Biomedical Nanosensors

Pan Stanford Series on Biomedical Nanotechnology

Series Editors

Vladimir Torchilin and Mansoor Amiji

Titles in the Series

Vol. 1
Handbook of Materials for Nanomedicine
Vladimir Torchilin and Mansoor Amiji, eds.
2010
978-981-4267-55-7 (Hardcover)
978-981-4267-58-8 (eBook)

Vol. 2
Nanoimaging
Beth A. Goins and William T. Phillips, eds.
2011
978-981-4267-09-0 (Hardcover)
978-981-4267-91-5 (eBook)

Vol. 3
Biomedical Nanosensors
Joseph Irudayaraj, ed.
2013
978-981-4303-03-3 (Hardcover)
978-981-4303-04-0 (eBook)

Vol. 4
Nanotechnology for Delivery of Therapeutic Nucleic Acids
Dan Peer, ed.
2013

Vol. 5
Inorganic Nanomedicine
Bhupinder Singh Sekhon, ed.
2014

Vol. 6
Nanotechnology for Cancer
Julia Ljubimova, ed.
2014

Vol. 7
Nanotechnology for Delivery of DNA and Related Materials
Bengt Fadeel, ed.
2015

Vol. 8
Translation Industrial Nanotechnology
Thomas Redelmeier, ed.
2015

Biomedical Nanosensors

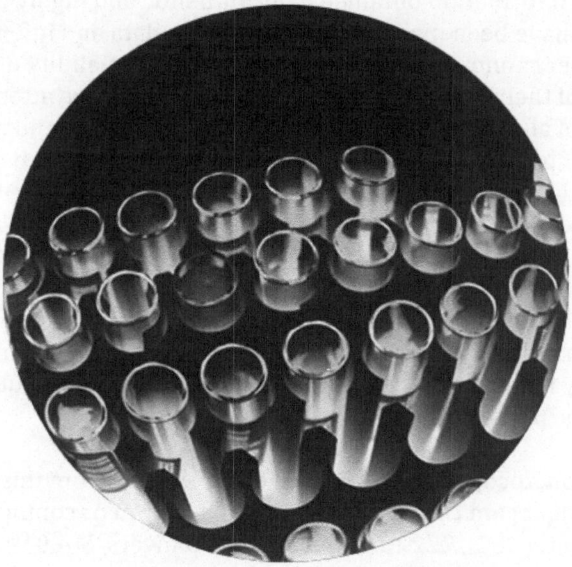

Joseph Irudayaraj

Editor

CRC Press
Taylor & Francis Group
Boca Raton London New York

CRC Press is an imprint of the
Taylor & Francis Group, an **informa** business

First published 2013 by Pan Stanford Publishing Pte. Ltd.

Published 2023 by CRC Press
Taylor & Francis Group
6000 Broken Sound Parkway NW, Suite 300
Boca Raton, FL 33487-2742

ISBN 13: 978-981-4303-03-3 (hbk)
ISBN 13: 978-981-4303-04-0 (ebk)

Visit the Taylor & Francis Web site at
http://www.taylorandfrancis.com

and the CRC Press Web site at
http://www.crcpress.com

British Library Cataloguing-in-Publication Data
A catalogue record for this book is available from the British Library.

Biomedical Nanosensors

Contents

Preface

The overall theme of *Biomedical Nanosensors* is to compile a comprehensive treatise on sensors for biomedicine. Specifically, we address the enthusiasm that nanotechnology and relevant tools have provided to medicine and biology in order to address diagnostics and therapeutics. The infusion of the nanoscale perspective to sensing is expected to allow the interrogation of biological systems, detection of ultrasmall concentrations of biomolecules, ions, and molecular interactions, exploration of cellular mechanisms, and integration of nanosensors into many other devices, including implantable devices and drug delivery vehicles. In this book, we will specifically address nanomaterials, constituting nanoparticles and nanotubes of inorganic as well as of polymeric composition of nano- and microscale dimensions, as these constitute a major direction in nanomedicine. Considering the diversity and the wealth of information in this field, we have attempted to focus on the basic concepts of fundamental and more recent materials and sensing paradigms leading to applications, including implantable devices, as these are central to micro- and nanoscale biomedical technology development.

In Chapter 1, we provide the fundamentals of biosensors, including the different transduction modes, biosensor elements, and bioconjugation strategies. A detailed account of molecular recognition is given in Chapter 2 in the context of molecular-imprinted polymer design, imprinting, and applications. Chapters 3 and 4 discuss the use of gold nanoparticles as plasmonic and Raman sensors. A comprehensive treatment of the fundamentals of gold nanoparticles as biosensors for ultrasensitive detection *in vitro* and *in vivo* can be found in these chapters. Chapters 5 and 6 discuss carbon nanotubes and graphene sensors. The physics of these versatile materials and their integration in sensor technology development, with a comprehensive survey of the work done in this field, is provided. The book will not be complete without a discussion on the use of single-molecule tools (Chapter 7), as these are becoming fundamental to understanding the mechanisms in biomedicine. In Chapter 8, recent advances in creating micro- and nanoparticles for inhalation drug delivery are addressed. Chapters 9 to 11 address *in vivo* therapeutics constituting bioelectric monitoring and stimulation, microelectromechanical systems, and implantable electrochemical biosensors. All of the chapters provide a detailed narrative on the future outlook.

Since the treatment is comprehensive, we expect this compilation to be valuable to scientists and engineers interested in the field of medicine, sensor development, and nanotechnology. Thus, this book could be used as an introductory text by graduate students and consultants wanting to obtain a quick introduction to biosensors for biomedicine with a focus on materials, properties, devices and technologies, and real-world applications.

Joseph Irudayaraj
Purdue University
June 2012

Chapter 1

Biomolecular Components of a Biosensor: Fundamentals

Kaustubh D. Bhalerao[a] **and Goutam J. Nistala**[b]

[a]*Agricultural and Biological Engineering, University of Illinois at Urb na-Champaign, 1304 West Pennsylvania Avenue, Urb na, IL 61801, USA*

[b]*Department of Chemical Engineering, Stauffer III, 381 North-South Mall, Stanford University, Stanford, CA 94305, USA*

bhalerao@illinois.edu

Biorecognition, the ability to recognize a specific biomolecular analyte from a complex sample, is key to the detection and quantification of biomolecules. Often, the physiological role of these analytes is well understood in terms of their interactions with other biomolecular components. This knowledge can be exploited to identify biomolecular elements that can serve as recognition components in a sensing system.

Analytes of interest are diverse in nature and include nucleic acids, proteins or peptide fragments, polysaccharides, or even small molecules. Several methodologies have evolved to recognize different kinds of analytes, some of which are based directly on the role of the analyte in the living system (in cases where the analyte is native to the living system) while others are based on imitating the immunological response of the analyte (in cases where the analyte is extraneous to the living system). Complementary to detection methodologies, there have been numerous developments in experimental and computational techniques to identify and produce such biomolecules in an efficient manner, as well as to conjugate these biomolecules with artificial sensing platforms.

While the field of biorecognition is not new in the strict sense of the term, it deserves to be recognized as an important subdiscipline of engineering. As biosensing and targeted therapeutics increase in relevance, the availability of a framework around organizing the information relevant to biorecognition will simplify and streamline the development process.

Biomedical Nanosensors
Edited by Joseph Irudayaraj

ISBN 978-981-4303-03-3 (Hardcover), 978-981-4303-04-0 (eBook)
www.panstanford.com

1.1 What Makes a Biosensor a Biosensor?

A sensor is a device that measures the presence or amount of substance and converts it into an interpretable signal. A biosensor is a class of sensor devices or systems in which a biochemical interaction is exploited as the basis of detecting the presence or level of a known analyte. This analyte may be an inorganic compound such as a salt, a small biological molecule such as a vitamin or sugar molecule, or even the concentration of larger biological macromolecules such as proteins or nucleic acids. The biochemical interaction between the sensor system and the analyte provides a biological context to the sensing event as often the sensing mechanism used by the biosensor is functionally analogous to the interaction between the analyte and its biological environment.

Consider the example of the commercially successful glucose monitors used by diabetes patients. These biosensors analyze droplet-sized blood samples for blood glucose level and provide an output in terms of milligrams of glucose per deciliter of blood. This information can be used for diagnostic purposes or to inform therapeutic actions. The level of glucose in the droplet is inferred through a biochemical reaction that occurs between the glucose in the sample and an enzyme (glucose oxidase or hexokinase) in the sampling strips. The progress of this reaction is measured either optically, as a change in the color of the strip, or electrically, by measuring the electrical current produced by the reaction. The biochemical reaction rate is converted into a number that can be interpreted by medical professionals. The glucose oxidase enzyme directly interacts with glucose in a manner and rate analogous to its interaction *in vivo*. Therefore, this biosensor system provides a way to measure the biologically relevant level of glucose in the sample and thus provides a more realistic assessment for diagnostic or therapeutic purposes.

The growing interest in the development of biosensors for biomedical applications is the result of two intertwined arcs of progress. Firstly, researchers in the life sciences continually push the boundaries of understanding life and disease at the molecular level. Cancer, cardiovascular diseases, diseases of the immune system such as AIDS and degenerative diseases such as Alzheimer's or Parkinson's are now described in terms of genetic mutations or the interplay between various biomolecules. Elucidating the molecular etiology of disease is a deterministic route to conceive of monitoring and intervention strategies directed specifically to detect or modulate the interaction between these biomolecules. Secondly, the progress in miniaturization of technology has allowed the manufacture of devices and systems comparable in size to biomolecules that can be leveraged as a therapeutic by directly intervening in the disease at the biomolecular level. Of course, these two developments are not altogether unrelated. The advances in understanding life at the molecular level have been enabled by advances in technology, while the drive of the life science community has been a motivating factor in developing these technologies.

Figure 1.1 shows an abstraction of a typical biosensor. Each biosensor can be said to be composed of three layers: recognition, transduction, and representation. The recognition layer, the subject of this chapter, is responsible for physically detecting the analyte by means of a biomolecular interaction. This recognition event is then transmitted to the transducer, which converts it into a machine-readable (typically electrical) signal. The representation layer takes this machine-readable signal and converts it into a form readily interpreted by humans.

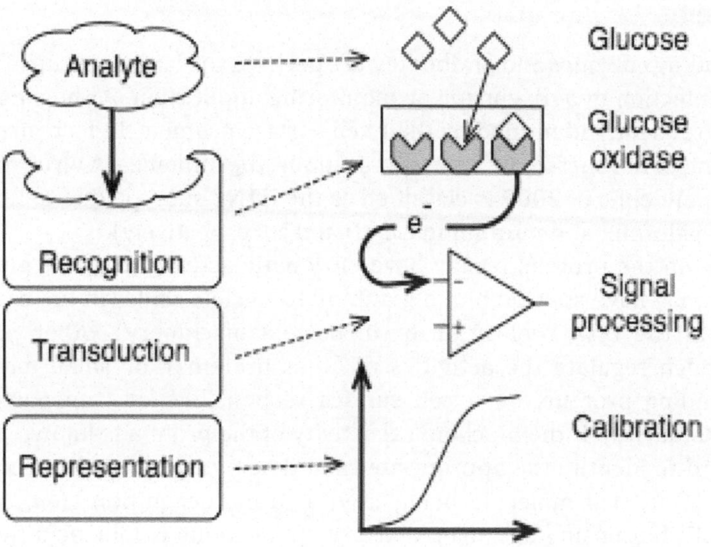

Figure 1.1 The three layers of the logical biosensor. The recognition layer binds to the analyte and leads to a measurable response in the transduction layer. The representation layer converts this response into a human-interpretable signal.

The recognition layer consists of a biomolecular system, which bears a known biochemical interaction with the analyte of interest. This biomolecular interaction either replicates the natural behavior of the analyte with the recognition layer or is closely analogous to it. This fidelity with the biological world provides the recognition event with a biological relevance and is the defining characteristic of a biosensor. If the affinity for the analyte borne by the recognition layer is highly specific, that is, it does not bind to anything other than the analyte, the recognition layer becomes especially advantageous as it can be used as a targeting system to extract and locally concentrate the analyte from a complex sample. In the example of the blood glucose sensor described above, the glucose oxidase enzyme embedded in the test strips bears a specific affinity only to glucose, which allows it to target glucose alone from the complex sample that is blood.

1.2 The Recognition Layer

Nature provides a vast array of biomolecules that may be utilized in a biosensor as part of its recognition apparatus. A biosensor may be designed for a variety of applications, including the detection of various proteins (e.g., the detection of a prostate-specific antigen in the blood serum as a biomarker for cancer) or antibodies, which are also proteins, but can specifically indicate the presence of an antigen. The analytes may also include nucleic acids as indicators of a virus, a genetic condition, or a carcinogenic mutation. In addition, the analytes may include small molecules such a polysaccharides, which are complex sugar molecules, or other small molecules such as glucose, urea, or in the case of blood diagnostics, lipids such as triglycerides or even blood gases such as acetone or ethanol. Each of these analytes requires a slightly different approach in the design of its cognate recognition layers as outlined in the following subsections:

1.2.1 Proteins

Proteins, including enzymes and antibodies, are perhaps the largest group of biomolecules targeted for detection in a diagnostic or monitoring application. Pathogens (viruses and bacteria) are frequently identified by their cell surface proteins. In fact, viruses are often typed according to cell surface proteins, for example, the influenza. A virus responsible for the swine flu pandemic of 2009 is classified as the H1N1 subtype due to its specific coat proteins hemagglutinin and nuraminidase, (H and N, respectively).

Depending on the protein, it may have enzymatic activity (e.g., the glucose oxidase enzyme) or it may have specific binding affinity to nucleic acids (in the case of proteins participating in the DNA replication or in protein machinery), other proteins (such as kinases, which regulate the activities of other proteins), or small molecules (such as vitamin-binding proteins on a cell surface responsible for transporting nutrients across cell boundaries). If the biochemical activity of the protein is known, it is relatively straightforward to identify the appropriate candidate moieties for the recognition layer. One example of a small molecule being used as a biorecognition element to detect a protein for both biosensing and drug delivery applications is folic acid (also known as folate or vitamin B9). Cancer cell lines such as the nasopharyngeal KB cells or ovarian cancer cells overexpress a protein known as folate receptor alpha (FOLR–α). The function of this protein is to bind to and sequester folic acid for subsequent transport into the cell,

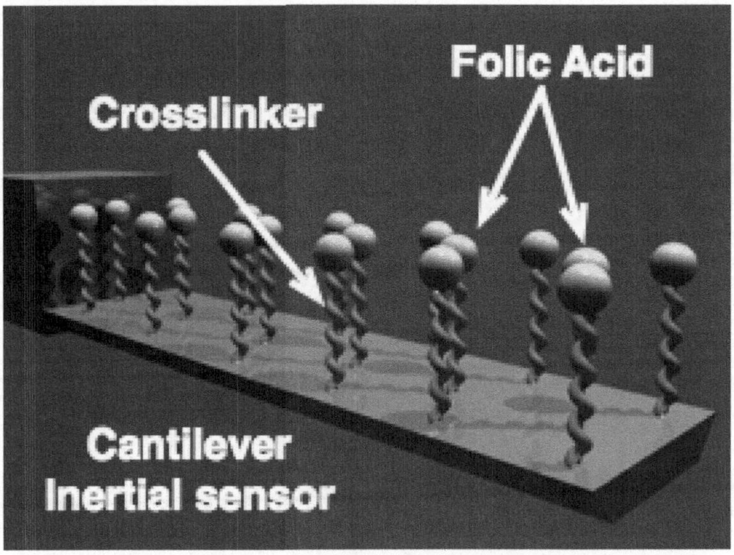

Figure 1.2 A cantilever-based biosensor. Folic acid molecules are chemically conjugated to the sensor surface. The transducer is a microcantilever that can be actuated electromechanically and whose resonance frequency can be measured using an electrooptical system. This setup is similar to microcantilevers used in atomic force microscopy. Changing the surface chemistry of the cantilever allows cancer cells' overexpressing folate receptors (•) to preferentially bind to the surface. Cancer cells binding to the sensor increase the inertia of the cantilever and result in a lowered resonance frequency, helping detect the presence of cancer.

where it is essential for DNA replication. Cancer cells, which replicate much faster than normal cells, face an increased demand of folic acid. In order to satisfy this demand, they upregulate the production of the folate receptor. In the KB or ovarian cancer cells, this overexpression can be as much as a 1,000-fold compared to normal tissue and serves as a prognostic marker for cancer.

Biosensors have been developed to exploit this property of increased affinity to folic acid. By applying a layer of folic acid to a suitable sensor platform, one can make the sensor stickier to cancerous KB cells in comparison with normal cells. If the sensor can measure the increase in mass due to the number of cells stuck to its surface, it can be used as a biosensor to detect cancer cells (Fig. 1.2).

Proteins with enzymatic activities can be very reliably detected and precisely quantified through the measurement of their activity in converting their respective substrates to products. The quantification of glucose oxidase is an example where the activity of the enzyme can be precisely quantified through its calibration against known concentrations of its cognate small-molecule substrate, glucose. A purified form of the analyte is necessary in the design of a biosensor. The proper testing and calibration of a biosensor can only be done when it is possible to purify the analyte using an independent procedure. The availability of a source of pure analyte allows for proper experimental controls in the testing procedure and is necessary to validate the claims of reliability, accuracy, precision, selectivity and specificity of a sensor.

1.2.2 Antibodies

Antibodies are proteins, and therefore, the preceding subsection is applicable to them. However, antibodies deserve a separate discussion for a number of reasons. Firstly, antibodies are components of animal immune systems, which are the natural biorecognition systems capable of recognizing and binding to trillions of foreign structural motifs that are not-self, that is, they do not resemble any part of the organism to which the immune system itself belongs. These structural motifs are referred to as antigenic epitopes, or simply epitopes.

Since the natural response of an immune system is to increase the number of antibodies directed to an epitope so as to enable their rapid sequestration and subsequent removal, the presence of significant quantities of antibodies to a certain antigen are a reasonable indicator of a past or continuing exposure to the antigen. For example, the test for tuberculosis in humans involves observing the reaction of the immune system to the attenuated tuberculin antigen injected just below the skin: a strong inflammatory response at the site of injection often indicates a latent infection suppressed by tuberculin antibodies. It follows then, that in the design of a biosensor to ascertain the presence of an infection of a pathogen, it is sometimes sufficient to detect the level of antibodies as a proxy indicator of that pathogen. This methodology is also commonly used to test whether a person has received vaccinations against different diseases.

Antibody complexes are called immunoglobulins. They are typically multivalent, that is, they have the capacity to bind multiple copies of their corresponding antigen. There are five major types of antibodies, classified as immunoglobulin types A, D, E, G, and M. (denoted as IgA, IgD, IgE, IgG, and IgM, respectively). Figure 1.3 shows the canonical form of the most common, IgG. It is a symmetric molecule comprising two identical halves joined to each other by means of disulfide linkages. The variable ends of the light and heavy chains create a pocket, which binds its cognate antigen in a manner similar to a lock-and-key mechanism. Permutations in the structure of the variable chain lead to the diversity of binding motifs in the antibody repertoire of the immune system. The heavy chains of the antibodies are identical within the organism but differ from organism to organism. This implies that one kind of organism, for example, a rabbit, can generate antibodies that specifically bind to the heavy region of the antibodies from another organism, for example, a cow. This property specifically enables enzyme ligated immunosorbent assays, (ELISAs) in which antibodies from one organism can be used to detect antibodies from another organism, which, in turn, are produced in response to exposure to an antigen.

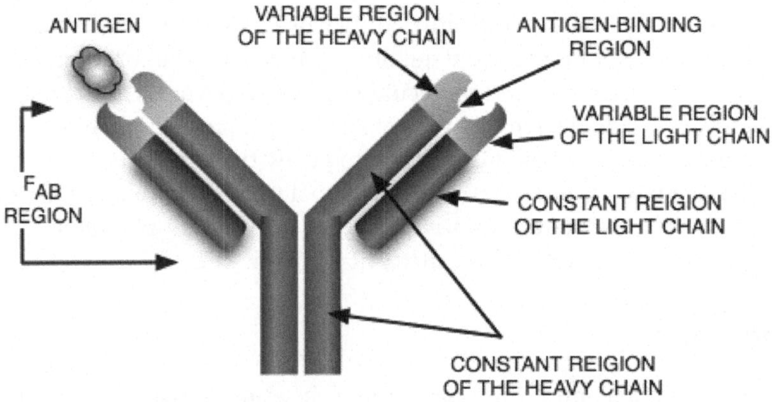

Figure 1.3 The structure of Immunoglobulin subtype G (IgG). Each IgG consists of a pair of peptide chains connected to each other by means of disulfide bonds. These provide structure to the antibody and are called heavy chains. A lighter peptide chain is connected to each arm of the heavy chain. Distal tips of the heavy and light chains are variable in sequence and form a pocket that can recognize and bind to an antigen. The fragment of the antibody denoted as the FAB region can be separated from the rest of the heavy chain.

As with protein analytes, in order to create a biosensor to detect the presence of an infectious pathogen, it is necessary to have a sample of the pure pathogenic antigen. The presence or absence can be inferred from the level of antibodies present in a sample that are capable of binding the antigen. These antibodies can be detected using secondary antibodies from a different organism, which bear affinity to the antibodies in the sample. This is the working principle of ELISA. The ELISA assays, in particular, are an inspiration for several biosensor concepts, which seek to miniaturize the conventional 96- or 384-well ELISA format to a size that can be accommodated on a microfluidic device [1–4].

1.2.3 Nucleic Acids

Although detecting the levels of proteins provide rich information about the biological process under investigation, sometimes it is more valuable to detect the nucleic acids responsible for producing those proteins. Consider a viral pathogen for instance. Detecting a protein produced from viral nucleic acids may not be an effective detection and diagnostic strategy; it could mean that if the proteins are already expressed, it may be too late to initiate therapy. Similarly, detecting nucleic acids (or mutations therein) that may indicate a proclivity to diseases such as cancer may have more value as a screening tool. For example, while genetic mutations in the breast cancer 1, early onset (BRCA1) gene do not immediately imply the existence of breast cancer in women, they may suggest a predisposition and provide valuable screening information to enable the early detection of the onset of the actual disease.

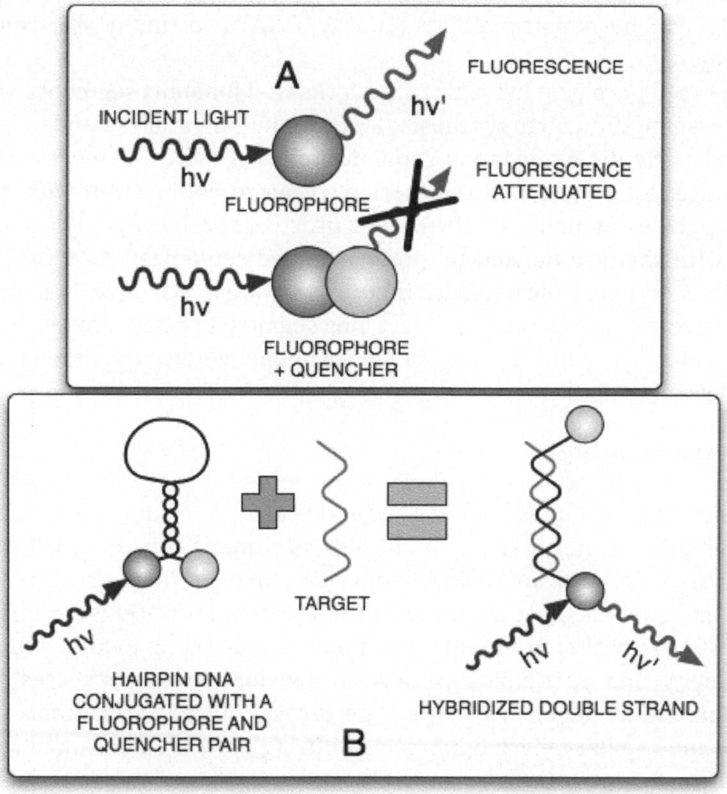

Figure 1.4 A DNA probe. A shows a fluorophore that fluoresces when light of a certain wavelength is incident upon it. When a quencher is in proximity with the fluorophore, the fluorescence is greatly attenuated. Such a scheme can be used for DNA detection. If a DNA segment is designed such that a portion of it bears affinity with a target DNA and the extreme base pairs are complementary to each other and if the ends are conjugated to a fluorophore and quencher, we get the hairpin-like structure shown in B. Such a structure will not fluoresce in the absence of the analyte DNA. However, in the presence of the target, the hairpin structure opens up and hybridizes with the target DNA. As such, the quencher can no longer suppress the fluorescence. Thus, the presence of the fluorescence can be interpreted as the detection of the target DNA. See also Color Insert.

The detection of specific strands of DNA is not as varied in scope as the detection of proteins. The detection relies primarily on the hybridization/dehybridization between two complementary strands of DNA. If the precise sequence of the analyte DNA strand is known, it is routine to synthetically create its complementary detector strands (for limited sizes of DNA). Furthermore, the sequence completely determines the energy released during the hybridization process or, conversely, the energy required to dehybridize the strand into its constituent single strands. Based on this principle, there exist several varieties of DNA-detecting probes combining single detector strands of DNA with optically active materials such as fluorescent nanoparticles. Some examples are shown in Fig. 1.4. The advantage of short DNA segments used for biorecognition is that they can be multiplexed. A single biosensor comprising a large number of detector strands placed in precise locations on a substrate, each bearing specificity to a different DNA segment, can be constructed to identify DNA from many different organisms simultaneously.

Detection based purely on hybridization works well for short segments of DNA since it becomes increasingly difficult to synthesize larger complementary strands of detector DNA. The gold standard for detecting larger sequences of DNA is the polymerase chain reaction (PCR), which uses a device called the thermal cycler in conjunction with a biochemical process to replicate segments of DNA. This process specifically replicates only those segments of DNA that are flanked by predetermined sequences. By amplifying specific segments of DNA, it is possible to enrich their proportion in a sample, making them easier to detect. Novel technologies used to detect long segments of DNA are generally validated against PCR, which is becoming increasingly cost effective through ease of availability of automated thermal cyclers.

1.2.4 Polysaccharides

Polysaccharides are complex natural biomolecules and are also referred to as glucosaminoglycans (GAGs). These molecules are made up of smaller mono- and disaccharides such as hyaluronic acid, chondroitin sulfate, or heparin. GAGs can be linear or branched and show a great diversity in their structure. GAGs were initially thought to function as structural components, but there is increasing evidence that they have biological activity and participate in several developmental processes such as cell-cell interaction and angiogenesis [5–8]. High throughput analytical tools such as mass spectrometry and capillary electrophoresis are being used to understand the structures, diversity, and function of GAGs in a new field called glycomics. There is also some evidence that some GAGs may serve as biomarkers for diseases such as cancer. It is expected that in the future, there may be significant diagnostic information to be had by profiling polysaccharides from a sample and dedicated biosensors may be created to aid this process.

Antibodies, which are versatile biorecognition elements, can be used to detect polysaccharides, but it is possible to generate DNA aptamers to specifically target polysaccharides as well. Aptamers are short segments of nucleic acids that fold upon themselves to create three-dimensional structures. These structures show binding affinity to polysaccharides in a manner similar to antibodies binding to their cognate antigenic epitopes. Aptamers can be created from DNA as well as RNA. Such structures exist naturally (as riboswitches) but can also be identified from a diverse pool of random nucleotide sequences by proper screening [9–12].

1.2.5 Other Small Molecules

In addition to polymeric biomarkers such as proteins, nucleotides, and polysaccharides, there exist a large number of smaller molecules produced in living tissue as a result of metabolism, whether normal or diseased. The science of diagnosis based on quantifying small molecules is in some cases older and more advanced than that based on larger biomolecules. There are numerous examples of small-molecule quantification for diagnostic purposes. Glucose is perhaps the most widely used measure as an indicator of diabetes. Levels of urea in blood are an indicator of kidney function. Breath analyzers routinely measure blood alcohol levels and can be designed to provide information on other volatile compounds such as ketones, which provide additional information on metabolism. In addition, the presence of nonnative small molecules such as toxins or illegal substances may also be informative in therapeutic or law enforcement situations.

Biosensors designed for detecting small molecules can be equally diverse. Biorecognition elements for small molecules can be as varied as proteins (e.g., vitamin-binding proteins can be used to detect vitamin levels), aptamers specifically identified to bind small molecules, or other small molecules that interact chemically with the analyte. This last strategy is by far the most common, as evidenced by numerous electrochemical assays designed for detecting lactic acid, ketones, and urea for medical diagnostics [13–17], as well as the sensors used in the food industry such as the biosensors used for the measurement of lactose, antibiotics, and animal drug residues in milk [18–20]. Given the huge diversity of analyte molecules, the choice of the appropriate recognition layer is not always obvious. It depends not only on the analyte but also on the ease of use of the recognition layer, its compatibility with the analyte of choice, available strategies to conjugate it to the sensor platform, and of course, its performance and reliability when used as a sensor component.

1.3 Bioconjugation

Nature has provided a vast array of biological molecules, which, by virtue of their biological function, can be used to recognize their analyte when they are coupled with a suitable transducer. The conjugation of biological materials with electromechanical ones, however, requires special consideration on the part of the biosensor designer. Technologies used for manufacturing microelectromechanical or microfluidic systems often use harsh chemicals, extreme temperatures, and abrasive mechanical procedures in their fabrication. Such procedures are usually not compatible with biological materials, which have evolved to function in relatively benign, usually aqueous, solvents, under moderate environmental conditions of temperature and pH. This implies that any conjugation procedure used to combine biological materials with artificial ones should provide the right environment for the biological materials to function as desired without damaging either the biological or the nonbiological components of the sensor.

Common sensor materials used to provide a platform for bioconjugation include silicon/silica-covered silicon; glass; metals such as titanium, gold, or nickel; polymeric (plastic) surfaces; or in some cases, even other biologically derived materials such as cellulose or paper.

Figure 1.5 shows the four primary methods by which biological materials may be conjugated with their artificial counterparts, depending upon the substrate and the biomolecule. If a substrate is porous, it is often simplest to entrap the biological material into the porous matrix. Examples include the blood glucose strips made of paper or disposable pregnancy tests, which use a porous ceramic matrix to deliver the sample to the absorbed reagents via capillary action. The advantages of this method include simplicity in manufacture and a relatively high surface-to-volume ratio, which enables a large amount of the biorecognition layer to be absorbed into the substrate. The primary disadvantage is poor control on the stoichiometry (i.e., the precise number of biomolecules per unit area/volume of substrate) if the pore size cannot be controlled, leading to variability in manufacturing and ultimately, variability in the accuracy and precision of the sensor performance. This method is also only suitable if the sample sizes analyzed are relatively small. For large volumes of samples, there is a possibility for the absorbed biomolecules to leach out into the sample and lead to further variability in operation.

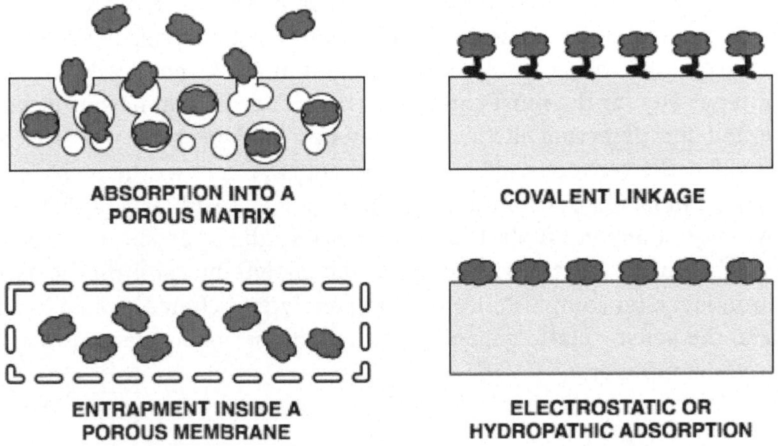

Figure 1.5 Four strategies to conjugate biomolecules with artificial surfaces.

A variation on this theme is for the biorecognition element to be trapped inside a porous membrane. Such a strategy can be advantageous in case of in-dwelling sensors, where the analyte is smaller in size than the biorecognition element. This strategy can overcome some of the limitations of pure absorption by allowing the analyte to freely pass across the membrane and by preventing the biorecognition element to leach out.

In reality, pure absorption or matrix entrapment probably does not exist by itself. Since larger biomolecules, whether proteins, nucleic acids, or polysaccharides, are composed of charged subunits (amino acids may have a positive or negative charge, while the phosphate-sugar backbone of DNA is negatively charged), there may exist a degree of interaction between the biomolecule and the substrate at the atomic level. Attractive van der Waals forces between the atoms of the biomolecule and the substrate can lead to the physical adsorption of the biomolecule on the substrate. For example, cysteine residues in proteins contain a sulfhydryl (-S-H) group that bears affinity for metal ions and readily binds to $Fe2^{+}$, $Pb2^{+}$, and $Hg2^{+}$ ions. Thus, if the substrate contains such ions, it may lead to a stronger bond with the biomolecule. However, this interaction can sometimes be undesirable if the bond disrupts or otherwise hinders the normal operation of the protein. Not coincidentally, the bond between cysteine and metal ions is often the molecular cause of heavy metal poisoning; any sensor that uses a biomolecule to detect heavy metals may reasonably be suspected to suffer the same fate.

The hydropathic interactions between a substrate and a biomolecule are also an important consideration in adsorptive processes, especially if the biosensor is operated in an aqueous environment (which is most often the case). If the substrate is hydrophobic, and if the biomolecule evolved in an aqueous environment (and is therefore hydrophilic), it may denature on the substrate and lose functionality. On the other hand, if the substrate is hydrophilic, the biomolecule may detach from it as soon as an aqueous environment is available.

Finally, a direct covalent linkage between the substrate and the biomolecule is often a relatively reliable bioconjugation strategy. Although more complex than absorption or adsorption, if a method to covalent bioconjugation procedure exists, it is often advisable to pursue it. The advantages of such a strategy include fine-grained control on the geometry of conjugation; in other words, it allows the orientation of the biomolecule relative to the substrate to be controlled if the precise locations of the covalent bonds are known. Secondly, it provides better control on the stoichiometry of the bioconjugation process since the number of binding interactions between the substrate and the biomolecule can be determined a priori. Thirdly, the knowledge of precise binding interactions allows predicting the properties, durability, and reliability under nominal operation environments for the recognition layer. For example, the bond determines its propensity for hydrolysis under different environmental conditions and, therefore, informs about the appropriate storage and sampling conditions necessary to ensure predictable behavior in storage and operation.

1.3.1 Cross-Linkers

Cross-linkers (Fig. 1.6), as the name suggests, are a class of molecular fasteners that allow coupling between multiple materials. Homobifunctional cross-linkers are the simplest instances of this class, which typically consist of an organic chain that ends in two identical functional groups. Such cross-linkers are used to create internal bonds within a homogenous matrix, thereby increasing its rigidity. Bioconjugation between two dissimilar materials, the substrate and the biomolecule, requires heterobifunctional cross-linkers, which support two different functional groups. A common subclass of heterobifunctional cross-linkers is the organometallic cross-linkers, which — as the name suggests — allows conjugation between an organic functional group and a metal. It is generally easy to deposit a thin layer of metals such as gold, nickel, chromium, or titanium on silicon-based microelectromechanical devices, allowing the conjugation of biomolecules using organometallic cross-linkers.

Examples of such linkers include thiol groups, which bind to gold, and phosphonic acid, which binds to titanium dioxide. The functional groups of the linker corresponding to the biomolecule typically employ covalent bonds to bind specifically to different residues on the biomolecule.

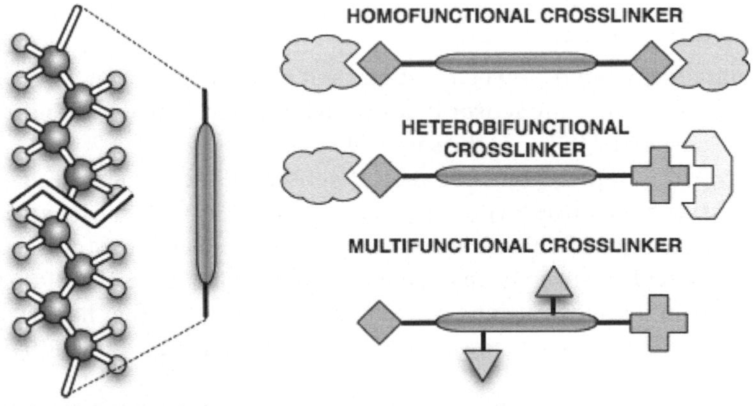

Figure 1.6 Different types of cross-linkers. An organic spacer chain with ends that bear affinity to different substances serves as a bridge to conjugate different materials. Depending upon the functional ends, cross-linkers may be classified as homofunctional, heterobifunctional, or multifunctional.

1.3.2 Development of Novel Recognition Elements

Beyond some well-publicized common cases, it is generally rare to have a recognition element readily available for any given analyte. In fact, the development of an appropriate (i.e., sufficiently specific and robust) recognition layer is a significant component of biosensor development. In some cases, the identification of the recognition layer is a trivial matter. For example, in the case of DNA detection, it is a straightforward matter of identifying the unique segment of DNA that needs to be recognized and determining the complement to it. Once the complement is manufactured, either synthetically or using a PCR-based technique, it can be conjugated to the biosensor by using suitable conjugation chemistry. When the analyte is a small molecule, it helps to know its biological role and interactions within living processes since those biomolecules that directly interact with the small molecule analyte are themselves leading candidates for recognition elements.

In order to generate protein-based recognition moieties to an analyte, there exist two main approaches: 1) a passive, screening approach, where the interaction with an analyte is tested against a library of proteins and strongly interacting candidates are identified and 2) an active approach, where a binding interaction between the analyte and protein is "rewarded" by allowing the interacting element to multiply and separate itself out from the rest of the library. The generation of antibodies to an analyte and the class of molecular "display" techniques are outlined next.

1.3.3 Antibody Generation and Processing

The search for an appropriate recognition element begins with a review of the systems biology of the analyte. The precise information of how an analyte interacts with the biological system may not always be known, or it may not be specific enough to help the designer identify a suitable recognition element. In such cases, it is useful to leverage the immune system in producing biorecognition elements to the analyte. Of course, this requires that the analyte be available as a pure sample, even if its biological effects are unknown.

Antibody generation to various analytes has been commonly utilized to create biorecognition reagents, which have been used in assays such as ELISA and Western blots. There are two types of antibodies, designated as polyclonal antibodies and monoclonal antibodies, which differ in the manner in which they are generated. Antibodies are produced by B lymphocytes, which are components of the immune system. The human repertoire of B lymphocytes contains billions of unique B cells, each displaying affinity for a specific epitope. In other words, the antibody production system consists of a library of recognition elements and an active process to select the specific B cell variant (or clone) that displays a high degree of affinity for the epitope. If the B cell clone binds to an analyte, biochemical cascades within the B cell induce massive proliferation of that clone and lead to an increase in the amount of antibodies for that epitope. These clones show a degree of variability around the specific binding affinity for the analyte. Those clones that show higher affinity than before proliferate further, while those that show lower affinity self-destruct. This positive feedback that identifies and promotes higher affinity is called "affinity maturation."

If an analyte is polyepitopic, that is, if there exist several distinct B cell clones capable of binding it and eliciting an antibody generation response and undergoing different trajectories of antibody maturation, then the set of distinct antibodies generated are called polyclonal antibodies since they arise from different B cell clones. Briefly, the method used to produce such antibodies is to inject a sample of the analyte into an animal, such as a mouse, a goat, or a horse, allow for the animal to generate an immune response to the analyte over the course of a few days, harvest the blood serum from the animal, and use an affinity-based separation process to harvest the antibodies generated.

Polyclonal antibodies are cheaper to produce than their monoclonal counterpart. However, because they consist of a diverse set, they are more variable in their binding characteristics for the original analyte. While polyclonal antibodies may offer sufficient selectivity in binding (i.e., they will bind the analyte when it is present), they may lack specificity (i.e., they may bind to things other than the analyte in a complex sample) owing to their diversity. This nonspecificity can negatively affect the operational reliability of the recognition layer of the biosensor by increasing the number of false positives in detection.

Monoclonal antibodies circumvent this problem, albeit at an additional cost. Instead of harvesting antibodies directly from the animal's serum in the case of polyclonal antibody production, the B cells are harvested. These healthy antibody-producing B cells are then fused with cancerous B cells, which have the property of unlimited cell division. This cell fusion is called a hybridoma. Each productive B cell clone transformed into a hybridoma is maintained in controlled *in vitro* conditions, where it replicates indefinitely (thanks to the cancer cell) and produces a stock of antibodies that are monoclonal, that is, the product of a single B cell clone. This results in increased costs due to the hybridoma technology; however, by virtue of being identical, monoclonal antibodies have more consistent binding characteristics and increased specificity. Therefore, the increase in costs is often justifiable as it can directly lead to better quality control measures for sensor products using monoclonal antibodies.

In applications like ELISA, a polyclonal or monoclonal antibody generated to the desired analyte may directly be used without further processing. In some biosensor applications, however, in which the transduction relies upon a change in mass (in quartz crystal microbalances or cantilever-based sensors), the mass of the antibody can be many times the mass of the analyte. In such cases, the change in mass due to the binding of an analyte to an antibody-consisting recognition layer can be extremely difficult to detect. Similarly, in cases where the biosensor transduction depends upon a change in the electrical properties on the surface of the sensor (such as in surface plasmon resonance or field effect transistor-based sensors), the biorecognition event, that is, the binding of the analyte to the antibody, may be too far away (greater than the Debye length) from the sensor surface owing to the size of the antibody. In these cases, a reduction in the size of the antibody leads to an increase in the signal-to-noise ratio.

Since the active component of the antibody, the substructure that actually participates in the binding, is the variable chain (see Fig. 1.3), it is possible to strip the heavy chain of the antibody while maintaining the activity of the antibody. Additional advantages of the antibody fragment (called Fabs) also include the ability for denser packing on a surface due to a smaller size and a more controlled structure. Given the advantages of Fabs over full-size antibodies, it may be that Fabs replace a significant portion of the antibody-based biorecognition mindshare in the near future.

1.3.4 Molecular Display Techniques

The active binding region in an antibody consists of no more than a few peptides, so positioned to create a pocket, which serves as a site to bind the antigen. The rest of the peptides in the antibodies serve as a scaffold to support the binding residues. Ergo, it can be possible to abstract only those few binding peptides and still retain some or all of the binding activity of the antibody. Techniques such as phage and cell surface display mimic the working of the immune system to identify small peptides (7–20-amino acid long) that display binding affinity to an analyte of choice. Phages are viruses that infect and replicate inside a bacterial host. They consist of a protein shell that encases the genetic material that encodes the coat protein as well as the program that enables the viruses to replicate inside the host. Specific proteins on the coat allow the virus to attach to the host cell's surface proteins. The technique of phage display works in the following way (Fig. 1.7): A library of phages is generated that is identical in every respect except for variations in a well-defined segment of the coat protein (caused, of course, by a corresponding variation in the genetic material that encodes it). This library can have billions of variants. When a solution containing the library of phages is exposed to a surface-immobilized analyte of choice, by some frequency, a small fraction of the library will contain phage variants, whose surface protein displays a binding affinity to the analyte. When the analyte-bound surface is washed, only the phages that are strongly bound to the analyte are retained on the surface. These phages are eluted from the analyte using a strong detergent, purified, and allowed to infect and replicate in suitable host cells.

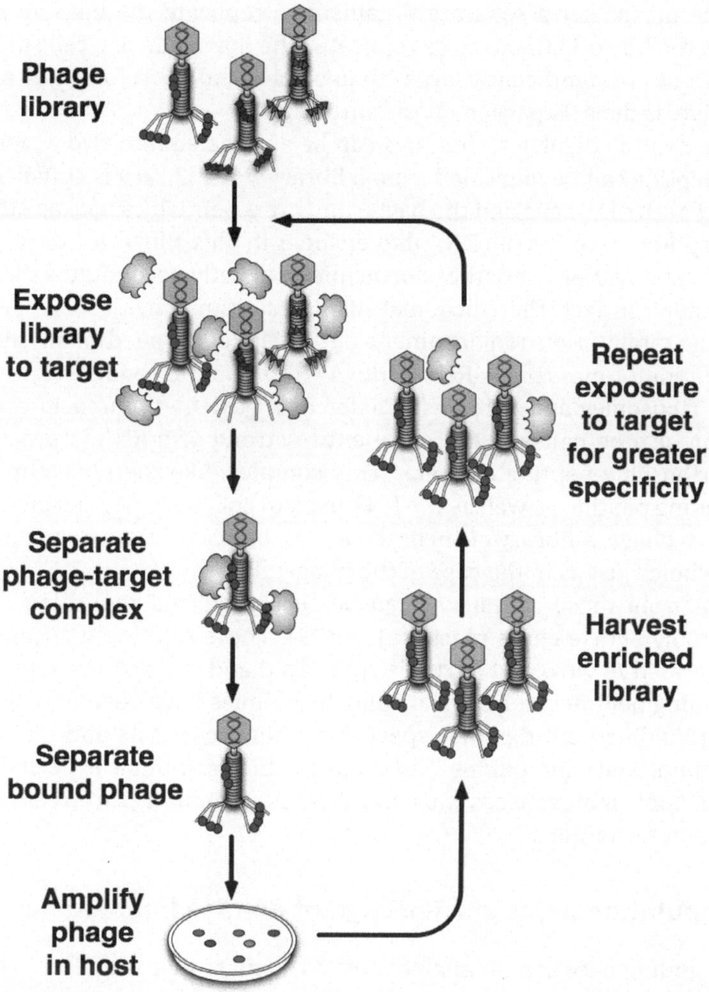

Figure 1.7 The phage display. A library of phages is exposed to a target. The subset of the library that binds to the target is separated and replicated inside the phage's native host. The subset is amplified and reexposed to the target. By increasing the aggressiveness with which the phage is separated from its target, the library can be enriched toward progressively higher binding affinities. The surface peptide that distinguishes the binding phage can be identified, synthesized, and used as a bioconjugation reagent.

Once the population of the phages is built back up, it has the effect of separating a small portion of the library and replicating it. This competitive binding process is called a panning round. Repeated panning with an increasingly aggressive washing process allows for isolating a progressively stronger binding phage. Sequencing the binding protein between pannings also produces information on any consensus sequences that emerge within the library and provide bioinformatic support for the increase in binding efficiency.

A cell surface display is similar to a phage display, with the difference that a bacterial cell is used to express the surface proteins, rather than a phage. This also simplifies the process a little bit in that a separate organism to replicate the binding protein is not required since the bacteria themselves replicate. The downside of a cell surface display is that bacterial cells are significantly larger than phages and, therefore, may not be suitable when the analyte is densely packed on the surface.

Both these peptide display techniques can be abstracted into two general principles: 1) a binding peptide can be identified from a library if the library is sufficiently large and diverse, and 2) both the phage and the bacterium are essentially a package that consists of the binding peptide as well as the DNA that encodes it. This allows for the easy replication of the binding peptide. Neither of these principles explicitly necessitates a living organism or a living system. In fact, the ribosomal display system provides this peptide display functionality in a totally *in vitro* environment and precludes the need for any living organism.

The translation of messenger RNA (mRNA) into peptide sequences can be achieved *in vitro* using ribosomes and suitable transfer RNA (tRNA)-amino acid complexes. This translation can also be paused before the entire strand of mRNA is processed, leading to an mRNA-ribosome-peptide complex. This complex, like the phage or bacterial cell, includes a binding peptide as well as the RNA used to encode it. Furthermore, it is of a size comparable to a phage. A library of such ribosomal complexes can be screened for binding an analyte of choice just as in the case of the phage display.

The mRNA from these complexes can be recovered and replicated using reverse-transcriptase polymerase chain reaction (rt-PCR). These replicated strands of DNA can then be transcribed *in vitro* and partially translated and the process repeated to obtain a strongly binding peptide. Molecular display techniques have been especially successful in developing binding peptides that specifically bind to metals and metal oxides such as gold, titanium, and zinc oxides. This makes this technique especially useful as a bioconjugation tool for biosensors and nanotechnology manufactured using conventional microfabrication techniques.

1.3.5 Computational Design/Redesign of Binding Interactions

The basis of binding between an analyte and its cognate biorecognition element is governed by thermodynamics. The degree of binding is a function of complementarity in the shape of the interface between the analyte and the recognition element, which, in turn, is dependent upon the configuration of the atoms and molecules that constitute the interface and its solvation environment. It follows that changes in the structure and environment of the interface can lead to manipulation of the binding characteristics. For example, the dissociation constants between two biomolecules can be altered significantly if the pH of its solvent environment is changed, a fact that is routinely used in various chromatographic techniques. Similarly, the hydrophobicity or hydrophilicity (and, therefore, its solubility in a solvent) of a molecule can be altered if it is cloaked in a suitable detergent. While it is possible, in theory, to manipulate these interfacial characteristics, in many cases, the rules that can predict such behavior precisely are not well understood. For example, while it is possible to get a sense of the solubility of a proteinic enzyme based on its amino acid composition and it is possible to surmise that a detergent could increase its solubility, it is rather difficult to predict the impact the detergent might have on its biological activity. Such information can only be obtained from experiments specifically designed for the purpose of determining structure-activity relationships.

However, if the detailed structure of a biorecognition element is known, such as in the case of an antibody, there are several ways that allow an engineer to fine-tune binding characteristics. The first method involves modifying the environment of the binding interaction by changing the temperature, pH, and the salt concentration of the solution. It is probable that biorecognition elements such as antibodies or the peptides generated from the various display techniques mentioned above are optimized for activity in the environmental/physiological conditions used in their production.

However, one may argue that this is only a local optimum depending upon conditions of competitive binding and may not necessarily hold in the case of noncompetitive binding in a pure sample. For instance, antibodies may be locally optimized for competitive binding under a whole host of nonimmunological components of blood serum (such as serum albumin or other proteins), and this local-binding optimum may not hold when such competitive action is not present.

Furthermore, if the biosensor operation requires environmental conditions that are necessarily nonphysiological (e.g., in nonaqueous solvents), the activity of any recognition layer that has been generated under dissimilar conditions should be reasonably considered suspect and should be experimentally validated.

Computational techniques also provide a route to understanding and tuning binding activity in the face of conditions that deviate from the nonphysiological. If the structure of the binding interface is well known and the molecular structure behind it well understood, it is possible to simulate modifications using computational tools based on thermodynamic principles. Using such tools, it is possible to both optimize a biomolecular interface as well as design new ones that are not seen natively. Computational methods provide a rational approach to designing biomolecular interactions, as opposed to discovering them through directed evolution techniques such as phage display. While such methods can drastically reduce the cost of designing novel biorecognition agents, they must be validated experimentally for use in any biosensor context.

Finally, one relatively facile way of modifying the binding activity of biomolecules is to simply control how they are conjugated to the sensor surface. As mentioned previously, the activity of a biomolecule is sensitive to how it is conjugated to the sensor surface. By limiting access of the analyte to the biorecognition element, one can effectively reduce its binding affinity. Access may be limited by encasing the biomolecule in a matrix or by conjugating it to the surface in a way that physically obstructs its binding residues. This is called stearic hindrance. It is important to note that stearic hindrance can only reduce the binding affinity. This reduction in sensitivity can be useful, however, in decreasing the instances of false positives through the reduction of nonspecific binding and is an interesting trick to keep in mind when designing biosensor surfaces.

1.4 Conclusions and Future Outlook

Biorecognition and bioconjugation are inextricably related when it comes to biosensor design. Bioconjugation is not exactly a new field. The material science for combining biological materials with artificial materials has accumulated over several decades. However, biorecognition, which has been motivated by biosensors and nanotechnology, has renewed the interest in developing, characterizing, and standardizing bioconjugation techniques. The science and engineering behind biorecognition will continue to evolve in the future. It is a broad field and is a necessary tool for several areas of technology besides biosensors for health, food, agriculture, and the environment, including tissue culture, pharmaceuticals, and industrial biotechnology. Biorecognition certainly deserves to be an important subdiscipline of materials science and engineering. Scientists and engineers actively involved in the development of novel biorecognition schemes should also consider the following propositions as a way to build an engineering framework around biorecognition:

1.4.1 Biorecognition Needs Better Integration with Informatics

The main step in designing a suitable biorecognition layer for a sensor application is to identify a biomolecule capable of binding to the analyte in consideration. Often, the choice of the recognition element is governed by, and also limited by, the knowledge of the analyte in its biological context. This knowledge often tends to be highly specialized and different from the knowledge of biotechnology required to generate the recognition element or the knowledge of materials science required to integrate the element into a workable device system. Having an informational resource that bridges these knowledge pools will help the biosensor designer immensely. Such a database, for instance, could, for a given analyte, provide a list of all known biomolecules that bear affinity for it and their sources, sequences, and structures; any methodologies available to produce, isolate, and purify them; lists of commercial suppliers; references to relevant literature; environmental conditions required for optimal performance; as well as any intellectual property rights associated with them. Such information is available today, but it is not organized in any central database and requires accessing disparate online and offline resources.

1.4.2 Standardization of Bioconjugate Chemistry

A consequence of the proposal for a better informatics resource would eventually lead to a degree of standardization of procedures that conjugate biorecognition elements to device surfaces. Such procedures would not only have detailed protocols but also specify the resulting performance of the biorecognition layer. There is great value in standardizing manufacturing processes as new regulations are established around manufacturing practices especially for medical devices.

1.4.3 Better Methodologies to Model Reliability and Environmental Impact

A specification on the reliability of a sensor must include considerations on its selectivity (the ability to detect the analyte when it is actually present), specificity (the ability to correctly infer the absence of the analyte), sensitivity (the lowest detectable level of the analyte), accuracy, precision, and aspects of durability and shelf life. Furthermore, significant research is also required on the potential for reusing and recycling it as well as the environmental impact resulting from its production and disposal. To the extent possible, such methodologies should be built into the development process and are a critical component of a systems-level approach to future biosensor design.

References

1. Babu, S., Mohapatra, S., Zubkov, L., and Murthy, S. (2009). A PMMA microcapillary quantum dot linked immunosorbent assay (QLISA), *Biosens. Bioelectron.*, **24**(12), 3467–3474.
2. Gomara, M. J., Fernandez, L., Perez, T., Ercilla, G., and Haro, I. (2009). Assessment of synthetic chimeric multiple antigenic peptides for diagnosis of GB virus C infection, *Anal. Biochem.*, **396**(1), 51–58.
3. Hartman, I., Gillies, A. R., Arora, S., Andaya, C., Royapet, N., Welsh, W. J., Wood, D. W., and Zauhar, R. J. (2009). Application of screening methods, shape signatures and engineered biosensors in early drug discovery process, *Pharm. Res.*, **26**(10), 2247–2258.
4. Wen, X., He, H., and Lee, L. J. (2009). Specific antibody immobilization with biotin-poly (l-lysine)-g-poly(ethylene glycol) and protein A on microfluidic chips, *J. Immunol. Methods*, **350**(1–2), 97–105.
5. Briganti, E., Spiller, D., Mirtelli, C., Kull, S., Counoupas, C., Losi, P., Senesi, S., Di Stefano, R., and Soldani. G. (2009). A composite fibrin-based scaffold for controlled delivery of bioactive pro-angiogenetic growth factors, *J. Control. Release*, **142**(1), 14–21.
6. Brule, S., Friand, V., Sutton, A., Baleux, F., Gattegno, L., and Charnaux, N. (2009). Glycosaminoglycans and syndecan-4 are involved in SDF-1/CXCL12-mediated invasion of human epitheloid carcinoma HeLa cells, *Biochim. Biophys. Acta.*, **1790**(12), 1643–1650.
7. Rusnati, M., and Urbinati, C. (2009). Polysulfated/sulfonated compounds for the development of drugs at the crossroad of viral infection and oncogenesis, *Curr. Pharm. Des.*, **15**(25), 2946–2957.
8. Wang, I. N., and Lu, H. H. (2006). Role of cell-cell interactions on the regeneration of soft tissue-to-bone interface, *Conf. Proc. IEEE Eng. Med. Biol. Soc.*, **1**, 783–786.
9. Fang, X., and Tan, W. (2009). Aptamers generated from cell-SELEX for molecular medicine: a chemical biology approach, *Acc. Chem. Res.*, **43**(1), 48–57.
10. Jeong, S., Lee, H. K., and Kim, M. Y. (2009). Use of RNA aptamers for the modulation of cancer cell signaling, *Methods. Mol. Biol.*, **542**, 363–377.
11. Kwak, H., Hwang, I., Kim, J. H., Kim, M. Y., Yang, J. S., and Jeong, S. (2009). Modulation of transcription by the peroxisome proliferator-activated receptor delta–binding RNA aptamer in colon cancer cells, *Mol. Cancer. Ther.*, **8**(9), 2664–2673.
12. Zhao, Z., Xu, L., Shi, X., Tan, W., Fang, X., and Shangguan, D. (2009). Recognition of subtype non-small cell lung cancer by DNA aptamers selected from living cells, *Analyst*, **134**(9), 1808–1814.
13. Forrow, N. J., Sanghera, G. S., Walters, S. J., and Watkin, J. L. (2005). Development of a commercial amperometric biosensor electrode for the ketone D-3-hydroxybutyrate, *Biosens. Bioelectron.*, **20**(8), 1617–1625.
14. Khan, A. S., Talbot, J. A., Tieszen, K. L., Gardener, E. A., Gibson, J. M., and New J. P. (2004). Evaluation of a bedside blood ketone sensor: the effects of acidosis, hyperglycaemia and acetoacetate on sensor performance, *Diab t. Med.*, **21**(7), 782–785.
15. Ozkan, M., Erhan, E., Terzi, O., Tan, I., and Ozöner, S. K. (2009). Thermostable amperometric lactate biosensor with Clostridium thermocellum L-LDH for the measurement of blood lactate, *Talanta*, **79**(5), 1412–1417.

16. Rawson, F. J., Purcell, W. M., Xu, J., Pemberton, R. M., Fielden, P. R., Biddle, N., and Hart, J. P. (2009). A microband lactate biosensor fabricated using a water-based screen-printed carbon ink, *Talanta*, **77**(3), 1149–1154.
17. Tiwari, A., Aryal, S., Pilla, S., and Gong, S. (2009). An amperometric urea biosensor based on covalently immobilized urease on an electrode made of hyperbranched polyester functionalized gold nanoparticles, *Talanta*, **78**(4–5), 1401–1407.
18. Rahman, M. M., Shiddiky, M. J., Rahman, M. A., and Shim Y. B. (2009). A lactate biosensor based on lactate dehydrogenase/nictotinamide adenine dinucleotide (oxidized form) immobilized on a conducting polymer/multiwall carbon nanotube composite film, *Anal. Biochem.*, **384**(1), 159–165.
19. Ravindranath, S. P., Mauer, L. J., Deb-Roy, C., and Irudayaraj, J. (2009). Biofunctionalized magnetic nanoparticle integrated mid-infrared pathogen sensor for food matrixes, *Anal. Chem.*, **81**(8), 2840–2846.
20. Suarez, G., Jin, Y. H., Auerswald, J., Berchtold, S., Knapp, H. F., Diserens, J. M., Leterrier, Y., Månson, J. A., and Voirin, G. (2009). Lab-on-a-chip for multiplexed biosensing of residual antibiotics in milk, *Lab Chip.*, **9**(11), 1625–1630.

Chapter 2

Macromolecule-Imprinted Polymers: Antibody/Receptor Mimics for Protein Recognition and Catalysis

Yun Peng,[a] David W. Britt,[b] Marie K. Walsh,[b] and Timothy Doyle[c]

[a]*Department of Physics, Utah State University, Logan, UT 84322, USA*

[b]*Department of Biological Engineering, Utah State University, Logan, UT 84322, USA*

[c]*Department of Nutrition Dietetics and Food Science, Utah State University, Logan, UT 84322, USA*

david.britt@usu.edu

2.1 Introduction

Molecular recognition, the preferential attraction and specific binding between ligand and receptor molecules, is a fundamental concept in nature. It occurs when two molecules are geometrically and chemically complementary. Example interactions include enzyme/substrate, antigen/antibody, RNA/protein, as well as a host of small molecule/protein interactions. Molecular recognition has traditionally been viewed as a lock-and-key mechanism, but many examples of induced fit are now recognized. Building upon these highly specific interactions, many approaches have been explored to synthesize materials that mimic natural molecular recognition. Among the various existing approaches, molecular imprinting has received considerable attention, offering several distinct advantages suitable for a broad range of applications.

Molecular imprinting is the synthesis of a highly cross-linked polymer network containing template molecule–induced binding cavities that are complementary to both the spatial configuration and the functionality of the target template molecules. The earliest attempts to synthesize polymers for molecular recognition were inspired by the theory of Pauling, formulated in the 1940s, on the complex question of antibody formation in the immune system (Ansell, Ramstrom *et al.* 1996; Mosbach and Ramstrom 1996). Pauling proposed that antibodies behaved like denatured proteins in the presence of chemical functionalities on the antigen, where their amino acid chains would be free to move to form

Biomedical Nanosensors
Edited by Joseph Irudayaraj

ISBN 978-981-4303-03-3 (Hardcover), 978-981-4303-04-0 (eBook)
www.panstanford.com

an antigen-guided shape via a mechanism he termed "molecular complementariness." Thus the antibody would then memorize a specific moiety of the antigen (Pauling 1940). While incorrect, this idea that a freely moving polymer chain could form a complementary mold around a structure inspired the field of molecularly imprinted polymers (MIPs).

Since MIP technology was first developed, the majority of molecules imprinted have been targets of low molecular weights (MWs), with many successful examples exhibiting high selectivity for the template over similar molecules. In contrast, macromolecule-based MIPs have seen much less success. The first report on protein imprinting was published in 1985 (Glad, Norrlow *et al.* 1985), and since this seminal work, many intrinsic limitations of protein (macromolecule) imprinting have been identified, including permanent entrapment, chemical complexity, conformational flexibility, and solubility incompatibility with many polymer solutions (Turner, Jeans *et al.* 2006; Ge and Turner 2008). Macromolecule imprinting still draws considerable attention from both scientific and industrial communities, given the unique advantages of MIPs over natural receptors, such as the low cost and robustness of polymers lending MIP technology to a wide range of applications. In this review, we highlight advances in developing "plastic antibodies." Applications of protein MIPs, which have been expanded to areas of analytical separation, biosensors, environmental monitoring, medical diagnostics, and targeted therapeutics, will also be discussed. We present our recent MIP work aiming to detect misfolded/pathogenic proteins using novel functional monomers that may also serve as catalysts exhibiting artificial chaperone activity. We conclude with a preview of our applications of MIP-coated nanoparticle therapeutics that can be activated through electromagnetic (EM) excitation.

2.1.1 Strategies for Synthesis of Molecularly Imprinted Polymers

The basic principle for MIP synthesis is relatively straightforward, and most published protocols follow the same general procedure. The functional monomers are mixed with the complementary template to form a prepolymerization complex, which is subsequently polymerized with a cross-linker in the presence of a proper solvent, most often an aprotic and nonpolar solvent. The mixture is cured to give a porous, highly cross-linked, and rigid material. The template molecules are then removed by washing with a solvent or a combination of chemical or enzymatic treatments, resulting in an imprinted matrix with molecular-scale cavities complementary in shape and functionality to the template molecules. Molecularly imprinted polymers posses a "permanent memory" of the template, most often demonstrated by enhanced rebinding of template molecule to the MIP versus a nonimprinted polymer equivalent (NIP). The selectivity of the MIP for the original template over structural analogues is a more challenging test for MIPs that work best for low-MW templates. The basic principles of molecular imprinting are illustrated in the scheme shown in Fig. 2.1.

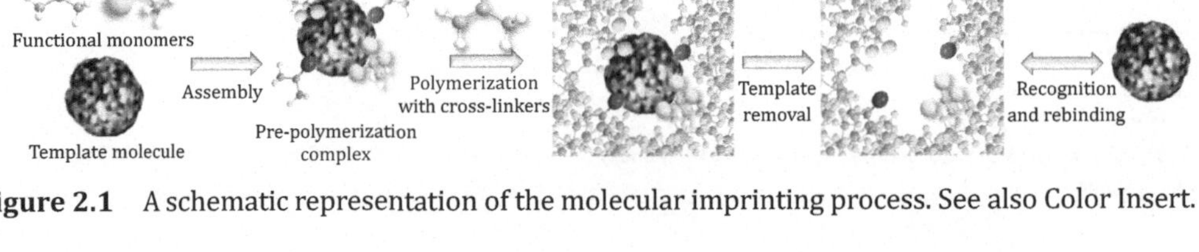

Figure 2.1 A schematic representation of the molecular imprinting process. See also Color Insert.

There are essentially two types of molecular imprinting strategies based on covalent and noncovalent interactions during the formation of the prepolymerization complex. For covalent MIPs, the functional monomers are covalently bonded with template molecules to form a well-defined template-monomer complex (prepolymerization complex). The chemical bonds ensure the correct spatial organization of the functional groups that is preserved during cross-linking. The template-functional monomer bonds are broken during template removal, leaving the functional groups in the binding cavities. The first successful covalent MIPs employed a simple sugar conjugated to a boronic acid derivative (Wulff 1972). The conjugated complex was then cross-linked with ethylene dimethacrylate. When the ester was cleaved, the rebinding of the template was found to be very selective. Due to the chemical linkage formed for the template-monomers complex, the binding sites will tend to be homogeneous and the formation of nontemplate-mediated binding sites in the polymer will be greatly decreased. However, the relatively specific structural requirements of covalent methods rather limit the variety of the templates that can be imprinted. Also, the relatively harsh conditions for the cleavage of the covalently bound template carry the risk of disrupting polymer rigidity.

Functional monomers oriented around templates using noncovalent interactions were explored in the decade following covalent MIPs and are used more extensively due to ease of formation of the template-monomer complex and template extraction. In this method, functional monomers and templates are mixed and allowed to form the prepolymerization complex by self-assembling. Typical template-monomer noncovalent interactions include hydrogen bonding, electrostatic interactions, π-π complexation, hydrophobic interactions, and metal-ligand complexation. After cross-linking with the matrix, the template can be removed via simple diffusion (for small molecules only), usually in a polar or acidic solvent that sufficiently overcomes the noncovalent interactions between the template and the imprinted polymer. Noncovalent imprinting allows for the introduction of a greater variety of functionalities into MIP binding sites and rapid rebinding kinetics. Thus, the majority of imprinting methods conducted recently, especially the imprinting of macromolecules, involves the use of the noncovalent technique. A summary of the characteristics of the covalent imprinting and noncovalent imprinting is given in Table 2.1.

Table 2.1 Characteristics of covalent imprinting and noncovalent imprinting methods

	Covalent imprinting	Noncovalent imprinting
Advantages	Low nonspecific rebinding Homogenous binding sites Strong and specific rebinding	Rapid and reversible rebinding Easy-to-remove template A wide variety of functional monomers
Disadvantages	Hard-to-remove template Slow rebinding kinetics Limited suitable template varieties	High nonspecific binding Heterogeneous binding sites

2.1.2 Small Molecule Imprinting: Validation of Mip Technology

The first imprinting attempts are attributed to the 1949 work of Dickey and involved the polymerizing of dye molecules in a sodium silicate matrix (Dickey 1949). The dye-imprinted silica demonstrated a pronounced selectivity for the template dye versus challenger dyes. In this case, no dye-specific functional monomers were employed. In 1972, a major advancement in molecular imprinting was achieved when template–"specific" functional monomers were employed in synthetic organic polymers (Takagishi 1972; Wulff 1972). Wulff proposed in his work that in addition to size and spatial complementarities, functional monomers were responsible for the molecular recognition by forming reversible covalent bonds with the template. Templates imprinted using this approach include glyceric acid (Wulff 1972), derivatives of mannose (Wulff 1977), galactose, fructose (Wulff and Haarer 1991), sialic acid (Kugimiya, Matsui *et al.* 1995), castasterone (Kugimiya, Matsui *et al.* 1998), and nucleotides (Sallacan, Zayats *et al.* 2002).

Following the demonstrations of covalent MIPs in the 1980s (Arshady 1981; Norrlow, Glad *et al.* 1984), noncovalent MIPs employing hydrogen bonds and electrostatic interactions have been extensively investigated. A variety of low-MW templates with high solubility in organic solvents have been imprinted in a range of polymers, resulting in a tremendous increase in the number of publications on MIPs in the last decade (Alexander, Andersson *et al.* 2006). A multitude of imaginative approaches to synthesize new imprint polymers has also been reported, such as the "sacrificial spacer approach" (Whitcombe, Rodriguez *et al.* 1995) or methods using metal coordination chemistry (Dhal 1991). Many of the imprinted polymers have been successively applied to electrophoresis and chromatographic analysis, solid phase extraction (Lai, Niessner *et al.* 2004; Caro, Marce *et al.* 2005; Pichon and Haupt 2006), enantiomer separation (Wang, Zhou *et al.* 2000; Sun, Yu *et al.* 2004), drug discovery (O'Connor, Paisner *et al.* 2007), controlled drug-release (Norell, Andersson *et al.* 1998; Hiratani and Alvarez-Lorenzo 2004; Kan, Geng *et al.* 2009), as well as sensor fabrication (Matsuguchi and Uno 2006; Uludag, Piletsky *et al.* 2007). Thus, molecular imprinting, after a half century of development, has emerged as a unique technology for the synthesis of polymeric receptors or artificial antibodies ("plastibodies"), offering a generic, robust, and cost-effective alternative to existing techniques such as monoclonal antibodies.

2.1.3 Macromolecule Imprinting: Toward Artificial Antibodies and Enzymes

Over the past 20 years, efforts to replicate macromolecular recognition behavior with synthetic polymers has received significant attention due to the success observed for many low-MW molecules (Ansell, Ramstrom *et al.* 1996; Ramstrom, Ye *et al.* 1996; Chianella, Lotierzo *et al.* 2002). Chemically synthesized imprints for a range of biological macromolecules and even microorganisms in aqueous solutions have been explored. Low-MW biological templates successfully imprinted include dipeptides (Kempe and Mosbach 1995) and endogenous neuropeptide (Leu-enkephalin) in acrylate-based polymers (Andersson, Muller *et al.* 1995). Hydrogel MIPs specific for the peptide hormone oxytocin have been demonstrated using only a portion of the hormone as the template, a process known as epitope imprinting (Rachkov and Minoura 2000). High-MW biological targets such as the tobacco mosaic virus have been imprinted using poly(allylamine hydrochloride) (Bolisay, Culver *et al.* 2006). More recently, *Escherichia coli*-imprinted gel granules synthesized by bulk imprinting were able to distinguish between different types and strains of bacteria by using free-zone electrophoresis (Bacskay, Takatsy *et al.* 2006).

Yeast imprints in sol-gels coupled to a quartz crystal microbalances (QCMs) showed high sensitivity for the detection of yeast and even the cell fragments in fluid conditions (Dickert and Hayden 2002). A similar process was employed for erythrocyte recognition (Dickert, Lieberzeit *et al.* 2004).

Among the many macromolecules, the imprinting of protein has received the most attention (Burow and Minoura, 1996; Bossi *et al.* 2001; Guo, Xia *et al.* 2004) owing to the growing demand for the production of cost-effective, robust, and reproducible protein recognition elements for protein-binding assays and associated applications (Mallik, Plunkett *et al.* 1994). Protein-selective MIPs are viewed as alternatives to the traditional antibodies and receptors used in diagnostics and biotechnology, which are often expensive to produce and subject to denaturation and aggregation with time and/or more extreme environments.

The standard "model proteins" for protein-selective MIPs such as albumin, lysozyme, and CyC have been intensively studied as templates (Liao, Wang *et al.* 1996; Hjerten, Liao *et al.* 1997; Tong, Hetenyi *et al.* 2001). However, the preparation of protein-selective MIPs in aqueous solvents faces a number of complications that are largely absent when targeting low-MW molecules. First of all, most well-established imprinting methods for small molecules cannot be directly applied to the imprinting of proteins due to their poor solubility in aprotic solvents, while the application of polar solvents weakens noncovalent linkages (e.g., electrostatic and hydrogen bonds) that are usually involved in the interaction of template proteins with the functional monomers. Second, reduced and restricted diffusion of high-MW protein molecules complicates template extraction and rebinding. Third, the flexible structures of proteins are prone to change with microenvironment conditions (e.g., temperature, pH, and ionic strength), making the imprinting of proteins difficult (Turner, Jeans *et al.* 2006). Novel methods for synthesis of MIPs addressing protein size and conformational flexibility are needed. Some distinct approaches will be discussed, including our approaches that leverage the protein flexibility and induced fit of molecular imprint–binding sites to develop conformation directing MIPs.

2.2 Molecular Imprinting of Proteins

2.2.1 Polymers for Protein Imprints

Due to the size, complexity, and susceptibility to denaturation and aggregation, the number of polymer systems suitable for protein MIPs is rather limited. Aqueous conditions, mild pH, and moderate temperatures are primary requirements. Secondary considerations include polymer porosity, integrity, and ease of protein-template removal.

2.2.1.1 Acrylamide-based MIPs

Acrylamide (AAm) gels were originally used in gel-electrophoresis for the separation and purification of high-MW proteins. For protein-selective MIPs, the hydrogen bond is often applied as a molecular recognition interaction. Owing to the fact that carboxyl groups can be both hydrogen donors and receptors, polymers formed by water-soluble acrylic monomers and their derivatives are among the most extensively used polymeric materials.

The polyacrylamide (pAAm) networks are relatively inert with low nonspecific protein binding (Tong, Hetenyi *et al.* 2001), and they reveal excellent biocompatibility *in vivo* (Saraydin, Unver-Saraydin *et al.* 2004). The polymerization mechanisms of the gel are well understood (Dunn 1993), and the gel properties can be easily manipulated to produce networks with a wide range of desired properties, including specific porosity, pore size, and even specific mechanical strengths.

Hjerten and coworkers demonstrated the synthesis of a pAAm gel and the imprinting of a large number of proteins. AAm and *N*, *N*-methylenebisacrylamide (MBisA) were used to produce a hydrogel with large pores. A series of proteins, including bovine hemoglobin (Hb), cytochrome C (CyC), and transferrin (Tf), as well as human growth hormone (HgH), RNase, and horse myoglobin (Mb) were evaluated as the target molecules (Liao, Wang *et al.* 1996; Hjerten, Liao *et al.* 1997). The synthesized bulk gel was granulated and packed into a column in which the template protein was washed out by acetic acid (HOAc) and sodium dodecyl sulfate (SDS). The protein recognition in this approach used affinity chromatography, in which various proteins passed through the column while only the target protein was selectively absorbed. By using this method, all protein templates were preferentially reabsorbed in the corresponding MIPed columns. A notable outcome from this work was that the horse Mb imprints were able to distinguish its target protein from whale-originated Mb, which has a very similar amino acid sequence and 3-D structure.

Following a similar approach, other proteins were recently imprinted in polymers from (mac)acrylic monomers. These proteins included bovine serum albumin (BSA) (Huang, Zhang *et al.* 2005; Pang, Cheng *et al.* 2006), lysozyme (Lyz) (Guo, Xia *et al.* 2004; Huang, Zhang *et al.* 2005; Kimhi and Bianco-Peled 2007), trypsin (Vaidya, Lele *et al.* 2001), and Hb (Guo, Xia *et al.* 2004). It is commonly accepted in these reports that steric factors as well as multiple-point electrostatic interactions contribute to the overall strong interactions between the polymers and the template proteins and thus directly influence the capacity and selectivity of the imprints. However, a recent study on the effect of charge density to the molecular recognition of imprinted hydrogel revealed that the interplay between hydrogel swelling, electrostatic interaction between the hydrogel and the template, and cavity shape recognition mechanisms that occur within imprinted polymers remain difficult to interpret (Janiak, Ayyub *et al.* 2009).

While high-template readsorption to AAm MIPs over corresponding NIPs are widely reported, the discrimination efficiency of the imprinted AAm-based polymer strongly depends on the ionic strength, cross-linking degree, and amount of template removal (Ou, Wu *et al.* 2004). In addition, the mobility of large molecules in acrylate chemistry–based imprints requires the gel form with a low degree of cross-linking, which inevitably leads to reduced stability of the material's mechanical properties (Conrad II 2005). To overcome this problem, Guo *et al.* prepared a protein-imprinted pAAm gel incorporating macroporous beads produced from chitosan (CS), a natural biopolymer composed of D-glucosamine and D-acetyl-glucosamine (Guo, Xia *et al.* 2004). Applied as a support, the CS beads allowed both the AAm and the template proteins (Hb) to diffuse into the pores for polymerization. The polymer beads obtained from the approach were hydrophobic and demonstrated good mechanical strength as well as sufficient chemical stability. Moreover, the Hb selectivity assay using BSA as a control showed a 20-fold difference in disassociation coefficients (K_D), attributed to the distinct configurations of these globular proteins.

2.2.1.2 Stimuli-sensitive hydrogel

Stimuli-sensitive hydrogels undergo reversible volume transitions upon triggering by template binding or external stimulus such as temperature (Tanaka 1978; Hirokawa 1984; Amiya, Hirokawa *et al.* 1987; Chen and Hoffman 1995; Yoshida, Uchida *et al.* 1995), pH (Tanaka 1980), solvent composition, electric field (Tanaka 1982; Osada, Okuzaki *et al.* 1992), ionic strength, light, and specific chemicals (Tanaka 1978; Tanaka 1980; Hirokawa 1984; Dusek 1993; Chen and Hoffman 1995; Okano 1998). These environmental-sensitive properties, high compatibility with aqueous solvents, and gel flexibility that may facilitate an induced fit provide versatile means for MIP fabrication in stimuli-sensitive hydrogels.

The employment of template-sensitive hydrogel for imprinting protein was conducted by Zhao and coworkers (Chen, Hua *et al.* 2008). The stimuli-responsive MIP, synthesized by using methacrylic acid (MAA) and AAm as functional monomers and N, N-MBisA as the cross-linker, responded to changes in temperature and solution ionic strength. More importantly, a notable shrinkage of the gel occurred upon its binding with imprinted proteins (lyz or CyC). The volume change was reversible: once the protein was released from the binding cavities, the hydrogel returned to its original size (Fig. 2.2). The process was considered to be the result of sequential adaptation steps: The protein-responsive MIP may first adapt a state favorable for target binding in the presence of proteins. The electrostatic forces that predominate the interactions between functional monomers and the protein favor adsorption of target proteins, which become partially fixed into the binding cavities. Subsequently, the flexibility of the polymer allows the functional groups to appropriately align around the protein to form a complementary structure, resulting in specific volume shrinking. While the mechanism for gel shrinkage is not clearly understood, the report provides a clear demonstration of "template induced fit" binding sites in protein MIPs.

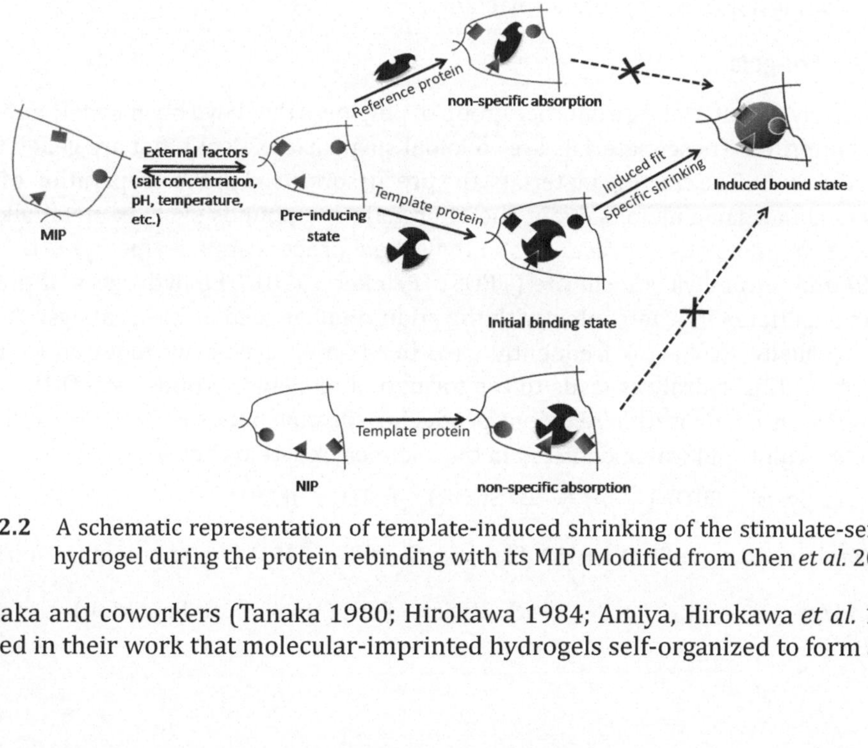

Figure 2.2 A schematic representation of template-induced shrinking of the stimulate-sensitive hydrogel during the protein rebinding with its MIP (Modified from Chen *et al.* 2008).

Tanaka and coworkers (Tanaka 1980; Hirokawa 1984; Amiya, Hirokawa *et al.* 1987) proposed in their work that molecular-imprinted hydrogels self-organized to form active sites with significantly different affinities to the target molecules in their swollen and collapsed states. This was later supported by Watanabe *et al.* (1998), who synthesized a temperature-sensitive hydrogel by copolymerizing N-isopropylacrylamide (NIPPAm) and acrylic acid (AA) with a cross-linker in the presence of template norephedrine. The polymer exhibited swelling at low temperatures and collapse at higher temperatures. Upon reintroducing the template molecules, the swollen gel at the low temperature exhibited no further volume change while the collapsed high-temperature hydrogel showed an increased swelling ratio. Cross-challenges with adrenaline demonstrated the imprinted hydrogel to be template sensitive and illustrated that the imprinted hydrogel in the collapsed state rather than swollen state memorized the template. It was also revealed from the report that the molecular memory function of the hydrogel was strongly affected by the gel preparation conditions, as demonstrated by their result that the polymer-specific volume did not respond to the imprint molecule when the solvent was changed from 1, 4-dioxane to water.

Demirel *et al.* also use a pH/temperature-sensitive hydrogel for imprinting BSA in poly(N-tert-butylAAm-coAAm/maleic acid) gels (Demirel, Ozcetin *et al.* 2005). Rebinding studies showed that pH, temperature, as well as the initial BSA concentration influenced the adsorption capacity of both MIP and NIP hydrogels. The maximum BSA binding to MIPed hydrogels occurred at pH 5.0, where BSA possesses a net negative charge. Although increased temperature did not favor BSA adsorption, the BSA-MIPed hydrogels exhibited a higher template-binding capability than nonimprinted polymers between 5°C and 40°C. Moreover, the binding capacity was also found to be directly proportional to the concentration of template protein during imprint formation. The BSA adsorption capacity for the hydrogel imprinted with 8.63 wt.% BSA was as high as ~59 mg of BSA per g of dry hydrogel in BSA solution. The adsorption for the reference protein casein was considerably low, indicating a degree of protein selectivity.

2.2.1.3 Sol-gels

Sol-gel-derived materials are another group of polymers that have been widely applied for protein imprints. These materials are colloidal suspensions ('sol') that are gelled to form a porous matrix. The starting materials, the precursors, used in the preparation of the sol are usually inorganic metal salts or metal organic compounds such as metal alkoxides. The most commonly used precursors in the sol-gel process are tetramethyl-orthosilicate (TMOS) and tetraethylorthosilicate (TEOS) (Brinker 1990). The hydrolysis of precursor regents is initiated in water, often with the addition of an acid or base catalyst. A mutual solvent (usually alcohol) is frequently used but is not necessarily required (Avnir and Kaufman 1987). Hydrolysis leads to the formation of silanol groups (≡Si–OH), followed by progressive condensation reactions producing siloxane bonds (≡Si–O–Si≡) and the by-products alcohol and water, depicted in the chemical reactions below:

$$\text{Hydrolysis:} \quad Si(OR)_4 + nH_2O \Rightarrow Si(OR)_{4-n}(OH)_n + nROH \tag{2.1}$$

$$\text{Condensation:} \quad \equiv Si{-}OH + HO{-}Si\equiv \Rightarrow \equiv Si{-}O{-}Si\equiv + H_2O \text{ (water condensation)} \tag{2.2}$$

$$\text{and/or} \quad \equiv Si{-}OR + HO{-}Si\equiv \Rightarrow \equiv Si{-}O{-}Si\equiv + ROH \text{ (alcohol condensation)} \tag{2.3}$$

During the sol-gel phase transition, the viscosity of the solution gradually increases as the colloidal sol becomes interconnected to form a rigid, porous network gel. Thus, the sol-gel process provides a facile method to incorporate numerous organic, organometallic, and biological molecules into a microporous network prepared under aqueous conditions. The physicochemical properties (e.g., the average pore size, pore distribution, the pore shape, and the surface area) of sol-gel-derived materials can be tuned by the process parameters such as acid or base catalysis, the addition of salts, concentration and chemistry of precursor (hydrophilic/hydrophobic/electrophilic), temperature, and aging (storage) conditions. Generally, a low-pH, low-water ratio results in denser materials with smaller average pore sizes, whereas high-pH, high-water preparations lead to more porous materials (Brinker 1990; Hench and West 1990; Buckley and Greenblatt 1994). Controlling these factors can be of significant importance for biological molecule activity in sol-gels as well as the molecular recognition MIP sol-gel process.

The earliest effort to prove that silane imprints could distinguish differences in the tertiary structure of proteins was made by Mosbach's group (Glad, Norrlow *et al.* 1985). A boronatesilane was used as the functional monomers for interaction with the carbohydrate segment of template glycoprotein Tf, followed by polymerization in a silicate matrix on the surface of porous silica beads. The selectivity of the Tf-imprinted silica beads was demonstrated by a high-performance liquid chromatography (HPLC) assay that showed higher affinity for the template Tf than for a BSA control. As a consequence of this work, protein imprints using silicates and polysiloxanes received considerable attention in the following years.

Particularly, Venton and coworkers (Venton and Gudipati 1995; Venton and Gudipati 1995) proposed in their reports that the protein-complementary binding pocket formed in the polysiloxane imprints results from the interaction of organo-functional side chains of silica-based monomers with template proteins (urease or BSA). The monolithic gel prepared with 3-aminopropyl triethoxysilane (APTES) as functional monomers and TEOS cross-linker (1:3 molar ratio) showed preferential rebinding of the target proteins. But they failed to show any imprinting effects when two more closely related proteins ($_{125}$I-Hb and $_{125}$I-Mb) were studied as the templates, suggesting that the chemistry and geometry of the binding pockets needed to be further modified to distinguish between closely related proteins. This was also supported from a recent report by Zhang *et al.* (2006), who combined the sol-gel and self-assembly technology to prepare a human serum albumin (HSA)-imprinted thin film on the surface of the piezoelectric quartz crystal Au-electrode modified with thioglycolic acid. By using the piezoelectric quartz crystal impedance and electrochemical impedance techniques, selective rebinding of the template protein to the imprinted surface was demonstrated; however, the binding capacity was affected by many factors, such as salts, solvents, the temperature, and the pH. In addition, Lee *et al.* (2007) recently correlated the expected relationship between the number of potential binding sites on macroporous polysiloxane scaffolds to the concentration of the template protein loaded into the sol before polymerization. The conclusion was based on their work on bulk imprinting of Lyz or RNase A via sol-gel processing with TEOS used as the cross-linker and APTES as the functional monomer. With their method, up to 62% of the loaded protein was surface accessible to the scaffolds. The imprinted scaffolds exhibited up to three times preference for binding with template proteins, even in the presence of a similar-sized competitor.

Sol-gel materials have a combination of advantages over many other polymers for protein imprinting. First, their unique mild aqueous preparation conditions favor the retention of the imprinted protein structure and biological properties inside the polymer matrix. Second, the rigid property of the sol-gel cross-linking network is favorable for retaining the shape of the cavities after the removal of the templates. In addition, sol-gel glasses are optically transparent and electro-, photo-, and chemically stable, the thickness, porosity, and surface area of the material are easily controlled and selectivity and diffusion are comparable and often better than acrylic polymer-based imprinted materials (Lahav, Kharitonov *et al.* 2001; Marx and Liron 2001).

2.2.1.4 3-Aminophenylboronic acid–based MIPs

3-aminophenylboronic acid (APBA) is an attractive "one-step" functional monomer/matrix polymer for protein imprinting because it presents a high content of hydroxyl groups that mediate the reversible recognition of various classes of biomacromolecules via electrostatic and hydrogen-bonding interactions and the mild polymerization conditions are nondisruptive to most biomacromolecules (Deore and Freund 2003; Rick and Chou 2005; Rick and Chou 2006). In addition, poly-APBA can be grafted tightly onto the surface of materials such as polystyrene (Piletsky, Piletska *et al.* 2001; Lu, Yan *et al.* 2009), glass (Chou, Rick *et al.* 2005) gold, and platinum (Rick and Chou 2006) by aromatic ring electron-pairing interactions (Piletsky, Piletska *et al.* 2000), offering the ease of integrating the polymer on a variety of sensor surfaces.

With the aim of constructing inexpensive sensors, Rick and Chou (2006) fabricated protein-imprinted electrodes by the cyclic voltammetric deposition of three polymer layers. An initial layer of polypyrrole was used as a supporting polymer, which was coated with two layers of poly-APBA, where only the outermost poly-APBA layer was imprinted with Lyz or CyC. Protein rebinding was determined via the current transmission reductions on the screen-imprinted platinum electrode attributed to the insertion of the nonconductive target proteins into imprinted binding pockets. Rebinding of Lyz or CyC to their respective imprints showed a distinct two-phase binding profile, while the target protein capture on the nonimprinted electrode only exhibited a progressive binding typical of nonspecific recognition. In the presence of 1 ppm proteins in solution, the current reductions for the Lyz- and CyC-imprinted electrodes were 30.3% and 66.2%, respectively, compared to 4.5% and 29.9% for their respective control electrodes.

Poly-APBA protein imprints have also been coupled to QCM gold chips, where discrimination between CyC and Lyz imprints was demonstrated (Rick and Chou 2005). Also employing QCM, Turner *et al.* employed APBA as a protein imprint matrix for discriminating distinct isoforms of bovine β-lactoglobulin (BLG) induced by temperature or fluoro-alcohols (Turner, Liu *et al.* 2007). The APBA MIPs in this report showed good steric specificity for the template conformation as demonstrated by reduced nonspecific binding of native BLG to MIPs formed against BLG imprinted in increasing concentrations of fluoroalcohol.

Although the APBA chemistry is a fairly simple protocol that conforms to the environmental constraints of protein imprinting, the interactions are highly dependent on environment with respect to rebinding and MIP stability. Lu *et al.* recently demonstrated that the adsorption of bovine Hb onto poly-APBA-grafted polystyrene microbeads was pH dependent (Lu, Yan *et al.* 2009), where the strongest binding occurred at pH 6.0. The enhanced binding capability was considered to be the result of increased electrostatic interactions between the positively charged target protein and the negatively charged poly-APBA at pH 6.0, whereas a higher or lower pH all led to repulsive electrostatic interactions. In addition, support material may also play a role for the stability of the APBA imprints, as poly-APBA film supported by polystyrene (core-shell microbeads) possessed high stability toward its model protein Hb while poly-APBA-grafted glass rapidly lost their affinity to template proteins. Additional examples of the influence of the supporting surface, pH, and protein isoelectric point on protein-poly-APBA MIPs are highlighted in the review by Turner *et al.* (2006).

2.2.2 Surface Imprinting

Imprinting a protein template in a bulk polymer matrix is the most straightforward approach for molecular imprinting. While this method works well for small templates, macromolecule templates may become permanently entrapped in a rigid-network polymer. The addition of a porogen during the imprinting process leads to the formation of macropores that facilitate the exposure of binding sites and improve mass transport; however, the decreased polymer density may compromise the matrix stability, resulting in a loss of recognition and binding efficiency after a few rounds of regeneration (Turner, Jeans *et al.* 2006). Harsh washing conditions with surfactants, acids, base, and/or protease may be necessary to remove the template, all of which may compromise the integrity of the imprinted binding site. Given the limitations associated with protein imprinting in bulk polymers, a number of surface-imprinting approaches have been developed for protein-selective MIPs.

A fundamental strategy of surface imprinting is to create a high surface-to-volume ratio and restrict the imprinted sites at or close to the surface of the MIP. Decreased selectivity may result because only part of the protein is bound and recognized, which in turn may lead to high binding-site heterogeneity. However, these limitations are often readily addressed by modifications of the polymerization protocols. Surface MIPs can often be constructed through the adaptation of a bulk-imprinting protocol combined with the appropriate selection of a support (Mirsky, Hirsch *et al.* 1999; Fang, Tan *et al.* 2005; Ouyang, Lei *et al.* 2008). Successful examples of surface imprinting on thin films and nanobeads and wires are discussed next.

2.2.2.1 Thin film

Poly-APBA was previously discussed as favorable for protein imprinting (Piletsky, Piletska *et al.* 2000), where Piletsy and coworkers described the technique for coating polystyrene microplates with a thin film of epinephrine-imprinted 3-amino-phenylboronic acid. The successful imprinting of terbumetone, a triazine herbicide, into APBA thin films has subsequently been reported (Sergeyeva, Matuschewski *et al.* 2001). These imprinted membranes introduced specific binding sites into porous membranes without reducing their high permeability and demonstrated remarkably increased affinity to template molecules. Bossi *et al.* (2001) applied the thin film for the fabrication of protein imprints. They coated the polystyrene surface of microplate wells with a thin polymer layer of poly-APBA and employed various proteins (lactoperoxidase, haemoglobin, horseradish peroxidase, CyC, and microperoxidase), covering a range of protein sizes and isoelectric points. The imprinting resulted in an increased affinity of the polymer toward the corresponding templates, while the protein desorption was size dependent, as demonstrated from the result that the dissociation constants of small proteins to the corresponding imprinted membrane were about 10^3 higher than that of large proteins. In addition, the imprint film revealed significant selectivity on recognizing Hb from its glycated isoforms (Hb-A1), which has only a small charge variance with Hb.

Microcontact imprinting is a classic example of surface imprinting developed by Ratner's group (Shi, Tsai *et al.* 1999). In this method, the template protein was preadsorbed and dried onto freshly cleaved mica, which was selected as it is atomically smooth. The adsorbed protein was then coated with disaccharide and then a thin film (10–30 nm) of hexafluoropropylene by radio frequency glow-discharge (RFGD) plasma deposition to "freeze" the disaccharide distribution. Molecular recognition was attributed to the multiple hydrogen bonds formed between the template protein and the hydroxyl groups of the disaccharides. The mica was peeled away, and the protein template was digested, leaving polysaccharide-decorated cavities complementary to the protein templates, as depicted in Fig. 2.3. Several proteins, including BSA, immunoglobulin G (IgG), Lyz, and RNase, were imprinted and cross-challenged by this approach (Ratner 1996; Ratner and Shi 1999; Shi, Tsai *et al.* 1999; Shi and Ratner 2000). Imprints were able to discriminate between Lyz and RNase, two proteins with comparable molecular dimensions and isoelectric points. The RNAse-MIP exhibited a 20-fold increase in selectivity over Lyz while the Lyz-MIP exhibited a 26-fold greater uptake of Lyz over RNAase, suggesting that the hydrogen bonding between the target protein and immobilized disaccharide functional monomers provides for high specificity.

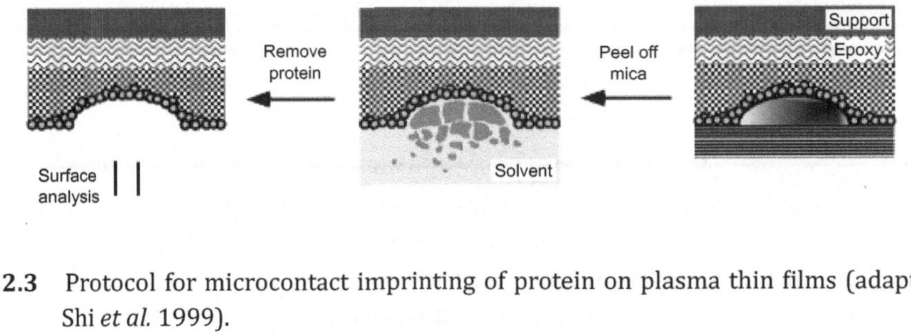

Figure 2.3 Protocol for microcontact imprinting of protein on plasma thin films (adapted from Shi *et al.* 1999).

Expanding from this technique, Chou and coworkers (2005) developed a method to create specific cavities for a multimeric protein. They prepared a single layer of a template protein together with functional monomers (O-4-nitrophenylphosphorylcholine, 4NPPC) on the surface of a glass cover slip. C-reactive protein (CRP), a protein with a pentameric structure assembled by five identical nonglycosylated subunits, was employed as the template protein. The protein layer was then deposited on the glass support and coated with a cross-linker (polyethylene glycol 400 dimethacrylate, PEG400MAA) and initiator for photo-polymerization. A thin film with CRP recognition sites was formed after the removal of the glass slide (Fig. 2.4). The rebinding experiments showed the CRP-binding capacity on the imprint to be 1 μg/cm^2, which was four times higher than when the MIP was cross-challenged with HSA and 346 times higher than for Lyz challenges. In the presence of competitive proteins, the polymer imprinted with CRP is able to bind 3.78 ng/cm^2 of CRP protein versus 0.08 ng/cm^2 of the competitor HSA. Advantages of microcontact imprinting include the use of only small amounts of monomer, solvent compatibility of the template, and the possibility to form template recognition sites in a solvent-free system (Chou, Rick *et al.* 2005; Rick and Chou 2006).

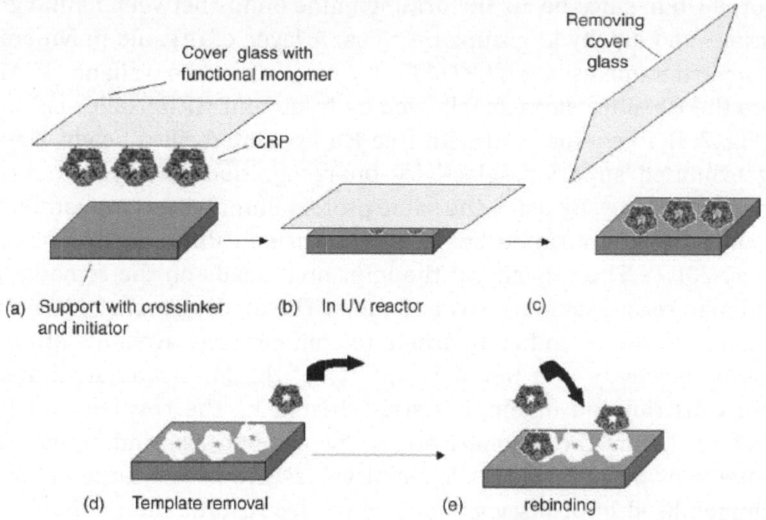

Figure 2.4 The schematic process of CRP microcontact imprinting and rebinding (adapted from Chou *et al.* 2005).

2.2.2.2 Core-shell microsphere imprinting

The first successful protein imprint on the surface of porous silica beads was developed by Mosbach's group (Glad, Norrlow *et al.* 1985). They polymerized a mixture of silanes on the surface of silica particles, leading to the formation of a thin layer of polymer network around the silica. Glycoprotein Tf was imprinted as the template, and boronatesilane functional monomers were added for interaction with the carbohydrate portion of Tf in the mixture. The enzymatic properties of the coated beads were about eight times higher than that from the ground bulk polymer produced from the same solution mixture. The competitive rebinding experiments analyzed by HPLC also revealed that Tf-imprinted microbeads have a higher affinity for Tf over BSA.

A number of proteins have been imprinted by a similar approach. For example, Burow imprinted glucose oxidase (GO) using acrylates on functionalized silica bead supports (Burow and Minoura 1996; Hirayama, Burow *et al.* 1998). Hirayama reported Lyz imprinting on silica beads with surface-modified AAm and AA or AAm and N, N-dimethylaminopropylacrylamide with $(NH4)_2S_2O_8$ (Hirayama, Sakai *et al.* 2001). Besides silica beads, agarose microsphare/beads, CS beads, high-density cross-linked pAAM beads, and mesoporous chloromethylated polystyrene beads have also been applied as the core and/or macroporuous support for surface imprinting.

For most protein imprinting, a template protein was usually added into the solution mixture before polymerization was initiated, where the protein interacts with polymer matrix via noncovalent interactions such as electrostatic, multipoint hydrogen bonds, and hydrophobic interaction as well as a shape conformation between the template protein and the polymer (Hirayama, Burow *et al.* 1998). Covalent immobilization of the template to beads has also been exploited by some groups for the synthesis of protein MIPs. Shiomi *et al.* anchored Hb to silica beads by forming imine bonds between amino groups on the protein surface and aldehyde groups on silica. A layer of organic polymers made from 3-aminopropyltrimethoxysilane (APTMS) and propyltrimethoxysilane (PTMS) was then formed onto the Hb-silica surface, followed by treatment with oxalic acid to remove the template (Fig. 2.5). In comparison with free-Hb-imprinted silica beads, Hb-immobilized imprinting exhibited superior selectivity on recognition of Hb, and the nonspecific adsorption was very low. By using the same protein immobilization method, Bonini *et al.* fabricated silica-based imprinted beads for HSA using APBA instead of silanes (Bonini, Piletsky *et al.* 2007). The capacity of the imprinted beads on the removal/depletion of HSA from human serum samples was evaluated. The binding capacity (1.4 mg/g) for the imprinted material was found comparable to commercially available affinity materials. The increased selectivity and binding capacity of the MIPs was attributed to a more homogeneous distribution of binding sites created by the covalent anchorage of the template before the matrix polymerization. The competitive binding with a mixture of neutral proteins or proteins with similar pI values showed that the recognition in the template-immobilized imprints was not due to electrostatic interactions and hydrogen bonds but mainly from the complementary shape of binding cavities.

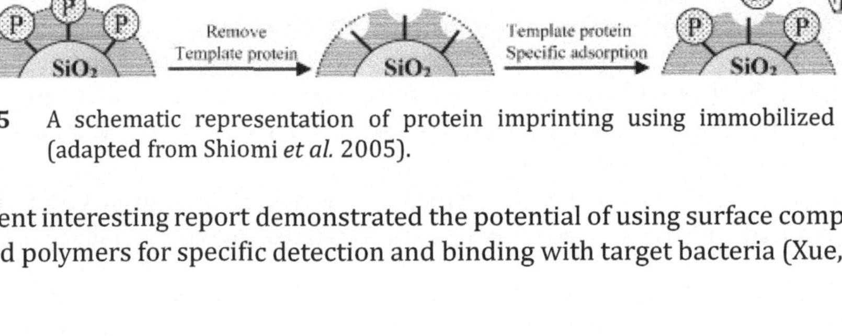

Figure 2.5 A schematic representation of protein imprinting using immobilized templates (adapted from Shiomi *et al.* 2005).

A recent interesting report demonstrated the potential of using surface components of imprinted polymers for specific detection and binding with target bacteria (Xue, Pan *et al.*

2009). In this work, *Staphylococcus aureus* (*S. aureus*) protein A (SpA), a specific surface protein for *S. aureus*, was utilized as the template protein and imprinted on pAAm gel beads, which were synthesized by inverse-phase suspension polymerization. The imprinted gel beads displayed a high adsorption capacity and specificity for SpA. In addition, the SpA imprints showed significant increased specificity and affinity for *S. aureus* among several types of competitive bacteria. The adsorbed amount of *S. aureus* was 10^3–10^4 CFU/g of gel beads. In contrast, nonimprinted polymer beads did not show any enhanced binding for either target or competitive bacteria.

An example with good utilization of sphere shapes of the support material is from the work of Tan and coworkers (2008). They employed a two-stage core-shell miniemulsion polymerization system for the fabrication of the BSA surface–imprinted submicrometer particles (0.5–0.6 μm) by using methyl methacrylate (MMA) and ethylene glycol dimethacrylate (EGDMA) as the functional and cross-linking monomers, respectively. Moreover, they made magnetic particles by encapsulating nanosized Fe_3O_4 magnetite into the polymeric support core in order to permit the isolation of imprinted particles using an external magnetic field (Fig. 2.6). Particles imprinted to recognize protein richly expressed in tumor tissues could be used either as affinity adsorbent in protein purification or as probe for the tumor imaging through magnetic resonance imaging. The report revealed that further modification of support material could provide the imprinting material with novel properties, which may play key roles for exploring the applications of the protein MIPs.

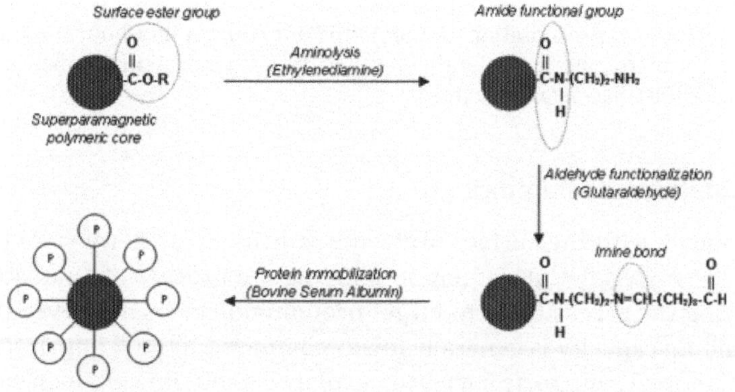

Figure 2.6 Surface imprinting of protein on magnetic particles by the covalent immobilization of template protein (Tan *et al.* 2008).

In a recent report, Li *et al.* (2008) prepared core-shell protein imprinting by using high-density cross-linked CS, a polysaccharide functional biopolymer, as the support sphere. Two organic silanes (APTMS and TEOS) were assembled and polymerized on the polysaccharide-protein surface via the sol-gel process in an aqueous solution. After the removal of the template (BSA), which was covalently immobilized by forming imine bonds with the functional amine groups of CS, large quantities of well-distributed pores were formed on the imprinted polymer shell surface. Due to the homogeneous binding sites formed by template immobilization, the protein-imprinted sol-gel surface exhibited a prevalent preference for the template protein over competitive proteins (Fig. 2.7). Moreover, template-immobilized imprints showed much higher BSA adsorption capacity than free-template-imprinted polymer beads.

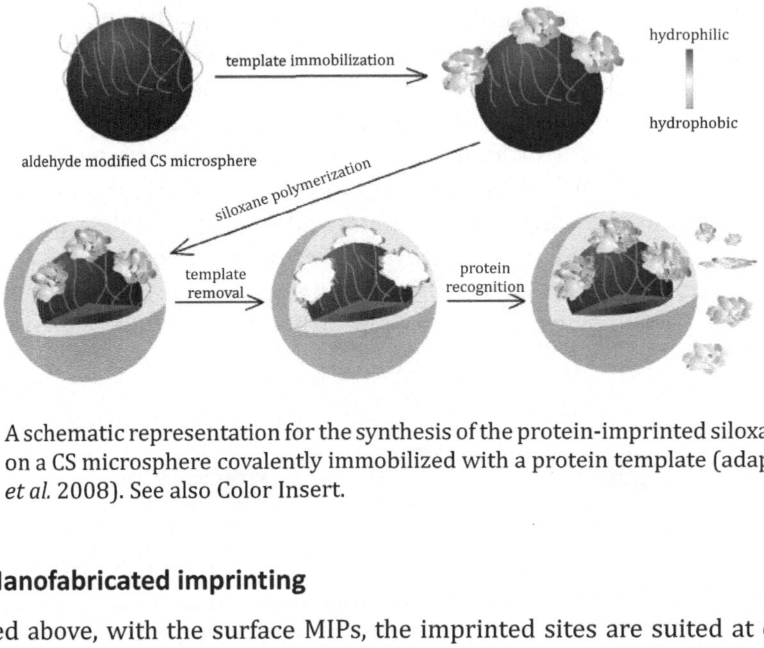

Figure 2.7 A schematic representation for the synthesis of the protein-imprinted siloxane polymer on a CS microsphere covalently immobilized with a protein template (adapted from Li *et al.* 2008). See also Color Insert.

2.2.2.3 Nanofabricated imprinting

As discussed above, with the surface MIPs, the imprinted sites are suited at or close to the surface of the material, thus largely simplifying the process of template removal and greatly enhancing the accessibility to target protein molecules. However, it is noted that the protein-binding capacity is generally lower compared to bulk imprinting. To address this problem, nanoscale supports have been explored to obtain very high surface-area-to-volume ratios.

Yilmaz *et al.* introduced a surface-imprinting method by immobilizing a theophyline template on porous silica beads. Different from conventional MIPs, the silica support is removed along with the template (Yilmaz, Haupt *et al.* 2000). Inspired by this idea, Li *et al.* (2006) developed a surface-imprinted polymer nanowire for protein recognition. The template protein was first immobilized on the pore walls of the nanoporous alumina and then filled with a mixture of AAm and N, N-MBisA. After polymerization in the nanopores, the template and the alumina membrane were subsequently removed by chemical dissolution, leaving behind polymer nanowires with protein-binding cavities situated on the surface (Fig. 2.8). The binding and competitive rebinding studies demonstrated the enlarged surface area on polymer nanowire results in a high protein-binding capacity and specificity.

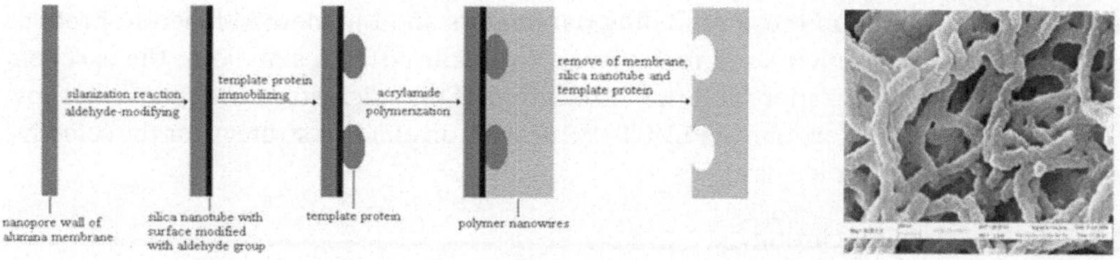

Figure 2.8 A schematic representation of the nanowire protein-imprinting approach by using an immobilized template and a sacrificial support (left) and SEM of pAAm nanowires (right) (adapted from Li *et al.* 2006).

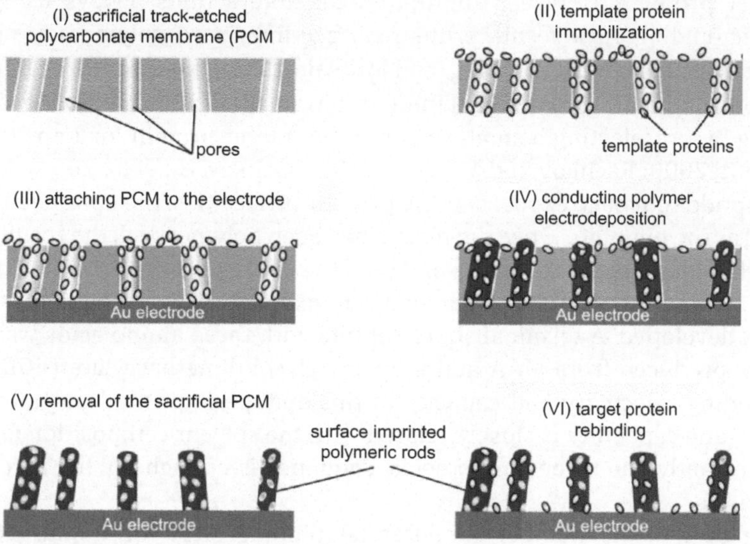

Figure 2.9 A schematic representation of the nanorods protein imprinting (adapted from Öpik *et al,* 2009).

However, the main disadvantage of this method is the weak adsorption of proteins on the alumina membrane requiring multistep treatment of the alumina before the proteins are attached. Also, the control of the polymerization is complicated (Öpik, Menaker *et al.* 2009). For this reason, Öpik adopted the technique employing nanorods for protein imprinting (Öpik, Menaker *et al.* 2009). In this report, fluorescently labelled avidin (AV-FITC), the template protein was allowed to interact with a polycarbonate membrane (PCM) with track-etched pores. Due to the hydrophobic properties of the membrane, the proteins attached to the pore wall simply by physical adsorption. The protein-modified PCM was located on the surface of a gold electrode, followed by electro-deposition polymerization of the conductive matrix, polystyrene sulphonate (PSS) doped poly (3, 4-ethylenedioxythiophene) (PEDOT). After removing the PCM membrane by dissolving in dichloromethane or chloroform, micro- to nanoscale PEDOT-PSS rods with complementary cavities for the target protein on the surface were synthesized (Fig. 2.9). The subsequent protein-rebinding assays revealed the method to be effective for protein recognition in terms of the binding capacity and selectivity. Although there is no direct comparison on those rebinding parameters and the method failed to produce monodispserse rods (microsized rods were also produced), the simplicity, the excellent control of the polymerization, and the localization of the microstructures as well as low nonspecific protein adsorption of PEDOT-PSS provide an efficient strategy for the selective recognition of macromolecules.

2.2.2.4 Epitope imprinting

The nature of surface imprinting results in the imprinting of only a portion of the target protein template. The recognition specificity depends on the steric properties and the complementary chemistries of the binding sites by forming numerous interaction points with target proteins. However, multiple weak interactions always lead to increased nonspecific binding. As a result, protein surface-imprinted polymers are susceptible to cross-reactivity, nonspecific binding, and low selectivity. Inspired by the inherent epitope recognition and binding for cell-cell interaction, an alternative method leverages epitope imprinting by preselecting a small region of the target protein for imprinting (Rachkov and Minoura 2000; Rachkov and Minoura 2001; Rachkov, Hu *et al.* 2004). In this method, a short peptide sequence exposed at the protein surface (i.e., the epitope of an antigen) is imprinted as the template. Once the matrix has been polymerized, the resultant imprinted material should be able to recognize and bind the entire protein. In this fashion, imprinted polymers for the recognition of nanopeptide oxytocin (a neurohypophyseal hormone) have been developed. A terminal short peptide with three amino acids was imprinted in the matrix produced from MAA and ethylene glycol dimethacrylate (EGDMA). The test on the binding selectivity and capacity of this epitope-imprinted polymer showed that the recognition capability is closely related with the solvent composition and the pH, and most importantly, the three amino acids' sequence is enough for the recognition of the whole peptide.

Owing to the low number of potential binding sites for nontarget protein, the nonspecific binding in epitope imprinting is largely reduced. The greatly enhanced specificity mainly results from the formation of cooperative and multivalent interactions between the target proteins and the imprinted cavities. Noncovalent linkages such as hydrogen bonds and hydrophobic interactions are the most widely applied interactions in epitope imprinting (Nishino, Huang *et al.* 2006). Other than this, Özcan *et al.* employed the metal-coordination interaction for CyC recognition (Ozcan, Say *et al.* 2006). The histidine residue exposed on the protein surface was imprinted as an epitope template, and a metal-chelating functional monomer, N-methacryloyl-(L)-histidine-Cu^{2+}, was integrated into the matrix polymer, which forms a metal-coordination interaction with the histidine. After the removal of template, the final imprinting material reveals even better stability, selectivity, and binding affinity. However, this approach is confined to protein templates (e.g., Hb and Mb) containing exposed residues that can form a metal coordination bond with the monomer.

More recently, a remarkable example of epitope imprinting for the capture of a protein has been reported by Shea and coworkers (Nishino, Huang *et al.* 2006). Three proteins, CyC, alcohol dehydrogenase (ADH) and BSA, were studied. The peptide sequences corresponding to the C-terminus domains of these three proteins were chemically synthesized and applied as the imprinting epitope. Each nonapeptide was covalently attached to a glass or oxidized silicon surface, followed by the polymerization of monomers containing AAm, N, N-ethylene-bis-acrylamide, and polyethylene glycol 200-diacrylate to produce a thin film (Fig. 2.10). The analysis of protein uptake revealed a roughly sevenfold higher affinity and capacity in the resulting film than nonimprinted film. Moreover, their epitope imprinting of BSA showed a significantly high selectivity for BSA when the correct peptide was employed as the epitope.

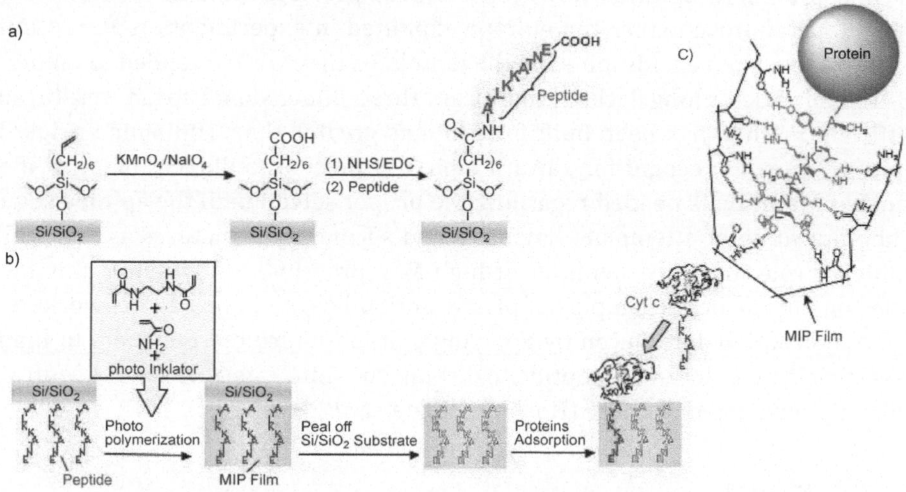

Figure 2.10 The protocol for epitope imprinting of proteins on glass or oxidized silicon surfaces. (a) *Glass surface modification and peptide attachment at N-terminal:* The vinylsilane surface was oxidized with $KMnO_4$, $NaIO_4$, and K_2CO_3, followed by the activation of the carboxylic acid-terminated monolayer with N-hydroxylsuccinimde/1-ethyl-3-[3-dimethylaminopropyl] carbodiimide (NHS/EDC). The peptide epitopes were subsequently attached to the substrate in a solution of phosphate buffer and NaCl. (b) *MIP film fabrication and evaluation:* The C-terminus nonapeptide epitope was attached through a tether to a glass or oxidized silicon surface by the N-terminal amino acid of the peptide. Monomers were photo-polymerized with a UV lamp; the glass substrate and/or silicon wafer was then removed by soaking the film and substrate in a buffer, whereupon the film and substrate spontaneously separated. (c) *Proposed target protein recognition of C-terminus peptide-sequence-imprinted surfaces by many cooperative weak interactions such as hydrogen bonds and hydrophobic interactions* (adapted from Nishino *et al.* 2006).

However, the determination of the effective sequence as epitope fragments was demonstrated to be complicated for higher-MW proteins. As Nishino *et al.* showed their epitope imprint for BSA lost most of its selectivity toward the target protein if a single amino acid mutant of the epitope peptide was used under the same conditions. Rachkov *et al.* imprinted an octapeptide analogue of angiotensin II (Rachkov, Hu *et al.* 2004). The synthesized material retained its binding with octapeptide after repeated use in aqueous solvent, while its parent protein angiotensin II was not bound at all to the polymer. Brown *et al.* (2008) synthesized a porous scaffold by sol-gel processing for the recognition of Lyz. The whole Lyz, as well as a 16-residue Lyz fragment, Lyz C peptide, was used as a template. The enzyme-imprinted scaffolds exhibited high selective binding of Lyz over the reference protein RNase, while the peptide-imprinted material, could not effectively recognize the enzyme.

The epitope imprinting method has provided successful examples for the imprinting of a variety of target molecules (Slinchenko, Rachkov *et al.* 2004; Tai, Lin *et al.* 2005). The chemical synthesis of peptide epitopes is inexpensive, and a commercial product based upon a concept similar to Rachkov's epitope approach is available (Huang 2005). Concerning protein recognition, however, it is still difficult to evaluate the true efficiency of the epitope approach since the epitope exploited in experiments reported thus far were up to a few amino acids long and the templates they were intended to mimic were just 8–10 amino acids long, lacking significant three-dimensional structures. No simple and efficient method has been built for proteins greater than 100 amino acids. Thus, the epitope imprints recognition toward high-MW proteins still faces many challenges and further attention is needed regarding the proper selection of the epitope sequence and the imprinting of a template having folded secondary structures as this will play a significant role in the recognition of high-MW proteins. Nonetheless, this method provides one of the best examples of plastic antibodies, and Shea and coworkers have gathered international attention by applying epitope-imprinted polymers to bind and sequester melittin (a low-MW peptide toxin) injected into rats at lethal concentrations, significantly reducing morbidity (Hoshino, Koide *et al.* 2010).

2.3 Applications

MIPs have been widely recognized as a very promising methodology for the preparation of tailor-made materials with selective adsorption. They have been intensively used for the investigation and development of sensors, catalysts, sorbents for solid-phase extraction, stationary phases for liquid chromatography, and enzyme, receptor, and antibody mimics. Although most of the examples are from the imprinting of small molecules (Tamayo, Turiel *et al.* 2007), protein-selective MIPs have shown significant development in recent years. Major application areas for protein imprints in bioseparation and biosensor fields, as well as protein imprints fabricated in our laboratory for use as artificial chaperones and gold nanoparticle supported enzyme mimics, will be discussed.

2.3.1 Analytical Separations

By far, the most popular application for molecular imprinted materials is for the analytical separation and purification of target molecules from mixed fluids or solutions. This consists of the synthesis of a monolithic polymer that has to be crushed and sieved to obtain particles of the desired size distribution with readily accessible imprint sites. The method is fast and simple in its practical execution and is widely reported in the literature for many different templates. For instance, Tong *et al.* developed an Hb-imprinted pAAm-agarose column by the entrapment of a soft gel in the pores of rigid, inert agarose beads (Tong, Hetenyi *et al.* 2001). The gel formed was cut into pieces, and the beads were freed from the surrounding pAAm gel by stirring. The chromatographic properties of these gels, including the protein-binding capacity and the flow rate, were demonstrated fourfold higher than nontemplate-imprinted polymers (NIPs). Recently, Qin *et al.* presented a facile method for Lyz-imprinted polymer beads (Qin, He *et al.* 2009). By surface-initiated radical polymerization (SIP), Lyz was copolymerized with AAm and MBisA in aqueous media and grafted on the surface of mesoporous chloromethylated polystyrene beads (MCP beads) containing dithiocarbamate iniferter (initiator transfer agent terminator). Polymeric beads formed were packed into a column for chromatographic separation. The results showed that the Lyz-MIP column exhibited good capability for separating Lyz from control proteins, including bovine Hb, BSA, ovalbumin, and CyC.

Some alternative polymerization strategies for the preparation of better analytical-grade imprinted materials have been proposed in the literature during recent years. Among many published papers, molecular-imprinted solid phase extraction (MISPE), is one of the fastest growing applications (Alexander, Andersson *et al.* 2006). However, the method is predominantly applied to the imprinting of small molecules, and reports on protein-imprinted MISPE are rare. In two closely related fields, affinity capillary electrophoresis and electrochromatography (CEC), protein imprinting has been conducted during the last decade. Takátsy *et al.* respectively copolymerized iron-free and iron-saturated Tf with AAm and N, N-methylene-bis-acrylamide with the goal of preparing imprinted artificial antibodies. The gel was granulated and freed from the selectively adsorbed template protein (as the antigen in this technique). By using the free-zone electrophoresis in a revolving capillary, the study revealed that the capillary-free zone electrophoresis cannot distinguish between iron-free and iron-saturated Tf due to their similar mobility in the gel, whereas significant differences in the mobility of the selective gel granules were observed depending on their interaction with iron-free or iron-saturated Tf (Takatsy, Kilar *et al.* 2006). By using a similar analysis approach, the same group also demonstrated the high selectivity of pAAm-based imprinting for human and bovine Hb species (Takatsy, Vegvari *et al.* 2007).

Recently, Denizli and coworkers presented a remarkable example of using protein-based MIPs for virus detection from biological fluids (Uzun, Say *et al.* 2009). They imprinted hepatitis B surface antibody (HBSA) in polymer particles prepared from poly (hydroxyethyl methacrylate-N-methacryloyl-l-tyrosine methyl ester) (PHEMAT). The developed material can be effectively applied to the purification of HBSA from human plasma. An HBSA-imprinted PHEMAT monolith column was synthesized by bulk polymerization using N-methacryloyl-l-tyrosine methyl ester as functional monomers for interaction with HBSA. After the polymer column was dried, ground, and sieved, particles 20~60 μm in size were obtained. Adsorption studies were performed on HBSA and antihepatitis A antibody positive human plasma. Competitive adsorption of HBSA, total antihepatitis A antibody, and total immunoglobulin E was investigated for demonstrating the selectivity. The particle imprints exhibited strong binding specificity for HBSA, as demonstrated from the results that the adsorption capability for HBSA was 18.3 times more than that for antihepatitis A antibody and 2.2 times more than that for immunoglobulin E.

Denizli's group also fabricated protein imprints via a metal-coordination method for Lyz purification from egg whites (Bereli, Andac *et al.* 2008) motivated by the high cost and sensitivity to methods associated with large-scale isolation and purification of Lyz, which usually leads to the loss of enzyme activity (Denizli and Piskin 2001). In this work, the Lyz template was imprinted into supermacroporous cryogels (Fig. 2.11), which are formed in frozen solutions of monomeric or polymeric precursors (Lozinsky, Galaev *et al.* 2003). Cu^{2+} was complexed with metal-coordinating monomer N-Methacryloyl-(l)-histidinemethylester (MAH), and the free radical polymerization of Lyz-MIPed cryogels were initiated by N,N,N,N-tetramethylene diamine (TEMED) in an ice bath. The relative selectivity of the Lyz-MIPed cryogel for target protein over BSA and CyC was 4.6 and 3.2 times greater than the nonimprinted cryogel, respectively. Moreover, the purity of Lyz desorbed from the Lyz-MIPed cryogel can be well maintained (~94%) and the recovery rate could reach up to 86%. In particular, the binding capacity retained good stability for target rebinding even after many adsorption-regeneration cycles, indicating Lyz imprints could be applied as an efficient, simple, and cost-effective alternative for large-scale Lyz separation.

2.3.2 Biosensors

As an analytical device, a chemical sensor is typically composed of a molecular recognition element and a transduction element to detect and analyze binding. Molecularly imprinted materials can be utilized as artificial sensing elements and combined with various physicochemical transducers to create biosensors.

While the mainstream of small-molecule imprinting is the bulk monolith, its incorporation into sensor platforms is not trivial. As protein-based MIPs are predominantly surface imprinted, the incorporation of the protein-MIPed film/membrane with sensor surfaces has become more and more popular for the development of novel types of biosensors in recent years (Cunliffe 2005; Ulbricht 2005). Generally, this is achieved by modifying a sensor surface with a MIP. Spin coating and other surface-grafting techniques are used for the controlled deposition of a thin layer of a protein-imprinted polymer on the sensor chips. After polymerization, the template protein is washed away, leaving an imprinted sensor surface for the recognition of specific protein analytes. For example, Rick and Chou deposited a thin layer of poly-APBA onto the gold surface of QCM, using Lyz or CyC as templates (Rick and Chou 2005). The modified sensor demonstrated good selectivity for the template proteins over competitive proteins, including albumin and Mb. A remarkable achievement of the experiment was that the MIPed polymer could recognize Lyz and CyC at the same time when they were used simultaneously as templates. In another report of the same group, cyclic voltammetric deposition was applied to provide more control of the protein-imprinted polymerization process on a screen-printed platinum support, resulting in the enhanced reproducibility of thin polymer films with a defined conductivity (Rick and Chou 2006).

Recently, some groups have explored the fabrication of protein MIPed biosensors for the specific detection of virus proteins and even whole viruses. Ramanaviciene prepared a polypyrrole-based MIP for the direct detection of the bovine leukaemia virus (BLV) glycoprotein gp51 (Ramanaviciene and Ramanavicius 2004). The polymer was prepared as a surface coating onto a platinum electrode in a potassium chloride solution. Using pulsed amperometric detection (PAD), they demonstrated binding, washing, and rebinding of the protein over several cycles. However, the recognition and affinity degraded with each repeat. Recently, Tai and coworkers fabricated an artificial receptor sensor for the specific recognition of the dengue virus by epitope imprinting on the surface of a QCM chip (Tai, Lin *et al.* 2006). Although an epitope of the dengue virus NS1 protein, pentadecapeptide, was employed as the protein template; the modified sensor surface exhibited increased polymer affinity toward both the pentadcapeptide and NS1 protein. Moreover, the quantitative detection of the dengue virus protein was achieved. In the presence of NS1 protein, the modified QCM chip even demonstrated a comparable frequency shift with control chips immobilized with anti-NS1 monoclonal antibodies.

Optical sensors fabricated by imprinting low-MW targets onto fluorescence responsive MIPs have been widely reported (Hillberg, Brain *et al.* 2005), but very few have been reported for protein imprinting. Recently, impressive work has been done by Bright's group (Tao, Tehan *et al.* 2006), who described protein-imprinted xerogels with integrated emission sites (PIXIES). The polymer was first prepared by protein imprinting in a sol-gel; after the template was washed away, the imprinted gel was treated with a noncovalent complex of the template protein and an arylazide-derivatized luminophore (BODIPY). Once the protein became bound with imprinted xerogels, the lunimophore reporter groups were also guided into the binding cavities. Upon UV irradiation, the azide functionality converted into an extremely reactive nitrene and inserted into the xerogel. Thus, the presence of specific protein at binding sites led to measurable changes of the imprint fluorescence. Bright's group prepared various PIXIES films out of 16 silane monomer formulations, including TEOS, APTES, n-octyltrimethoxysilane (C8-TMOS), and bis(2-hydroxyethyl)-aminopropyltriethoxysilane (HAPTS). By using ovalbumin as the template and fluorescent arylazidederivatized BODIPYs FL as the luminophore, all silane formulations showed increased fluorescence upon rebinding with ovalbumin, among which the highest emission was about 10 times higher than control.

The protein-responsive PIXIES films also demonstrated strong selectivity, with responses to competing proteins (HSA and chemically modified ovalbumin) being over 200-fold smaller. In addition, a remarkable property of the PIXIES system is the high stability of the quantum yield of BODIPY upon the changes in the pH or the solvent polarity (Karolin, Johansson *et al.* 1994), thus excluding the possibility that the increased fluorescence results from microenvironment changes induced by analyte binding, which has been reported in other sol-gel fluorescence imprints (Carlson, Lloyd *et al.* 2006).

The efficiency of the fluorescent protein–imprinting system was further demonstrated by additional results using human interleukin-1 alpha (IL-1α) and beta (IL-1β) as templates. The two interleukins, which have 27% homology, can be identified by the highly responsive PIXIES, demonstrating that the PIXIES system could be a promising technology to integrate with MIP-based optical sensors for sensitive, high-throughput screening using protein molecular imprints with enhanced binding properties.

2.3.3 Artificial Chaperones for Conformational Diseases

The proper folding of protein structures is of vital importance to their biological functions. The well-defined three-dimensional conformation of protein is also coupled essentially with many biological processes ranging from molecular recognition and transportation to cell growth and regulation (Radford and Dobson 1999). In addition, the long-term stability and the capability to interact specifically with their natural partner are also determined by correct folded structures of proteins. Thus, it is reasonable that protein failure to fold into the intended structure is linked to a wide variety of pathological conditions termed as "conformational diseases," including some of the currently most perplexing medical problems, such as Alzheimer's disease, Parkinson's disease, and the prion encephalopathies (Collinge and Rossor 1996; Walker and LeVine 2000). In spite of the diverse clinic presentations, the molecular basis of the neurodegenerative disorders are all involved in the structure transition of proteins from helices to β-sheets (Oltersdorf, Fritz *et al.* 1989; Huang, Prusiner *et al.* 1996), leading to multimerization of misfolded proteins into insoluble, extracellular aggregates and/or intracellular inclusions, including cross-beta sheet amyloid fibrils. Thus, blocking the induction of the β-structure linkages and the subsequent self-association of different pathological macrostructures are most important strategies for the prospective therapy of the conformational diseases (Carrell and Lomas 1997).

MIP technology is well poised to develop global catalytic protein conformational imprinting (CPC-MIPs), which can be applied as "artificial chaperones" that bind specifically with target proteins and, most importantly, catalyze β-sheet to α-helix structure transitions in target proteins (Fig. 2.11). It is anticipated that combining the synthetic chaperone system with suitable platforms will permit the analysis, diagnosis, and even the possible treatment of conformational diseases.

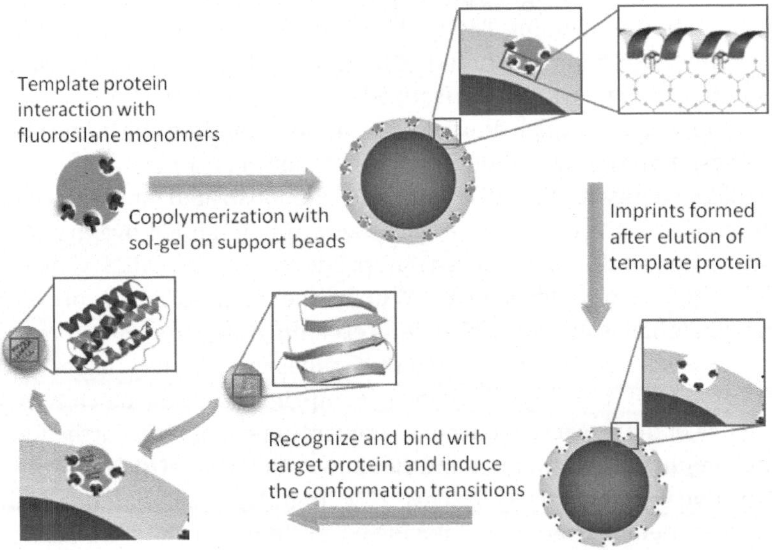

Figure 2.11 A schematic representation of catalytic protein conformational imprinting (CPC-MIPs).

Peng and Britt have employed TEOS as the precursor for the preparation of a polymeric matrix via sol-gel processing in which 3-(heptafluoroisopropoxy)propyltrie thoxysilane (7F) and (3,3,3-trifluoropropyl)trimethoxysilane (3F) were incorporated as functional monomers for the fabrication of the artificial chaperone imprinting (Peng, 2011). The fluorosilanes were selected because they are structure analogues of two fluorine substitute alcohols — 1,1,1,3,3,3-hexafluoro-2-propanol (HFIP) and 2,2,2-trifluoromethanol (TFE) — which have been identified showing remarkable effects on inducing and stabilizing α-helical structures in various peptides and proteins (Kumaran and Roy 1999; Naeem, Khan *et al.* 2004; Naseem and Khan 2005; Fatima, Ahmad *et al.*

2006). Preliminary results by Peng and Britt demonstrate that 3F copolymerization with TEOS under select sol-gel reaction conditions, such as pH and water/silane ratio, led to the formation of transparent silica gels with a uniform distribution of encapsulated protein while the more hydrophobic 7F, on the other hand, tended to self-aggregate and led to protein aggregation in the silica gel (Peng, 2011).

To investigate the 3F-induced conformational transitions, the interactions between the silane and two model proteins, β-lactoglobulin (BLG) and bovine carbonic anhydrase (BCA), were investigated mainly by fluorescence spectroscopy measurements. The protein unfolding caused by the fluoro-alcohols TFE and HFIP under the same conditions was applied as the controls. In contrast to TFE, which showed almost no structure changes on template proteins, HFIP led to significant conformation transitions in BLG and BCA, introducing the formation of a molten globule intermediate with the accumulation of α-helical structures. The hydrolyzed free monomer from 3F behaved similar to HFIP, partially unfolding the BLG and leading to the production of a stable intermediate but at significantly lower concentrations than for HFIP. Although the partially unfolded BCA intermediate induced by the free monomer from 3F was more susceptible to pH conditions, the maximum fluorescence intensity of 1,8-ANS after binding with the protein encapsulated in the 3F/TEOS gel was much higher than that with the free 3F monomer in the hydrolyzed solution, indicating a greatly enhanced structure-inducing effect of the 3F silane molecule after its copolymerization with TEOS matrix (Peng, 2011).

In addition, the mechanism involved in 3F-induced conformation transitions was also demonstrated via comparing the results with the structure effect from less hydrophobic n-propyltrimethoxysilane (nPM), which shares with 3F the same structure minus the tri-fluoro groups. From this study, nPM exhibited much weaker α-inducing effects on BLG and BCA. Therefore, the significant effects of 3F on protein conformations are believed to be the addictive contribution from fluoro-alkyl groups, particularly the tri-fluoro moieties. The strong activity of the fluorosilanes on the protein secondary structure is hypothesized to arise from both the hydrophobic interaction of 3F with the protein as well as silanol-protein interactions (e.g., H-bonding) that favor α-helix propensities.

2.3.4 Enzyme Mimics Using Electromagnetically Activated Au-MIP Nanoparticles

Humans have been augmenting the immune system for centuries to fight toxins and microbial diseases through the use of vaccines and antimicrobial agents. Novel approaches to enhance the immune system are needed, however, to counter new diseases and pandemics — particularly those that will spread and flourish with global warming — to treat existing diseases more effectively, to counter bioterrorism threats, and to fight the increase in chronic, age-related diseases in the population. Such advancements will most likely include the development of artificial immune systems, as predicted by Holland (2002), or the invention of artificial phagocytes that consume a broad spectrum of toxins and microbes in the body similar to the "panphages" envisioned by Vinge (1986). These artificial immune systems and cells will be based on advances in nano- and molecular biotechnology. For example, artificial antigen-presenting cells for stimulating T-cell production have recently been developed that use biodegradable polymer micro- and nanoparticles with encapsulated cytokines and surface recognition ligands (Steenblock and Fahmy 2008). These new technologies will also include the development of enzyme mimics that are artificial nanoscale materials designed to facilitate the recognition and neutralization of harmful agents in the body.

The application of nanoparticles to medical therapies has exploded in the past decade and offers a promising avenue for the development of enzyme mimics and artificial immune cells. By functionalizing the nanoparticle with the appropriate ligand, the nanoparticle can be targeted to a specific site or cell type in the body that has an affinity for the ligand. Cancer has been a particularly successful target for nanoparticles due to the overexpression of certain types of receptors on the plasma membrane (e.g., the Her-2/neu receptor in many types of breast cancers). Tumors additionally exhibit enhanced uptake of nanoparticles due to the leakiness of tumor vasculature caused by abnormally large gaps in the epithelial cell junctions in the vascular walls. Finally, nanoparticles made of gold, carbon nanotubes, and magnetic compounds have been shown to efficiently absorb EM radiation ranging from radio frequencies (RF) to x-rays. EM absorption mechanisms can be used to either directly heat the targeted tissue or control the release of drugs and have been explored for new therapeutic approaches such as the RF ablation of cancer cells (Gannon, Cherukuri *et al.* 2007; Gannon, Patra *et al.* 2008), the microwave dissolution of beta amyloid plaques (Bastus, Kogan *et al.* 2007), the near-infrared (NIR) phototherapy of cancer (Huang, El-Sayed *et al.* 2006), the x-ray ablation of tumors (Hainfeld, Slatkin *et al.* 2004), and the selective release of multiple drugs using ultrafast laser melting of gold nanoparticles (Wijaya, Schaffer *et al.* 2009).

A current obstacle to the widespread implementation of nanoparticle methods to disease is the need to find an appropriate ligand or selection mechanism for targeting the nanoparticles to noncellular, disease-causing agents. Such agents typically lack plasma membrane receptors for targeting and include macromolecules such as misfolded proteins, prions, toxins, and advanced glycation end products (AGEs). They may also comprise pathogens, including nonenveloped viruses such as rhinoviruses, rotaviruses, enteroviruses, adenoviruses, and papillomaviruses. The "macromolecular" diseases are particularly responsible for many chronic health problems associated with aging, such as protein conformational diseases (Alzheimer's, Huntington's, etc.), diabetes-related complications (microangiopathy and macrovascular disease), and cardiovascular disease.

The ability to imprint macromolecules such as proteins onto polymers provides a new and unique opportunity to target MIP-nanoparticle constructs to these disease-causing agents (Turner, Jeans *et al.* 2006). For example, a MIP has been used to detect the dengue virus NS1 protein (Tai, Lin *et al.* 2005). To date, MIP and MIP-coated nanoparticles have been primarily studied for use as receptor mimics for drug development, biosensors, biochemical separation, and drug delivery (Vaihinger, Landfester *et al.* 2002; Tovar, Krauter *et al.* 2003; Matsui, Akamatsu *et al.* 2004; Ye and Haupt 2004; Matsui, Akamatsu *et al.* 2005; Schillemans and van Nostrum 2006; Perez-Moral and Mayes 2007; Kan, Geng *et al.* 2009; Li, He *et al.* 2009; Wang, Wang *et al.* 2009). Combining the targeting mechanism of a MIP with the EM excitation properties of conducting, semiconducting, or magnetic nanoparticles, however, offers new opportunities for creating enzyme mimics for the noninvasive treatment of disease.

The basic concept for an enzyme mimic is to either coat an EM-absorbing nanoparticle with a MIP or to embed many nanoparticles into a larger MIP particle. The affinity between the MIP and the imprinted molecule allows the nanoparticle composite to selectively target harmful proteins and macromolecules either *in vivo* or *ex vivo*. EM radiation in the tissue or solution is then preferentially absorbed by the nanoparticles, producing thermal, acoustic, or chemical energy that transforms, destroys, or inactivates the target molecule. This concept has been applied in a nonbiomedical context to the photodegradation of small pollutant molecules such as phenol, 2-chlorophenol, and 4-chlorophenol using MIP-coated TiO_2 nanoparticles as the photocatalysts and UV light as the EM source (Shen, Zhu *et al.* 2007).

Since UV and visible light attenuate rapidly in biological tissue, other EM wavelengths have been explored for activating nanoparticles. One such band that has elicited strong interest is the NIR, where gold-shelled nanoparticles and nanorods exhibit a strong absorption due to surface plasmon resonances. A plasmon is a collective electron oscillation in the metal that strongly absorbs light at the resonant frequency. However, tissues are still moderately opaque to NIR light, which will only have an effective penetration of 2–3 cm for soft tissue and less for bone. The RF band has also been explored as an excitation source with both gold nanoparticles and carbon nanotubes (Gannon, Cherukuri *et al.* 2007; Gannon, Patra *et al.* 2008). The dominant heating mechanism in the RF band has been proposed and modeled as ohmic (resistive) heating (Moran, Wainerdi *et al.* 2009). Heating effects have also been produced at RF frequenies using magnetic particles (Hilger, Hiergeist *et al.* 2002). The primary advantage of RF frequencies is their ability to penetrate deep tissue. The demonstration of the break-up of beta amyloid aggregates under low-power microwave irradiation (100 MW at 12 GHz for 2–3 min.) using 10 nm gold nanoparticles suggests that the microwave band may also be efficient for activating enzyme mimics (Bastus, Kogan *et al.* 2007). Since the observed dissolution of the beta amyloid was significantly greater than what could be accounted for by heating effects, Bastus *et al.* (2007) speculated that an electron-phonon coupling mechanism was responsible for converting the microwaves to acoustic vibrations in the nanoparticle. The phonons subsequently transferred to the proteins as molecular vibrations.

A particularly interesting and beneficial application for enzyme mimics would be the targeting and breakdown of AGEs. The protein cross-links that result from glycation are responsible for many of the complications of diabetes, including cardiovascular disease, microvascular disease, neuropathy, and kidney failure. Pharmacological methods are being developed to reverse AGE cross-linking (Ulrich and Zhang 1997), but a principal problem with a pharmacological approach is the ability to design drugs with a high selectivity for specific AGEs that do not damage normal functional macromolecules. The molecular imprinting of AGEs onto polymer surfaces has been demonstrated (Sode, Yanai *et al.* 2001; Sode, Ohta *et al.* 2003; Yamazaki, Saito *et al.* 2006; Rajkumar, Warsinke *et al.* 2007), as well as their binding to gold nanoparticles (GhoshMoulick, Bhattacharya *et al.* 2007), for the development of AGE sensors. Molecular imprinting provides a high degree of specificity for AGE molecules that may not be attainable with pharmaceuticals. Combined with EM activation methods or controlled drug release for the cleaving of AGE cross-links, a MIP-nanoparticle construct may provide a new and potentially unparalleled tool in nanomedicine for the treatment of diabetes and other chronic illnesses.

Further development of enzyme mimics and artificial immune cells based on MIP-nanoparticle composites faces many challenges. These include finding new fabrication methods for MIP nanoconstructs, developing biocompatible MIPs for *in vivo* or *ex vivo* applications, gaining a more quantitative understanding of the physical mechanisms underlying the EM activation of nanoparticles, and designing clinically practical and effective therapies. Overcoming these challenges, however, will result in new and more versatile nanoparticle-based therapies for medicine.

2.4 Perspective

Due to the enormous potential for applications in fundamental science and industrial fields, the molecular imprinting of proteins has been and continues to be an attractive theme. The imprinting of proteins is much more challenging than that for small molecules simply because of the unique properties of the proteins, which are of essential importance for their biological functions but are still not fully compatible with current molecular imprinting systems. The main problems resulting directly from template proteins include poor protein solubility in organic solvents, impeded mass transportation due to large molecular sizes, and protein structure flexibility and instability. For these reasons, while numerous molecular imprints for low-MW molecules have been prepared and some have yielded commercial success, the development of imprints for proteins has been slow. Recently, however, some remarkable advances have been achieved. Varieties of monomers addressing the high density of recognition sites have been employed for specific interactions with proteins, and many novel synthetic protocols have been developed to overcome various obstacles associated with protein imprinting. On the basis of these current trends, the future of protein imprinting will continue to grow and transition toward more commercial applications. It is anticipated that these fully synthetic molecular recognition techniques will evolve to effectively compete with biological antibodies and receptors and bring about reliable applications in drug delivery, disease diagnosis, therapeutic monitoring, chemical analysis, and environmental control.

Surface imprinting offers superior properties over bulk monoliths for the imprinting of proteins, and the prevailing strategy usually involves the formation of multiple and specific interactions between the template proteins and the imprinting polymers. However, it is difficult to determine at this time which approach will provide the most universal and effective method for imprinting across a wide range of proteins. A number of protocols have been reviewed here, and each has associated advantages and disadvantages with regard to "generic protein imprinting." The dependence of MIP affinity for target protein on polymer composition is convoluted with the microenvironmental conditions (solvent, buffer composition, ionic strength, pH, temperature, etc.). Further examples demonstrating efficacy against a wider range of proteins (and recognition in complex milieus) for each technique are necessary. Other questions regarding applications, such as the long-term stability of the imprints and the generic application of a method, as well as the robustness, flexibility, signal-generation capabilities (for biosensors), and activation efficiency (for therapeutic nanoparticles) of the material, require additional study and are the driving forces behind development of plastic antibodies and enzymes that will not only rival, but exceed their natural counterparts.

References

1. Alexander, C., Andersson, H. S., Andersson, L. I., Ansell, R. J., Kirsch, N., Nicholls, I. A., O'Mahony, J., and Whitcombe, M. J. (2006). Molecular imprinting science and technology: a survey of the literature for the years up to and including 2003, *J. Mol. Recognit.*, **19**(2), 106–180.
2. Amiya, T., Hirokawa, Y., Hirose, Y., Li, Y., and Tanaka, T. (1987). Reentran phase-transition of N-isopropyllacrylamide gels in mixed-solvents, *J. Chem. Phys.*, **86**(4), 2375–2379.
3. Andersson, L. I., Muller, R., Vlatakis, G., and Mosbach, K. (1995). Mimics of the binding-sites of opioid receptors obtained by molecular imprinting of enkephalin and morphine, *Proc. Natl. Acad. Sci. USA*, **92**(11), 4788–4792.
4. Ansell, R. J., Ramstrom, O., and Mosbach, K. (1996). Towards artificial antibodies prepared by molecular imprinting, *Clin. Chem.*, **42**(9), 1506–1512.
5. Arshady, R., and Mosbach, K. (1981). Synthesis of substrateselective polymers by host-guest polymerization, *Macromol. Chem. Phys.*, **182**, 687–692.
6. Avnir, D., and Kaufman, V. R. (1987). Alcohol is an unnecessary additive in the silicon alkoxide sol-gel process, *J. Non-Cryst. Solids*, **92**(1), 180–182.
7. Bacskay, I., Takatsy, A., Vegvari, A., Elfwing, A., Balllagi-Pordany, A., Kilar, F., and Hjerten, S. (2006). Universal method for synthesis of artificial gel antibodies by the imprinting approach combined with a unique electrophoresis technique for detection of minute structural differences of proteins, viruses, and cells (bacteria). III: gel antibodies against cells (bacteria), *Electrophoresis*, **27**(23), 4682–4687.
8. Bastus, N. G., Kogan, M. J., Amigo, R., Grillo-Bosch, D., Araya, E., Turiel, A., Labarta, A., Giralt, E., and Puntes, V. F. (2007). Gold nanoparticles for selective and remote heating of beta-amyloid protein aggregates, *Mater. Sci. Eng., C*, **27**(5–8), 1236–1240.
9. Bereli, N., Andac, M., Baydemir, G., Say, R., Galaev, I. Y., and Denizli, A. (2008). Protein recognition via ion-coordinated molecularly imprinted supermacroporous cryogels, *J. Chromatogr., A*, **1190**(1–2), 18–26.
10. Bolisay, L. D., Culver, J. N., and Kofinas, P. (2006). Molecularly imprinted polymers for tobacco mosaic virus recognition, *Biomaterials*, **27**(22), 4165–4168.
11. Bossi, A., Piletsky, S. A., Piletska, E. V., Righetti, P. G., and Turner, A. P. F. (2001). Surface-grafted molecularly imprinted polymers for protein recognition, *Anal. Chem.*, **73**(21), 5281–5286.
12. Brinker, C. J., and Scherer, G. W. (eds) (1990). *Sol-Gel Science: The Physics and Chemistry of Sol-Gel Processing*, Academic Press, New York.
13. Brown, M. E., and Puleo, D. A. (2008). Protein binding to peptide-imprinted porous silica scaffolds, *Chem. Eng. J.*, **137**(1), 97–101.
14. Buckley, A. M., and Greenblatt, M. (1994). The sol-gel preparation of silica-gels, *J. Chem. Educ.*, **71**(7), 599–602.
15. Burow, M., and Minoura, N. (1996). Molecular imprinting: synthesis of polymer particles with antibody-like binding characteristics for glucose oxidase, *Biochem. Biophys. Res. Comm.*, **227**(2), 419–422.
16. Carlson, C. A., Lloyd, J. A., Dean, S. L., Walker, N. R., and Edmiston, P. L. (2006). Sensor for fluorene based on the incorporation of an environmentally sensitive fluorophore proximal to a molecularly imprinted binding site, *Anal. Chem.*, **78**(11), 3537–3542.

17. Caro, E., Marce, R. M., Cormack, P. A. G., Sherrington, D. C., and Borrull, F. (2005). Synthesis and application of an oxytetracycline imprinted polymer for the solid-phase extraction of tetracycline antibiotics, *Anal. Chim. Acta*, **552**(1–2), 81–86.

18. Carrell, R. W., and Lomas, D. A. (1997). Conformational disease, *Lancet*, **350**(9071), 134–138.

19. Chen, G. H., and Hoffman, A. S. (1995). Graft-copolymers that exhibit temperature-induced phase-transitions over a wide-range of pH, *Nature*, **373**(6509), 49–52.

20. Chen, Z. Y., Hua, Z. D., Xu, L., Huang, Y., Zhao, M. P., and Li, Y. Z. (2008). Protein-responsive imprinted polymers with specific shrinking and rebinding, *J. Mol. Recognit.*, **21**(1), 71–77.

21. Chianella, I., Lotierzo, M., Piletsky, S. A., Tothill, I. E., Chen, B. N., Karim, K., and Turner A. P. F. (2002). Rational design of a polymer specific for microcystin-LR using a computational approach, *Anal. Chem.*, **74**(6), 1288–1293.

22. Chou, P. C., Rick, J., and Chou, T. C. (2005). C-reactive protein thin-film molecularly imprinted polymers formed using a micro-contact approach, *Anal. Chim. Acta*, **542**(1), 20–25.

23. Collinge, J., and Rossor, M. (1996). A new variant of prion disease, *Lancet*, **347**(9006), 916–917.

24. Conrad II, P. G., and Shea, K. J. (2005). Use of metal co-ordination for controlling imprinted polymers, in *Molecularly Imprinted Materials: Science and Technology* (ed. Yan, M., and Ramstreom, O.), Marcel Dekker, New York, pp. 123–180.

25. Cunliffe, D., and Alexander, C. (2005). Surface imprinting, in *Molecularly Imprinted Materials: Science and Technology* (ed. Yan, M., and Ramstreom, O.), Marcel Dekker, New York, pp. 249–283.

26. Demirel, G., Ozcetin, G., Turan, E., and Caykara, T. (2005). pH/temperature-sensitive imprinted ionic poly (N-tert-butylacrylamide-co-acrylamide/maleic acid) hydrogels for bovine serum albumin, *Macromol. Biosci.*, **5**(10), 1032–1037.

27. Denizli, A., and Piskin, E. (2001). Dye-ligand affinity systems, *J. Biochem. Biophys. Methods*, **49**(1–3), 391–416.

28. Deore, B., and Freund, M. S. (2003). Saccharide imprinting of poly(aniline boronic acid) in the presence of fluoride, *Analyst*, **128**(6), 803–806.

29. Dhal P. K., and Arnold, F. H. (1991). Template-mediated synthesis of metal-complexing polymers for molecular recognition, *J. Am. Chem. Soc.*, **113**, 7417–7418.

30. Dickert, F. L., and Hayden, O. (2002). Bioimprinting of polymers and sol-gel phases. Selective detection of yeasts with imprinted polymers, *Anal. Chem.*, **74**(6), 1302–1306.

31. Dickert, F. L., Lieberzeit, P., Miarecka, S. G., Mann, K. J., Hayden, O., and Palfinger, C. (2004). Synthetic receptors for chemical sensors - subnano- and micrometre patterning by imprinting techniques, *Biosens. Bioelectron.*, **20**(6), 1040–1044.

32. Dickey, F. (1949). The preparation of specific adsorbents, *Proc. Natl. Acad. Sci. USA*, **35**, 227–229.

33. Doyle, T. E. (2006). Iterative simulation of elastic wave scattering in arbitrary dispersions of spherical particles, *J. Acoust. Soc. Am.*, **119**(5), 2599–2610.

34. Dunn, M. J. (ed) (1993). *Gel Electrophoresis: Proteins*, Alden Press, Oxford, UK.

35. Fang, G. Z., Tan, J., and Yan, X. P. (2005). An ion-imprinted functionalized silica gel sorbent prepared by a surface imprinting technique combined with a sol-gel process for selective solid-phase extraction of cadmium(II), *Anal. Chem.*, **77**(6), 1734–1739.

36. Fatima, S., Ahmad, B., and Khan, R. H. (2006). Fluoroalcohols induced unfolding of succinylated Con A: native like beta-structure in partially folded intermediate and alpha-helix in molten globule like state, *Arch. Biochem. Biophys.*, **454**(2), 170–180.

37. Gannon, C. J., Cherukuri, P., Yakobson, B. I., Cognet, L., Kanzius, J. S., Kittrell, C., Weisman, R. B., Pasquali, M., Schmidt, H. K., Smalley, R. E., and Curley, S. A. (2007). Carbon nanotube-enhanced thermal destruction of cancer cells in a noninvasive radiofrequency field, *Cancer*, **110**(12), 2654–2665.

38. Gannon, C. J., Patra, C. R., Bhattacharya, R., Mukherjee, P., and Curley, S. A. (2008). Intracellular gold nanoparticles enhance non-invasive radiofrequency thermal destruction of human gastrointestinal cancer cells, *J. Nanobiotechnol.*, **6**(2).

39. Ge, Y., and Turner, A. P. F. (2008). Too large to fit? Recent developments in macromolecular imprinting, *Trends Biotechnol.*, **26**(4), 218–224.

40. GhoshMoulick, R., Bhattacharya, J., Mitra, C. K., Basak, S., and Dasgupta, A. K. (2007). Protein seeding of gold nanoparticles and mechanism of glycation sensing, *Nanomed. Nanotechnol. Biol. Med.*, **3**(3), 208–214.

41. Glad, M., Norrlow, O., Sellergren, B., Siegbahn, N., and Mosbach, K. (1985). Use of silane monomers for molecular imprinting and enzyme entrapment in polysiloxane-coated porous silica, *J. Chromatogr.*, **347**(1), 11–23.

42. Greiner, W., and Maruhn, J. A. (eds) (1996). *Nuclear Models*. Springer-Verlag, Berlin.

43. Guo, T. Y., Xia, Y. Q., Hao, G. J., Song, M. D., and Zhang, B. H. (2004). Adsorptive separation of hemoglobin by molecularly imprinted chitosan beads, *Biomaterials*, **25**(27), 5905–5912.

44. Hainfeld, J. F., Slatkin, D. N., and Smilowitz, H. M. (2004). The use of gold nanoparticles to enhance radiotherapy in mice, *Phys. Med. Biol.*, **49**(18), N309–N315.

45. Hench, L. L., and West, J. K. (1990). The sol-gel process, *Chem. Rev.*, **90**(1), 33–72.

46. Hilger, I., Hiergeist, R., Hergt, R., Winnefeld, K., Schubert, H., and Kaiser, W. A. (2002). Thermal ablation of tumors using magnetic nanoparticles – an in vivo feasibility study, *Invest. Radiol.*, **37**(10), 580–586.

47. Hillberg, A. L., Brain, K. R., and Allender, C. J. (2005). Molecular imprinted polymer sensors: implications for therapeutics, *Adv. Drug Delivery Rev.*, **57**(12), 1875–1889.

48. Hiratani, H., and Alvarez-Lorenzo, C. (2004). The nature of backbone monomers determines the performance of imprinted soft contact lenses as timolol drug delivery systems, *Biomaterials*, **25**(6), 1105–1113.

49. Hirayama, K., Burow, M., Morikawa, Y., and Minoura, N. (1998). Synthesis of polymer-coated silica particles with specific recognition sites for glucose oxidase by the molecular imprinting technique, *Chem. Lett.*, (8), 731–732.

50. Hirayama, K., Sakai, Y., and Kameoka, K. (2001). Synthesis of polymer particles with specific lysozyme recognition sites by a molecular imprinting technique, *J. Appl. Polym. Sci.*, **81**(14), 3378–3387.

51. Hjerten, S., Liao, J. L., Nakazato, K., Wang, Y., Zamaratskaia, G., and Zhang, H. X. (1997). Gels mimicking antibodies in their selective recognition of proteins, *Chromatographia*, **44**(5–6), 227–234.

52. Holland, J. H. (2002). What is to come and how to predict it, in *The Next Fifty Years: Science in the First Half of the Twenty-First Century* (ed. Brockman, J.), Vintage, New York.

53. Hoshino, Y, Koide, H., Urakami, T., Kanazawa, H., Kodama, T., Oku, N., and Shea, K. J. (2010). Recognition, neutralization, and clearance of target peptides in the bloodstream of living mice by molecularly imprinted polymer nanoparticles: a plastic antibody, *J. Am. Chem. Soc.*, **132** (19), 6644–6645.

54. Huang, C.-S. (2005). Compositions and methods for capturing, isolating, detecting, analyzing and quantifying macromolecules. U.S. Patent. B. Aspira Biosystems, CA. 6,979,573.

55. Huang, J. T., Zhang, J., Zhang, J. Q., and Zheng, S. H. (2005). Template imprinting amphoteric polymer for the recognition of proteins, *J. Appl. Polym. Sci.*, **95**(2), 358–361.

56. Huang, X. H., El-Sayed, I. H., Qian, W., and El-Sayed, M. A. (2006). Cancer cell imaging and photothermal therapy in the near-infrared region by using gold nanorods, *J. Am. Chem. Soc.*, **128**(6), 2115–2120.

57. Huang, Z. W., Prusiner, S. B., and Cohen, F. E. (1996). Scrapie prions: a three-dimensional model of an infectious fragment, *Fold. Des.*, **1**(1), 13–19.

58. Janiak, D. S., Ayyub, O. B., and Kofinas, P. (2009). Effects of charge density on the recognition properties of molecularly imprinted polymeric hydrogels, *Macromolecules*, **42**(5), 1703–1709.

59. K. Dusek (ed) (1993). *Responsive Gels: Volume Transitions I, Advances applications as the next generation of materials for in Polymer Science*, Springer, Berlin.

60. Kan, X. W., Geng, Z. R., Zhao, Y., Wang, Z. L., and Zhu, J. J. (2009). Magnetic molecularly imprinted polymer for aspirin recognition and controlled release, *Nanotechnology*, **20**(16).

61. Karolin, J., Johansson, L. B. A., Strandberg, L., and Ny, T. (1994). Fluorescence and absorption spectroscopic properties of dipyrrometheneboron difluoride (BODIPY) derivatives in liquids, lipid-membranes, and proteins, *J. Am. Chem. Soc.*, **116**(17), 7801–7806.

62. Kempe, M., and Mosbach, K. (1995). Receptor-binding mimetics – a novel molecularly imprinted polymer, *Tetrahedron Lett.*, **36**(20), 3563–3566.

63. Kimhi, O., and Bianco-Peled, H. (2007). Study of the interactions between protein-imprinted hydrogels and their templates, *Langmuir*, **23**(11), 6329–6335.

64. Kugimiya, A., Matsui, J., Abe, H., Aburatani, M., and Takeuchi, T. (1998). Synthesis of castasterone selective polymers prepared by molecular imprinting, *Anal. Chim. Acta*, **365**(1–3), 75–79.

65. Kugimiya, A., Matsui, J., Takeuchi, T., Yano, K., Muguruma, H., Elgersma, A. V., and Karube, I. (1995). Recognition of sialic-acid using molecularly imprinted polymer, *Anal. Lett.*, **28**(13), 2317–2323.

66. Kumaran, S., and Roy, R. P. (1999). Helix-enhancing propensity of fluoro and alkyl alcohols: influence of pH, temperature and cosolvent concentration on the helical conformation of peptides, *J. Pept. Res.*, **53**(3), 284–293.

67. Lahav, M., Kharitonov, A. B., Katz, O., Kunitake, T., and Willner, I. (2001). Tailored chemosensors for chloroaromatic acids using molecular imprinted TiO2 thin films on ion-sensitive field-effect transistors, *Anal. Chem.*, **73**(3), 720–723.

68. Lai, J. P., Niessner, R., and Knopp, D. (2004). Benzo[a]pyrene imprinted polymers: synthesis, characterization and SPE application in water and coffee samples, *Anal. Chim. Acta*, **522**(2), 137–144.

69. Lee, K., Itharaju, R. R., and Puleo, D. A. (2007). Protein-imprinted polysiloxane scaffolds, *Acta Biomater.*, **3**(4), 515–522.

70. Li, F., Li, J., and Zhang, S. S. (2008). Molecularly imprinted polymer grafted on polysaccharide microsphere surface by the sol-gel process for protein recognition, *Talanta*, **74**(5), 1247–1255.

71. Li, L., He, X. W., Chen, L. X., and Zhang, Y. K. (2009). Preparation of core-shell magnetic molecularly imprinted polymer nanoparticles for recognition of bovine hemoglobin, *Chem. Asian J.*, **4**(2), 286–293.

72. Li, Y., Yang, H. H., You, Q. H., Zhuang, Z. X., and Wang, X. R. (2006). Protein recognition via surface molecularly imprinted polymer nanowires, *Anal. Chem.*, **78**(1), 317–320.

73. Liao, J. L., Wang, Y., and Hjerten, S. (1996). Novel support with artificially created recognition for the selective removal of proteins and for affinity chromatography, *Chromatographia*, **42**(5–6), 259–262.

74. Lozinsky, V. I., Galaev, I. Y., Plieva, F. M., Savinal, I. N., Jungvid, H., and Mattiasson, B., (2003). Polymeric cryogels as promising materials of biotechnological interest, *Trends Biotechnol.*, **21**(10), 445–451.

75. Lu, Y., Yan, C. L., and Gao, S. Y. (2009). Preparation and recognition of surface molecularly imprinted core-shell microbeads for protein in aqueous solutions, *Appl. Surf. Sci.*, **255**(12), 6061–6066.

76. Mallik, S., Plunkett, S. D., Dhal, P. K., Johnson, R. D., Pack, D., Shnek, D., and Arnold, F. H. (1994). Towards materials for the specific recognition and seperation of proteins, *New J. Chem.*, **18**(3), 299–304.

77. Marx, S., and Liron, Z. (2001). Molecular imprinting in thin films of organic-inorganic hybrid sol-gel and acrylic polymers, *Chem. Mater.*, **13**(10), 3624–3630.

78. Matsuguchi, M., and Uno, T. (2006). Molecular imprinting strategy for solvent molecules and its application for QCM-based VOC vapor sensing, *Sens. Actuators, B*, **113**(1), 94–99.

79. Matsui, J., Akamatsu, K., Hara, N., Miyoshi, D., Nawafune, H., Tamaki, K., and Sugimoto, N. (2005). SPR sensor chip for detection of small molecules using molecularly imprinted polymer with embedded gold nanoparticles, *Anal. Chem.*, **77**(13), 4282–4285.

80. Matsui, J., Akamatsu, K., Nishiguchi, S., Miyoshi, D., Nawafune, H., Tamaki, K., and Sugimoto, N. (2004). Composite of Au nanoparticles and molecularly imprinted polymer as a sensing material, *Anal. Chem.*, **76**(5), 1310–1315.

81. Mirsky, V. M., Hirsch, T., Piletsky, S. A., and Wolfbeis, O. S. (1999). A spreader-bar approach to molecular architecture: formation of stable artificial chemoreceptors, *Angew. Chem., Int. Ed.*, **38**(8), 1108–1110.

82. Moran, C. H., Wainerdi, S. M., Cherukuri, T. K., Kittrell, C., Wiley, B. J., Nicholas, N. W., Curley, S. A., Kanzius, J. S., and Cherukuri, P. (2009). Size dependent Joule heating of gold nanoparticles using capacitively coupled radiofrequency fields, *Nano Res.*, **2**, 400–405.

83. Mosbach, K., and Ramstreom, O. (1996). The emerging technique of molecular imprinting and its future impact on biotechnology, *Bio-Technology*, **14**(2), 163–170.

84. Naeem, A., Khan, K. A., and Khan, R. H. (2004). Characterization of a partially folded intermediate of papain induced by fluorinated alcohols at low pH, *Arch. Biochem. Biophys.*, **432**(1), 79–87.

85. Naseem, F., and Khan, R. H. (2005). Characterization of a common intermediate of pea lectin in the folding pathway induced by TFE and HFIP, *Biochim. Biophys. Acta-Gen. Subjects*, **1723**(1–3), 192–200.

86. Nishino, H., Huang, C. S., and Shea, K. J. (2006). Selective protein capture by epitope imprinting, *Angew. Chem., Int. Ed.*, **45**(15), 2392–2396.

87. Norell, M. C., Andersson, H. S., and Nicholls, I. A. (1998). Theophylline molecularly imprinted polymer dissociation kinetics: a novel sustained release drug dosage mechanism, *J. Mol. Recognit.*, **11**(1–6), 98–102.

88. Norrlow, O., Glad, M., and Mosbach, K. (1984). Acrylic polymer preparations containing recognition sites obtained by imprinting with substrates, *J. Chromatogr.*, **299**(1), 29–41.

89. O'Connor, N. A., Paisner, D. A., Huryn, D., and Shea, K. J. (2007). Screening of 5-HT1A receptor antagonists using molecularly imprinted polymers, *J. Am. Chem. Soc.*, **129**(6), 1680–1689.

90. Okano, T., (ed) (1998). *Biorelated Polymers and Gels*, Academic Press, Boston.

91. Oltersdorf, T., Fritz, L. C., Schenk, D. B., Lieberburg, I., Johnsonwood, K. L., Beattie, E. C., Ward, P. J., Blacher, R. W., Dovey, H. F., and Sinha, S. (1989). The secreted form of the alzheimers amyloid precursor protein with the kunitz domain is protease nexin-ii, *Nature*, **341**(6238), 144–147.

92. Öpik, A., Menaker, A., Reut, J., and Syritski, V. (2009). Molecularly imprinted polymers: a new approach to the preparation of functional materials, *Proc. Est. Acad. Sci.*, **58**(1), 3–11.

93. Osada, Y., Okuzaki, H., and Hori, H. (1992). A polymer gel with electrically driven motility, *Nature*, **355**(6357), 242–244.

94. Ou, S. H., Wu, M. C., Chou, T. C., and Liu, C. C. (2004). Polyacrylamide gels with electrostatic functional groups for the molecular imprinting of lysozyme, *Anal. Chim. Acta*, **504**(1), 163–166.

95. Ouyang, R. Z., Lei, J. P., and Ju, H. X. (2008). Surface molecularly imprinted nanowire for protein specific recognition, *Chem. Commun.*, (44), 5761–5763.

96. Ozcan, A. A., Say, R., Denizli, A., and Ersoz, A. (2006). L-histidine imprinted synthetic receptor for biochromatography applications, *Anal. Chem.*, **78**(20), 7253–7258.

97. Pang, X. S., Cheng, G. X., Lu, S. L., and Tang, E. J. (2006). Synthesis of polyacrylamide gel beads with electrostatic functional groups for the molecular imprinting of bovine serum albumin, *Anal. Bioanal. Chem.*, **384**(1), 225–230.

98. Pauling, L. (1940). A theory of the structure and process of formation of antibodies, *J. Am. Chem. Soc.*, **62**(10), 2643–2657.

99. Peng, Y. (2011). *Fluoro-Silane as a Functional Monomer for Protein Conformational Imprinting*, Paper 906. All Graduate Theses and Dissertations, Utah State University. http://digitalcommons.usu.edu/etd/906

100. Perez-Moral, N., and Mayes, A. G. (2007). Molecularly imprinted multi-layer core-shell nanoparticles – a surface grafting approach, *Macromol. Rapid Commun.*, **28**(22), 2170–2175.

101. Pichon, V., and Haupt, K. (2006). Affinity separations on molecularly imprinted polymers with special emphasis on solid-phase extraction, *J. Liq. Chromatogr. Relat. Technol.*, **29**(7–8), 989–1023.

102. Piletsky, S. A., Piletska, E. V., Bossi, A., Karim, K., Lowe, P., and Turner, A. P. F. (2001). Substitution of antibodies and receptors with molecularly imprinted polymers in enzyme-linked and fluorescent assays, *Biosens. Bioelectron.*, **16**(9–12), 701–707.

103. Piletsky, S. A., Piletska, E. V., Chen, B. N., Karim, K., Weston, D., Barrett, G., Lowe, P., and Turner, A. P. F. (2000). Chemical grafting of molecularly imprinted homopolymers to the surface of

microplates. Application of artificial adrenergic receptor in enzyme-linked assay for beta-agonists determination, *Anal. Chem.*, **72**(18), 4381–4385.

104. Qin, L., He, X. W., Zhang, W., Li, W. Y., and Zhang, Y. K. (2009). Surface-modified polystyrene beads as photografting imprinted polymer matrix for chromatographic separation of proteins, *J. Chromatogr., A,* **1216**(5), 807–814.

105. Rachkov, A., Hu, M. J., Bulgarevich, E., Matsumoto, T., and Minoura, N. (2004). Molecularly imprinted polymers prepared in aqueous solution selective for [Sar(1), Ala(8)] angiotensin II, *Anal. Chim. Acta*, **504**(1), 191–197.

106. Rachkov, A., and Minoura, N. (2000). Recognition of oxytocin and oxytocin-related peptides in aqueous media using a molecularly imprinted polymer synthesized by the epitope approach, *J. Chromatogr., A,* **889**(1–2), 111–118.

107. Rachkov, A., and Minoura, N. (2001). Towards molecularly imprinted polymers selective to peptides and proteins. The epitope approach, *Biochim. Biophys. Acta-Protein Str. Mol. Enzymol.*, **1544**(1–2), 255–266.

108. Radford, S. E., and Dobson, C. M. (1999). From computer simulations to human disease: emerging themes in protein folding, *Cell,* **97**(3), 291–298.

109. Rajkumar, R., Warsinke, A., Mohwald, H., Scheller, F. W., and Katterle, M. (2007). Development of fructosyl valine binding polymers by covalent imprinting, *Biosens. Bioelectron.*, **22**(12), 3318–3325.

110. Ramanaviciene, A., and Ramanavicius, A. (2004). Molecularly imprinted polypyrrole-based synthetic receptor for direct detection of bovine leukemia virus glycoproteins, *Biosens. Bioelectron.*, **20**(6), 1076–1082.

111. Ramstrom, O., Ye, L., and Mosbach, K. (1996). Artificial antibodies to corticosteroids prepared by molecular imprinting, *Chem. Biol.*, **3**(6), 471–477.

112. Ratner, B. D. (1996). The engineering of biomaterials exhibiting recognition and specificity, *J. Mol. Recognit.*, **9**(5–6), 617–625.

113. Ratner, B. D., and Shi, H. Q. (1999). Recognition templates for biomaterials with engineered bioreactivity, *Curr. Opin. Solid. St. M.*, **4**(4), 395–402.

114. Rick, J., and Chou, T. C. (2005). Enthalpy changes associated with protein binding to thin films, *Biosens. Bioelectron.*, **20**(9), 1878–1883.

115. Rick, J., and Chou, T. C. (2005). Imprinting unique motifs formed from protein-protein associations, *Anal. Chim. Acta*, **542**(1), 26–31.

116. Rick, J., and Chou, T. C. (2006). Amperometric protein sensor – fabricated as a polypyrrole, poly-aminophenylboronic acid bilayer, *Biosens. Bioelectron.*, **22**(3), 329–335.

117. Rick, J., and Chou, T. C. (2006). Using protein templates to direct the formation of thin-film polymer surfaces, *Biosens. Bioelectron.*, **22**(4), 544–549.

118. Sallacan, N., Zayats, M., Bourenko, T., Kharitonov, A. B., and Willner, I. (2002). Imprinting of nucleotide and monosaccharide recognition sites in acrylamidephenylboronic acid-acrylamide copolymer membranes associated with electronic transducers, *Anal. Chem.*, **74**(3), 702–712.

119. Saraydin, D., Unver-Saraydin, S., Karadag, E., Koptagel, E., and Guven, O. (2004). In vivo biocompatibility of radiation crosslinked acrylamide copolymers, *Nucl. Instrum. Meth. B.*, **217**(2), 281–292.

120. Schillemans, J. P., and van Nostrum, C. F. (2006). Molecularly imprinted polymer particles: synthetic receptors for future medicine, *Nanomedicine*, **1**(4), 437–447.

121. Sergeyeva, T. A., Matuschewski, H., Piletsky, S. A., Bendig, J., Schedler, U., and Ulbricht, M. (2001). Molecularly imprinted polymer membranes for substance-selective solid-phase extraction from water by surface photo-grafting polymerization, *J. Chromatogr., A*, **907**(1–2), 89–99.

122. Shen, X. T., Zhu, L. H., Li, J., and Tang, H. Q. (2007). Synthesis of molecular imprinted polymer coated photocatalysts with high selectivity, *Chem. Commun.*, (11), 1163–1165.

123. Shi, H. Q., and Ratner, B. D. (2000). Template recognition of protein-imprinted polymer surfaces, *J. Biomed. Mater. Res.*, **49**(1), 1–11.

124. Shi, H. Q., Tsai, W. B., Garrison, M. D., Ferrari, S., and Ratner, B. D. (1999). Template-imprinted nanostructured surfaces for protein recognition, *Nature*, **398**(6728), 593–597.

125. Shiomi, T., Matsui, M., Mizukami, F., and Sakaguchi, K. (2005). A method for the molecular imprinting of hemoglobin on silica surfaces using silanes, *Biomaterials*, **26**(27), 5564–5571.

126. Slinchenko, O., Rachkov, A., Miyachi, H., Ogiso, M., and Minoura, N. (2004). Imprinted polymer layer for recognizing double-stranded DNA, *Biosens. Bioelectron.*, **20**(6), 1091–1097.

127. Sode, K., Ohta, S., Yanai, Y., and Yamazaki, T. (2003). Construction of a molecular imprinting catalyst using target analogue template and its application for an amperometric fructosylamine sensor, *Biosens. Bioelectron.*, **18**(12), 1485–1490.

128. Sode, K., Yanai, Y., Ota, S., Tsugawa, W., and Yamazaki, T. (2001). Molecular recognition of frutosyl amine compounds by molecularly imprinted polymers, *Nippon Kagakkai Koen Yokoshu*, **79** (2), 840.

129. Steenblock, E. R., and Fahmy, T. M. (2008). A comprehensive platform for ex vivo T-cell expansion based on biodegradable polymeric artificial antigen-presenting cells, *Mol. Ther.*, **16**(4), 765–772.

130. Sun, R., Yu, H., Luo, M., and Shen, Z. (2004). Construction and application of a stoichiometric displacement model for retention in chiral recognition of molecular imprinting, *J. Chromatogr. A.*, **1055**(1–2), 1–9.

131. Tanaka, T., Filmore, D., Sun, S.-T., Nishio, I., Swislow, G., and Shah, A. (1980). Phase transitions in ionic gels, *Phys. Rev. Lett.*, **45**, 1636–1639.

132. Tanaka, T., Nishio, I., Sun, S.-T., and Ueno-Nishio, S. (1982). Collapse of temperature-sensitive polymers can also lead to the gels in an electric field, *Science*, **218**, 467–469.

133. Tai, D. F., Lin, C. Y., Wu, T. Z., and Chen, L. K. (2005). Recognition of dengue virus protein using epitope-mediated molecularly imprinted film, *Anal. Chem.*, **77**(16), 5140–5143.

134. Tai, D. F., Lin, C. Y., Wu, T. Z., Huang, J. H., and Shu, P. Y. (2006). Artificial receptors in serologic tests for the early diagnosis of dengue virus infection, *Clin. Chem.*, **52**(8), 1486–1491.

135. Takagishi, T., and Klotz, I. M. (1972). Macromolecule-small molecule interactions; introduction of additional binding sites in polyethyleneimine by disulfide cross-linkages, *Biopolymers*, **11**, 483–491.

136. Takatsy, A., Kilar, A., Kilar, F., and Hjerten, S. (2006). Universal method for synthesis of artificial gel antibodies by the imprinting approach combined with a unique electrophoresis technique for detection of minute structural differences of proteins, viruses and cells (bacteria), Ia. Gel antibodies against proteins (transferrins), *J. Sep. Sci.*, **29**(18), 2802–2809.

137. Takatsy, A., Vegvari, A., Hjerten, S., and Kilar, F. (2007). Universal method for synthesis of artificial gel antibodies by the imprinting approach combined with a unique electrophoresis technique

for detection of minute structural differences of proteins viruses and cells, (bacteria). lb. Gel antibodies against proteins (hemoglobins), *Electrophoresis*, **28**(14), 2345–2350.

138. Tamayo, F. G., Turiel, E., and Martin-Esteban, A. (2007). Molecularly imprinted polymers for solid-phase extraction and solid-phase microextraction: recent developments and future trends, *J. Chromatogr., A*, **1152**(1–2), 32–40.
139. Tan, C. J., Chua, M. G., Ker, K. H., and Tong, Y. W. (2008). Preparation of bovine serum albumin surface-imprinted submicrometer particles with magnetic susceptibility through core-shell miniemulsion polymerization, *Anal. Chem.*, **80**(3), 683–692.
140. Tanaka., T. (1978). Collapse of gels and the critical endpoint, *Phys. Rev. Lett.*, **40**, 820–823.
141. Tao, Z. Y., Tehan, E. C., Bukowski, R. M., Tang, Y., Shughart, E. L., Holthoff, W. G., Cartwright, A. N., Titus, A. H., and Bright, F. V. (2006). Templated xerogels as platforms for biomolecule-less biomolecule sensors, *Anal. Chim. Acta*, **564**(1), 59–65.
142. Tong, D., Hetenyi, C., Bikadi, Z., Gao, J. P., and Hjerten, S. (2001). Some studies of the chromatographic properties of gels ('artificial antibodies/receptors') for selective adsorption of proteins, *Chromatographia*, **54**(1–2), 7–14.
143. Tovar, G. E. M., Krauter, I., and Gruber, C. (2003). Molecularly imprinted polymer nanospheres as fully synthetic affinity receptors, *Colloid Chem. Ii.*, **227**, 125–144.
144. Turner, N. W., Jeans, C. W., Brain, K. R., Allender, C. J., Hlady, V., and Britt, D. W. (2006). From 3D to 2D: a review of the molecular imprinting of proteins, *Biotechnol. Prog.*, **22**(6), 1474–1489.
145. Turner, N. W., Liu, X., Piletsky, S. A., Hlady, V., and Britt, D. W. (2007). Recognition of conformational changes in, beta-lactoglobulin by molecularly imprinted thin films, *Biomacromolecules*, **8**(9), 2781–2787.
146. Ulbricht, M. (ed) (2005). Molecularly imprinted polymer films and membranes, in *Molecularly Imprinted Materials: Science and Technology*, Marcel Dekker, New York.
147. Ulrich, P., and Zhang, X. (1997). Pharmacological reversal of advanced glycation end-product-mediated protein crosslinking, *Diabetologia*, **40**, S157–S159.
148. Uludag, Y., Piletsky, S. A., Turner, A. P. F., and Cooper, M. A. (2007). Piezoelectric sensors based on molecular imprinted polymers for detection of low molecular mass analytes, *FEBS J.*, **274**(21), 5471–5480.
149. Uzun, L., Say, R., Unal, S., and Denizli, A. (2009). Hepatitis B surface antibody purification with hepatitis B surface antibody imprinted poly(hydroxyethyl methacrylate-N-methacryloyl-L-tyrosine methyl ester) particles, *J. Chromatogr., B: Anal. Technol. Biomed. Life Sci.*, **877**(3), 181–188.
150. Vaidya, A. A., Lele, B. S., Kulkarni, M. G., and Mashelkar, R. A. (2001). Creating a macromolecular receptor by affinity imprinting, *J. Appl. Polym. Sci.*, **81**(5), 1075–1083.
151. Vaihinger, D., Landfester, K., Krauter, I., Brunner, H., and Tovar, G. E. M. (2002). Molecularly imprinted polymer nanospheres as synthetic affinity receptors obtained by miniemulsion polymerisation, *Macromol. Chem. Phys.*, **203**(13), 1965–1973.
152. Venton, D. L., and Gudipati, E. (1995). Entrapment of enzymes using organo-functionalized polysiloxane copolymers, *Biochim. Biophys. Acta-Protein Str. Mol. Enzymol.*, **1250**(2), 117–125.
153. Venton, D. L., and Gudipati, E. (1995). Influence of protein on polysiloxane polymer formation – evidence for induction of complementary protein-polymer interactions, *Biochim. Biophys. Acta-Protein Str. Mol. Enzymol.*, **1250**(2), 126–136.

154. Vinge, V. (ed) (1986). *Marooned in Realtime*, Simon and Schuster, New York.

155. Walker, L. C., and LeVine, H. (2000). The cerebral proteopathies – neurodegenerative disorders of protein conformation and assembly, *Mol. Neurobiol.*, **21**(1–2), 83–95.

156. Wang, J. F., Zhou, L. M., Liu, X. L., Wang, J. D., and Shang, Z. H. (2000). Simultaneous chiral separation of two pairs of enantiomers using series column of molecular imprinted phase, *Chin. J. Anal. Chem.*, **28**(9), 1096–1098.

157. Wang, X., Wang, L. Y., He, X. W., Zhang, Y. K., and Chen, L. X. (2009). A molecularly imprinted polymer-coated nanocomposite of magnetic nanoparticles for estrone recognition, *Talanta*, **78**(2), 327–332.

158. Watanabe, M., Akahoshi, T., Tabata, Y., and Nakayama, D. (1998). Molecular specific swelling change of hydrogels in accordance with the concentration of guest molecules, *J. Am. Chem. Soc.*, **120**(22), 5577–5578.

159. Whitcombe, M. J., Rodriguez, M. E., Villar, P., and Vulfson, E. N. (1995). A new method for the introduction of recognition site functionality into polymers prepared by molecular imprinting – synthesis and characterization of polymeric receptors for cholesterol, *J. Am. Chem. Soc.*, **117**(27), 7105–7111.

160. Wijaya, A., Schaffer, S. B., Pallares, I. G., and Hamad-Schifferli, K. (2009). Selective release of multiple DNA oligonucleotides from gold nanorods, *ACS Nano*, **3**(1), 80–86.

161. Wulff, G., and Haarer, J. (1991). Enzyme-analog built polymers. 29. The preparation of defined chiral cavities for the racemic-resolution of free sugars, *Makromol. Chemie-Macromol. Chem. Phys.*, **192**(6), 1329–1338.

162. Wulff G., and Sarhan, A. (1972). The use of polymers with enzyme-analogous structures for the resolution of racemates, *Angew. Chem., Intl. Ed. Engl.*, **11**, 341.

163. Wulff, G., Vesper, W., Grobe-Einsler, R., and Sarhan, A. (1977). Enzyme analogue built polymers: 4. On the synthesis of polymers containing chiral cavities and their use for the resolution of racemates, *Makromol. Chem.*, **178**, 2799–2816.

164. Xue, X. H., Pan, J., Xie, H. M. M., Wang, J. H., and Zhang, S. A. (2009). Specific recognition of staphylococcus aureus by staphylococcus aureus protein A-imprinted polymers, *React. Funct. Polym.*, **69**(3), 159–164.

165. Hirokawa, Y., and Tanaka, T. (1984). Volume phase transition in a nonionic gel, *J. Chem. Phys.*, **81**, 6379–6380.

166. Yamazaki, T., Saito, A., and Sode, K. (2006). Molecular recognition of AGEs using molecularly imprinted polymer, *Nippon Kagakkai Koen Yokoshu*, **86**(2), 944.

167. Ye, L., and Haupt, K. (2004). Molecularly imprinted polymers as antibody and receptor mimics for assays, sensors and drug discovery, *Anal. Bioanal. Chem.*, **378**(8), 1887–1897.

168. Yilmaz, E., Haupt, K., and Mosbach, K. (2000). The use of immobilized templates – a new approach in molecular imprinting, *Angew. Chem., Int. Ed.*, **39**(12), 2115–2118.

169. Yoshida, R., Uchida, K., Kaneko, Y., Sakai, K., Kikuchi, A., Sakurai, Y., and Okano, T. (1995). Comb-type grafted hydrogels with rapid de-swelling response to temperature-changes, *Nature*, **374**(6519), 240–242.

170. Zhang, Z. H., Long, Y. M., Nie, L. H., and Yao, S. Z. (2006). Molecularly imprinted thin film self-assembled on piezoelectric quartz crystal surface by the sol-gel process for protein recognition, *Biocatal. Biotransform.*, **21**(7), 1244–1251.

Chapter 3

Chemical and Biological Sensing and Imaging Using Plasmonic Nanoparticles and Nanostructures

Qi Wang and Chenxu Yu
Department of Agricultural and Biosystems Engineering, Iowa State University, Ames, IA 50010, USA
chenxuyu@iastate.edu

3.1 Introduction

Surface plasmons (SPs) are coherent electron oscillations that are excited by electromagnetic (EM) radiation at the metal-dielectric interface. The characteristics of such light-metal interaction are determined by the properties of the metal, the dielectric environment, and the frequency of the EM radiation. It is represented by the phenomenon called surface plasmon resonance (SPR), where the frequency of the SP resonates with that of the excitation radiation. In recent years, study of this phenomenon has grown into a new field called plasmonics [1–3]. Much attention has been paid to plasmonics because of its potential in miniaturized optical devices, sensors for chemical and biological targets, and photonic circuits that could be used in next-generation medical diagnostics and therapeutics [4–8]. Advances in nanofabrication technology over the past decade are the key driving force in plasmonics study. A variety of nanomaterials and nanostructures have been synthesized and fabricated, including metallic nanoparticles [9] and nanoholes [10] with precisely controlled shape, size, and/or spacings [11, 12]. Advances in EM field computation and modeling have also greatly improved the theoretical understanding of the optical properties of these nanomaterials and nanostructures [13, 14]; combining theoretical modeling with precise synthetic control, it is now possible to design and fabricate nanomaterials and/or nanostructures that possess desired optical properties. In addition to the progress in nanofabrication, a variety of techniques and procedures have been developed to modify the surfaces of the nanomaterials to introduce specific

Biomedical Nanosensors
Edited by Joseph Irudayaraj

ISBN 978-981-4303-03-3 (Hardcover), 978-981-4303-04-0 (eBook)
www.panstanford.com

binding sites that allow the selective binding and detection of specific targets for chemical and biological sensing [15, 16]. Quantitative assays are also being developed for sensing applications [17, 18].

Detection schemes often utilize the effect of enhancement on the local EM field surrounding the nanomaterials caused by SPR. These enhancements, surface-enhanced Raman scattering (SERS) being one example, can lead to extraordinary sensitivity in chemical sensing. As being discussed in the chapter by Chula *et al.* the detection level of a single target molecule has been reached in SERS [19]. Further development in nanofabrication methodologies and sensor instrumentation will lead to much wider applications of these extremely sensitive nanosensors outside research laboratories [20].

Two types of SPRs are used in surface-based sensing [20] — (i) propagating surface plasmonpolaritons (SPPs), whichare evanescently confined SPs at the metal-dielectric interface, and (ii) nonpropagating localized SPRs (LSPRs) or nanoSPRs, which are SPs confined within nano-sized particles and/or structures. SPPs can be excited on thin metal films using grating or prism couplers [21]. They propagate along the metal surface, and their evanescent fields decay exponentially from the surface (normal to the metal-dielectric surface) [22]. Changes in the dielectric properties above the metal (i.e., refractive index) shift the plasmon resonance wavelength, which is detected as a change in intensity, wavelength, or angle shift in sensing applications [23]. Most commercial SPR sensors (i.e., Biacore® SPR sensors) are based on SPP design.

NanoSPR is nonpropagatingplasmon excitation that can be resonantly excited on metal nanoparticles and around nanoholes or nanowells in thin metal film [20]. The spectral characteristics of nanoSPR depend on the size, shape, chemical composition, and local dielectric environment [20, 24, 25]. The dependence on local dielectric environment is the base for design and implementation of label-free optical-sensing schemes. Binding/absorption of target molecules onto the surfaces of plasmonic nanoparticles/nanostructures introduces changes in the surface dielectric properties, which subsequently lead to shift of the nanoSPR band and/or changes in the nanoSPR magnitude. Monitoring these changes will provide the capacity to detect the target molecules at high sensitivity in real time [17, 18, 26, 27]. NanoSPR also creates enhanced EM fields surrounding the plasmonic nanostructures, which are used for performing surface-enhanced spectroscopies, that is, SERS [28]. In this chapter, we will review the latest progress on nanoSPR and their applications in chemical [29, 30] and biological sensing [15, 16, 29, 31].

3.2 Theoretical Background of nanoSPR

Gold (Au) and silver (Ag) are the most commonly used materials in nanoSPR since they exhibit strong SPRs in the visible-near-infrared (NIR) wavelength range [32]. At these wavelengths, the dielectric properties of the metals are described by a complex, wavelength-dependent dielectric coefficient, ε

$$\varepsilon(\lambda) = \varepsilon_r(\lambda) + i\varepsilon_i(\lambda) \tag{3.1}$$

where $\varepsilon = m^2$ and $m = n + ik$ is the complex refractive index given as a function of the refractive index n and the absorption coefficient k[20]. Noble metal nanoparticles can support nanoSPR [33] when the excitation photon frequency resonates with the collective oscillation of the conduction electrons confined inside the nanoparticles (Fig. 3.1) [34]. In the case of nanoSPR, the size of the particles (*d*) is much smaller than the wavelength of the incident light ($d << \lambda$). The conduction electrons inside the nanoparticles move in phase-up plane-wave excitation. This leads to the buildup of polarization charges on the nanoparticle surface, allowing a resonance to occur at a specific frequency (the nanoparticle dipole plasmon frequency) [33, 35]. A homogeneous field builds up in the interior of the nanoparticle, while a dipolar field is produced at the exterior, resulting in strong light scattering, the appearance of intense SPR bands, and the enhancement of the nearfield in the immediate vicinity of the nanoparticles. The spectroscopic responses of larger metallic nanoparticles are different due to the excitation of higher-order modes such as quadrupoles and retardation and skin-depth effects [33, 35]. The bandwidth, peak height, and peak wavelength all depend on the nanoparticle chemical composition, size, and geometrical shape (Fig. 3.2) and the dielectric property of the surrounding environment [33, 36, 37].

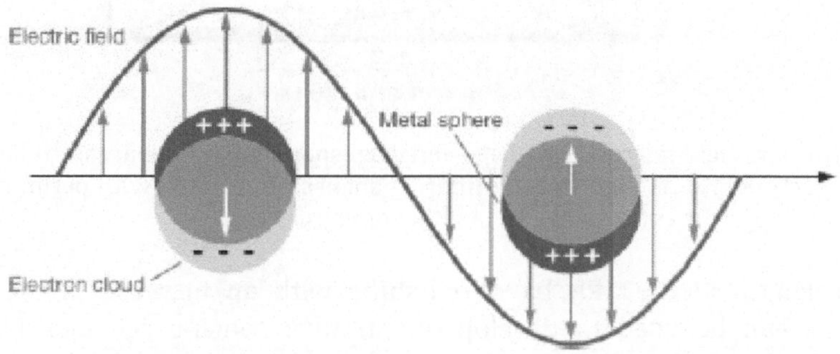

Figure 3.1 A schematic illustration of a localized SP of a metal sphere showing the displacement of the electron charge cloud relative to the nuclei. (Reprinted with permission from ref. 34. Copyright 2007 Annual Reviews.)

Optical properties of spherical nanoparticles can be modeled using the classical Mie theory [33, 35, 38]. When the wavelength of light is greater than the size of nanoparticles, the extinction coefficient $E(\lambda)$ — the sum of absorption and scattering — of a spheroid metallic nanoparticle is given by:

$$E(\lambda)=\frac{9}{c}\varepsilon_{\text{med}}^{3/2}VN\left[\frac{\varepsilon_{\text{i}}}{(\varepsilon_{\text{r}}+\chi\varepsilon_{\text{med}})^2+\varepsilon_{\text{i}}^2}\right] \tag{3.2}$$

where $\omega(=2\omega v)$ is the angular frequency of the incident light, c is the speed of light, V is the nanoaprticle volume, and N is the number density of the nanoparticles. ε_{med} is the dielectric constant of the surrounding medium, λ is the excitation wavelength, and χ is a form factor that describes the nanoparticle's aspect ratio (χ = 2 for spheres and increases directly with the aspect ratio of the nanoparticle). ε_{r} and ε_{i} are the real and imaginary parts of the dielectric function of the nanoparticle, respectively. For a spherical particle with $|\varepsilon_{\text{r}}| >> \varepsilon_{\text{i}}$, $E(\lambda)$ has a maximum when $\varepsilon_{\text{r}}(\omega) = -2\varepsilon_{\text{med}}$, and this is the LSPR or nanoSPR condition.

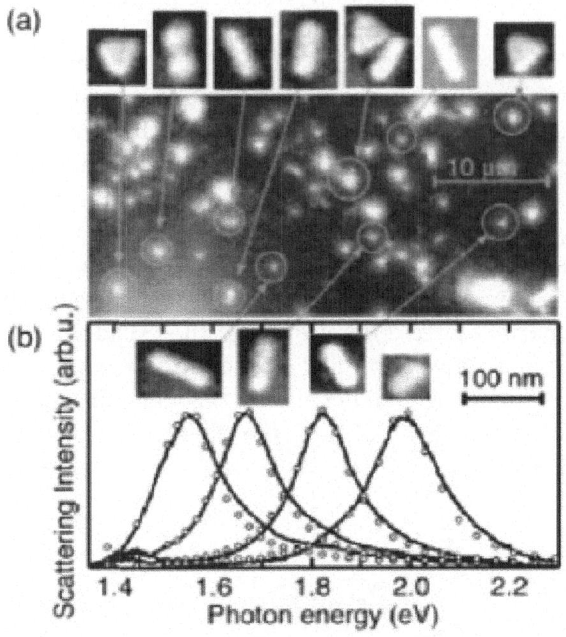

Figure 3.2 (a) Dark-field microscopy image and corresponding SEM images; (b) light-scattering spectra of Au nanocrystals of different shapes. (Reprinted with permission from ref. 39. Copyright 2003 American Institute of Physics.)

The wavelength of the SPR band red-shifts with an increase in the magnitude of ε_{med}, which can be used to develop quantitative sensing schemes [17, 18]. For nonsphericalnanoparticles, the SPs are unevenly distributed around them, manifesting in a shape dependence of the SPR spectra [40]. For example, the SPR band of Au nanorods splits into two peaks — (i) a strongly red-shifted longitudinal mode and (ii) a slightly blue-shifted transverse mode. As the aspect ratio of a nanorod increases, the separation between the two plasmonic peaks becomes more pronounced [12, 41–49]. As the aspect ratio increases, the wavelength of the longitudinal band also increases in a almost linear correlation, as shown in Fig. 3.3 [42], accompanied by a visible color change of the nanoparticle suspension (Fig. 3.4). The wavelengths of the longitudinal bands are also strongly correlated to the refractive index of the surrounding dielectric, and a quantitative correlation could be established (Fig. 3.5), which could be utilized for quantitative biological sensing [17].

Triangular metallic nanoparticles (or nanoprism) exhibit multiple plasmon resonances, a longitudinal (bulk) plasmon mode, and a very large field enhancement at their sharp tips [50, 51]. In general, anisotropic nanoparticles display very high field enhancement around their sharp tips/corners and edges, which leads to strong SERS effect [52].

To theoretically calculate fields and optical properties of nanoparticles with more complex shapes and compositions, a generalized quasistatic approach needs to be combined with numerical methods such as T-matrix [53], the discrete dipole approximation [54, 55], or finite difference time-domain simulations [56, 57]. Important insights have been acquired by applying these methods to calculate the resonant frequency, local field enhancement, absorption, scattering, and extinction efficiency of complex nanostructures.

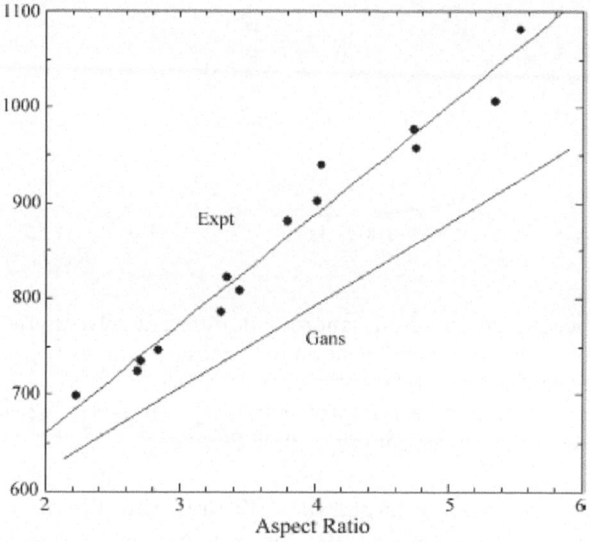

Figure 3.3 An experimental and theoretical plot of the SP longitudinal band position versusthe aspect ratio for pure Au rods in water. (Expt: experimental results; Gans: theoretical prediction by Gans' model [Eq. 3.2]). (Reprinted with permission from ref. 42. Copyright 2005 Elsevier.)

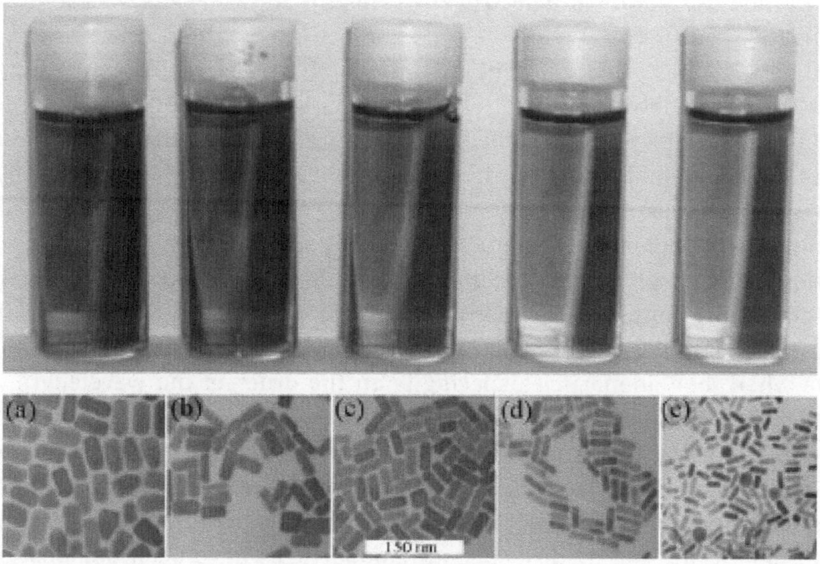

Figure 3.4 The color of Au rods and the respective micrographs. The color changes take place for very small changes in the mean aspect ratio. (Reprinted with permission from ref. 45. Copyright 2005 Elsevier.) See also Color Insert.

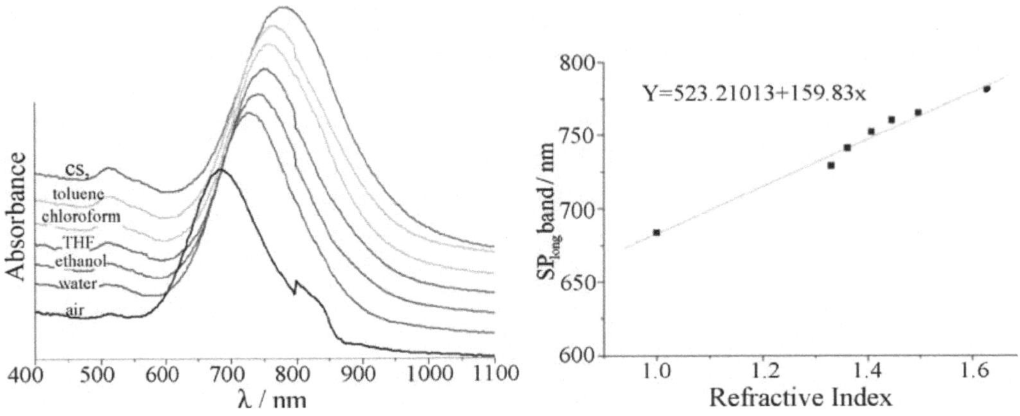

Figure 3.5 Experimental spectra of Au nanorods in different solvents (left) and change in the longitudinal plasmon band position with refractive index (right). To obtain $n = 1$, the particles were coated on glass. However, the effective refractive index for small particles in air supported on glass is approximately $n_{eff} = (n_{air} + n_{glass})/2 \sim 1.25$ (Reprinted with permission from ref. 45. Copyright 2005 Elsevier.)

As the nanoparticles become larger (d > 30 nm), the dipolar resonance red-shifts and suffers a substantial broadening [20]. The red-shift arises due to a reduction of the depolarization field caused by the retardation effects [58] where the conduction electrons no longer all move in phase. Radiative losses [59] also start to contribute significantly to the plasmon damping, eventually dominating it totally for Au and Ag nanoparticles with d >100 nm. The broadening negatively impacts the sensitivity of these nanoparticles to changes in their surrounding dielectric environment and weakens their applicability as highly sensitive sensors.

Mie theory applies to only noninteracting nanoparticles well separated in the solid state or present in low concentrations in a solution [20]. Recently, an assembly of ordered arrays of interacting metallic nanoparticles and nanostructures like nanohole arrays created on metallic thin films has been acquired, and they exhibit interesting new properties that are also being explored for sensing applications [60, 61]. In such cases, each nanoparticle or nanohole is much smaller in size compared to the wavelength of the exciting light and acts as electric dipole. Two types of EM interactions prevail in this context, depending on the spacing between adjacent spots — (i) near-field coupling and (ii) far-field dipolar interactions [62]. Far-field dipolar interactions with a d^{-1} dependence dominate when then anoparticles spacing is on the order of the wavelength (λ) of the exciting light [63], while the near-field interactions with a d^{-3} dependence dominate for spacings much smaller than λ [64]. These dependences on spacings have been exploited in colorimetric-sensing schemes [65]. Nanoholes created in metallic thin films are also being explored for nanoSPR sensing. Strong correlations between the light scattering of holes and light scattering by disk-shaped nanoparticles have been demonstrated [66]. Nanoholes tend to exhibit broader scattering features than nanoparticles and can also serve as point sources for SPP waves in the thin films [20, 60, 61].

3.3 Synthesis, Fabrication, and Functionalization of Plasmonic Nanostructures

Synthesis and fabrication of metallic nanomaterials and nanostructures has been a very active research field for many years due to their importance in many applications, including photonics and optoelectronics [1, 3, 7, 67–71], electronics [72–75], chemical and biological sensing [76–81], and medical diagnosis and therapies [82–86]. As of today, the chemical and physical properties of metallic nanomaterials and nanostructures can be fine-tuned by controlling their sizes, shapes, and compositions [23, 35, 87–90]. Many methods have been developed to make isotropic and anisotropic nanostructures with well-controlled sizes and shapes from a variety of materials [11, 12, 83, 88, 90–93]. These approaches can be summarized into two categories, top-down and bottom-up [20]. Top-down techniques involve using various forms of conventional lithographic techniques to pattern nanostructures (e.g., onto planar substrate) [94–97], whereas in bottom-up methods, nanostructures are "built" or "assembled" from atoms, molecules, and/or more complex mesoscale objects [18, 20, 66, 98–102]. In the following section, we discuss the variety of synthesis and fabrication methods in more detail.

3.3.1 Wet-Chemistry Synthesis

A typical bottom-up wet-chemistry synthesis of spherical nanoparticles of noble metals involves the reduction of its metal salt in a solution and proceeds in two stages [103, 104]. In the first stage, metal ions are reduced to metal atoms, which then undergo rapid collisions to form stable icosahedral nuclei of 1–2 nm in size. The initial concentration of nuclei depends on the concentration of the reducing agent, the solvent, temperature, and reduction potential of the reaction. This stage typically gets completed in a few seconds and determines the heterogeneity in the sizes and shapes of the nanoparticles. Further growth of the nanoparticles occurs in the second stage by the reduction of metal ions on the surface of the nuclei until all ions are consumed. Simultaneously, a stabilization reaction is occurring, whereas a stabilizing agent is added onto the surface of the nanoparticles to prevent them from aggregation either through Coulombic repulsion [16, 105, 106] or steric hindrance [103, 107, 108]. For example, Au nanoparticles are often stabilized by thiol-capping agents [109] that impart negative or positive charges to the particles.

The reduction of the metal salt can be carried out electrochemically [110–115], photochemically [106, 116–119], sonochemically [120–122], or using chemical reductants such as citrates [106], hydrides [123, 124], alcohols [125, 126], hydrogen [92], hydroxylamine [127, 128], or hydrazine [129, 130]. By choosing a specific reductant, stabilizer, temperature, and concentrations of the reactants, the shapes and sizes of the nanoparticles can be controlled.

Nanorods and nanoshells are often produced using wet-chemistry approaches [90, 91, 131–135]. Typically, for making Au nanorods, a growing solution is prepared containing Au salt, a mild reducing agent (e.g., ascorbic acid), a surfactant stabilizer (e.g., cetyltrimethylammonium bromide [CTAB]), and other additives (e.g., $AgNO_3$); a nanoparticle "seed" (2~4 nm) is then added to the growth solution, and the Au salt is reduced on the surface of the seeds [131]. Because CTAB binds to the side faces of the growing nanoparticle (due to favorable packing of Au atoms) more strongly and slows down the growth along the transverse axis, the growth along the longitudinal axis outpaces it and the nanoparticles eventually become anisotropic and grow into a rodshape [131]. The morphology, aspect ratio, and yield of rods are controlled by the composition and concentrations of the surfactant(s), additives, seeds, Au salts, and the reducing agent [42, 44, 90, 131, 132, 136–140]. Because the longitudinal plasmon mode of nanorods can be tuned by adjusting the aspect ratio, they are particularly useful in photonics and biotechnology [12, 17, 41–49, 133, 141]. An Au nanoshell is another type of nanostructure with tunable plasmonic properties [86, 142]. Halas and coworkers formed such core-shell structures by seeded metallization of colloidal silica spheres [134]: the silica particles are functionalized with a monolayer of an amine-terminated silane (aminopropyltriethoxysilane, APTES), which is then used to bind small Au particles. After the initial seeding, more Au is deposited by an electroless plating procedure [134, 135], and the thickness of the Au shell is controlled by adjusting the initial Au seed coverage and the amount of reductant using in plating. By controlling the size and thickness of the shell-core structure, the plasmonic properties can be fine-tuned, as shown in Fig. 3.6.

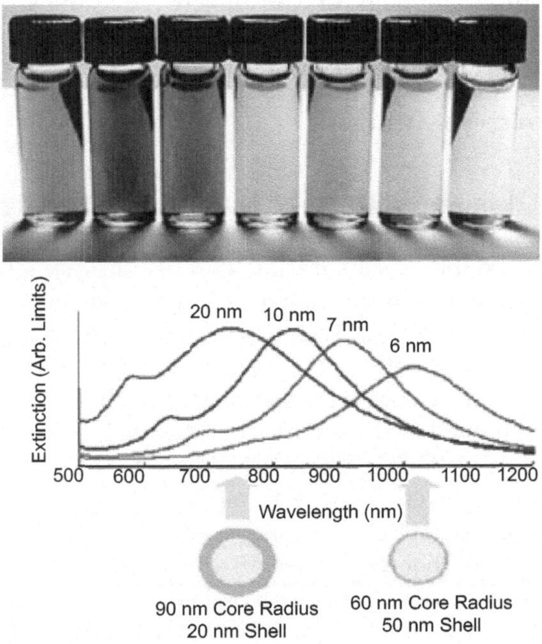

Figure 3.6 A visual demonstration of the tunability of metal nanoshells (top) and optical spectra of Au shell-silica core nanoshell. (Reprinted with permission from ref. 86. Copyright 2004 www.tcrt.org.) See also Color Insert.

Galvanic replacement reaction has also been explored for the production of Au-Ag alloy nanocages and nanoboxes [143, 144]. Firstly, Ag nanocubes are made through the polyol process [11, 88], in which ethylene glycol is used both as a solvent and as a reducing agent to reduce Ag salt into well-defined Ag nanoparticles with a variety of sizes and shapes [11, 88]. Then, utilizing a galvanic replacement reaction, hollow or porous Au nanocages or Au-Ag alloy nanoboxes are made from the Ag nanocubetemplates [143, 144] simply by adding Au salt into a suspension of Ag nanocubes. The nanoSPR of these nanostructures can be tuned from the visible to NIR by varying the amount of Au salt added. More detailed reviews on nanoparticle syntheses through wet-chemistry approaches can be found elsewhere [9, 45, 145, 146].

3.3.2 Lithography

3.3.2.1 Top-down lithography

Surface-bound metallic nanostructures with precisely controlled size, shape, and spacing can be obtained using scanning beam lithography, such as electron beam lithography (EBL) [96, 97] and focused ion beam (FIB) lithography [147]. The substrate for SERS [148–150] and plasmonic-enhanced fluorescence [151] have been made using these methods. In EBL, a tightly focuses electron beam is used to pattern-scan a thin layer of resist (a radiation-sensitive polymer), which makes it either more or less soluble in an organic developer solution. The patterned resist is then used as a mask in subsequent etching or deposition processes to generate nanostructures with well-defined geometry down to 20 nm resolution [29]. In FIB, a focused beam of ions (typically Ga^{2+}) is used to perform the patterning in physically or chemically assisted processes [147, 152–157]. FIB is capable of forming patterns with ~10 nm resolution [20]. Various metallic nanostructures such as circular silts [152], nanoholes [158, 159], slit grating [160, 161], and V-grooves [162] have been fabricated by FIB.

3.3.2.2 Colloidal lithography

Colloidal lithography has been used to form random arrays of nanoholes[163], nanodisks [164], and nanorings [102]. Polystyrene particles (PS) are absorbed onto a substrate by electrostatic assembly; the patterning of the PS is controlled governed by particle-particle repulsion, which can be controlled by adjusting the electrolyte concentration of the colloidal solution [99]. The randomly absorbed PS particles are then used as a mask for subsequent deposition and/or etching processes wherein the size of asphere dictates the size of the resulting nanostructure.

3.3.2.3 Nanosphere lithography

Nanosphere lithography (NSL) is invented by Van Duyne and coworkers [18, 165] for the fabrication of periodic arrays of metallic nanostructures. It utilizes a nanosphere colloidal etching mask to create nanostructures with controlled size and shape [18]. In

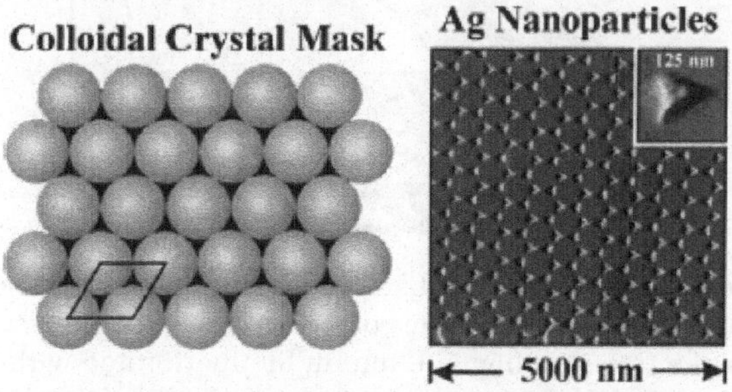

Figure 3.7 A schematic illustration of a hexagonally close-packed colloidal crystal mask (left) and a representative AFM image of a triangular nanoparticle array (right). (Reprinted with permission from ref. 18. Copyright 2001 American Chemical Society.)

NSL, a single-layer colloidal crystal etching mask of hexagonally close-packed latex or silica spheres is deposited on a glass substrate. The colloidal crystal contains triangular-shaped void spaces between the neighboring particles. Ag is deposited on top of the mask layer and fills the voids; when the mask is finally removed by sonication, arrays of surface-bound triangular nanoparticles with a P6*mm* symmetry are produced, and these structures have been used for nanoSPR sensing and SERS (Fig. 3.7) [18, 166–169]. The main difference between colloidal lithography and NSL is that the colloidal spheres do not form a hexagonal close-packed (hcp) monolayer on the substrate as NSL does [20].

3.3.2.4 Soft lithography

In soft lithography, a structured elastomer is used as a stamp or mold to pattern a material of interests [170–172]. Siloxane-based elastomers such as hard poly(dimethylsiloxane) (*h*-PDMS) and UV-curable PDMS are often used in nano-scaled soft lithography due to their large moduli [173–176]. A soft lithography procedure used by Odom and coworkers to create two-dimensional nanohole arrays in Au is schematically illustrated in Fig. 3.8. [177, 178]. In this process, an array of PDMS posts is deposited on a silicon wafer. A thin layer of chromium is then deposited by electron beam evaporation followed by the removal of the PDMS posts. The Cr layer now serves as an etching mask. The exposed Si is then an isotropically etched using a KOH/isopropyl (IPA) solution to produce pyramidal-shaped voids. A layer of Au is then deposited by e-beam evaporation to form a two-dimensional Au nanoholes array [178] and pyramidal-shaped Au nanoparticles [179].

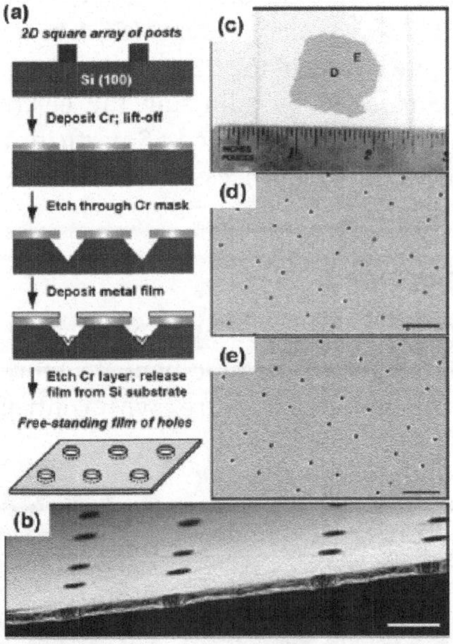

Figure 3.8 Fabrication and structural characterization of large-area hole arrays; (a) the preparation of free-standing films of subwavelength hole arrays; (b) the SEM image of a portion of a free-standing 100 nm Au film perforated with 250 nm holes, scale bar 500 nm; (c) an optical micrograph of a free-standing film placed on a glass substrate; (d) and (e) an SEM image of representative areas of the film illustrating the uniformity of the nanohole array; the holes are spaced 1.6 × 2.4 μm. Scale bar: 2 μm. (Reprinted with permission from ref. 178. Copyright 2005 American Chemical Society.)

3.3.3 Functionalization of Plasmonic Nanostructures

To make plasmonic nanostructures into useful nanosensors for chemical and biological probes, it is necessary to add functionality to them through a functionalization process. This is particularly true for nanoparticles made through wet-chemistry synthesis: the surfactant stabilizer needs to be replaced by a chemicallyactive layer to which further attachment of biologically functional molecules such as antibodies and DNAs can be performed.

The first step in a functionalization process is to chemically modify the surface of the nanoparticles and/or nanostructures to add a "dock" onto each particle to which biological molecules can attach subsequently. Thiol-containing compounds are widely used for Au particle modification to utilize the easily formed S-Au bond. After the thiol ends of the molecules are anchored onto the Au particle surface, the chemical groups at the other end (i.e., -COOH and -NH_2) can be made to react with the functional molecules (i.e., the antibody).

Thiolatedpolyethylene glycol (PEG) has been used to replace CTAB on Au nanorods (Pierrat 2007 *Nano Letts.*). Small thiol-containing molecules, such as 3-animo, 5-mercapto, 1, 2, 4 triazole (AMTAZ), and 11-mercaptoundecaonic acid (MUDA) have also been used to modified Au nanorods, and subsequent attachment of antibodies were demonstrated [140, 141]. In this approach, the Au nanorods were kept at an elevated temperature to facilitate CTAB removal and subjected to constant sonication to minimize particle aggregation. The Au nanorod probes functionalized with antibodies were used for multiplex characterization of antibody-antigen bindings, and quantitative assay of the binding kinetics was reported [17]. The process to make molecular probes using Au nanorods is illustrated in Fig. 3.9.

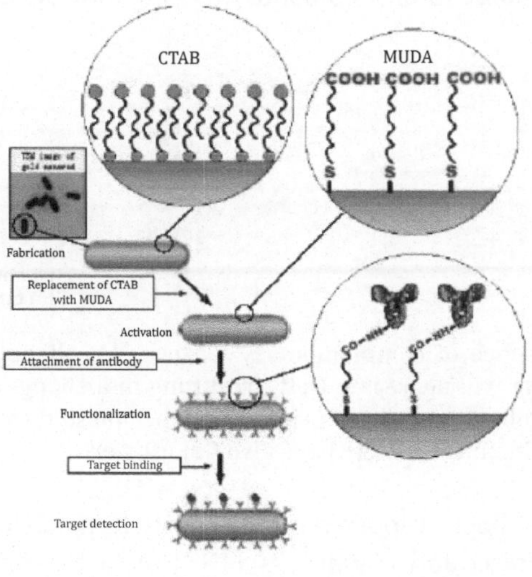

Figure 3.9 An illustration of the fabrication of gold nanorod molecular probes (GNrMPs); Steps shown are: (a) the synthesis of Au nanorods; (b) the activation step to replace the CTAB coating with MUDA; (c) functionalization and attachment of antibodies to the MUDA anchor; (d) target detection, depicting the binding of GNrMPs to their molecular targets. (Reprinted with permission from ref. 141. Copyright 2007 American Chemical Society.) See also Color Insert.

In a recent approach, Hamad-Schifferli and coworkers developed a round-trip modification process to replace CTAB on Au nanorods with mercaptohexanoic acid (MHA) and then attached single-strand DNA to them [180, 181]. In this process, CTAB-coated Au nanorods were firstly subjected to an aqueous-to-organic phase transfer, in which CTAB was replaced by dodecanethiol (DDT). Then an organic-to-aqueous phase transfer was performed, in which DDT was replaced by MHA, resulting in an MHA-coated Au nanorods aqueous solution. Then, thiolated single-strand DNA was attached to the Au nanorods through further ligand exchange with MHA [180, 181]. In any functionalization efforts with DNA and DNA hybridization, charge screening (salt-aging) was necessary to compensate for the electrostatic repulsion between the negatively charged DNA molecules [182].

3.4 Application of nanoSPR in Chemical and Biological Sensing

3.4.1 Solution-Based nanoSPR Sensing

3.4.1.1 Colorimetric sensing based on particle-particle coupling

The colorimetric detection of biological targets using nanoSPR induced by particle-particle coupling is pioneered by Mirkin and coworkers [65, 183]. Au nanoparticles were functionalized using two batches of single-strand nucleotides; each strand is complementary to part of the target strand; upon hybridization to the targets, a controlled aggregation of Au nanoparticles occurred and exhibited strong red-shift (Fig. 3.10) and a visible color change was observed. The temperature profile of DNA hybridization detected by particle-coupling nanoSPR sensing showed very sharp transition behavior, and single base-pair mismatch could be identified using this detection scheme [183].

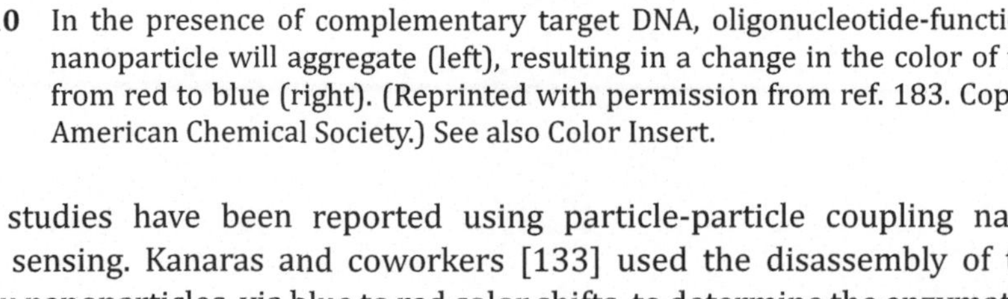

Figure 3.10 In the presence of complementary target DNA, oligonucleotide-functionalized Au nanoparticle will aggregate (left), resulting in a change in the color of the solution from red to blue (right). (Reprinted with permission from ref. 183. Copyright 2005 American Chemical Society.) See also Color Insert.

Many studies have been reported using particle-particle coupling nanoSPR for biological sensing. Kanaras and coworkers [133] used the disassembly of the closely coupled Au nanoparticles, via blue to red color shifts, to determine the enzymatic cleavage of DNA, where in a recognition site for the restriction endonulease Eco*RI* was designed to monitor the enzymatic cleavage activity. Han *et al.* used oligonucleotide-functionalized Au nanoparticles to measure the relative binding strengths of duplex and triplex DNA-binding molecules [184, 185]. Peptide nucleic acid (PNA) and aptamer-modified Au nanoparticles were also used to control assembly rates and aggregate sizes of Au nanoparticles for sensing applications [77, 186–189].

Cation-specific functionalized Au nanoparticles have also been developed for the detection of metal ion targets. Pb^{2+} (Fig. 3.11) [190] and Hg^{2+} are among the metal ion targets that have been reportedly detected using Au nanoparticle-based nanoSPR sensing schemes [76, 190–193].

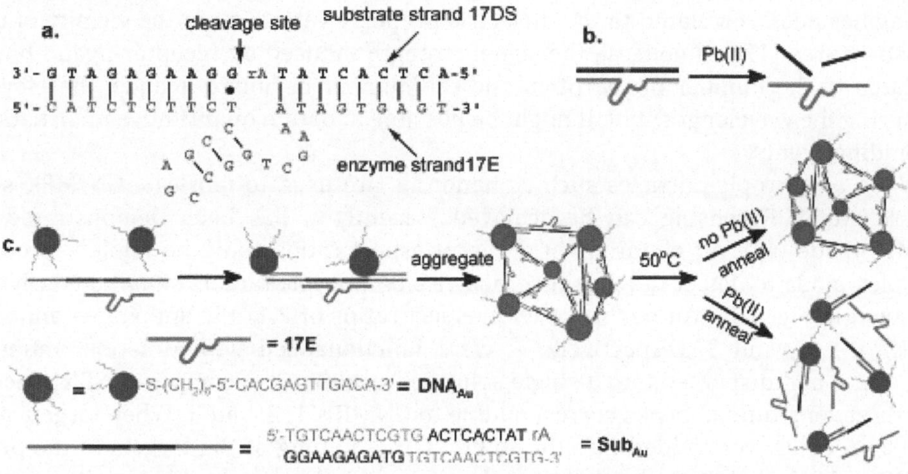

Figure 3.11 (a) The secondary structure of the "8–17" DNAzyme, which consists of an enzyme strand (17E) and a substrate strand (17DS). The cleavage site is indicated by a black arrow. Except for a ribonucleoside adenosine at the cleavage site (rA), all other nucleosides are deoxyribonucleosides. (b) The cleavage of 17DS by 17E in the presence of Pb^{2+}. (c) Schematics of DNAzyme-directed assembly of Au nanoparticles and their application as biosensors for metal ions such as Pb^{2+}. In this system, the 17DS has been extended on both the 3' and 5' ends for 12 bases, which are complementary to the 12-mer DNA attached to the 13-nm Au nanoparticles (DNA^{Au}). (Reprinted with permission from ref. 190. Copyright 2003 American Chemical Society.)

All of the solution-based colorimetric methods employ the plasmonic coupling in aggregated nanoparticle systems. They are in general sensitive and do not require sophisticated instrumentation.

3.4.1.2 Single-particle nanoSPR sensing

Plasmons are exceptionally sensitive reporters for chemical or physical phenomena that influence the refractive index of the local environment of a nanoSPR probe. SPR spectroscopy [194–196] and imaging [22, 26, 197–200] are well-known label-free optical detection methods for real-time molecular-binding events. Early SPR sensing was based on plasmon produced in macroscopic metallic film; as nanoparticles being exploited for sensing application, particle-based nanoSPR label-free sensing also became available [17, 27].

In aqueous solutions, Au nanostructures exhibit strong plasmon bands depending on their geometric shapes and sizes. For spherical particles, a strong absorption band around 520 nm due to the excitation of plasmons by incident light can be readily observed [45].

For nanorods, two distinct plasmon bands, one associated with the transverse (~520 nm) mode and the other with the longitudinal mode (usually >600 nm), could be observed [45]. Quadrupoleplasmon modes have also been reported for more complex structures such as prisms [201]. The wavelengths of these bands are affected by changes in the dielectric properties in the close vicinity of these nanostructures due to the binding of ligands to the corresponding receptor molecules (i.e., antibodies) immobilized onto the nanostructures through chemi- or physisorption. The magnitude of wavelength shift induced by target binding has been correlated to the dielectric property variations in the vicinity of these nanostructures [17]. In general, the signal strength induced by receptor-ligand binding is related to the number of receptors and the number of bound ligands; therefore, by measuring the wavelength shift, it might be possible to obtain quantitative information of the binding events.

When anisotropic particles such as nanorods are used to fabricate GNrMPs, single particle nanoSPR sensing can be achieved. Recently, it has been demonstrated that GNrMPs made with Au nanorods of different aspect ratios could be implemented in a multiplex mode to detect the presence/absence of multiple targets simultaneously [27]. As shown in Fig. 3.12, Au nanorods with aspect ratios of 2.1, 4.5, and 6.5 — annotated as GNrMP1, 2, and 3, respectively — were functionalized with different antibodies (Fig. 3.12a) and dispersed into a single solution, resulting in a composite SPR spectrum with three longitudinal peaks corresponding to GNrMPs 1, 2, and 3. When targets of the three antibodies were added into the solution, the binding of the targets to the proper GNrMPs selectively initiated red-shift to the corresponding longitudinal peaks, reporting the presence of the targets. The easily distinguishable multibands of GNrMPs supported a multiplex-detection scheme.

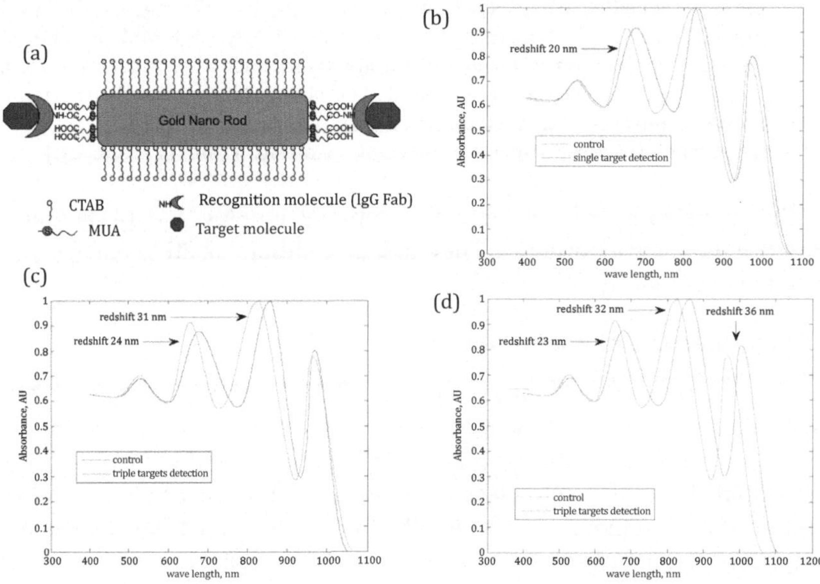

Figure 3.12 (a) Aschematic representation of a GNR molecular probe; multiplexing detection of various targets using GNR molecular probes; (b) one target; (c) two targets; (d) three targets; (Reprinted with permission from ref. 27. Copyright 2007 American Chemical Society.) See also Color Insert.

Treating each GNrMP as a single sensing platform, the equilibrium surface excess, Γ_{targets} expresses as the number of molecules/cm^2, for 1:1 binding of anti-IgG to particle-bound IgG, is given by the Langmuir isotherm [18]:

$$\frac{\Gamma_{\text{target}}}{\Gamma_{\text{target}}^{\max}} = \frac{K_{\text{a}}[T]}{(1+K_{\text{a}}[T])} \tag{3.3}$$

where $\Gamma_{\text{target}}^{\max}$ is the saturation value of Γ_{targets} (i.e., when every capturing agent is saturated by the target), $[T]$ is the concentration of targets, and K_{a} is the apparent affinity constant for 1:1 binding. Adopting the model developed by Campbell and coworkers [202] to analyze SPR responses to target binding, the GNrMP response (R) to target binding, where anti-IgG is used as capturing agents and their complementary IgG is used as targets, is given by [17],

$$R = S(n_{\text{eff}} - n_{\text{ext}}) \tag{3.4}$$

where S is the sensitivity factor of GNrMPs, defined as [17]:

$$S = \frac{d\lambda_{\text{res}}}{dn_{\text{s}}} \propto \lambda_{\text{p}}\sqrt{Y} \propto \lambda_{\text{p}} AR \tag{3.5}$$

where λ_{res} is the plasma wavelength of nanostructures, n_{s} is the refractive index of the surroundings, λ_{p} is the plasmon wavelength of bulk metal, and Y is a geometric parameter proportional to the square of the aspect ratio (AR) of the nanorods.

In Eq. 3.4, n_{ext} is the bulk refractive index of the external medium ($n_{\text{ext}} = n_{\text{water}} = 1.33$) and n_{eff} is the effective refractive index of the quadralayer structure (layer 1 = MUA SAM, layer 2 = anti-IgG, layer 3 = IgG, and layer 4 = water) surrounding each GNrMP. S is determined experimentally for each GNrMP. The effective refractive index of the quadra-layer structure is determined by integrating the distance-dependent local refractive index, $n(z)$, weighted by the square of the local EM field, $E(z)$, as [17]:

$$n_{\text{eff}} = \frac{2}{l_{\text{d}}}\int_0^{\infty} n(z)E^2(z)dz \tag{3.6}$$

where

$$n(z) = \begin{cases} n_{\text{sam}},\ 0 \le z \le d_{\text{sam}} \\ n_{\text{anti-igG}},\ d_{\text{sam}} \le z \le d_{\text{sam}} + d_{\text{anti-igG}} \\ n_{\text{IgG}},\ d_{\text{sam}} + d_{\text{anti-igG}} \le z \le d_{\text{sam}} + d_{\text{IgG}} + d_{\text{anti-IgG}} \\ n_{\text{water}},\ d_{\text{sam}} + d_{\text{IgG}} + d_{\text{anti-IgG}} \le z < \infty \end{cases} \tag{3.7}$$

Here $n_{\text{IgG}} = n_{\text{anti-IgG}}$; d_{sam} is the thickness of the MUDA SAM; d_{IgG} is the thickness of the IgG layer, which is the same as $d_{\text{anti-IgG}}$; $E(z)$ is assumed to be only dependent on the local surface normal z; and l_{d} is the characteristic decay length.

The measured response of GNrMPs ($\Delta R = R_{\text{layer3}} - R_{\text{layer2}}$) is given by:

$$\Delta R = S(n_{\text{IgG}} - n_{\text{water}})e^{-\frac{2d_{\text{sam}}}{l_{\text{d}}}}(1 - e^{-\frac{4d_{\text{IgG}}}{l_{\text{d}}}}) \tag{3.8}$$

Plotting the normalized GNrMP responses ($\Delta R/\Delta R_{max}$) versus [T] gives the binding affinity of the targets to the GNrMPs [17], as shown in Fig. 3.13.

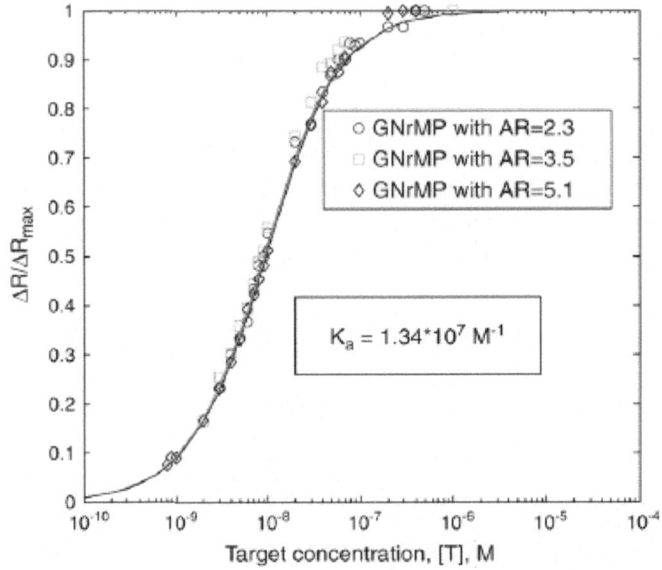

Figure 3.13 Normalized GNrMP response, $\Delta R/\Delta R_{max}$, versus the target concentration [T]. (Reprinted with permission from ref. 17. Copyright 2007 Biophysical Society.)

Using GNrMPs in a multiplexing detection experiment, three targets were quantitatively detected at nM levels [17], as shown in Table 3.1.

Table 3.1 Comparison of measured target concentration to real values [17]

	Target 1	Target 2	Target 3
Real value	100 nM	20 nM	20 nM
Measured value	92.32 nM	19.14 nM	15.86 nM

The binding affinity between GNR-bound anti-IgGs and IgGs is two orders of magnitude higher than that between surface-bound anti-IgGs and IgGs, showing that using nanoparticles as SPR sensors can significantly increase the sensitivity compared to macroscopic SPR sensors [17].

Single particle–based GNrMPs were also used to detect multiple-cell-surface biomarkers to identify cancer stem cells in living cells [141].

3.4.2 Chip-Based nanoSPR Sensing

3.4.2.1 Raleigh scattering of nanostructures

Raleigh scattering of metallic nanoparticles is also being explored for biosensing [203, 204]. It has been suggested that one single 80 nm Au nanoparticle, which does not photo bleach or blink, exhibits light-scattering power equivalent to the emission of 10^6 fluorescein molecules [205]. These optical properties have been exploited for labeling in immunoassays and DNA microarrays [206–209], imaging contrast agents [86, 210], and molecular rulers [211–213].

When nanoparticles undergo aggregation, their light-scattering properties change. Therefore, target-induced aggregation of Au nanoparticles has been used to measure trace amounts of DNAs at picomolar concentrations [214]. Muller and coworkers demonstrated an even lower detection limit: in their approach, 50 nm Au nanoparticles functionalized with segments of ssDNA were mixed with a target ssDNA solution to initiate hybridization. The hybridization led to the aggregation of the Au nanoparticles. An aliquot of the solution containing the aggregated Au nanoparticles was then spotted onto a glass slide (a planar waveguide) that was illuminated using a planar fiber-optic illuminator (Fig. 3.14a). The evanescent light was scattered from the aggregated Au nanoparticles and imaged with a complementary metal oxide semiconductor (CMOS) camera. Isolated individual Au nanoparticles scattered green light, while aggregated Au nanoparticles, which indicated the existence of the target ssDNAs, scattered orange light (Fig. 3.14b) due to plasmon coupling between aggregated particles. This approach allowed the detection of target DNA at femtomolar concentration [215]. The high sensitivity achieved without having to use polymerase chain reaction (PCR) amplification makes this approach a promising method for the detection of trace amounts of DNA in point-of-care molecular diagnosis.

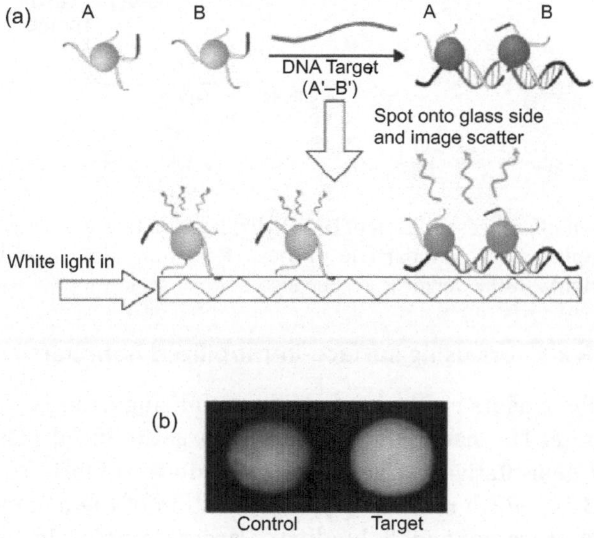

Figure 3.14 Colorimetric detection of nucleic acids using scattered light. (a) Step 1: DNA-Au nanoparticle probes (A and B) are hybridized to a DNA target in solution. Step 2: the samples are spotted onto a glass slide, which is illuminated with white light in the plane of the slide. (b) The evanescent-induced scatter from the Au nanoparticles is visually observed. Individual 40 nm to 50 nm diameter Au probes scatter green light, whereas complexed probes scatter yellow to orange light because of a plasmon band red-shift. (Reprinted with permission from ref. 215. Copyright 2004 Nature Publishing Group.) See also Color Insert.

In another approach, the ratio of the scattered plasmon light intensity at two wavelength is used to quantify a target analyte, such as glucose [216]. The ratio of scattered light intensity [217] or polarization [218] at two different angles as the function of target-induced aggregation or dissociation of plasmonic nanoparticles was also used for biosensing. The ratiometric methods are independent of nanoparticle concentration, and the stability of the assay is thus improved compared to single wavelength measurements.

Plasmonic nanoparticles are also being used as labels to replace fluorescent tags in microarray-based analyses to improve sensitivity [206–209]. Muller *et al.* [207] recently used 15 nm oligonucleotide-functionalized Au nanoparticles to label captured target DNA on a sandwich microarray assay (Fig. 3.15). Once the Au nanoparticles were fixed, an Ag amplification step was used to increase the target signal by electroless reduction of Ag^+ into Ag at the Au nanoparticle surfaces [209]. The glass slide was then imaged. The Ag amplification was reported to yield a 1,000-fold increase in sensitivity compared to the commonly used Cy3 dye [209], and DNA was detected at femtomolar concentration [207]. Through theimplementation of a bio-bar-code technique to further amplify the signal, attomolar and zeptomolar sensitivity was reported [219].

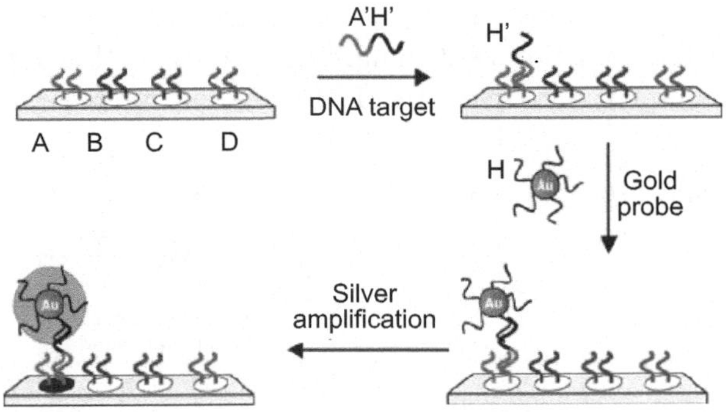

Figure 3.15 A schematic representation of DNA hybridization to microarrays and detection using Ag-amplified Au nanoparticle probes. (Reprinted with permission from ref. 209. Copyright 2000 Elsevier B. V.) See also Color Insert.

3.4.2.2 NanoSPR sensors using surface-immobilized nanoparticles

Chip-based nanoSPR sensors can be made by immobilizing nanoparticles on an optically transparent substrate. The nanoSPR response of target-induced plasmon red-shift can then be measured quantitatively. The typical procedure to fabricate a nanoSPR chip is illustrated in Fig. 3.16, which involves (a) the activation of the nanoparticle surface and immobilization of the receptor and (b) binding of target biomolecules and characterization of the sensor performance [220]. The nanoSPR of the Ag or Au monolayer on the chip can be measured by UV-Vis spectrometer [221, 222], and binding events can be monitored in real time.

Changes in the refractive index occurring near the surface of the nanoSPR chips (induced by target binding to the chip) shift the nanoSPR peak position and magnitude [4, 221, 223–226]. Wavelength shifts of ~167, ~76, and 252 nm/RIU were observed for spherical Ag [224], Au nanoparticles [4], and Au nanorods[227], respectively. The binding of antibodies to BSA and human serum albumin (HSA) has been studied using nanoSPR chips [225], and selective detection of phosphopeptides at nanomolar concentration has been reported [226]. Higher-sensitivity nanoSPR chips made by immobilizing spherical silica-Au shell and rice-shaped hematite-Au shell have also been developed [228, 229].

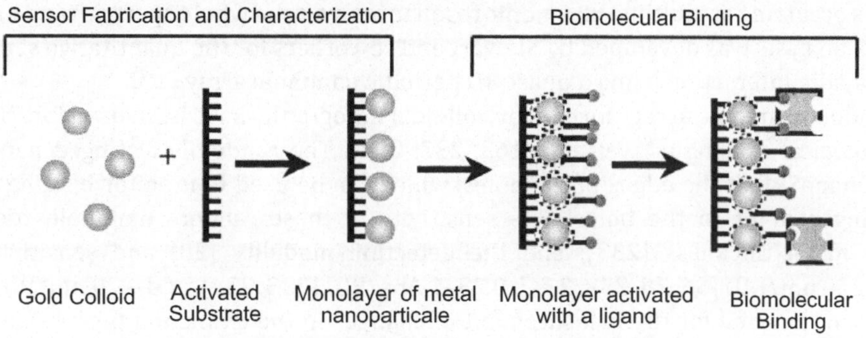

Figure 3.16 The steps involved in the fabrication of a nanoSPR biosensor and its use for the quantification of biomolecular interactions. (Reprinted with permission from ref. 220. Copyright 2004 Elsevier.)

NanoSPR chips have also been made by directly forming nanostructures on a solid substrate through NSL [6, 203, 230, 231], colloidal lithography [163, 232–234], soft lithography [13], or metal thin-film evaporation [235–237]. For example, arrays of triangular Ag NOs formed by NSL (Fig. 3.7) were exploited for sensing applications. The quantitative detection of concanavalin A [230], streptavidin [6], and antibiotin [238], as well as their binding affinity, was obtained. In a recent work, biomarkers for Alzheimer's disease, amyloid-β-derived ligands (ADDL), were detected at femtomolar concentrations using a triangular Ag nanoSPR chip in a sandwich assay [79]. The sensitivity of nanoSPR chips can be further improved if single-particle spectroscopic techniques are used to monitor nanoSPR response from individual nanoparticles. McFarland and coworkers have shown that a nanoSPR peak shift of ~40 nm could be observed upon the binding of ~100 zeptomoles of 1-hexadecanethiol to a single triangular Ag nanoparticle (as measured by resonant Rayleigh scattering spectroscopy). This high sensitivity and the small transducer size suggest that single nanoparticles could be useful for the analysis of trace amounts of targets in precious samples.

3.4.2.3 Nanohole arrays

The enhanced transmission of light through periodic arrays of subwavelength nanoholes in metal film has been actively explored for sensing applications since it was first discovered in 1998 by Ebbesen *et al.* [239]. Although a variety of mechanisms may be operative in bringing about the enhanced transmission, in most cases involving thin Au films, it is the SP that plays the key role [240, 241]. The optical properties of plasmonic quasi-3D periodic nanohole arrays are extremely sensitive to the changes in the refractive index of the solution that passes through it. Both plasmonic peak positions and intensities were observed to be linearly correlated to the RI. The most sensitive peak at

~1023 nm linearly red-shifted with a sensitivity of 700–800 nm/RIU and linearly increased in intensity with a sensitivity of ~2.5–3.5 Abs/RIU [13].

SPR sensing is typically performed by following the response of an individual peak to binding events at the sensor surface. However, it does not fully capture the sensitivity of the quasi-3D plasmonic sensor. To take full advantage of all the peak shifts and intensity changes occurring at multiple plasmonic resonances over the whole spectral range, a multi spectral analysis was developed by Stewart and coworkers for the quantitative sensing of biotin-avidin interaction using a quasi-3D periodic nanohole array [13].

Random nanohole arrays formed by colloidal lithography have been used for chemical and biological sensing as well [78, 163, 233, 234]. The randomly arranged nanoholes exhibit nanoSPR at the edges of the holes, which can be used to monitor binding events occurring near or in the holes. The sensitivity of these random nanoholes depends on the metal thickness [233]; and the detection modality [20] and sensitivities of ~100–270 nm/RIU [66, 78, 233, 234], 0.23–1 Abs/RIU [233, 234], and ~70 nm/RIU [163] have been reported for the measurement of changes in the extinction peak wavelength, extinction peak magnitude, and scattering peak wavelength, respectively.

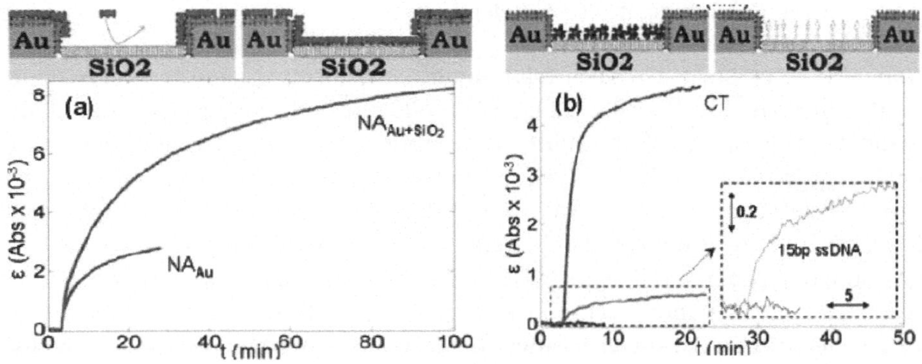

Figure 3.17 The detection of lipid-membrane-mediated binding events using SPB in random arrays of Au nanohole. (a) Temporal variation in extinction measured at the longer wavelength inflection point (725 nm) of the LSPR peak upon addition of ~0.3 μM NA. In both cases, Au is modified with biotin-BSA. In one case, biotin-modified SPB patches cover the SiO_2 regions (NA_{Au+SiO_2}), while in the other — the top-right illustration in (a) — unmodified SPB patches cover the SiO_2 regions (NA_{Au}) — the top-left illustration in (a). (b) Variation in extinction measured upon addition of ~0.5 μM cholera toxin (CT) to GM1-modified (5 wt%) SPBs — purple curve, illustration of SPB-modified nanohole shown in the upper left of (b). (Inset) Magnification of changes in extinction versus time upon the addition of a 15-base-long noncomplementary (0.2 μM, blue) and a fully complementary strand (0.2 μM, red) to SPB patches modified with a DNA construct carrying two cholesterol moieties at its one end and a 15-base-long single strand available for hybridization at the other. (Reprinted with permission from ref. 234. Copyright 2005 American Chemical Society.) See also Color Insert.

One advantage of a nanohole-based sensor is that a phospholipid bilayer (SPB) can be put to cover the hole and mimic cytoplasmic membrane; thus, interactions between membrane-embedded proteins and other molecules can be studied potentially. Dahlinand coworkers constructed such a system, as shown in Fig. 3.17. The SPBs were supported on the silica surface inside a nanohole. The left panel of Fig. 3.17 shows the interaction between protein NA with biotin-modified lipids in the SPB: without biotin-modification, NA only binds to bBSA on the Au surface (left); with biotin modification of the SPB, NA binds to the SPB as well as the Au surface and a threefold increase of nanoSPR response at 725 nm is observed. The right panel of Fig. 3.17 shows the binding of cholera toxin to SPBs containing ganglioside G_{M1} glycolipids (left) and the hybridization of 15-base single-strand DNA (ssDNA) to SPBs presenting a complementary strand (right). Further development of this technique will enable investigation of interaction between signals and membrane proteins and greatly advance our understanding of the cell-cell signaling processes.

3.4.3 Surface-Enhanced Spectroscopies

The nanoSPR phenomenon has also been utilized to enhance spectroscopic sensitivity in chemical- and molecular-specific sensing and imaging, most noticeable in Raman and fluorescence spectroscopies [20]. Surface plasmons are able to create a very strong localized EM field surrounding the nanostructures, which, in turn, leads to remarkable enhancement of spectroscopic signals from molecules that are interacting with the local field. In Raman spectroscopy, the weak Raman signal (due to the small Raman cross sections [10^{-30} to 10^{-25} cm^2] for most molecules) can be enhanced up to 10^{13}~10^{14} times and Raman scattering from a single molecule has been observed [28]. Such enhancement has enabled many analytical applications for SERS, ranging from DNA sequencing to forensics [203, 242–262]. More details on SERS-based biosensing are discussed in a separate chapter by Chula *et al.*

NanoSPR has also been used to enhance fluorescence. The enhanced EM field surrounding plasmonic nanostructures will lead to enhanced fluorescent signals from fluorophores interacting with the field [20]. Plasmon-enhanced fluorescence (PEF) has been applied to the study of complex biomolecular systems [151, 218, 263, 264]. Fluorophores quench in close proximity to metallic surfaces (typically < 5 nm) [265]; however, at longer distances (5–200 nm), the coupling between the fluorophores and the enhanced local field increases the absorption cross section and the radiative decay rate of the fluorophores [264], which leads to increased photostability, enhanced wavelength-tunable emission, and FRET efficiency [264].

A number of different metallic nanostructures have been used for PEF, such as nanohole arrays, nanoparticles, thin film, and deposited nano islands [151, 266–268]. Cu^{2+} was detected using fluorophores-functionalized Au nanoparticles at 1 μM [269], whereas normal fluorescence-based detection for Cu^{2+} is difficult because of the quenching of fluorescence by Cu^{2+}. Au nanoparticles were also used to detect the presence of Hg^{2+} in pond water at 2 ppb [266]. Enhancement to standard fluorophores such as Cy3 and fluorescein by nanoparticles immobilized onto a solid substrate was found to be dependent on the widths and heights of the nanoparticles, with a maximum enhancement (20 times) being obtained with a particle size of 85–90 nm [130]. The PEF is a relatively new field, and the optimization of PEF is yet to be established.

3.4.4 Plasmonic Sensing Beyond the Diffraction Limit

The resolution of optical detection methods is often dictated by the wave nature of light, or the diffraction limit. Plasmonic nanostructures provide a possibility to go beyond this limit. Two approaches have been proposed to push detection beyond diffraction limits — apertureless near-field optical microscopy (a-NSOM) and tip-enhanced Raman spectroscopy (TERS); with these techniques, spatial resolution that is only achievable by electron microscopy and surface probe microscopy (SPM) is now possible for optical detections [20].

In a-NSOM, an apertureless metalized tip is capable of supporting SP and brings about a 10–1,000 times enhancement of the local EM field; the spatial resolution is limited only by the physical size of the tip, which can be ~10 nm 270]. Tips coated with 15–25 nm Ag were used to detect Alexa 532 dye molecules with a resolution of 15 nm [271].

In TERS, the sharp metallic tip creates SP that enhances local EM field, which, in turn, leads to a significant enhancement in Raman scattering. Enhancement factors of 10^2 to 10^4 were observed in experiments [272]; with a theoretically predicted enhancement factor upto 10^{10}, there is still much room for improvement[272]. Tips of silicon nanowires with Au coating have been shown to be good Raman enhancers for chemical sensing [273]. The combination of nm-scale spatial resolution and chemical imaging capability of TERS constitutes an area of enormous interests to biomedical and biomolecular sensing.

3.5 Conclusions and Future Perspectives

Over the past decade, advances in nanomaterial synthesis and fabrication have led to an explosion of both efforts and reports on the emerging field of plasmonics. Significant progresses have been made in creating novel nano-sized sensors for chemical and biological sensing and imaging. Researchers all over the world are actively exploring two lines of researches — (i) new materials are being fabricated and new theories and models are being developed and investigated to expand our understanding of plasmonic phenomena at the nanoscales, and (ii) new sensing instruments and/or platforms are being designed and exploited based on the growing knowledge bases to provide unprecedented sensitivity and resolution for chemical and biological sensing and imaging. These two lines of research are merging to yield next-generation nanosensors.

One of the emerging trends in plasmonic sensing is the effort to use plasmonic devices for subwave length "imaging" [20]. The spatial resolution in optical microscopy is limited to $\lambda/2$ due to the Rayleigh criterion. This limit is caused by the loss of the evanescent field intensity in the far field, which carries high spatial frequency information [20]. Complex methods that overcome this limit, such as near-field scanning optical microscopy, have been developed [270]. However, the complexity of the methods has limited their accessibility to a limited group of people and their widespread use for biological imaging has not been realized. Nanoplasmonics provides another alternative. It has been shown that metallic nanohole arrays give anomalous high transmission of light inherently [239] and theoretically can be the base for subwavelength resolution microscopy in conjunction with "perfect lens" made from negative index metamateials [274]. In theory, such a system can provide resolution as small as 25 nm [20]. Recent reports have demonstrated that SPR near-field dimensions are smaller than free-space radiation at the same frequency [20]. Nanostructures with sizes of 90 nm with 170 nm spacings have been constructed by exposing photoresist through a nanohole array [275]. Fields with dimensions as small as 25 nm were obtained using plasmonic structures as well [276]. It has been proposed that wide-field optical microscopy can be combined with plasmon-coupled transmission and optical near-field to yield subwavelength resolution [277–279]. Recently, two different types of magnifying metamaterial-based suprelenses that can be integrated into a conventional far-field optical microscope were reported [280, 281]. Liu *et al.* demonstrated a magnifying optical metamaterial-based superlens consisting of a curved periodic stack of Ag (35 nm) and Al_2O_3 (35 nm) deposited on a half-cylindrical cavity in quartz [280]. This setup was able to image a pair of 35 nm lines spaced 150 nm apart. In another report, Smolyaninov *et al.* demonstrated a superlens design that utilized multilayer photonic metamaterial consisting of alternating layers of positive and negative refractive indexes [281]. With this design, a resolution of 70 nm was obtained. New subwavelength resolution near-field microscopy implementing nanoplasmonic designs that will facilitate biological imaging, especially subcellular imaging, is about to become a reality in the near future.

References

1. Maier, S. A., and Atwater, H. A. (2005). Plasmonics: localization and guiding of electromagnetic energy in metal/dielectric structures, *Appl. Phys. Lett.*, **98**, 01101.
2. Atwater, H. A. (2007). The promise of plasmonics, *Sci. Am.*, **296**(4), 56–63.
3. Ozbay, E. (2006). Plasmonics: merging photonics and electronics at nanoscale dimensions, *Science*, **311**(5758), 189–193.
4. Nath, N., and Chilkoti, A. (2002). A colorimetric gold nanoparticle sensor to interrogate biomolecular interactions in real time on a surface, *Anal. Chem.*, **74**(3), 504–509.
5. Hu, M., Chen, J., Li, Z.-Y., Au, L., Hartland, G. V., Li, X., Marquez, M., and Xia, Y. (2006). Gold nanostructures: engineering their plasmonic properties for biomedical applications, *Chem. Soc. Rev.*, **35**(11), 1084–1094.
6. Haes, A. J., and Van Duyne, R. P. (2002). A nanoscale optical biosensor: sensitivity and selectivity of an approach based on the localized surface plasmon resonance spectroscopy of triangular silver nanoparticles, *J. Am. Chem. Soc.*, **124**(35), 10596–10604.
7. Bozhevolnyi, S. I., Volkov, V. S., Devaux, E., Laluet, J.-Y., and Ebbesen, T. W. (2006). Channel plasmon subwavelength waveguide components including interferometers and ring resonators, *Nature*, **440**(7083), 508–511.
8. Malyarchuk, V., Hua, F., Mack, N. H., Velasquez, V. T., White, J. O., Nuzzo, R. G., and Rogers, J. A. (2005). High performance plasmonic crystal sensor formed by soft nanoimprint lithography, *Opt. Express*, **13**(15), 5669–5675.
9. Daniel, M.-C., and Astruc, D. (2004). Gold nanoparticles: assembly, supramolecular chemistry, quantum-size-related properties, and applications toward biology, catalysis, and nanotechnology, *Chem. Rev.*, **104**(1), 293–346.
10. Brolo, A. G., Gordon, R., Leathem, B., and Kavanagh, K. L. (2004). Surface plasmon sensor based on the enhanced light transmission through arrays of nanoholes in gold films, *Langmuir*, **20**(12), 4813–4815.

11. Wiley, B., Sun, Y., Mayers, B., and Xia, Y. (2005). Shape-controlled synthesis of metal nanostructures: the case of silver, *Chem.-Eur. J.*, **11**(2), 454–463.
12. Murphy, C. J., Sau, T. K., Gole, A. M., Orendorff, C. J., Gao, J., Gou, L., Hunyadi, S. E., and Li, T. (2005). Anisotropic metal nanoparticles: Synthesis, assembly, and optical applications, *J. Phys. Chem. B*, **109**(29), 13857–13870.
13. Stewart, M. E., Mack, N. H., Malyarchuk, V., Soares, J. A. N. T., Lee, T.-W., Gray,S. K., Nuzzo, R. G., and Rogers, J. A. (2006). Quantitative multispectral biosensing and 1D imaging using quasi-3D plasmonic crystals, *PNAS.*, **103**(46), 17143–17148.
14. Grady, N. K., Halas, N. J., and Nordlander, P. (2004). Influence of dielectric function properties on the optical response of plasmon resonant metallic nanoparticles, *Chem. Phys. Lett.*, **399**(1–3), 167–171.
15. Wang, C., Ma, Z., Wang, T., and Zu, S. (2006). Synthesis, assembly, and biofunctionalization of silica-coated gold nanorods for colorimetric biosensing, *Adv. Funct. Mater.*, **16**(13), 1673–1678.
16. Genevieve, M., Vieu, C., Carles, R., Zwick, A., Briere, G., Salome, L., and Trevisiol, E. (2007). Biofunctionalization of gold nanoparticles and their spectral properties, *Microelectron. Eng.*, **84**(5–8), 1710–1713.
17. Yu, C., and Irudayaraj, J. (2007). Quantitative evaluation of sensitivity and selectivity of multiplex nanoSPR biosensor assays, *Biophys. J.*, **93**(10), 3684–3692.
18. Haynes, C. L., and Van Duyne, R. P. (2001). Nanosphere lithography: aversatile nanofabrication tool for studies of size-dependent nanoparticle optics, *J. Phys. Chem. B*, **105**(24), 5599–5611.
19. Kneipp, K., Kneipp, H., and Bohr, H. G. (2006). Single-molecule SERS spectroscopy, *Top. Appl. Phys.*, **103**, 261–278.
20. Stewart, M. E., Anderton, C. R., Thompson, L. B., Maria, J., Gray, S. K., Rogers, J. A., and Nuzzo, R. G. (2008). Nanostructured plasmonic sensors, *Chem. Rev.*, **108**(2), 494–521.
21. Raether, H. (1988). *Surface Plasmons on Smooth and Rough Surfaces and On Gratings*, Springer.
22. Wark, A. W., Lee, H. J., and Corn, R. M. (2005). Long-range surface plasmon resonance imaging for bioaffinity sensors, *Anal. Chem.*, **77**(13), 3904–3907.
23. Lee, K.-S., and El-Sayed, M. A. (2006). Gold and silver nanoparticles in sensing and imaging: sensitivity of plasmon response to size, shape, and metal composition, *J. Phys. Chem. B*, **110**(39), 19220–19225.
24. Hutter, E., and Fendler, J. (2004). Exploitation of localized surface plasmon resonance, *Adv. Mater.*, **16**(19), 1685–1706.
25. Ghosh, S. K., Nath, S., Kundu, S., Esumi, K., and Pal, T. (2004). Solvent and ligand effects on the localized surface plasmon resonance (LSPR) of gold colloids, *J. Phys. Chem. B*, **108**(37), 13963–13971.
26. Ly, N., Foley, K., and Tao, N. (2007). Integrated label-free protein detection and separation in real time using confined surface plasmon resonance imaging, *Anal. Chem.*, **79**(6), 2546–2551.
27. Yu, C., and Irudayaraj, J. (2007). Multiplex biosensor using gold nanorods, *Anal. Chem.*, **79**(2), 572–579.
28. Stockman, M. I. (2006). Electromagnetic theory of SERS, *Top. Appl. Phys.*, **103**, 47–66.
29. Stewart, M. E., Anderton, C. R., Thompson, L. B., Maria, J., Gray, S. K., Rodgers, J. A., and Nuzzo, R. G. (2008). Nanostructured plasmonic sensors, *Chem. Rev.*, **108**, 494–521.
30. Chau, L.-K., Lin, Y.-F., Cheng, S.-F., and Lin, T.-J. (2006). Fiber-optic chemical and biochemical probes based on localized surface plasmon resonance, *Sens. Actuators B*, **113**(1), 100–105.

31. Zhao, J., Zhang, X., Yonzon, C. R., Haes, A. J., and Van Duyne, R. P. (2006). Localized surface plasmon resonance biosensors, *Nanomedicine*, **1**(2), 219–228.

32. Bohren, C. F., and Huffman, D. R. (1983). *Ab orption and Scattering of Light by Small Particles*, John Wiley and Sons, New York.

33. Kreibig, U., and Vollmer, M. (eds) (1995). *Optical Properties of Metal Clusters*, vol. 25, Springer, Berlin.

34. Willets, K. A., and Van Duyne, R. P. (2006). Localized surface plasmon resonance spectroscopy and sensing, *Annu. Rev. Phys. Chem.*, **58**, 267–297.

35. Kelly, K. L., Coronado, E., Zhao, L. L., and Schatz, G. C. (2003). The optical properties of metal nanoparticles: the influence of size, shape, and dielectric environment, *J. Phys. Chem. B*, **107**(3), 668–677.

36. Mulvaney, P. (1996). Surface plasmon spectroscopy of nanosized metal particles, *Langmuir*, **12**(3), 788–800.

37. Underwood, S., and Mulvaney, P. (1994). Effect of the solution refractive index on the color of gold colloids, *Langmuir*, **10**(10), 3427–3430.

38. Mie, G. (1908). Contributions to the optics of turbid media, especially colloidal metal solutions, *Annu. Phys.*, **25**, 377–445.

39. Kuwata, H., Tamaru, H., Esumi, K., and Miyano, K. (2003). Resonant light scattering from metal nanoparticles: practical analysis beyond Rayleigh approximation, *Appl. Phys. Lett.*, **83**(22), 4625–4627.

40. Sun, Y. G., and Xia, Y. N. (2003). Synthesis and characterization of metal nanostructures with hollow interiors, *Proc. SPIE.*, **5221**, 164–173.

41. El-Sayed, M. A. (2001). Some interesting properties of metals confined in time and nanometer space of different shapes, *Acc. Chem. Res.*, **34**(4), 257–264.

42. Perez-Juste, J., Liz-Marzan, L. M., Carnie, S., Chan, D. Y. C., and Mulvaney, P. (2004). Electric-field-directed growth of gold nanorods in aqueous surfactant solutions, *Adv. Funct. Mater.*, **14**(6), 571–579.

43. Link, S., and El-Sayed, M. A. (2003). Optical properties and ultrafast dynamics of metallic nanocrystals, *Annu. Rev. Phys. Chem.*, **54**, 331–366.

44. Kou, X., Zhang, S., Tsung, C.-K., Yang, Z., Yeung, M. H., Stucky, G. D., Sun,L., Wang, J., and Yan, C. (2007). One-step synthesis of large-aspect-ratio single-crystalline gold nanorods by using CTPAB and CTBAB surfactants, *Chem.-Eur. J.*, **13**(10), 2929–2936.

45. Perez-Juste, J., Pastoriza-Santos, I., Liz-Marzan, L. M., and Mulvaney, P. (2005). Gold nanorods: synthesis, characterization and applications, *Coord. Chem. Rev.*, **249**(17–18), 1870–1901.

46. Hu, M., Wang, X., Hartland, G. V., Mulvaney, P., Perez-Juste, J., and Sader, J. E. (2003). Vibrational response of nanorods to ultrafast laser induced heating: theoretical and experimental analysis, *J. Am. Chem. Soc.*, **125**(48), 14925–14933.

47. Lee, K. S., and El-Sayed, M. A. (2005). Dependence of the enhanced optical scattering efficiency relative to that of absorption for gold metal nanorods on aspect ratio, size, end-cap shape, and medium refractive index, *J. Phys. Chem. B*, **109**(43), 20331–20338.

48. Link, S., Mohamed, M. B., and El-Sayed, M. A. (1999). Simulation of the optical absorption spectra of gold nanorods as a function of their aspect ratio and the effect of the medium dielectric constant, *J. Phys. Chem. B*, **103**(16), 3073–3077.

49. Yu, Y.-Y., Chang, S.-S., Lee, C.-L., and Wang, C. R. C. (1997). Gold nanorods: electrochemical synthesis and optical properties, *J. Phys. Chem. B*, **101**(34), 6661–6664.

50. Hao, E., Bailey, R. C., Schatz, G. C., Hupp, J. T., and Li, S. (2004). Synthesis and optical properties of "branched" gold nanocrystals, *Nano Lett.*, **4**(2), 327–330.

51. Kottmann, J. P., Martin, O. J. F., Smith, D. R., and Schultz, S. (2000). Field polarization and polarization charge distributions in plasmon resonant nanoparticles, *Opt. Express*, **6**, 213.

52. McLellan, J. M., Li, Z.-Y., Siekkinen, A. R., and Xia, Y. (2007). The SERS activity of a supported Ag nanocube strongly don its orientation relative to laser polarization, *Nano Lett.*, **7**(4), 1013–1017.

53. Barber, P. W., and Hill, S. C. (1990). *Light Scattering by Particles: Computational Methods*, World Scientific Publishing, Singapore.

54. Yang, W.-H., Schatz, G. C., and Van Duyne, R. P. (1995). Discrete dipole approximation for calculating extinction and Raman intensities for small particles with arbitrary shapes, *J. Chem. Phys.*, **103**(3), 869–875.

55. Sosa, I. O., Noguez, C., and Barrera, R. G. (2003). Optical properties of metal nanoparticles with arbitrary shapes, *J. Phys. Chem. B*, **107**(26), 6269–6275.

56. Shlager, K. L., and Schneider, J. B. (1995). A selective survey of the finite-difference time-domain literature, *IEEE Antenn. Propag. Mag.*, **37**(4), 39–57.

57. Gray, S. K., and Kupka, T. (2003). Propagation of light in metallic nanowire arrays: finite-difference time-domain studies of silver cylinders, *Phys. Rev. B*, **68**(4), 045415.

58. Meier, M., and Wokaun, A. (1983). Enhanced fields on large metal particles: dynamic depolarization, *Opt. Lett.*, **8**(11), 581–583.

59. Wokaun, A., Gordon, J. P., and Liao, P. F. (1982). Radiation damping in surface-enhanced Raman scattering, *Phys. Rev. Lett.*, **48**(14), 957.

60. Yin, L., Vlasko-Vlasov, V. K., Rydh, A., Pearson, J., Welp, U., Chang,S. H., Gray,S. K., Schatz, G. C., Brown, D. B., and Kimball, C. W. (2004). Surface plasmons at single nanoholes in Au films, *Appl. Phys. Lett.*, **85**(3), 467–469.

61. Wannemacher, R. (2001). Plasmon-supported transmission of light through nanometric holes in metallic thin films, *Opt. Commun.*, **195**(1–4), 107–118.

62. Maier, S. A., Brongersma, M. L., Kik, P. G., and Atwater, H. A. (2002). Observation of near-field coupling in metal nanoparticle chains using far-field polarization spectroscopy, *Phys. Rev. B*, **65**(19), 193408.

63. Lamprecht, B., Schider, G, Lechner, R. T., Ditlbacher, H., Krenn, J. R., Leitner, A., and Aussenegg, F. R. (2000). Metal nanoparticle gratings: influence of dipolar particle interaction on the plasmon resonance, *Phys. Rev. Lett.*, **84**(20), 4721.

64. Maier, S. A., Kik, P. G., and Atwater, H. A. (2002). Observation of coupled plasmon-polariton modes in Au nanoparticle chain waveguides of different lengths: estimation of waveguide loss, *Appl. Phys. Lett.*, **81**(9), 1714–1716.

65. Mirkin, C. A. L., Robert, L., Mucic, R. C., and Storhoff, J. J. (1996). A DNA-based method for rationally assembling nanoparticles into macroscopic materials, *Nature*, **382**, 607–609.

66. Prikulis, J., Hanarp, P., Olofsson, L., Sutherland, D., and Kall, M. (2004). Optical spectroscopy of nanometric holes in thin gold films, *Nano Lett.*, **4**(6), 1003–1007.

67. Weeber, J. C., Gonzalez, M. U., Baudrion, A. L., and Dereux, A. (2005). Surface plasmon routing along right angle bent metal strips, *Appl. Phys. Lett.*, **87**(22), 221101–221103.

68. Dionne, J. A., Lezec, H. J., and Atwater, H. A. (2006). Highly confined photon transport in subwavelength metallic slot waveguides, *Nano Lett.*, **6**(9), 1928–1932.

69. Yin, L., Vlasko-Vlasov, V. K., Pearson, J., Hiller, J. M., Hua, J., Welp, U., Brown, D. E., and Kimball, C. W. (2005). Subwavelength focusing and guiding of surface plasmons, *Nano Lett.*, **5**(7), 1399–1402.

70. Satuby, Y., and Orenstein, M. (2007). Surface-plasmon-polariton modes in deep metallic trenches- measurement and analysis, *Opt. Express*, **15**(7), 4247–4252.

71. Bozhevolnyi, S. I., Volkov, V. S., Devaux, E., and Ebbesen, T. W. (2005). Channel plasmon-polariton guiding by subwavelength metal grooves, *Phys. Rev. Lett.*, **95**(4), 046802.

72. Ko, S. H., Pan, H., Grigoropoulos, C. P., Luscombe, C. K., Frechet, J. M. J., and Poulikakos, D. (2007). Air stable high resolution organic transistors by selective laser sintering of ink-jet printed metal nanoparticles, *Appl. Phys. Lett.*, **90**(14), 141103.

73. Ko, S. H., Park, I., Pan, H., Grigoropoulos, C. P., Pisano, A. P., Luscombe, C. K., and Frechet, J. M. J. (2007). Direct nanoimprinting of metal nanoparticles for nanoscale electronics fabrication, *Nano Lett.*, **7**(7), 1869–1877.

74. Herderick, E. D., Tresback, J. S., Vasiliev, A. L., and Padture, N. P. (2007). Template-directed synthesis, characterization and electrical properties of Au–TiO_2–Au heterojunction nanowires, *Nanotechnology*, **18**, 155204.

75. Blümel, A., Klug, A., Eder, S., Scherf, U., Moderegger, E., and List, E. J. W. (2007). Micromolding in capillaries and microtransfer printing of silver nanoparticles as soft-lithographic approach for the fabrication of source/drain electrodes in organic field-effect transistors, *Org. Electron.*, **8**(4), 389–395.

76. Lee, J.-S., Han, M. S., and Mirkin, C. A. (2007). Colorimetric detection of mercuric ion (Hg^{2+}) in aqueous media using DNA-functionalized gold nanoparticles, *Angew. Chem. Int. Edit.*, **46**(22), 4093–4096.

77. Liu, J., and Lu, Y. (2006). Fast colorimetric sensing of adenosine and cocaine based on a general sensor design involving aptamers and nanoparticles, *Angew. Chem. Int. Edit.*, **45**(1), 90–94.

78. Gao, D., Chen, W., Mulchandani, A., and Schultz, J. S. (2007). Detection of tumor markers based on extinction spectra of visible light passing through gold nanoholes, *Appl. Phys. Lett.*, **90**(7), 073901–073903.

79. Haes, A. J., Chang, L., Klein, W. L., and Van Duyne, R. P. (2005). Detection of a biomarker for Alzheimer's disease from synthetic and clinical samples using a nanoscale optical biosensor, *J. Am. Chem. Soc.*, **127**(7), 2264–2271.

80. Storhoff, J. J., Elghanian, R., Mucic, R. C., Mirkin, C. A., and Let singer, R. L. (1998). One-pot colorimetric differentiation of polynucleotides with single base imperfections using gold nanoparticle probes, *J. Am. Chem. Soc.*, **120**(9), 1959–1964.

81. Haes, A. J., Haynes, C. L., McFarland, A. D., Schatz, G. C., Van Duyne, R. P., and Zou, S. (2005). Plasmonic materials for surface-enhanced sensing and spectroscopy, *MRS Bull.*, **30**(5), 368–375.

82. Huang, X., El-Sayed, I. H., Qian, W., and El-Sayed, M. A. (2006). Cancer cell imaging and photothermal therapy in the near-infrared region by using gold nanorods, *J. Am. Chem. Soc.*, **128**(6), 2115–2120.

83. Chen, J., Wang, D., Xi, J., Au, L., Siekkinen, A., Warsen, A., Li, Z.-Y., Zhang, H., Xia, Y., and Li, X. (2007). Immuno gold nanocages with tailored optical properties for targeted photothermal destruction of cancer cells, *Nano Lett.*, **7**(5), 1318–1322.

84. Durr, N. J., Larson, T., Smith, D. K., Korgel, B. A., Sokolov, K., and Ben-Yakar, A. (2007). Two-photon luminescence imaging of cancer cells using molecularly targeted gold nanorods, *Nano Lett.*, **7**(4), 941–945.

85. Loo, C., Lowery, A., Halas, N., West, J., and Drezek, R. (2005). Immunotargeted nanoshells for integrated cancer imaging and therapy, *Nano Lett.*, **5**(4), 709–711.

86. Loo, C., Lin, A., Hirsch, L., Lee, M.-H., Barton, J., Halas, N., West, J., and Drezek, R. (2004). Nanoshell-enabled photonics-based imaging and therapy of cancer, *Technol. Cancer Res. Treat.*, **3**(1), 33–40.
87. Lu, L., Kobayashi, A., Tawa, K., and Ozaki, Y. (2006). Silver nanoplates with special shapes: controlled synthesis and their surface plasmon resonance and surface-enhanced Raman scattering properties, *Chem. Mat.*, **18**(20), 4894–4901.
88. Wiley, B. J., Im, S. H., Li, Z.-Y., McLellan, J., Siekkinen, A., and Xia, Y. (2006). Maneuvering the surface plasmon resonance of silver nanostructures through shape-controlled synthesis, *J. Phys. Chem. B*, **110**(32), 15666–15675.
89. Chen, H. M., Hsin, C. F., Liu, R.-S., Lee, J.-F., and Jang, L.-Y. (2007). Synthesis and characterization of multi-pod-shaped gold/silver nanostructures, *J. Phys. Chem. C*, **111**(16), 5909–5914.
90. Murphy, C. J., Sau, T. K., Anand, G., and Orendorff, C. J. (2005). Surfactant-directed synthesis and optical properties of one-dimensional plasmonic metallic nanostructures, *MRS Bull.*, **30**(5), 349–355.
91. Xiong, Y., Cai, H., Wiley, B. J., Wang, J., Kim, M. J., and Xia, Y. (2007). Synthesis and mechanistic study of palladium nanobars and nanorods, *J. Am. Chem. Soc.*, **129**(12), 3665–3675.
92. Evanoff, D. D., and Chumanov, G. (2004). Size-controlled synthesis of nanoparticles. 1. "silver-only" aqueous suspensions via hydrogen reduction, *J. Phys. Chem. B*, **108**(37), 13948–13956.
93. Nehl, C. L., Liao, H., and Hafner, J. H. (2006). Optical properties of star-shaped gold nanoparticles, *Nano Lett.*, **6**(4), 683–688.
94. Su, K.-H., Wei, Q.-H., Zhang, X., Mock, J. J., Smith, D. R., and Schultz, S. (2003). Interparticle coupling effects on plasmon resonances of nanogold particles, *Nano Lett.*, **3**(8), 1087–1090.
95. Gunnarsson, L., Rindzevicius, T., Prikulis, J., Kasemo, B., Kall, M., Zou, S., and Schatz, G. C. (2005). Confined plasmons in nanofabricated single silver particle pairs: experimental observations of strong interparticle interactions, *J. Phys. Chem. B*, **109**(3), 1079–1087.
96. Rechberger, W., Hohenau, A., Leitner, A., Krenn, J. R., Lamprecht, B., and Aussenegg, F. R. (2003). Optical properties of two interacting gold nanoparticles, *Opt. Commun.*, **220**(1–3), 137–141.
97. Grand, J., Adam, P.-M., Grimault, A.-S., Vial, A., de la Chapelle, M. L., Bijeon, J.-L., Kostcheev, S., and Royer, P. (2006). Optical extinction spectroscopy of oblate, prolate and ellipsoid shaped gold nanoparticles: experiments and theory, *Plasmonics*, **1**, 135–140.
98. Whitesides, G. M., and Love, C. J. (2001). The art of building small, *Sci. Am.*, **285**(3), 38.
99. Hanarp, P., Sutherland, D. S., Gold, J., and Kasemo, B. (2003) Control of nanoparticle film structure for colloidal lithography, *Colloids Surf. A*, **214**(1–3), 23–36.
100. Haes, A. J., Zhao, J., Zou, S., Own, C. S., Marks, L. D., Schatz, G. C., and Van Duyne, R. P. (2005). Solution-phase, triangular Ag nanotriangles fabricated by nanosphere lithography, *J. Phys. Chem. B*, **109**(22), 11158–11162.
101. Hicks, E. M., Lyandres, O., Hall, W. P., Zou, S., Glucksberg, M. R., and Van Duyne, R. P. (2007). Plasmonic properties of anchored nanoparticles fabricated by reactive ion etching and nanosphere lithography, *J. Phys. Chem. C*, **111**(11), 4116–4124.
102. Aizpurua, J., Hanarp, P., Sutherland, D. S., Kall, M., Bryant, G. W., and Garcia de Abajo, F. J. (2003) Optical properties of gold nanorings, *Phys. Rev. Lett.*, **90**(5), 057401.
103. Khomutov, G. B., and Koksharov, Y. A. (2006). Effects of organic ligands, electrostatic and magnetic interactions in formation of colloidal and interfacial inorganic nanostructures, *Adv. Collid. Interface Sci.*, **122**(1–3), 119–147.
104. Roucoux, A., Schulz, J., and Patin, H. (2002). Reduced transition metal colloids: a novel family of reusable catalysts?, *Chem. Rev.*, **102**(10), 3757–3778.

105. Ott, L. S., and Finke, R. G. (2006). Nanocluster formation and stabilization fundamental studies: investigating "solvent-only" stabilization en route to discovering stabilization by the traditionally weakly coordinating anion BF4 plus high dielectric constant solvents, *Inorg. Chem.*, **45**(20), 8382–8393.

106. Pillai, Z. S., and Kamat, P. V. (2004). What factors control the size and shape of silver nanoparticles in the citrate ion reduction method?, *J. Phys. Chem. B*, **108**(3), 945–951.

107. Anand, M., Bell, P. W., Fan, X., Enick, R. M., and Roberts, C. B. (2006). Synthesis and steric stabilization of silver nanoparticles in neat carbon dioxide solvent using fluorine-free compounds, *J. Phys. Chem. B*, **110**(30), 14693–14701.

108. Hunter, R. J. (2001). *Foundations of Colloid Science*, Oxford University Press, Oxford.

109. Yu, C., and Irudayaraj, J. (2007). Multiplexing biosensor for molecular fingerprinting using gold nano rods, *Anal. Chem.*, **79**(2), 572–579.

110. Lee, W., Scholz, R., Nielsch, K., and Gösele, U. (2005). A template-based electrochemical method for the synthesis of multisegmented metallic nanotubes13, *Angew. Chem. Int. Edit.*, **44**(37), 6050–6054.

111. Huang, C.-J., Chiu, P.-H., Wang, Y.-H., Yang, C.-F., and Feng, S.-W. (2007). Electrochemical formation of crooked gold nanorods and gold networked structures by the additive organic solvent, *J. Colloid Interface Sci.*, **306**(1), 56–65.

112. Starowicz, M., Stypula, B., and Banas, J. (2006). Electrochemical synthesis of silver nanoparticles, *Electrochem. Commun.*, **8**(2), 227–230.

113. Huang, C.-J., Chiu, P.-H., Wang, Y.-H., and Yang, C.-F. (2006). Synthesis of the gold nanodumbbells by electrochemical method, *J. Colloid Interface Sci.*, **303**(2), 430–436.

114. Vosgröne, T., Meixner, A. J., Anders, A., Dietz, H., Sandmann, G., and Plieth, W. (2005). Electrochemically deposited silver particles for surface enhanced Raman spectroscopy, *Surf. Sci.*, **597**(1–3), 102–109.

115. Plieth, W., Dietz, H., Anders, A., Sandmann, G., Meixner, A., Weber, M., and Kneppe, H. (2005). Electrochemical preparation of silver and gold nanoparticles: characterization by confocal and surface enhanced Raman microscopy, *Surf. Sci.*, **597**(1–3), 119–126.

116. Sakamoto, M., Tachikawa, T., Fujitsuka, M., and Majima, T. (2007). Photochemical formation of Au/Cu bimetallic nanoparticles with different shapes and sizes in a poly(vinyl alcohol) film, *Adv. Funct. Mater.*, **17**(6), 857–862.

117. Kim, F., Song, J. H., and Yang, P. (2002). Photochemical synthesis of gold nanorods, *J. Am. Chem. Soc.*, **124**(48), 14316–14317.

118. McGilvray, K. L., Decan, M. R., Wang, D., and Scaiano, J. C. (2006). Facile photochemical synthesis of unprotected aqueous gold nanoparticles, *J. Am. Chem. Soc.*, **128**(50), 15980–15981.

119. Zhu, J., Shen, Y., Xie, A., Qiu, L., Zhang, Q., and Zhang, S. (2007). Photoinduced synthesis of anisotropic gold nanoparticles in room-temperature ionic liquid, *J. Phys. Chem. C*, **111**(21), 7629–7633.

120. Park, J.-E., Atobe, M., and Fuchigami, T. (2006). Synthesis of multiple shapes of gold nanoparticles with controlled sizes in aqueous solution using ultrasound, *Ultrason. Sonochem.*, **13**(3), 237–241.

121. Jiang, L.-P., Wang, A.-N., Zhao, Y., Zhang, J.-R., and Zhu, J.-J. (2004). A novel route for the preparation of monodisperse silver nanoparticles via a pulsed sonoelectrochemical technique, *Inorg. Chem. Com.*, **7**(4), 506–509.

122. Su, C.-H., Wu, P.-L., and Yeh, C.-S. (2003). Sonochemical synthesis of well-dispersed gold nanoparticles at the ice temperature, *J. Phys. Chem. B*, **107**(51), 14240–14243.

123. Pal, A., Shah, S., and Devi, S. (2007). Preparation of silver, gold and silver-gold bimetallic nanoparticles in w/o microemulsion containing triton X-100, *Colloids Surf., A*, **302**(1–3), 483–487.

124. Kariuki, N. N., Luo, J., Maye, M. M., Hassan, S. A., Menard, T., Naslund, H. R., Lin, Y., Wang, C., Engelhard, M. H., and Zhong, C.-J. (2004). Composition-controlled synthesis of bimetallic gold-silver nanoparticles, *Langmuir*, **20**(25), 11240–11246.

125. Toshima, N., and Yonezawa, T. (1998). Bimetallic nanoparticles: novel materials for chemical and physical applications, *New J. Chem.*, **22**, 1179–1201.

126. Wiley, B., Herricks, T., Sun, Y., and Xia, Y. (2004). Polyol synthesis of silver nanoparticles: use of chloride and oxygen to promote the formation of single-crystal, truncated cubes and tetrahedrons, *Nano Lett.*, **4**(9), 1733–1739.

127. Yuan, H., Ma, W., Chen, C., Zhao, J., Liu, J., Zhu, H., and Gao, X. (2007). Shape and SPR evolution of thorny gold nanoparticles promoted by silver ions, *Chem. Mat.*, **19**(7), 1592–1600.

128. Srnova-Sloufova, I., Vlckova, B., Bastl, Z., and Hasslett, T. L. (2004). Bimetallic (Ag) Au nanoparticles prepared by the seed growth method: two-dimensional assembling, characterization by energy dispersive x-ray analysis, x-ray photoelectron spectroscopy, and surface enhanced Raman spectroscopy, and proposed mechanism of growth, *Langmuir*, **20**(8), 3407–3415.

129. Youk, J. H., Park, M.-K., Locklin, J., Advincula, R., Yang, J., and Mays, J. (2002). Preparation of aggregation stable gold nanoparticles using star-block copolymers, *Langmuir*, **18**(7), 2455–2458.

130. Shi, X., Ganser, T. R., Sun, K., Balogh, L. P., and Baker, J., Jr. (2006). Characterization of crystalline dendrimer-stabilized gold nanoparticles, *Nanotechnology.*, **17**, 1072–1078.

131. Jana, N. R., Gearheart, L., and Murphy, C. J. (2001). Seed-mediated growth approach for shape-controlled synthesis of spheroidal and rod-like gold nanoparticles using a surfactant template, *Adv. Mater.*, **13**(18), 1389–1393.

132. Iqbal, M., Chung, Y.-I., and Tae, G. (2007). An enhanced synthesis of gold nanorods by the addition of pluronic (F-127) via a seed mediated growth process, *J. Mater. Chem.*, **17**, 335–342.

133. Kanaras, A. G., Wang, Z., Brust, M., Cosstick, R., and Bates, A. D. (2007). Enzymatic disassembly of DNA-gold nanostructures, *Small*, **3**(4), 590–594.

134. Oldenburg, S. J., Averitt, R. D., Westcott, S. L., and Halas, N. J. (1998). Nanoengineering of optical resonances, *Chem. Phys. Lett.*, **288**(2–4), 243–247.

135. Halas, N. (2005). Playing with plasmons: tuning the optical resonant properties of metallic nanoshells, *MRS Bull.*, **30**(5), 362–367.

136. Gole, A., and Murphy, C. J. (2004). Seed-mediated synthesis of gold nanorods:role of the size and nature of the seed, *Chem. Mat.*, **16**(19), 3633–3640.

137. Sau, T. K., and Murphy, C. J. (2004). Seeded high yield synthesis of short Au nanorods in aqueous solution, *Langmuir*, **20**(15), 6414–6420.

138. Nikoobakht, B., and El-Sayed, M. A. (2003). Preparation and growth mechanism of gold nanorods (NRs) using seed-mediated growth method, *Chem. Mater.*, **15**(10), 1957–1962.

139. Ha, T. H., Koo, H.-J., and Chung, B. H. (2007). Shape-controlled syntheses of gold nanoprisms and nanorods influenced by specific adsorption of halide ions, *J. Phys. Chem. C*, **111**(3), 1123–1130.

140. Yu, C., Vahgese, L., and Irudayaraj, J. (2007). Surface modification of CTAB capped gold nanorods, *Langmuir*, **23**(17), 9114–9119.

141. Yu, C., Nakshatri, H., and Irudayaraj, J. (2007). Identity profiling of cell surface markers by multiplex gold nanorod probes, *Nano Lett.*, **7**(8), 2300–2306.

142. Neeves, A. E., and Birnboim, M. H. (1989) Composite structures for the enhancement of nonlinear-optical susceptibility, *J. Opt. Soc. Am. B*, **6**(4), 787–796.

143. Sun, Y., Mayers, B., and Xia, Y. (2003). Metal nanostructures with hollow interiors, *Adv. Mater.*, **15**(7–8), 641–646.

144. Lu, X., Tuan, H.-Y., Chen, J., Li, Z.-Y., Korgel, B. A., and Xia, Y. (2007). Mechanistic studies on the galvanic replacement reaction between multiply twinned particles of Ag and $HAuCl_4$ in an organic medium, *J. Am. Chem. Soc.*, **129**(6), 1733–1742.

145. Cushing, B. L., Kolesnichenko, V. L., and O'Connor, C. J. (2004). Recent advances in the liquid-phase syntheses of inorganic nanoparticles, *Chem. Rev.*, **104**(9), 3893–3946.

146. Dahl, J. A., Maddux, B. L. S., and Hutchison, J. E. (2007). Toward greener nanosynthesis, *Chem. Rev.*, **107**(6), 2228–2269.

147. Ohno, T., Bain, J. A., and Schlesinger, T. E. (2007). Observation of geometrical resonance in optical throughput of very small aperture lasers associated with surface plasmons, *J. App. Phy.*, **101**(8), 083107–083110.

148. Wang, S., Pile, D. F. P., Sun, C., and Zhang, X. (2007). Nanopin plasmonic resonator array and its optical properties, *Nano Lett.*, **7**(4), 1076–1080.

149. De Jesús, M. A., Giesfeldt, K. S., Oran, J. M., Abu-Hatab, N. A., Lavrik, N. V., and Sepaniak, M. J. (2005). Nanofabrication of densely packed metal-polymer arrays for surface-enhanced Raman spectrometry, *Appl. Spec.*, **59**, 1501–1508.

150. Billot, L., Lamy de la Chapelle, M., Grimault, A. S., Vial, A., Barchiesi, D., Bijeon, J. L., Adam, P. M., and Royer, P. (2006). Surface enhanced Raman scattering on gold nanowire arrays: evidence of strong multipolar surface plasmon resonance enhancement, *Chem. Phys. Lett.*, **422**(4–6), 303–307.

151. Corrigan, T. D., Guo, S., Phaneuf, R. J., and Szmacinski, H. (2005). Enhanced fluorescence from periodic arrays of silver nanoparticles, *J. Fluoresc.*, **15**(5), 777–784.

152. Chang, C.-K., Lin, D.-Z., Yeh, C.-S., Lee, C.-K., Chang, Y.-C., Lin, M.-W., Yeh, J.-T., and Liu, J.-M. (2007). Experimental analysis of surface plasmon behavior in metallic circular slits, *Appl. Phys. Lett.*, **90**(6), 061113–061115.

153. Fu, Y., Ngoi, B. K. A., Zhou, W., and Loh, T. F. (2006). Direct formation of nanopore array via focused ion beam fine milling and surface coating techniques, *Int. J. Nanosci.*, **5**(4–5), 529–534.

154. Taniguchi, J., Ohno, N., Takeda, S., Miyamoto, I., and Komuro, M. (1998). Focused-ion-beam-assisted etching of diamond in XeF_2, *J. Vac. Sci. Technol. B*, **16**(4), 2506–2510.

155. Cabrini, S., Carpentiero, A., Kumar, R., Businaro, L., Candeloro, P., Prasciolu, M., Gosparini, A., Andreani, C., De Vittorio, M., Stomeo, T., and Di Fabrizio, E. (2005). Focused ion beam lithography for two dimensional array structures for photonic applications, *Microelectron. Eng.*, **787 9**, 11–15.

156. Reyntjens, S., and Puers, R. (2000). Focused ion beam induced deposition: fabrication of three-dimensional microstructures and Young's modulus of the deposited material, *J. Micromech. Microeng.*, **10**(2), 181–188.

157. Ishida, M., Fujita, J., Ichihashi, T., Ochiai, Y., Kaito, T., and Matsui, S. (2003). Focused ion beam-induced fabrication of tungsten structures, *J. Vac. Sci. & Tech. B*, **21**(6), 2728–2731.

158. Dintinger, J., Klein, S., and Ebbesen, T. W. (2006). Molecule-surface plasmon interactions in hole arrays: enhanced absorption, refractive index changes, and all-optical switching, *Adv. Mater.*, **18**(10), 1267–1270.

159. Przybilla, F., Genet, C., and Ebbesen, T. W. (2006). Enhanced transmission through Penrose subwavelength hole arrays, *Appl. Phys. Lett.*, **89**(12), 121115–121117.

160. Lopez-Tejeira, F., Rodrigo, S. G., Martin-Moreno, L., Garcia-Vidal, F. J., Devaux, E., Ebbesen, T. W., Krenn, J. R., Radko, I. P., Bozhevolnyi, S. I., Gonzalez, M. U., Weeber, J. C., and Dereux, A. (2007). Efficient unidirectional nanoslit couplers for surface plasmons, *Nat. Phys.*, **3**(5), 324–328.

161. Gonzalez, M. U., Weeber, J. C., Baudrion, A. L., Dereux, A., Stepanov, A. L., Krenn, J. R., Devaux, E., and Ebbesen, T. W. (2006). Design, near-field characterization, and modeling of 45 [degree] surface-plasmon Bragg mirrors, *Phys. Rev. B*, **73**(15), 155416–155428.

162. Volkov, V. S., Bozhevolnyi, S. I., Devaux, E., Laluet, J.-Y., and Ebbesen, T. W. (2007). Wavelength selective nanophotonic components utilizing channel plasmon polaritons, *Nano Lett.*, **7**(4), 880–884.

163. Rindzevicius, T., Alaverdyan, Y., Dahlin, A., Hook, F., Sutherland,D. S., and Kall, M. (2005). Plasmonic sensing characteristics of single nanometric holes, *Nano Lett.*, **5**(11), 2335–2339.

164. Hanarp, P., Kall, M., and Sutherland, D. S. (2003). Optical properties of short range ordered arrays of nanometer gold disks prepared by colloidal lithography, *J. Phys. Chem. B*, **107**(24), 5768–5772.

165. Duval Malinsky, M., Kelly, K. L., Schatz, G. C., and Van Duyne, R. P. (2001). Nanosphere lithography: effect of substrate on the localized surface plasmon resonance spectrum of silver nanoparticles, *J. Phys. Chem. B*, **105**(12), 2343–2350.

166. Riboh, J. C., Haes, A. J., McFarland, A. D., Ranjit Yonzon, C., and Van Duyne, R. P. (2003). A nanoscale optical biosensor: real-time immunoassay in physiological buffer enabled by improved nanoparticle adhesion, *J. Phys. Chem. B*, **107**(8), 1772–1780.

167. Schmidt, J. P., Cross, S. E., and Buratto, S. K. (2004). Surface-enhanced Raman scattering from ordered Ag nanocluster arrays, *J. Chem. Phys.*, **121**(21), 10657–10659.

168. Haynes, C. L., and Van Duyne, R. P. (2003). Plasmon-sampled surface-enhanced Raman excitation spectroscopy, *J. Phys. Chem. B*, **107**(30), 7426–7433.

169. Haes, A. J., Hall,W. P., Chang, L., Klein,W. L., and Van Duyne, R. P. (2004). A localized surface plasmon resonance biosensor: first steps toward an assay for Alzheimer's disease, *Nano Lett.*, **4**(6), 1029–1034.

170. Xia, Y., and Whitesides, G. M. (1998). Soft lithography, *Annu. Rev. Mat. Sci.*, **28**(1), 153–184.

171. Rogers, J. A., and Nuzzo, R. G. (2005). Recent progress in soft lithography, *Mater. Today*, **8**(2), 50–56.

172. Weibel, D. B., DiLuzio, W. R., and Whitesides, G. M. (2007). Microfabrication meets microbiology, *Nat. Rev. Micro.*, **5**(3), 209–218.

173. Odom, T. W., Love, J. C., Wolfe, D. B., Paul, K. E., and Whitesides, G. M. (2002). Improved pattern transfer in soft lithography using composite stamps, *Langmuir*, **18**(13), 5314–5320.

174. Schmid, H., and Michel, B. (2000). Siloxane polymers for high-resolution, high-accuracy soft lithography, *Macromolecules.*, **33**(8), 3042–3049.

175. Hua, F., Sun, Y., Gaur, A., Meitl, M. A., Bilhaut, L., Rotkina, L., Wang, J., Geil, P., Shim, M., Rogers, J. A., and Shim, A. (2004). Polymer imprint lithography with molecular-scale resolution, *Nano Lett.*, **4**(12), 2467–2471.

176. Choi, K. M., and Rogers, J. A. (2003). A photocurable poly(dimethylsiloxane) chemistry designed for soft lithographic molding and printing in the nanometer regime, *J. Am. Chem. Soc.*, **125**(14), 4060–4061.

177. Henzie, J., Barton, J. E., Stender, C. L., and Odom, T. W. (2006). Large-area nanoscale patterning: chemistry meets fabrication, *Acc. Chem. Res.*, **39**(4), 249–257.

178. Kwak, E.-S., Henzie, J., Chang, S.-H., Gray, S. K., Schatz, G. C., and Odom, T. W. (2005). Surface plasmon standing waves in large-area subwavelength hole arrays, *Nano Lett.*, **5**(10), 1963–1967.

179. Henzie, J., Kwak, E.-S., and Odom, T. W. (2005). Mesoscale metallic pyramids with nanoscale tips, *Nano Lett.*, **5**(7), 1199–1202.

180. Wijaya, A., and Hamad-Schifferli, K. (2008). Ligand customization and DNA functionalization of gold nanorods via round-trip phase transfer ligand exchange, *Langmuir,* **24**(18), 9966–9969.

181. Wijaya, A., Schaffer, S. B., Pallares, I. G., and Hamad-Schifferli, K. (2009). Selective release of multiple DNA oligonucleotides from gold nanorods, *ACS Nano.*, **3**(1), 80–86.

182. Hill, H. D., and Mirkin, C. A. (2006). The bio-barcode assay for the detection of protein and nucleic acid targets using DTT-induced ligand exchange, *Nature Protocols.*, (1), 324–336.

183. Rosi, N. L., and Mirkin, C. A. (2005). Nanostructures in biodiagnostics, *Chem. Rev.*, **105**(4), 1547–1562.

184. Han, M. S., Lytton-Jean, A. K. R., Oh, B.-K., Heo, J., and Mirkin, C. A. (2006). Colorimetric screening of DNA-binding molecules with gold nanoparticle probes, *Angew. Chem. Int. Edit.*, **45**(11), 1807–1810.

185. Han, M. S., Lytton-Jean, A. K. R., and Mirkin, C. A. (2006). A gold nanoparticle based approach for screening triplex DNA binders, *J. Am. Chem. Soc.*, **128**(15), 4954–4955.

186. Chakrabarti, R., and Klibanov, A. M. (2003). Nanocrystals modified with peptide nucleic acids (PNAs) for selective self-assembly and DNA detection, *J. Am. Chem. Soc.*, **125**(41), 12531–12540.

187. Ellington, A. D., and Szotak, J. W. (1990) In vitro selection of RNA molecules that bind specific ligands, *Nature,* **346**(6287), 818–822.

188. Huang, C.-C., Huang, Y.-F., Cao, Z., Tan, W., and Chang, H.-T. (2005). Aptamer-modified gold nanoparticles for colorimetric determination of platelet-derived growth factors and their receptors, *Anal. Chem.,* **77**(17), 5735–5741.

189. Tuerk C., and Gold, L. (1990) Systematic evolution of ligands by exponential enrichment: RNA ligands to bacteriophage T4 DNA polymerase, *Science,* **249**(4968), 505–510.

190. Liu, J., and Lu, Y. (2003). A colorimetric lead biosensor using DN Azyme-directed assembly of gold nanoparticles, *J. Am. Chem. Soc.*, **125**(22), 6642–6643.

191. Liu, J., and Lu, Y. (2004). Accelerated color change of gold nanoparticles assembled by DNAzymes for simple and fast colorimetric $Pb2^{+}$ detection, *J. Am. Chem. Soc.*, **126**(39), 12298–12305.

192. Lin, S.-Y., Wu, S.-H., and Chen, C.-h. (2006). A simple strategy for prompt visual sensing by gold nanoparticles: general applications of interparticle hydrogen bonds, *Angew. Chem. Int. Edit.*, **45**(30), 4948–4951.

193. Nakashima, H., Furukawa, K., Kashimura, Y., and Torimitsu, K. (2007). Anisotropic assembly of gold nanorods assisted by selective ion recognition of surface-anchored crown ether derivatives, *Chem. Commun.*, (10), 1080–1082.

194. Homola, J. (2003). Present and future of surface plasmon resonance biosensors, *Anal. Bioanal. Chem.*, **377**(3), 528–539.

195. Pattnaik, P. (2005). Surface plasmon resonance, *Appl. Biochem. Biotechnol.*, **126**, 79–92.

196. Myszka, D. G. (1997). Kinetic analysis of macromolecular interactions using surface plasmon resonance biosensors, *Curr. Opin. Biotech.*, **8**(1), 50–57.

197. Li, Y., Lee,H. J., and Corn, R. M. (2007). Detection of protein biomarkers using RNA aptamer microarrays and enzymatically amplified surface plasmon resonance imaging, *Anal. Chem.*, **79**(3), 1082–1088.

198. Fang, S., Lee, H. J., Wark, A. W., and Corn, R. M. (2006). Attomole microarray detection of microRNAs by nanoparticle-amplified SPR imaging measurements of surface polyadenylation reactions, *J. Am. Chem. Soc.*, **128**(43), 14044–14046.

199. Brockman, J. M., Nelson, B. P., and Corn,R. M. (2000). Surface plasmon resonance imaging measurements of ultrathin organic films, *Annu. Rev. Phys. Chem.*, **51**(1), 41–63.

200. Piliarik, M., Vaisocherová, H., and Homola, J. (2007). Towards parallelized surface plasmon resonance sensor platform for sensitive detection of oligonucleotides, *Sens. Actuators B*, **121**(1), 187–193.

201. Millstone, J. E., Park, S., Shuford, K. L., Qin, L., Schatz, G. C., and Mirkin, C. A. (2005). Observation of a quadrupole plasmon mode for a colloidal solution of gold nanoprisms, *J. Am. Chem. Soc.*, **127**(15), 5312–5313.

202. Jung, L. S., Campbell, C. T., Chinowsky, T. M., Mar, M. N., and Yee, S. S. (1998). Quantitative interpretation of the response of surface plasmon resonance sensors to adsorbed films, *Langmuir*, **14**(19), 5636–5648.

203. Stuart, D. A., Haes, A. J., Yonzon, C. R., Hicks, E. M., and Van Duyne, R. P. (2005). Biological applications of localized surface plasmonic phenomena, *IEE Proc.-Nanobo technol.*, **152**(1), 13–32.

204. Aslan, K., Lakowicz, J. R., and Geddes, C. D. (2005). Plasmon light scattering in biology and medicine: new sensing approaches, visions and perspectives, *Curr. Opin. Chem. Biol.*, **9**(5), 538–544.

205. Yguerabide, J., and Yguerabide, E. E. (1998). Light-scattering submicroscopic particles as highly fluorescent analogs and their use as tracer labels in clinical and biological applications: I. theory, *Anal. BioChem.*, **262**(2), 137–156.

206. Bally, M., Halter, M., Vörös, J., and Grandin, H. M. (2006). Optical microarray biosensing techniques, *Surf. Interface Anal.*, **38**(11), 1442–1458.

207. Storhoff, J. J., Marla, S. S., Bao, P., Hagenow, S., Mehta, H., Lucas, A., Garimella, V., Patno, T., Buckingham, W., Cork, W., and Müller, U. R. (2004). Gold nanoparticle-based detection of genomic DNA targets on microarrays using a novel optical detection system, *Biosens. Bioelectron.*, **19**(8), 875–883.

208. Bao, Y. P., Huber, M., Wei, T.-F., Marla, S. S., Storhoff, J. J., and Müller, U. R. (2005). SNP identification in unamplified human genomic DNA with gold nanoparticle probes, *Nucl. Acids Res.*, **33**(2), e15.

209. Taton, A. T., Mirkin, C. A., and Letsinger, R. L. (2000). Scanometric DNA array detection with nanoparticle probes, *Science*, **289**(5485), 1757.

210. Alivisatos, P. (2004). The use of nanocrystals in biological detection, *Nat. Biotechnol.*, **22**(1), 47–52.

211. Reinhard, B. M., Siu, M., Agarwal, H., Alivisatos, A. P., and Liphardt, J. (2005). Calibration of dynamic molecular rulers based on plasmon coupling between gold nanoparticles, *Nano Lett.*, **5**(11), 2246–2252.

212. Liu, G. L., Yin, Y., Kunchakarra, S., Mukherjee, B., Gerion, D., Jett, S. D., Bear, D. G., Gray, J. W., Alivisatos, A. P., Lee, L. P., and Chen, F. F. (2006). A nanoplasmonic molecular ruler for measuring nuclease activity and DNA footprinting, *Nat. Nanotechnol.*, **1**(1), 47–52.

213. Reinhard, B. R. M., Sheikholeslami, S., Mastroianni, A., Alivisatos, A. P., and Liphardt, J. (2007). Use of plasmon coupling to reveal the dynamics of DNA bending and cleavage by single EcoRV restriction enzymes, *PNAS.*, **104**(8), 2667–2672.

214. Du, B. A., Li, Z.-P., and Liu, C.-H. (2006). One-step homogeneous detection of DNA hybridization with gold nanoparticle probes by using a linear light-scattering technique, *Angew. Chem. Int. Edit.*, **45**(47), 8022–8025.

215. Storhoff, J. J., Lucas, A. D., Garimella, V., Bao, Y. P., and Müller, U. R. (2004). Homogeneous detection of unamplified genomic DNA sequences based on colorimetric scatter of gold nanoparticle probes, *Nat. Biotechnol.*, **22**, 883–887.

216. Aslan, K., Lakowicz, J. R., and Geddes, C. D. (2005). Nanogold plasmon resonance-based glucose sensing. 2. wavelength-ratiometric resonance light scattering, *Anal. Chem.*, **77**(7), 2007–2014.

217. Aslan, K., Holley, P., Davies,L., Lakowicz, J. R., and Geddes, C. D. (2005). Angular-ratiometric plasmon-resonance based light scattering for bioaffinity sensing, *J. Am. Chem. Soc.*, **127**(34), 12115–12121.

218. Aslan, K., Lakowicz, J. R., and Geddes, C. D. (2005). Angular-dependent polarization-based plasmon light scattering for bioaffinity sensing, *Appl. Phys. Lett.*, **87**(23), 234108–234110.

219. Nam, J.-M., Stoeva, S. I., and Mirkin, C. A. (2004). Bio-bar-code-based DNA detection with PCR-like sensitivity, *J. Am. Chem. Soc.*, **126**(19), 5932–5933.

220. Nath, N., and Chilkoti, A. (2004). Label free colorimetric biosensing using nanoparticles, *J. Fluoresc.*, **14**(4), 377–389.

221. Nath, N., and Chilkoti, A. (2004). Label-free biosensing by surface plasmon resonance of nanoparticles on glass: optimization of nanoparticle size, *Anal. Chem.*, **76**(18), 5370–5378.

222. Taue, S., Nishida, K., Sakaue, H., and Takahagi, T. (2007). Immobilization of gold nanoparticles on silanized substrate for sensors based on localized surface plasmon resonance, *Surf. Sci. Nanotechnol.*, **5**, 74–79.

223. Frederix, F., Friedt, J.-M., Choi, K.-H., Laureyn, W., Campitelli, A., Mondelaers, D., Maes, G., and Borghs, G. (2003). Biosensing based on light absorption of nanoscaled gold and silver particles, *Anal. Chem.*, **75**(24), 6894–6900.

224. Malynych, S., and Chumanov, G. (2006). Coupled planar silver nanoparticle arrays as refractive index sensors, *J. Opt. A*, **8**(4), S144–S147.

225. Fujiwara, K., Watarai, H., Itoh, H., Nakahama, E, and Ogawa, N. (2006). Measurement of antibody binding to protein immobilized on gold nanoparticles by localized surface plasmon spectroscopy, *Anal. Bioanal. Chem.*, **386**(3), 639–644.

226. Lin, H.-Y., Chen, C.-T., and Chen, Y.-C. (2006). Detection of phosphopeptides by localized surface plasma resonance of titania-coated gold nanoparticles immobilized on glass substrates, *Anal. Chem.*, **78**(19), 6873–6878.

227. Marinakos, S. M., Chen, S., and Chilkoti, A. (2007). Plasmonic detection of a model analyte in serum by a gold nanorod sensor, *Anal. Chem.*, **79**(14), 5278–5283.

228. Tam, F., Moran, C., and Halas, N. (2004). Geometrical parameters controlling sensitivity of nanoshell plasmon resonances to changes in dielectric environment, *J. Phys. Chem. B*, **108**(45), 17290–17294.

229. Wang, H., Brandl, D. W., Le, F., Nordlander, P., and Halas, N. J. (2006). Nanorice: a hybrid plasmonic nanostructure, *Nano Lett.*, **6**(4), 827–832.

230. Yonzon, C. R., Jeoung, E., Zou, S., Schatz, G. C., Mrksich, M., and Van Duyne, R. P. (2004). A comparative analysis of localized and propagating surface plasmon resonance sensors: the binding of concanavalin A to a monosaccharide functionalized self-assembled monolayer, *J. Am. Chem. Soc.*, **126**(39), 12669–12676.

231. Jensen, T. R., Duval, M. L., Kelly, K. L., Lazarides, A. A., Schatz, G. C., and Van Duyne, R. P. (1999). Nanosphere lithography: effect of the external dielectric medium on the surface plasmon

resonance spectrum of a periodic array of silver nanoparticles, *J. Phys. Chem. B,* **103**(45), 9846–9853.

232. Larsson, E. M., Alegret, J., Kall, M., and Sutherland, D. S. (2007). Sensing characteristics of NIR localized surface plasmon resonances in gold nanorings for application as ultrasensitive biosensors, *Nano Lett.*, **7**(5), 1256–1263.
233. Dahlin, A. B., Tegenfeldt, J. O., and Hook, F. (2006). Improving the instrumental resolution of sensors based on localized surface plasmon resonance, *Anal. Chem.*, **78**(13), 4416–4423.
234. Dahlin, A., Zach, M., Rindzevicius, T., Kall, M., Sutherland, D. S., and Hook, F. (2005). Localized surface plasmon resonance sensing of lipid-membrane-mediated biorecognition events, *J. Am. Chem. Soc.*, **127**(14), 5043–5048.
235. Kalyuzhny, G., Vaskevich, A., Schneeweiss, M. A., and Rubinstein, I. (2002). Transmission surface-plasmon resonance (T-SPR) measurements for monitoring adsorption on ultrathin gold island films, *Chem. Eur. J.*, **8**(17), 3849–3857.
236. Lahav, M., Vaskevich, A., and Rubinstein, I. (2004). Biological sensing using transmission surface plasmon resonance spectroscopy, *Langmuir,* **20**(18), 7365–7367.
237. Kalyuzhny, G., Schneeweiss, M. A., Shanzer, A., Vaskevich, A., and Rubinstein, I. (2001). Differential plasmon spectroscopy as a tool for monitoring molecular binding to ultrathin gold films, *J. Am. Chem. Soc.*, **123**(13), 3177–3178.
238. Riboh, J. C., Haes, A. J., McFarland, A. D., Ranjit Yonzon, C., and Van Duyne, R. P. (2003). A nanoscale optical biosensor: real-time immunoassay in physiological buffer enabled by improved nanoparticle adhesion, *J. Phys. Chem. B,* **107**(8), 1772–1780.
239. Ebbesen, T. W., Lezec, H. J., Ghaemi, H. F., Thio, T., and Wolff, P. A. (1998). Extraordinary optical transmission through sub-wavelength hole arrays, *Nature,* **391**(6668), 667–669.
240. Gao, H., Henzie, J., and Odom, T. W. (2006). Direct evidence for surface plasmon-mediated enhanced light transmission through metallic nanohole arrays, *Nano Lett.*, **6**(9), 2104–2108.
241. Chang, S.-H., Gray, S., and Schatz, G. (2005). Surface plasmon generation and light transmission by isolated nanoholes and arrays of nanoholes in thin metal films, *Opt. Express,* **13**(8), 3150–3165.
242. Kneipp, K., and Kneipp, H. (2006). Single molecule Raman scattering, *Appl. Spec.*, **60**, 322A–334A.
243. Beckman, R., Beverly, K., Boukai, A., Bunimovich, Y., Choi, J. W., DeIonno, E., Green, J., Johnston-Halperin, E., Luo, Y., Sheriff, B., Stoddart, J. F., and Heath, J. R. (2006). Molecular mechanics and molecular electronics, *Faraday Disc.*, **131**, 9–22.
244. Albrecht, M. G., and Creighton, J. A. (1977). Anomalously intense Raman spectra of pyridine at a silver electrode, *J. Am. Chem. Soc.*, **99**(15), 5215–5217.
245. Jeanmaire, D. L., and Van Duyne, R. P. (1977). Surface Raman spectroelectrochemistry: PartI. heterocyclic, aromatic, and aliphatic amines adsorbed on the anodized silver electrode, *J. Electroanal. Chem.*, **84**(1), 1–20.
246. Otto, A. (2006). On the significance of Shalaev's lsquohot spotsrsquo in ensemble and single-molecule SERS by adsorbates on metallic films at the percolation threshold, *J. Raman Spec.*, **37**(9), 937–947.
247. Schatz, G. C., Young, M. A., and Van Duyne, R. P. (2006). Electromagnetic mechanism of SERS, *Top. Appl. Phys.*, **103**, 19–46.
248. Lim, J. K., Kim, I.-H., Kim, K.-H., Shin, K. S., Kang, W., Choo, J., and Joo, S.-W. (2006). Adsorption of dimethyl sulfide and methanethiolate on Ag and Au surfaces: surface-enhanced Raman scattering and density functional theory calculation study, *Chem. Phys.*, **330**(1–2), 245–252.

249. Zhang, X., Young, M. A., Lyandres, O., and Van Duyne, R. P. (2005). Rapid detection of an anthrax biomarker by surface-enhanced Raman spectroscopy, *J. Am. Chem. Soc.*, **127**(12), 4484–4489.

250. Kuncicky, D. M., Prevo, B. G., and Velev, O. D. (2006). Controlled assembly of SERS substrates templated by colloidal crystal films, *J. Mater. Chem.*, **16**, 1207–1211.

251. Anderson, D. J., and Moskovits, M. (2006). A SERS-active system based on silver nanoparticles tethered to a deposited silver film, *J. Phys. Chem. B*, **110**(28), 13722–13727.

252. Li, Z., Tong, W. M., Stickle, W. F., Neiman, D. L., Williams, R. S., Hunter, L. L., Talin, A. A., Li, D., and Brueck, S. R. J. (2007). Plasma-induced formation of Ag nanodots for ultra-high-enhancement surface-enhanced Raman scattering substrates, *Langmuir*, **23**(9), 5135–5138.

253. Brolo, A. G., Arctander, E., Gordon, R., Leathem, B., and Kavanagh, K. L. (2004). Nanohole-enhanced Raman scattering, *Nano Lett.*, **4**(10), 2015–2018.

254. Sun, L., Yu, C., and Irudayaraj, J. (2008). Raman multiplexers for alternative gene splicing, *Anal. Chem.*, **80**(9), 3342–3349.

255. Sun, L., Yu, C., and Irudayaraj, J. (2007). Surface-enhanced Raman scattering based non fluorescent probe for multiplex DNA detection, *Anal. Chem.*, **79**(11), 3981–3988.

256. Bell, S. E. J., and Sirimuthu, N. M. S. (2006). Surface-enhanced Raman spectroscopy (SERS) for sub-micromolar detection of DNA/RNA mononucleotides, *J. Am. Chem. Soc.*, **128**(49), 15580–15581.

257. Tang, H.-W., Yang, X. B., Kirkham, J., and Smith, D. A. (2007). Probing intrinsic and extrinsic components in single osteosarcoma cells by near-infrared surface-enhanced Raman scattering, *Anal. Chem.*, **79**(10), 3646–3653.

258. Shamsaie, A., and Irudayaraj, J. (2007). Intracellularly grown gold nanoparticles (IGAuN) as potential SERS probes. *J. Biomed. Opt.*, **12**(2), 020502-1-3.

259. Zhang, J., Gao, Y., Alvarez-Puebla, R. A., Buriak, J. M., and Fenniri, H. (2006). Synthesis and SERS properties of nanocrystalline gold octahedra generated from thermal decomposition of $HAuCl_4$ in block copolymers FNR, *Adv. Mater.*, **18**(24), 3233–3237.

260. Orendorff, C. J., Gearheart, L., Jana, N. R., and Murphy, C. J. (2006). Aspect ratio dependence on surface enhanced Raman scattering using silver and gold nanorod substrates, *Phys. Chem. Chem. Phys.*, **8**(1), 165–170.

261. Wang, T., Hu, X., and Dong, S. (2006). Surfactantless synthesis of multiple shapes of gold nanostructures and their shape-dependent SERS spectroscopy, *J. Phys. Chem. B*, **110**(34), 16930–16936.

262. Delfino, I., Bizzarri, A. R., and Cannistraro, S. (2006). Time-dependent study of single-molecule SERS signal from yeast cytochrome C, *Chem. Phys.*, **326**(2–3), 356–362.

263. Hsieh, B.-Y., Chang, Y.-F., Ng, M.-Y., Liu, W.-C., Lin, C.-H., Wu, H.-T., and Chou, C. (2007). Localized surface plasmon coupled fluorescence fiber-optic biosensor with gold nanoparticles, *Anal. Chem.*, **79**(9), 3487–3493.

264. Lakowicz, J. (2006). Plasmonics in biology and plasmon-controlled fluorescence, *Plasmonics.*, **1**(1), 5–33.

265. Lakowicz, J., Geddes, C., Gryczynski, I., Malicka, J., Gryczynski, Z., Aslan, K., Lukomska, J., Matveeva, E., Zhang, J., Badugu, R., and Huang, J. (2004). Advances in surface-enhanced fluorescence, *J. Fluoresc.*, **14**(4), 425–441.

266. Huang, C.-C., and Chang, H.-T. (2006). Selective gold-nanoparticle-based "turn-on" fluorescent sensors for detection of mercury(II) in aqueous solution, *Anal. Chem.*, **78**(24), 8332–8338.

267. Garrett, S. H., Smith, L. H., and Barnes, W. L. (2005). Fluorescence in the presence of metallic hole arrays, *J. Mod. Opt.*, **52**(8), 1105–1122.

268. Nagamura, T., Yamamoto, M., Terasawa, M., and Shiratori, K. (2003). High performance sensing of nitrogen oxides by surface plasmon resonance excited fluorescence of dye-doped deoxyribonucleic acid thin films, *Appl. Phys. Lett.*, **83**(4), 803–805.

269. He, X., Liu, H., Li, Y., Wang, S., Li, Y., Wang, N., Xiao, J., Xu, X., and Zhu, D. (2005). Gold nanoparticle-based fluorometric and colorimetric sensing of copper^{2+} ions, *Adv. Mater.*, **17**(23), 2811–2815.

270. Kim, J., and Song, K.-B. (2007). Recent progress of nano-technology with NSOM, *Micron.*, **38**(4), 409–426.

271. Frey, H. G., Bolwien, C.,Brandenburg, A., Ros, R., and Anselmetti, D. (2006). Optimized apertureless optical near-field probes with 15 nm optical resolution, *Nanotechnology.*, **17**(13), 3105–3110.

272. Pettinger, B. (2006). Tip-enhanced Raman spectroscopy (TERS), *Top. Appl. Phys.*, **103**, pp. 217–240.

273. Becker, M., Sivakov, V., Andra, G., Geiger, R., Schreiber, J., Hoffmann, S., Michler, J., Milenin, A. P., Werner, P., and Christiansen, S. H. (2007). The SERS and TERS effects obtained by gold droplets on top of Si nanowires, *Nano Lett.*, **7**(1), 75–80.

274. Pendry, J. B. (2000). Negative refraction makes a perfect lens, *Phys. Rev. Lett.*, **85**(18), 3966.

275. Srituravanich, W., Fang, N., Sun, C., Luo, Q., and Zhang, X. (2004). Plasmonic nanolithography, *Nano Lett.*, **4**(6), 1085–1088.

276. Liu, Z.-W., Wei, Q.-H., and Zhang, X. (2005). Surface plasmon interference nanolithography, *Nano Lett.*, **5**(5), 957–961.

277. Smolyaninov, I. I., Elliott, J., Zayats, A. V., and Davis, C. C. (2005). Far-field optical microscopy with a nanometer-scale resolution based on the in-plane image magnification by surface plasmon polaritons, *Phys. Rev. Lett.*, **94**(5), 057401.

278. Garini, Y., Kutchoukov, V. G., Bossche, A., Alkemade, P. F. A., Docter, M. W., Verbeek, P. W., van Vliet, L. J., and Young, I. T. (2004). Toward the development of a three-dimensional mid-field microscope, *Proc. SPIE.*, **5327**, 115–122.

279. Docter, M. W., Young, I. T., Kutchoukov, V. G., Bossche, A., Alkemade, P. F., and Garini, Y. (2005). A novel concept for a mid-field microscope, *Proc. SPIE.*, **5703**, 118–126.

280. Liu, Z., Lee, H., Xiong, Y., Sun, C., and Zhang, X. (2007). Far-field optical hyperlens magnifying sub-diffraction-limited objects, *Science*, **315**(5819), 1686.

281. Smolyaninov, I. I., Hung, Y.-J., and Davis, C. C. (2007). Magnifying superlens in the visible frequency range, *Science*, **315**(5819), 1699–1701.

Chapter 4

Surface-Raman Scattering for Medical Applications

Mustafa Çulha
Yeditepe University, Faculty of Engineering and Architecture, Department of Genetics and Bioengineering, 34755 Ataşehir-Istanbul, Turkey
mculha@yeditepe.edu.tr or mustafa.cullha@gmail.com

4.1 Introduction

There is an enormous interest to utilize surface-enhanced Raman scattering (SERS) in a variety of applications. This is mostly due to its sensitivity and power to provide very specific chemical information about the molecule or molecular structure of interest in a short time without significant sample preparation steps. Due to the fact that this information can be used for detection and identification besides characterization, it has become an important analytical technique in recent years and it is considered one of the most sensitive diagnostic tools available to the scientist [1]. This chapter covers the recent SERS-based biomolecular detection and identification of approaches and assays from the biological application perspective. First, a brief introduction to Raman scattering and SERS is presented. In the later section of the chapter, the detection of proteins and DNA; the characterization, identification, and detection of microorganisms; single-cell studies; and tissue differentiation are discussed.

4.2 Introduction to Raman Scattering and Surface-Enhanced Raman Scattering

The Raman scattering process is the result of the inelastic collision of photons with molecules [2, 3]. During the scattering process, photons may gain or lose energy while

Biomedical Nanosensors
Edited by Joseph Irudayaraj

ISBN 978-981-4303-03-3 (Hardcover), 978-981-4303-04-0 (eBook)
www.panstanford.com

interacting with the molecules. A change in the photons energy means a change in the photons frequency relative to the excitation laser light's frequency. A Raman spectrum is composed of several lines arising from different vibrational modes of the bonds in a molecule. Since this chemical information is specific for the molecule, it is, therefore, considered that the Raman spectrum of a molecule is the "fingerprint" for that molecule. Raman scattering is inherently weak, and a typical cross section is between 10^{-30} and 10^{-25} cm^2 per molecule. When compared to a fluorescence cross section, whose typical value is between 10^{-17} and 10^{-16} cm^2, Raman scattering is rather weak. Therefore, a large number of molecules in Raman experiments are necessary to achieve the same level of signal in fluorescence. Therefore, the measurement of Raman scattering requires a powerful light source and very sensitive detectors. It is indeed the development of more sensitive and cost-effective Raman instruments that was delayed until the advancements in laser and detector technology. With the recent advances in laser and detector technology and optics, the cost and size of Raman spectrometers have significantly reduced without the sacrifice of sensitivity. For the fundamentals and instrumentation of Raman scattering, a number of textbooks and tutorial reviews are available and the readers are encouraged to address these sources for further details [4–6].

The major breakthrough with the use of Raman scattering as a characterization and detection technique came with the discovery of the enhancement in scattering on roughened noble metal surfaces. In 1974, Fleischman observed the enhancement in Raman scattering of pyridine molecules on silver electrode surfaces [7]. However, this enhancement was first explained with the increased pyridine concentration on a roughed electrode surface that resulted with an increased surface area. In 1977, Jeanmarie and Van Duyne [8] and Albrecht and Creighton [9] proposed two separate mechanisms, electromagnetic (EM) and charge transfer, respectively, to explain the increased Raman scattering. In 1978, Moskovits proposed a relationship between enhancement in Raman scattering and surface plasmons [10]. In the following years, it was demonstrated that enhancement in Raman scattering did not occur only on the surface of silver, gold, and copper but also on other conductors, such as aluminum, indium, platinum, and transition metals [11]. It is now commonly accepted that the enhancement mechanism has two components — EM, which is related to the surface plasmons, and chemical, which is related to the charge transfer between the metal surface and adsorbed molecule [12, 13]. The contribution of the EM enhancement is much larger than the charge transfer that depends on the chemical structure of the molecule adsorbed onto the nanostructured noble metal surface.

Roughed electrode surfaces and colloidal nanoparticles of silver and gold have been employed as SERS substrates since the early days of the discovery. The development of SERS substrates from historical perspectives has recently been reviewed [14]. The relationship between the enhancement and surface roughness was noticed in the early days and then later confirmed to be an important factor with the better understanding of the plasmonic properties of the metals. Realizing that the junctions and groves of the nanostructures, where surface plasmons are localized, provided the optimal enhancement; the investigations have been moved to the generation of such structures using lithographic or other approaches and techniques. As the advancements in plasmonics progressed, a better understanding of the concept helped the manipulation of the EM enhancement component for its optimal utility. Although the lithographic techniques are not sophisticated enough to locate the plasmonic structures with nanometer precision, the developments in this area positively contributed to the way of making customized novel structures to test their performances as SERS-active structures [15]. The SERS substrates were prepared using e-beam lithography [16, 17], nanosphere lithography [18, 19, 20, 21], nanoimprint lithography [22], and on-wire lithography [23, 24]. An excellent tutorial review is available for detailed discussion of the preparation of SERS substrates from the design perspective [15].

It should also be noted that the early reports claiming that SERS is capable of detecting single molecules also contributed to the development of the technique due to the enormous interest for its use in single-molecule detection [25, 26]. Although the sensitivity down to single molecule can be achieved in only isolated cases where fractal silver aggregates are present, the recent studies proved that single-molecule detection can be achieved on rationally designed nanostructures as well [27]. Even with all these advances, the problem from the analytical perspective remains as concrete as in early days. The success of the nanostructured surfaces using several different lithographic techniques seems limited with the current technology. Although very high sensitivity can be achieved with these structures and surfaces, a number of problems such as reproducible generation of the surfaces and high cost still remain a major obstacle.

In SERS experiments, the detection scheme can be designed based on the nature of the molecule whose detection is pursued. If the chemical "fingerprint" information can be reproducibly obtained from the target molecule, the molecule's intrinsic signature can be used for detection. Otherwise, a reporter molecule must be used to perform the detection indirectly. The requirements for the use of a molecule's own SERS spectrum for detection purposes are to have a distinct and recognizable spectrum. Although the requirements are more easily established by small molecules with an aromatic ring or electron rich functional group, it can be rather difficult to achieve this goal with large biomolecules and molecular structures such as microorganisms. Therefore, the use of a reporter molecule is preferred, especially for the detection of DNA and proteins that are composed of similar repeating units. In the indirect detection scheme, the reporter molecule is usually a small and Raman-active molecule that is chemically or physically linked to one of the molecules in the detection system. The presence or absence of the SERS spectrum from this reporter molecule as a result of the molecular interaction between the molecules in question indirectly establishes the detection.

The other important consideration is the fact that the molecule of interest must be in the close vicinity of the plasmonic structures. When a small molecule is considered, enhancement is possible as long as the molecule remains in the same location and position near the plasmonic structure. However, when the size of molecule increases, special attention must be given to the size relationship between the plasmonic structure and the target molecule. A large molecule such as a protein or DNA fragment may not come into contact from all points with the nanostructure and the obtained SERS spectra remain limited to the parts of the molecule interacting with the surface of the plasmonic structure.

Although numerous ways of preparation of new substrates for several applications have been reported to date, the fundamental issues regarding the control of the plasmonic properties of the substrates are preventing the construction of universally applicable substrates. The fundamentals and application of plasmonic structures have been explored in a separate chapter of this book, and the reader can also find excellent books and reviews in the literature regarding the fundamentals and properties of plasmonics [28, 29]. With all these considerations in mind, the best approach for the design of a new substrate is still case specific. The detection or characterization scheme can be rationally designed with the knowledge of the target molecule or molecular structure. As explained above, this is due to the fact that the molecule or molecular structure must be in contact with or in close vicinity to a noble metal structure. This condition cannot be satisfied for all types of molecules and molecular structures, especially biological macromolecules and higher organizations such as bacteria and viruses and is discussed in detail later in the chapter.

With the recent increasing momentum for the detection of chemical and biological species, SERS has been proven to be a powerful technique not only for the detection of environmentally [30] and biological relevant molecules [31] but also for the characterization and identification of a number of higher organizations such as viruses, bacteria, and eukaryotic cells, which are discussed later in the chapter. This power can be attributed to its several distinguished features, such as high sensitivity, narrow spectral bandwidths, the ability to provide molecular-level information, and minimum interference from water. The SERS substrates can be classified into several groups from different perspectives. However, they are here examined from the perspective of the biomolecules whose detection and characterization is targeted due to the diversity of SERS substrates and for the sake of clarity in this chapter.

4.3 Applications of SERS in Biomedical Sciences and Medicine

4.3.1 Protein Detection

Proteins are essential components of living systems and may have structural and/or regulatory functions. Since they are the end products of gene expression and are responsible for a number of functions, any failure in cellular processes is mostly reflected as a change in the protein profile of a living system. Therefore, the detection of proteins has clinical significance. Protein detection is usually performed with electrophoresis-mass spectroscopy hyphenated systems. However, the instrumentation for this technique is expensive and requires highly trained personnel.

Label-free detection of proteins is highly desired using SERS. However, proteins are difficult molecules to acquire comparable SERS spectra. This difficulty can be explained with their molecular structure. Since they are folded from long peptide chains, they have flexible structures prone to environmental changes such as pH and temperature. They are also composed of similar molecular units, amino acids, making the identification difficult due to the similarity in their SERS spectra. However, SERS is still powerful enough to serve in the detection of specific proteins. In this regard, two approaches can be followed. The first involves label-free detection, where the intrinsic protein SERS signature is used, and the second is where a Raman reporter molecule is included in the assay.

As mentioned, as long as the SERS spectrum from a molecule can be reproducibly obtained, it can be used for identification and detection. However, it is not always straightforward to obtain reproducible SERS spectra from a large biological molecule such as protein. However, in a recent report, it was claimed that an assay for label-free detection of proteins in solutions was developed [32]. The detection of lysozyme, ribonuclease B, avidin, catalase, and hemoglobin was performed in citrate-reduced silver colloidal suspension. The only difference in the protocol was the use of sodium sulfate to initiate the aggregation of silver nanoparticles rather than a halide ion such as chloride. The detection limits for lysozyme and catalase as low as 5-μg/ mL and 50-ng/mL, respectively, were achieved.

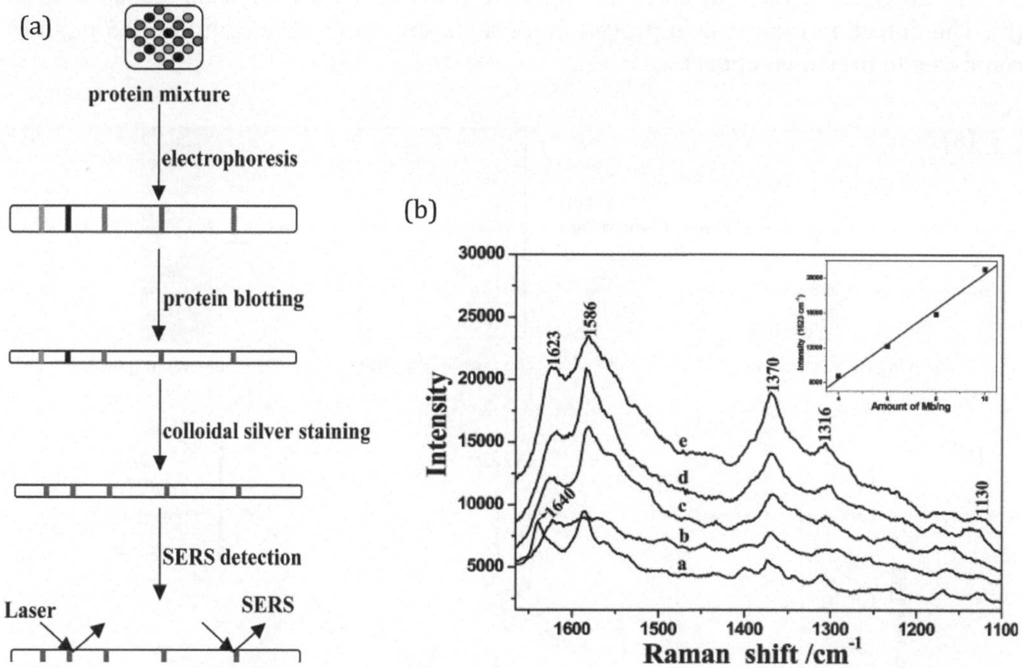

Figure 4.1 (a) Procedure for SERS protein detection based on a Western blot; (b) SERRS spectra of Mb with different amounts on a nitrocellulose membrane — **a:** solid Mb, **b:** 4 ng, **c:** 6 ng, **d:** 8 ng, and **e:** 10 ng. The inset shows an intensity of a SERRS peak at 1370 cm^{-1} versus the amount of Mb on a nitrocellulose membrane. (Reprinted with permission from ref. 33. Copyright 2008 American Chemical Society.)

In another approach, the proteins were detected after a Western blot [33]. Western blotting is an analytical method used to detect proteins in a mixture. First, the protein mixture is separated with gel electrophoresis. The separated proteins are then transferred onto a membrane and targeted with an antibody or recently have been analyzed using mass spectroscopy [34–36]. In that proposed technique, the bands on a slab gel were transferred onto a nitrocellulose membrane and then stained with a silver colloidal suspension. The SERS experiment was performed with a Raman instrument equipped with a 514.5 nm laser, where its use contributes to further enhancement of Raman scattering (surface-enhanced resonance Raman scattering [SERRS]). Figure 4.1 shows the procedure for SERS protein detection based on a Western blot. Myoglobin (Mb) and bovine serum albumin

(BSA) were used as model proteins, and the SERRS detection limit of Mb is found to be 4 ng/band. In a more recent study, Culha and coworkers demonstrated a novel approach for label-free protein detection [37]. In that study, proteins and colloidal AgNPs were convectively assembled on glass surfaces and the SERS spectra of proteins regardless of the protein size and surface charge properties were successfully acquired. The convective assembly process, which is assembly of particles from a moving and drying droplet, allows the proteins and AgNPs to assemble into ordered structures in a reproducible manner. Figure 4.2 shows the assembly process using "convective assembly" and the thin film formed on glass surface (a) and SERS spectra of avidin with decreasing concentration (b). The detection limit was improved more than one order of magnitude (0.5 μg/mL) compared to previous reports.

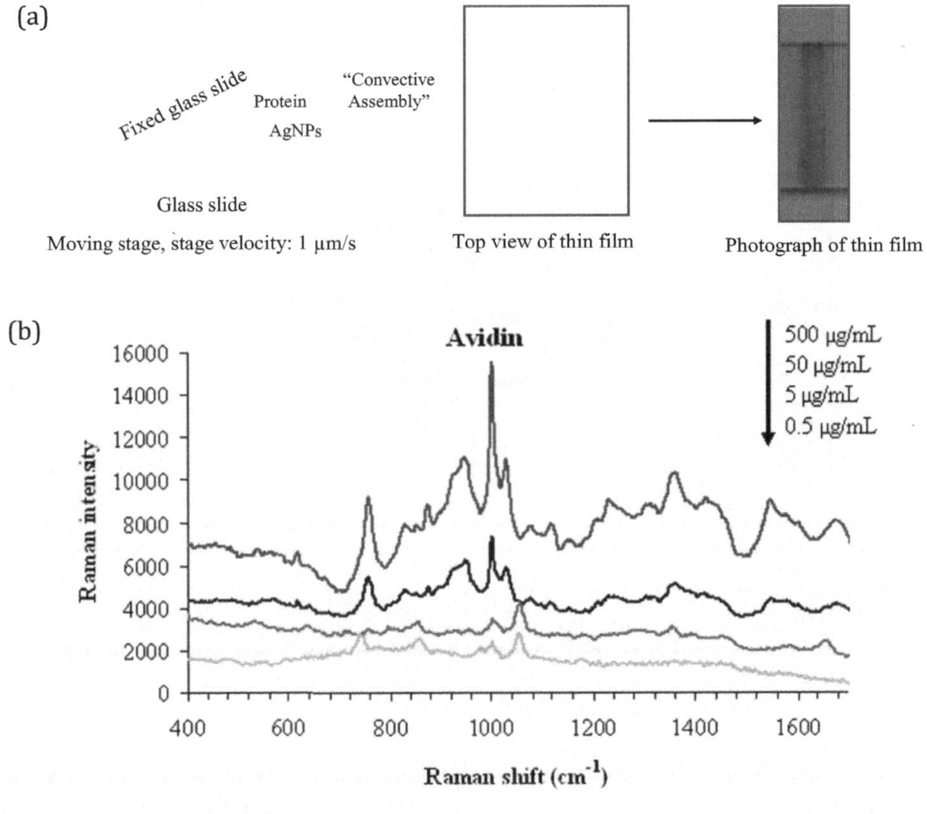

Figure 4.2 (a) The convective assembly process of proteins and AgNPs; (b) SERS spectra of avidin with decreasing concentrations. (Reprinted with permission from ref. 37. Copyright 2010 American Chemical Society.)

Immunoassays can be given as good examples to the reporter molecule–based SERS detection schemes. In traditional immunoassay-based systems, the detection is established using a detection technique such as radioactivity, electrochemical chemical, or fluorescence. However, all of these techniques have certain drawbacks that limit their use. For example, the radioactivity-based techniques may pose a danger to the user upon long-term exposure, electrochemical methods need a species that can be oxidized or reduced in the system, and finally fluorescence suffers from inherently broad bandwidths limiting the multiplex detection.

In SERS-based immunoassays, a noble metal element has to be in place for the SERS enhancement. This is usually achieved with the use of gold or silver colloidal nanoparticles or surfaces. In the first example presented here, the AgNPs are used and in the other examples, gold nanoparticles are used after surface modifications.

An enzyme-linked immunosorbent assay (ELISA) was reported for protein detection using SERRS [38]. In that assay, fluorescein isothiocyanate (FITC) was used as the Raman reporter molecule. The assay is illustrated in Fig. 4.3. The whole process is the same as the conventional ELISA procedure. The bottom of the microwells was first coated with the antigens. After blocking the well bottom to prevent nonspecific binding, the antibodies labeled with FITC treated the antigens at the bottom of the wells. Finally, the well plates were treated with colloidal silver nanoparticles for SERRS enhancement. The authors claim that this approach can be beneficial in several ways. First, the concentration of the antigens can be determined from the SERRS signal without the need for an enzymatic reaction. Second, SERRS spectra can be obtained from the bottom of the plate without displacement. Third, a detection limit as low as 0.2 ng/mL can be achieved and this is comparable to the conventional ELISA based detection.

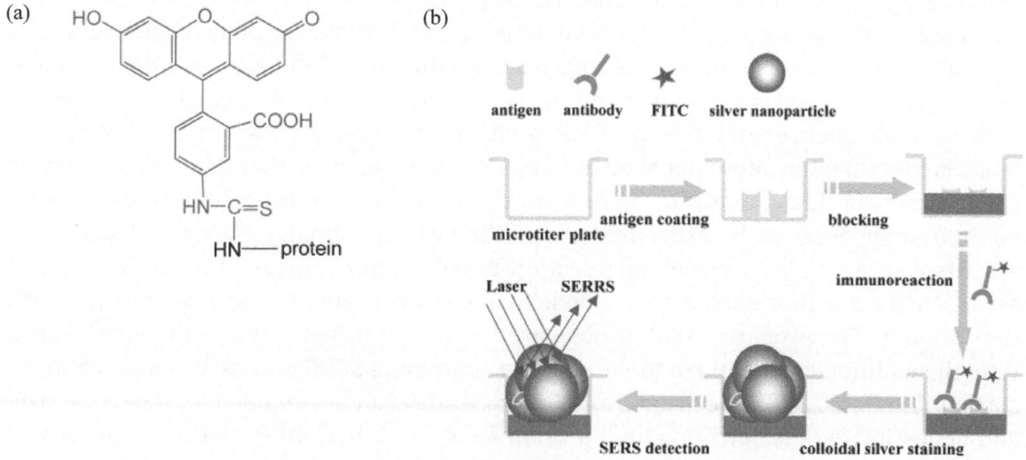

Figure 4.3 (a) The structure of FITC; (b) the proposed procedure for SERRS-based immunoassay on a microtiter plate. (Reprinted with permission from ref. 38. Copyright 2008 American Chemical Society.)

In a recent report, Porter *et al.* reviewed the detection of proteins, viruses, and microorganisms in heterogeneous immunoassays [39]. In thereview, the fundamentals and applications of SERS-based immunoassays were outlined. The authors also described their own system in a three-step procedure. In the first step, a capture antibody-immobilized substrate is used to capture and concentrate the antigens from the solution. In the second step, the captured antigens are treated with the AuNPs coated with the detection antibody. In the final step, SERS spectra are acquired from the surface. The antibodies can be covalently linked to the gold-coated substrates through the thiol groups of a linker group such as dithiobis (succinimidyl undecanoate) (DSU) or dithiobis (succinimidyl propionate) (DSP). Proteins and other amine-containing substances can be easily attached to the carboxyl group of these molecules through amide linkages. The prepared surface is treated with BSA to block the residual succinimidyl groups. It should be noted that the gold surface was prepared using template-stripped gold (TSG) that generates subnanometer surface roughness and atomic force microscope (AFM) and scanning electron microscope (SEM) can be used for the characterization and validation of SERS enhancement [40].

In most SERS-based immunoassays, the AuNPs were modified with a reporter molecule and capture antibodies. For surface modifications of AuNPs, three approaches were explored. In the first approach, while the antibodies were adsorbed onto AuNPs, reporter Raman molecules were chemically attached [41]. It is claimed that this approach has cross-talk problems due to desorption of some of the antibodies from the AuNP surface and the reabsorption of the desorbed antibodies by the second SERS probe in a two-probe system. In the second approach, a bifunctional Raman reporter with a large scattering cross section and a reactive moiety was covalently coupled to an antibody to prevent cross talk. 5,5′ dithiobis (succinimidyl-2-nitrobenzoate) (DSNB) gave the best results among the tested Raman reporters (see Fig. 4.3 b). This procedure covalently links both the antibody and the Raman reporter molecule. With this approach, the detection limit was improved due to better enhancement in SERS scattering, and the femtomolar detection limit for f-PSA directly in human serum was reported [42].

In the third approach, AuNPs were modified with two different thiolated molecules. In an analogous manner to the second approach, a bifunctional molecule similar to DSNB-carrying disulfide and succinimidyl groups is attached to the AuNPs. The second molecule, such as 4-nitrobenzenethiol, is preferably chosen because its Raman cross-section is much higher than the bifunctional molecule [43]. It was reported that this approach yields similar limits of detection to the second approach. However, it provides the opportunity for multi-analyte assays by using different antibodies and Raman reporter molecules.

It is also possible to design aggregation-based immunoassays [44]. In this type of assay, AuNPs are first coated with thioalted Raman reporter molecules and then with the antibody. The prepared SERS probes are treated with BSA to block the empty sites. With the addition of an antigen to the solution containing SERS probes, the aggregation is started. A suitable SERS spectrum from the aggregates can be obtained for detection. The authors estimated the detection limit for human IgG as 0.1-μg mL^{-1}, which is much lower than the UV/V-based and light-scattering-based aggregation immunoassays.

4.3.2 DNA Detection

With the enormous progress in genomic research, the relationship between gene mutations and disease formation has never been this clear. The diversity and large number of mutations and variations in DNA have forced scientists to develop novel mutation-detection schemes. Among the number of mutation-detection techniques, fluorescence-based ones are the most popular due to their sensitivity. However, fluorescence suffers from inherently broadbands (50–100 nm bandwidths) and does not allow the detection of more than one target at a time. When multiple fluorescence molecules are used for the detection of multiple targets, it is necessary to use complex algorithms to resolve the significantly overlapped spectra. In addition, the necessity to use a different laser source for each fluorescence molecule is another drawback of the technique. With its "fingerprint" property, very narrow spectral bands features (~ 1 nm bandwidth), and competing sensitivity with fluorescence, SERS may be an alternative technique.

The use of SERS labels for DNA probing was first reported by Vo-Dinh in 1994 [45]. In this pioneering study, cresyl fast violet (CFV) was chemically attached to an 18-base-long oligonucleotide and this labeled oligonucleotide was hybridized with the target DNA before SERS measurement [46]. Later, the same group demonstrated the detection of the HIV virus after the amplification of the HIV (gag) gene through the use of a CFV-labeled forward primer and the capture of this labeled oligonucleotide with an immobilized complementary oligonucleotide before SERS measurement. Vo-Dinh's group also demonstrated the detection of HIV DNA and the breast cancer 1, early onset (BRCA1) gene by immobilizing the capture oligonucleotides on surfaces and hybridizing with the Raman-active molecule labeled "target oligonucleotides" [46, 47].

The detection of six viruses, hepatitis A virus Vall7 polyprotein gene (HVA), hepatitis B virus surface antigen gene (HVB), human immunodeficiency virus (HIV), Ebola virus (EV), variola virus (smallpox, VV), and Bacillus anthracis (BA) protective antigen geneDNA, down to 20-femtomolar level was demonstrated using six spectrally distinct and commercially available Raman labels. The AuNPs with an average of 13-nm size were modified with oligonucleotides complementary to a portion of the viruses' 30-base-long DNA and Raman-active dyes [48]. A complementary segment of oligonucleotide targeting the virus DNA was immobilized onto gold surfaces. Both dye labeled "oligonucleotides attached to gold nanoparticles" and "oligonucleotides immobilized on gold surfaces" were hybridized from the two ends of the 30-base-long complementary target DNA. In the final step, a silver staining procedure was performed through the hydroqinone reduction of silver ions. With this approach, the detection of viral DNA was successfully achieved.

A DNA-detection SERS probe operating similar to the working principle of molecular beacons was demonstrated [49]. The probe is composed of two parts — a stem and a loop (see Fig. 4.7, top). While the stem part is complementary to each other, the loop is complementary to the target DNA. The whole structure consisted of a 42-base-long hairpin and a 7-base-pair stem. The stem was modified with rhodamine 6G and hexanethiol at the 3' end and the 5' end, respectively. When the oligonucleotide gets chemically attached to citrate-reduced silver nanoparticles through an Ag-S bond, the rhodamine 6G is brought close to the surface of silver nanoparticles since the stem part is in a naturally hybridized form. Fig. 4.4 summarizes the working principle of the SERS probe. In the absence of the target DNA, a SERS signal from rhodamine 6G is recorded. However, when a target DNA is present, the stem part is forced to open because the loop part is complementary to the target DNA; rhodamine 6G is removed from the silver surface as the stem part is opened and there is no SERS signal generated. The authors demonstrated the detection of HIV-1 virus with this approach.

In later years, other interesting applications of DNA detection using SERS were reported. For example, alternative splice junctions were detected with SERS using non-fluorescent labels [50]. In that study, the proof of the principle of multiplex detection of pre-mRNA splicing of BRCA1 gene, which was linked to malignancy in breast tissue, was demonstrated [51, 52]. In another interesting recent report, Qin *et al.* demonstrated

DNA detection down to 100 fM on SERS-active arrays of nanodisks prepared by on-wire lithography and modified with Raman-active reporter molecules [24]. Irudayaraj and coworkers later extended their work to quantify splice variants [53, 54].

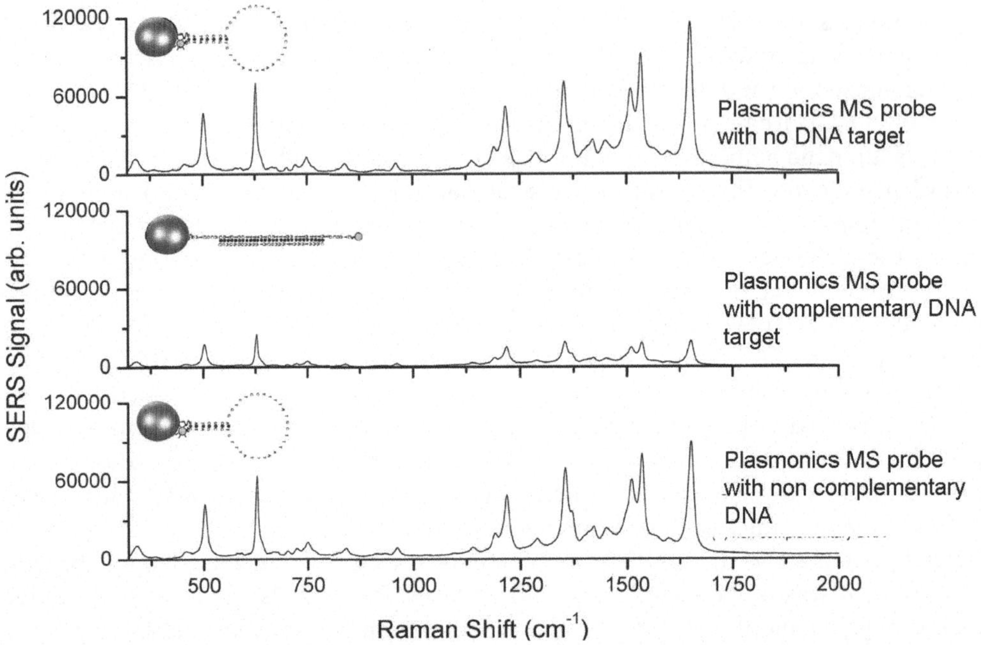

Figure 4.4 Detection of HIV-1: The SERS spectrum from the detection system. (a) No target DNA sequence (top curve); (b) a complementary HIV-1 DNA target (positive diagnostic: middle curve); (c) in the presence of a noncomplementary DNA target sequence (negative diagnostic: bottom curve). (Reprinted with permission from ref. 49. Copyright 2008 American Chemical Society.)

The detection of DNA without using a label is highly desired. Braun *et al.* reported a SERS-based sensing methodology that detects single-stranded DNA without labeling the DNA with a reporter molecule [55]. The sensing platform is seen in Fig. 4.5. A 24-mer oligonucleotide shown as a half-blue (a) and a half-red (b) oligonucleotide in Fig. 4.5 was used as a model. First, 15 nm size AgNPs were functionalized with the complement of one-half of the target oligonucleotide, and second a 100 nm thick silver thin film was used to immobilize with the complement of the remaining part of the target. The oligonucleotide-immobilized silver thin-film surface was treated with metcaptohexanathiol to prevent nonspecific binding to the silver surface. Then, 5-((2-(and-3)-S (acetylmercapto)suc cinoyl)amino) fluorescein (SAMSA fluorescein, F), which contains a protected thiol, was chemically attached to the silver surface as a SERS-active molecule. When the oligonucleotide-and-SERS-active-molecule-attached silver surface was incubated with the AgNPs labeled with the complement of one-half of the target oligonucleotide, a sandwich structure was formed and SERS spectra from the label were acquired due to the presence of closely located AgNPs. The authors claimed that they constructed hot-spots in the sandwich structure; therefore, the increase in the enhancement factor additional to the overall enhancement is about 5×10^5.

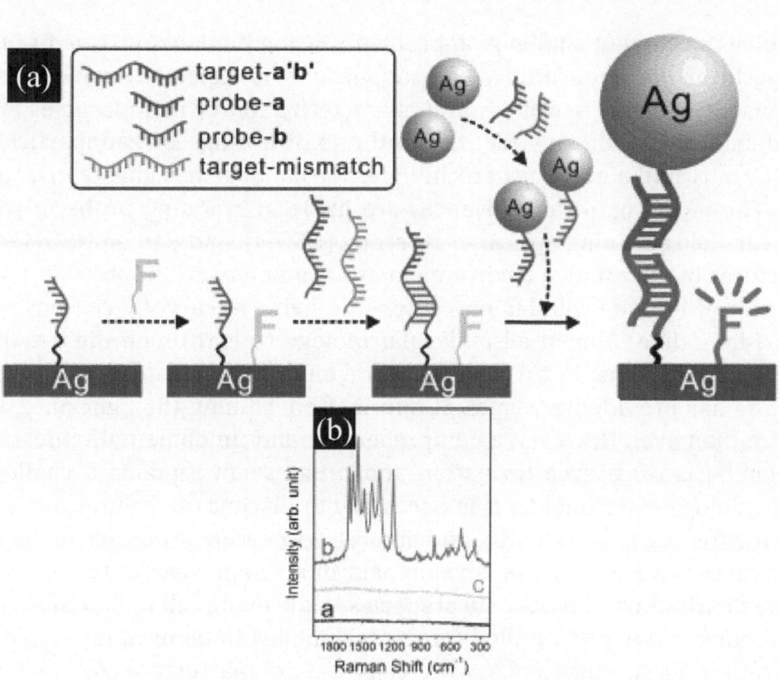

Figure 4.5 (a) Immobilization of half-capture probes (red on the surface and blue on nanoparticles), a Raman-active label (F) on a 100 nm thick Ag surface and 15 nm size Ag nanoparticles, and incubation with target oligonucleotide and half-capture-probe-coated Ag nanoparticles; (b) SERS spectra of an Ag thin film — **a**; in the presence of the Raman-active label (F) — **b**; and in the presence of target-mismatch — **c**. (Reprinted with permission from ref. 55. Copyright 2008 American Chemical Society.) See also Color Insert.

SERRS is another mode of Raman scattering, and it increases the sensitivity of the technique even further. In SERRS-based DNA detection assays, oligonucleotides are labeled with a Raman-active dye that has absorption in the UV/Vis region of the spectrum; it preferably absorbs the light at the wavelength of the laser light used in the experiment. The enhancement factor can increase up to 10^3 in a SERS experiment under proper conditions, and this enhancement factor is additional to SERS enhancement. The instrumentation for SERRS is the same as the standard Raman instrumentation. In order to benefit from both resonance enhancement and SERS enhancement, the surface plasmon excitation of the SERS substrate and the absorption of the molecule of interest must be overlapped. Graham and Faulds have recently reviewed the fundamentals of DNA detection using SERRS, and Graham's group reported the synthesis and use of a number of molecules that are utilized in quantitative and qualitative DNA detection [56]. Therefore, the use of AgNPs is preferred due to the absorption maxima of their colloidal suspensions at around 420 nm.

4.3.3 Cellular Applications

The use of nanoparticles as novel tools in drug and gene delivery, cellular and biomedical imaging, and therapeutics are under intense investigations. This is mostly due to their promise to resolve the current challenging problems facing humankind. One of the most significant impacts of the applications of nanoparticles is expected in medicine. The problems associated with the synthesis and characterization of nanoparticles are now partly resolved, and new protocols for the synthesis of a range of nanoparticles with reasonable uniformity and a number of techniques for nanoparticle characterization are now available. The most significant advances are made in scanning probe microscopy such as AFM, near-field scanning optical microscopy (NSOM), and Raman NSOM to study biological structures in their native environment at nanometer resolution.

As a basic unit of life, the cellular processes in a living eukaryotic cell have been a focal point for a long time. Almost all molecular biology runs through the use of living cells, and the recent advances in this field have revolutionized medicine. The progress made in medicine has provided a wealth of information, shining the light on a number of diseases more than ever. However, a comprehensive and fundamental understanding of the molecular basis of disease formation and progression remains a challenge. In most molecular biology experiments, it is necessary to disrupt the cell for analysis and characterization. After lysis, the integrity and naturally occurring processes of the cell are lost. Therefore, there is a need for the development of new approaches to study cellular processes or the distribution of biochemical species in the living cell in real time.

Three requirements must be fulfilled for the techniques to be used for single-living-cell characterization. First, since eukaryotic cells are compartmentalized and several biochemical processes are underway simultaneously in different parts of the cell, the technique must have optimal spatial resolution. Second, since the cellular processes are very fast, the technique must have the temporal resolution that is compatible with the cellular processes. Third, since the concentration of cellular species is very low, the sensitivity of the technique must be as low as a single molecule level.

If not wholly, SERS can partly be a solution to these challenges. Since the SERS measurement is limited with the close vicinity of noble nanoparticles and their aggregates, the spatial resolution is limited with the nanometer scale independent from the laser spot. As long as the biochemical processes proceed in the vicinity of the nanoparticles or their aggregates, the changes can be monitored. Finally, with its competing sensitivity with fluorescence and high immunity to water, it is an excellent technique to study living cells.

Although SERS might comply with requirements for single-living-cell studies, the main limitation of the technique is the necessity of the introduction of colloidal nanoparticles into living cells. This adds further complexity to a cellular system that is already chaotic in nature. In addition to noble metal colloidal nanoparticles, the microelectrodes and fiber-optic tips coated with a noble metal are also used for cellular SERS investigations [57, 58].

The major obstacle to the use of noble metal nanoparticles in living cells is their fate in the cell and the molecular species in contact with the nanoparticles. Although the cellular uptake of nanoparticles is fairly well understood, their fate in the cell is not completely understood yet. Most of the recent studies indicate that the majority of nanoparticles remain in endosomes in the form of aggregates. Both transmission electron microscope (TEM) and confocal microcopy studies confirm that this is the case. However, it is possible that some of the nanoparticles trapped in the endosomes may escape and come into contact with other cellular organelles and molecular species in the cytosol.

Most of the reported applications of intracellular SERS use non-modified noble metal nanoparticles. A small volume of AgNP or AuNP suspension is added into the growth medium. The aggregates of these nanoparticles are taken into the cells via endocytosis or phagocytosis [59]. The surface charge properties of nanoparticles play a critical role from the point they are added to the growth media. AuNPs or AgNPs prepared with citrate reduction possess negative surface charge due to the adsorption of citrate ions on the nanoparticle surface. When a solution possessing ionic species is added to the colloidal suspension, a change in surface charge properties of nanoparticles is inevitable and ionic layers from the nanoparticles surface into the solution with decreasing density (Debye-Hückel ionic layers) are formed. This change in the surface charge of nanoparticles alters the behavior of nanoparticles in the solution, and they form aggregates and precipitate as repulsion forces are diminished as a result of the formation of a charge double layer on the surface.

When a non-modified nanoparticle suspension is added to the growth media, a change in the behavior of nanoparticles is expected due to the presence of ionic species and proteins such as BSA. The major route for the cellular uptake of nanoparticles is now well understood and is mostly through endocytosis [60]. As the cell takes nanoparticles or their aggregates into the cell, they are transported to the endosome and remain there for some time. Before being able to identify the origin of the bands on the intracellular SERS spectra, the final destination of the gold nanoparticles must be identified. The feasibility of using SERS for single-cell analysis goes back to the early 1990s [61, 62, 63]. In those early studies, the distribution of antitumor drugs such as doxorubicin was investigated in cancerous cell lines using SERS. Citrate-reduced AgNPs were used in the SERS experiments. The use of AuNPs along with a near-infrared (NIR) laser was later demonstrated [64]. A more recent study combined SERS and TEM for better understanding of the localization and behavior of AuNPs in immortalized rat renal proximal tubule cells (IRPT) and in the mouse macrophage cell line J774 [65]. This study concluded that AuNPs entering cells through endocytosis remained in the endosomes and as the exposure time of the cells to nanoparticles increased, the aggregate size and morphology changed. Analysis of the SERS spectra obtained from AuNPs aggregates in endosomes was performed. It should be noted here that when AuNPs are exposed to cells, they might escape from endosomes and interact with other organelles [66, 67].

When nanoparticles are introduced into a cell through endocytosis, they seem to prefer to stay in endosomes, at least for a certain period of time. Since the SERS signal is acquired from their location, most of the signal originates from the inside of endosomes. An alternative approach could be intercellularly grown nanoparticles [68]. Since gold ions are introduced into the cell, when the gold ions are reduced, their localization will be quite different compared to the endosomal uptake. Since the gold nanoparticles are more diversely distributed in the cell, the acquired SERS spectra contain much richer information.

The use of microtips or fiber-optic bundles is another way of acquisition of SERS spectra from a single living cell. A tapered optical fiber down to 100 nm covered with a thin layer of silver film using physical vapor deposition was used for the analysis of intracellular regions [69]. A fiber-optic bundle capable of measuring as many as 30,000 individual subdiffraction-limited locations on a sample's surface simultaneously was constructed. The bundle surface was coated with roughened silver thin film to allow the SERS measurements [70]. The authors claim that these probes can be used in cellular biology, medicine, food science, and materials science.

An Au core-Ag shell nanoparticle was synthesized to target cancer cells for cell differentiation [71]. Rhodamine 6G, a highly Raman-active dye, was first adsorbed onto 13 nm average-size gold nanoparticles, and then BSA was adsorbed to stabilize the nanoparticles. Finally, these nanoparticles were coated with a thin silver shell. Then antibodies targeting the surface proteins of HEK293 cancer cells were chemically attached to the silver surface. The prepared SERS probes were used for the selected targeting of cancer cells.

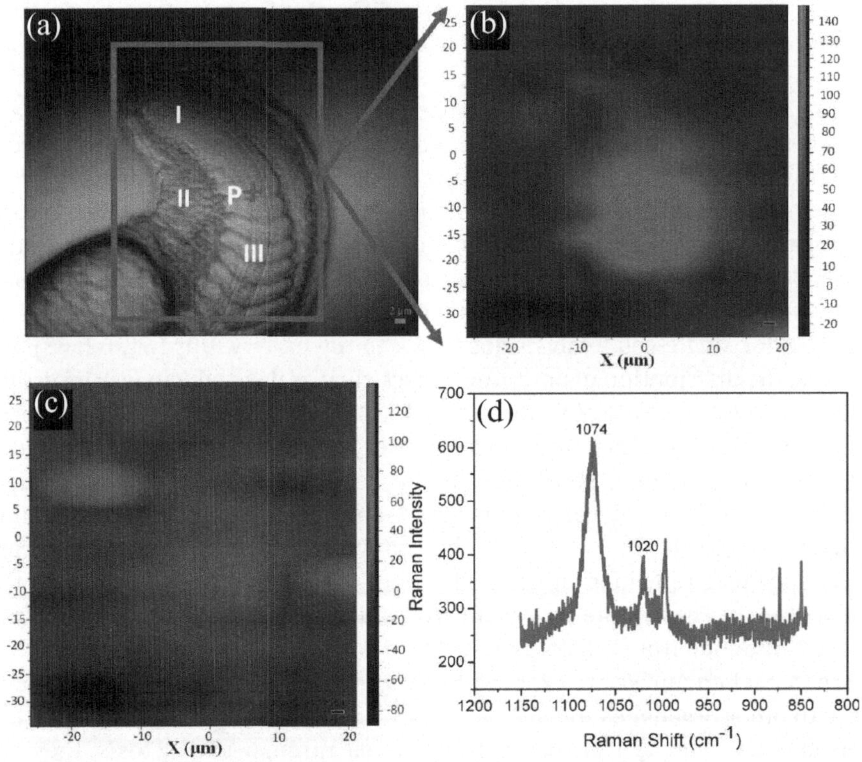

Figure 4.6 Raman-mapping images constructed from the SERS spectra of gold nanoprobes labeled with two different reporter molecules in the posterior region of a zebrafish embryo. (a) The optical image of the zebrafish embryo (18–20 hpf); the areas marked as I, II, and III representing presomitic mesoderim, blood island, and somite, respectively; (b) SERS-intensity map of the C-C vibrational band from the SERS of mercaptobenzoic acid at 1078 cm^{-1}; (d) a SERS-intensity map of the ring-breathing vibration band from the SERS of mercaptopyridine at 1020 cm^{-1}; (d) a typical SERS spectrum obtained from the point P seen in panel A. (Reprinted with permission from ref. 72. Copyright 2010 American Chemical Society).

In a more recent study, the distribution of gold nanoprobes in zebrafish embryos was demonstrated by Wang *et al.* [72]. The mercaptobenzoic acid-modified AuNPs were injected into the cell region of the embryo and the distribution of labeled AuNPs was monitored during the growth stages of the embryo by Raman mapping. The changes in the SERS spectra of the labeled AuNP probes were monitored in several locations in the embryo, and the SERS spectra of labeled AuNPs were used to construct the mapping images. Finally, the multiplex distribution of AuNPs labeled with mercaptobenzoic acid and mercaptopyridine was demonstrated in different organs of the embryo. Figure 4.6 shows the Raman-mapping image of the distribution of labeled Au nanoprobes in zebrafish embryo. This study is an important application demonstrating the capability of the technique in cellular applications.

4.3.4 Detection and Identification of Microorganisms

Fast and accurate identification of microorganisms has clinical importance. SERS has been proven to be a technique that can provide fingerprint information from viruses [3, 74] and bacteria. Upon reproducible acquisition of the SERS spectra from microorganisms, identification can be performed without the need for extra effort. The first SERS spectrum of bacteria was reported by Holt and Cotton [75]. In the light of successive studies, the feasibility of identification and classification of bacteria causing urinary tract infection was demonstrated by combining statistical methods [76–86]. Later, reproducible SERS spectra acquisition from bacterial samples was demonstrated by Culha *et al.* [85–86]. The critical component in a bacterial SERS experiment is the sample preparation. The problem that causes the irreproducibility of the SERS spectra in molecular detection worsens the bacterial detection and identification. The most suitable SERS substrate in bacterial SERS experiments is the colloidal suspensions of AuNPs or AgNPs. However, when the colloidal suspension and bacterial samples are mixed, a heterogeneous mixture is generated. When the laser spot is moved on the sample, variations in SERS spectra are observed and this makes the comparison of bacterial SERS difficult. The two sample-preparation methods, one of mixing the bacterial samples with increased colloidal suspension concentration (4x) and the second of "convective-assembly" of both bacteria and silver nanoparticles on glass surfaces, were developed by Culha *et al.* [86]. The SERS spectra obtained from such samples are highly reproducible spectra that can be used for bacterial identification without extensive statistical data treatment. Figure 4.7 shows the self-assembled *S. Chonii* and silver nanoparticles and the SERS spectra obtained from such a surface. Instead of acquiring several spectra from the same sample and averaging, only one SERS spectrum can be used for the identification. The feasibility of using the same sample-preparation method was also tested for yeast [87]. It was found that the increased colloidal suspension concentration helped to acquire reproducible SERS spectra that can be used for identification. In a most recent report, the identification of bacteria in bacterial mixtures was also reported [88].

The first report concerning the SERS of viruses appeared in 1993 [89]. In that study, the characterization of turnip virus coat protein on colloidal silver nanoparticles was demonstrated. The characterization of polydrin of *Trabala vishnou* nuclear polyhedrosis virus and *Dendrolimus punctatus* nuclear polyhedrosis virus by adsorbing onto colloidal silver nanoparticle surface was also reported [90]. Recently, Ag silver nanorods array constructed with the oblique angle deposition (OAD) method were used for viral detection using SERS [91]. In this method, there is no need for a capture probe. The extracted viruses are left onto the nanorod arrays, and the SERS spectra are acquired. These spectra are used for the identification and discrimination of viruses by comparing the SERS spectra obtained from standard viruses.

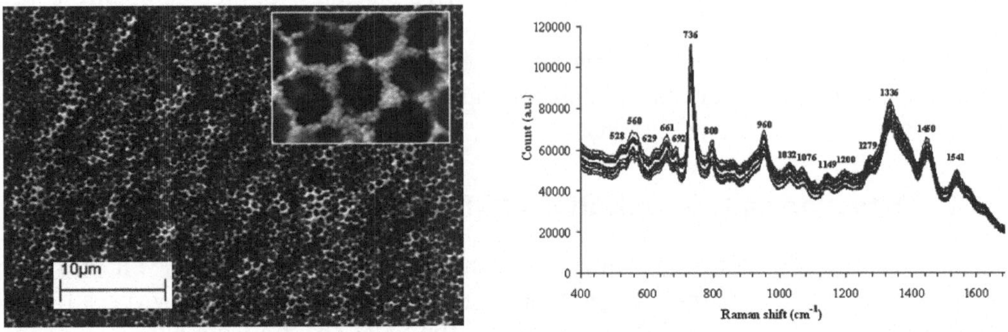

Figure 4.7 (a) "Convective-assembly" of *S. Cohnii* and silver nanoparticles; (b) SERS spectra obtained from the sample seen in (a). (Reprinted with permission from ref. 86. Copyright 2008 American Chemical Society.)

SERS-based immunoassays for the detection of hepatitis B virus were proposed [92]. The detection of feline calicivirus with a sandwich-type immunoassay was also reported [93]. The detection limit of the assay was claimed to be $\sim 1 \times 10^6$ viruses/mL. The immunoassay-based SERS-detection schemes are very similar to the immunoassays discussed earlier in this chapter.

4.3.5 Tissue Differentiation

Tissues are complex structures composed of several different types of specialized cells. In each organ, tissues are composed of a collection of cells that perform the specific tasks dictated by their specialization. A disease or disorder starts in a cell at the molecular level first, and it goes undetected until the individual feels the symptoms, which indicates the spread of the problem in one or more organs. SERS is powerful enough to detect the chemical changes in a given location. Therefore, it can be used to detect the change in the molecular composition of tissue due to a disease or a disorder.

Tissue differentiation with spectroscopic techniques such as infrared (IR) and bulk-Raman has been widely investigated for a long time [94–98]. Raman spectroscopy is especially advantageous over IR spectroscopy with compatibility with water and minimal sample-preparation steps. However, Raman spectroscopy can suffer from autofluorescence with the use of a laser line in the UV/visible region of the spectrum. A laser line in the NIR or IR region can help to diminish the fluorescence and also increases the penetration depth of light into the tissue.

The use of SERS for tissue differentiation is a new concept, and there is only one report for its use for this purpose [99]. In that study, a slice of tissue was placed on a surface and silver colloidal suspension was spotted on the top of the tissue and SERS spectra acquired. The utility of SERS as tissue differentiation has been investigated by Culha and coworkers [100]. In their studies, a millimeter-size tissue was frozen in liquid nitrogen and crushed with a pestle. The concentrated silver colloidal suspension was added and mixed with the crushed tissue. This mixture was spotted on a substrate and SERS spectra were acquired. Figure 4.8 shows the SERS spectra obtained from the different sections of rat brain. As it is seen, the spectral features are significant enough to differentiate their locations in the brain. This indicates that SERS is sensitive enough to observe the molecular differences in tissue. The SERS spectrum from each organ has different spectral features, indicating the specialization of the organs that the tissue is obtained from. This study is further expanded for the differentiation of several tumors and healthy tissue samples [101].

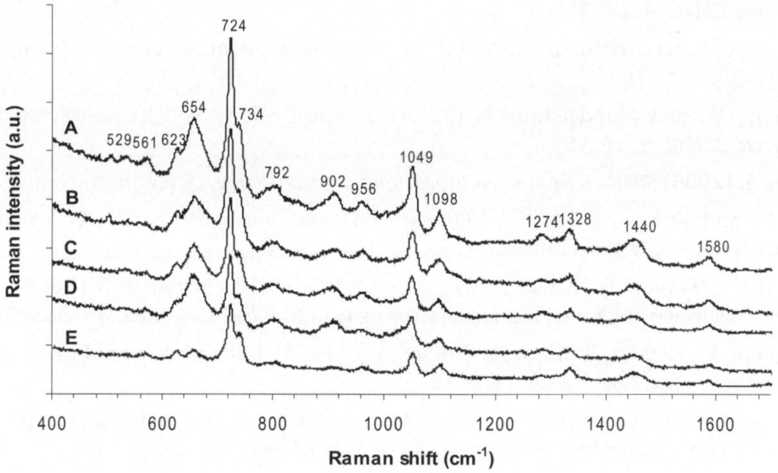

Figure 4.8 SERS spectra of **A** the cortex, **B** the brain stem, **C** the glandular lobe of hypophysis, **D** subcortial tissue, and **E** the cerebellum.

4.4 Concluding Remarks and Future Outlook

SERS has experienced a dramatic development in the last decade. The advances in plasmonics and a better understanding of its fundamentals are the major factors for this advancement. Since it provides high sensitivity and "fingerprint" information about the molecule of interest, it is becoming an important technique not only for detection but also for characterization. Low water interference is another important factor that makes the technique suitable for applications in biological samples.

The biggest challenge in the application of SERS as an analytical technique is the difficulty in the reproducible preparation of substrates with high sensitivity. This challenge becomes more apparent in biological applications because of the nature of the biological molecules. The best approach for biological applications is to design and construct problem-specific substrates. This especially becomes an important consideration in the detection of molecular species in biological systems such as in living cells. As the technology to fine-tune plasmonic structures is advanced and becomes more readily available, it will become possible to construct novel SERS-active structures not only for the detection of several biologically important molecules and species but also for the monitoring of the cellular processes in a single living cell.

Acknowledgments

The author gratefully acknowledges the contribution of Associate Professors Andrew J. Harvey, İlknur Sur, and Mehmet Kahraman during the preparation of this chapter.

References

1. Petry, R., Schmitt, M., Popp, J. (2003). Raman spectroscopy-a prospective tool in the life sciences, *Chem. Phys. Chem.*, **4**, 14–30.
2. Raman, C. V., and Krishnan, K. S. (1928). A new type of secondary radiation, *Nature*, **121**, 501–502.
3. Landsberg, G., and Mandelstam, L. (1928). A novel effect of light scattering in crystals, *Naturwissenschaften*, **16**, 557.
4. Ricardo, A. (2006) *Surface-Enhanced Vibrational Spectroscopy*, Wiley-Interscience, U. S.
5. Lasch, P., and Kneipp, J. (2008). *Biomedical Vibrational Spectroscopy*, Wiley-Interscience, Canada.
6. Maher, R. C., Galloway, C. M., Le Ru, E. C., Cohena, L. F., and Etchegoin, P. G. (2008). Vibrational pumping in surface enhanced Raman scattering (SERS), *Chem. Soc. Rev.*, **37**, 965–979.
7. Fleischman, M., Hendra, P. J., and McQuillan, A. J. (1974). Raman spectra of pyridine adsorbed at a silver electrode, *Chem. Phys. Lett.*, **26**, 163–166.
8. Jeanmaire, D. L., and Van Duyne, R. P. (1977). Heterocyclic, aromatic, and aliphatic-amines absorbed on anodized silver electrode, *J. Electroanal. Chem.*, **84**, 1–20.
9. Albrecht, M. G., and Creighton, J. A. (1977). Anomalously intense Raman spectra of pyridine at a silver electrode, *J. Am. Chem. Soc.*, **99**, 5215.
10. Moskovits, M. (1978). Surface roughness and the enhanced intensity of Raman scattering by molecules adsorbed on metals, *J. Chem. Phys.*, **69**, 4159.
11. Tian, Z. Q., Yang, Z. L., Yen, B., and Wu, D. Y. (2006). SERS from transition metals and excited by ultraviolet light, *Top. Appl. Phys.*, **103**, 125–146.
12. Campion, A., and Kambhampati, P. (1998). Surface-enhanced Raman scattering, *Chem. Soc. Rev.*, **27**, 241–250.
13. Moskovits, M. (2005). Surface-enhanced *Raman spectroscopy*: a brief retrospective, *J. Raman Spectrosc.*, **36**, 485–496.
14. Baker, G. A., and Moore, D. S. (2005). Progress in plasmonic engineering of surface-enhanced Raman-scattering substrates toward ultra-trace analysis. *Anal. Bioanal. Chem.*, **382**, 1751–1770.
15. Banholzer, M. J., Millstone, J. E., Qin, L., and Mirkin, C. A. (2008). Rationally designed nanostructures for surface-enhanced Raman spectroscopy, *Chem. Soc. Rev.*, **37**, 885–897.
16. Fromm, D. P., Sundaramurthy, A., Kinkhabwala, A., Schuck, P. J., Kino, F. S., and Moerner, W. E. (2006). Exploring the chemical enhancement for surface-enhanced Raman scattering with Au bowtie nanoantennas, *J. Chem. Phys.*, **124**, 061101.
17. Félidj, N., Aubard, J., Lévi, G., Krenn, J. R., Hohenau, A., Schider, G., Leitner, A., and Aussenegg, F. R. (2003). Optimized surface-enhanced Raman scattering on gold nanoparticle arrays, *Appl. Phys. Lett.*, **82**, 3095–3097.
18. Dick, L. A., McFarland, A. D., Haynes, C. L., and Van Duyne, R. P. (2002). Metal film over nanosphere (mfon) electrodes for surface-enhanced Raman spectroscopy (SERS): improvements in surface nanostructure stability and suppression of irreversible loss, *J. Phys. Chem. B*, **106**, 853–860.

19. Haes, A. J., Haynes, C. L., McFarland, A. D., Schatz, G. C., Van Duyne, R. P., and Zou, S. (2005). Plasmonic materials for surface-enhanced sensing and spectroscopy, *MRS Bull.*, **30**, 368–375.

20. McFarland, A. D., Young, M. A., Dieringer, J. A., and Van Duyne, R. P. (2005). Wavelength-canned surface-enhanced Raman excitation spectroscopy, *J. Phys. Chem. B*, **109**, 11279–11285.

21. Zhang, X., Zhao, J., Whitney, A. V., Elam, J. W., and Van Duyne, R. P. (2006). Ultrastable substrates for surface-enhanced Raman spectroscopy: Al_2O_3 overlayers fabricated by atomic layer deposition yield ımproved anthrax biomarker detection, *J. Am. Chem. Soc.*, **128**, 10304–10309.

22. Alvarez-Puebla, R., Cui, B., Bravo-Vasquez, J.-P., Veres, T., and Fenniri, H. (2007). Nanoimprinted SERS-active substrates with tunable surface plasmon resonances, *J. Phys. Chem. C*, **111**, 6720–6723.

23. Qin, L., Zou, S., Xue, C., Atkinson, A., Schatz, G. C., and Mirkin, C. A. (2006)., Designing, fabricating and imaging Raman hot spots, *Proc. Natl. Acad. Sci. USA*, **103**, 13300–13303.

24. Qin, L., Banholzer, M. J., Millstone, J. E., and Mirkin, C. A. (2007). Nanodisk codes, *Nano Lett.*, **7**, 3849–3853.

25. Nie, S. M., and Emory, S. R. (1997). Probing single molecules and single nanoparticles by surface enhanced Raman scattering, *Science*, **275**, 1102–1106.

26. Kneipp, K., Wang, Y., Kneipp, H., Perelman, L. T., Itzkan, I., Dasari, R., and Feld, M. S. (1997). *Single molecule detection using surface-enhanced Raman* scattering (SERS)., *Phys. Rev. Lett.*, **78**, 1667–1670.

27. Ward, D. R., Grady, N. K., Levin, C. S., Halas, N. J., Wu,Y., Nordlander, P., and Natelson, D. (2007). Electromigrated nanoscale gaps for surface-enhanced Raman spectroscopy, *Nano Lett.*, **7**, 1396–1400.

28. Kelly, K. L., Coronado, E., Zhao, L. L., and Schatz, G. C. (2003). The optical properties of metal nanoparticles: the influence of size, shape, and dielectric environment, *J. Phys. Chem. B*, **107**, 668–677.

29. Maier, S. A. (2007). *Plasmonics: Fundamentals and Applications*, Springer, New York.

30. Bhandari, D., Walworth, M. J., and Sepaniak, M. J. (2009). Dual function surface-enhanced Raman active extractor for the detection of environmental contaminants, *Appl Spectrosc.*, **63**, 571–578.

31. Kneipp, J., Kneipp, H., and Kneipp, K. (2008). SERS- a single-molecule and nanoscale tool for bioanalytics, *Chem Soc Rev.*, **37**(5), 1052–1060.

32. Han, X. X., Huang, G. G., Zhao, B., and Ozaki, Y. (2009). Label-free highly sensitive detection of proteins in aqueous solutions using surface-enhanced Raman scattering, *Anal. Chem.*, **81**, 3329–3333.

33. Han, X. X., Jia, H. Y., Wang, Y. F., Lu, Z. C., Wang, C. X., Xu, W. Q., Zhao, B., and Ozaki Y. (2008). Analytical technique for label-free multi-protein detection based on western blot and surface-enhanced Raman scattering, *Anal. Chem.*, **80**, 2799–2804.

34. Methogo, R. M., Dufresne-Martin, G., Leclerc, P., Leduc, R., and Klarskov, K. J. (2005). Mass spectrometric peptide fingerprinting of proteins after western blotting on polyvinylidene fluoride and enhanced chemiluminescence detection, *Proteome Res.*, **4**, 2216–2224.

35. Nakanishi, T., Ohtsu, I., Furuta, M., Ando, E., and Nishimura, O. (2005). Direct MS/MS analysis of proteins blotted on membranes by a matrix-assisted laser desorption/ionization-quadrupole ion trap-time-of-flight tandem mass spectrometer, *J. Proteome Res.*, **4**, 743–747.

36. Ohtsu, I., Nakanisi, T., Furuta, M., Ando, E., and Nishimura, O. (2005). *Direct matrix-assisted laser desorption/ionization time-of-flight mass spectrometric identification of proteins on membrane detected by Western blotting and lectin blotting, J. Proteome Res.*, **4**, 1391–1396.

37. Kahraman, M., Sur, I., and Culha, M. (2010). Label-free detection of proteins from self-assembled protein-silver nanoparticle structures using surface-enhanced Raman scattering, *Ana. Chem.*, **82**(18), 7596–7602.

38. Han, X. X., Cai, L. J., Guo, J., Wang, C. X., Ruan, W. D., Han, W. Y., Xu, W. Q., Zhao, B., and Ozaki, Y. (2008). Fluorescein isothiocyanate linked immunoabsorbent assay based on surface-enhanced resonance Raman scattering, *Anal. Chem.*, **80**, 3020–3024.

39. Porter, M. D., Lipert, R. J., Siperko, L. M., Wang, G., and Narayanan, R. (2008). SERS as a bioassay platform: fundamentals, design, and applications, *Chem. Soc. Rev.*, **37**, 1001–1011.

40. Stamou, D., Gourdon, D., Liley, M., Burnham, N. A., Kuik, A., Vogel, H., and Duschl, C. (1997). Uniformly flat gold surfaces: imaging the domain structure of organic monolayers using scanning force microscopy, *Langmuir*, **13**, 2425–2428.

41. Ni, J., Lipert, R. J., Dawson, G. B., and Porter, M. D. (1999). Immunosassay readout method using extrinsic Raman labels adsorbed on immunogold colloids, *Anal. Chem.*, **71**, 4903–4908.

42. Grubisha, D. S., Lipert, R. J., Park, H. Y., Driskell, J., and Porter, M. D. (2003). Femtomolar detection of prostate-specific antigen: an immunoassay based on surface-enhanced Raman scattering and immunogold labels, *Anal. Chem.*, **75**, 5936–5943.

43. Park, H.-Y. (2005). *Chip-Scale Bioassays Based on Surface-Enhanced Raman Scattering: Fundamentals and Applications*, Iowa State University, Ames.

44. Chen, J.-W., Lei, Y., Liu, X.-J., Jiang, J.-H., Shen, G.-L., and Yu, R.-Q. (2008). Immunoassay using surface-enhanced Raman scattering based on aggregation of reporter-labeled immunogold nanoparticles, *Anal. Bioanal. Chem.*, **392**,187–193.

45. Vo-Dinh, T., Houck, K., and Stokes, D. L. (1994). surface-enhanced Raman gene probes, *Anal. Chem.*, **66**, 3379–3383.

46. Vo-Dinh, T. L., Allain, R., and Stokes, D. L. (2002). Cancer gene detection using surface-enhanced Raman scattering (SERS), *J. Raman Spectrosc.*, **33**, 511–516.

47. Culha, M., Stokes, D., Allain, L. R., and Vo-Dinh, T. (2003). Surface-enhanced Raman scattering substrate based on a self-assembled monolayer for use in gene diagnostics, *Anal. Chem.*, **75** (22), 6196–6201.

48. Cao, Y. W. C., Jin, R., and Mirkin, C. A. (2002). Nanoparticles with Raman spectroscopic fingerprints for DNA and RNA detection, *Science*, **297**, 1536–1540.

49. Wabuyele, M., and Vo-Dinh, T. (2005). Detection of HIV type 1 DNA sequence using plasmonics nanoprobes, *Anal. Chem.*, **77**, 7810–7815.

50. Sun, L., Yu, C., and Irudayaraj, J. (2008). Raman multiplexers for alternative gene splicing, *Anal. Chem.*, **80**, 3342–3349.

51. Orban, T. I., and Olah, E. (2001). Expression profiles of BRCA1 splice variants in asynchronous and in G1/S synchronized tumor cell lines, *Biochem. Biophys. Res. Commun.*, **280**, 32–38.

52. Mixon, M., Kittrell, F., and Medina, D. (2000). Expression of Brca1 and splice variant Brca1delta11 RNA levels in mouse mammary gland during normal development and tumorigenesis, *Oncogene*, **19**, 5237–5243.

53. Sun, L., and Irudayaraj, J. (2009). Quantitative surface-enhanced Raman for gene expression estimation, *Biophys. J.*, **96**(11), 4709–4716.

54. Sun, L., and Irudayaraj, J. (2009). PCR-free quantification of multiple splice variants in a cancer gene by surface-enhanced Raman spectroscopy, *J. Phys. Chem. B*, **113**(42), 14021–14025.

55. Braun, G., Lee, S. J., Dante, M., Nguyen, T.-Q., Moskovits, M., and Reich, N. O. (2007). Surface-enhanced Raman spectroscopy for DNA detection by nanoparticle assembly onto smooth metal films, *J. Am. Chem. Soc.*, **129**, 6378–6379.

56. Graham, D., and Faulds, K. (2008). Quantitative SERRS for DNA sequence analysis, *Chem. Soc. Rev.*, **37**, 1042–1051.

57. Beljebbar, A., Sockalingum, G. D., Morjani, H., and Manfait, M. (1999). Raman and SERS microspectroscopy on living cells: a promising tool toward cellular drug response and medical diagnosis, *Proc. SPIE.*, **3608**, 175–184.

58. Kasili, P.M., Song, J. M., and Vo-Dinh, T. (2004). Optical sensor for the detection of caspase-9 activity in a single cell, *J. Am. Chem. Soc.*, **126**, 2799–2806.

59. Osaki, F., Kanamori, T., Sando, S., Sera, T., and Aoyana, Y. (2004). A quantum dot conjugated sugar ball and its cellular uptake on the size effects of endocytosis in the subviral region, *J. Am. Chem. Soc.*, **126**, 6520–6521.

60. Chithrani, B. D., Ghazani, A. A., and Warren, C. W. C. (2006). Determining the size and shape dependence of gold nanoparticle uptake into mammalian cells, *Nano Lett.*, **6**(4), 662–668.

61. Manfait, M., Morjani, H., Millot, J. M., Debal, V., Angiboust, J. F., and Nabiev, I. (1990). Drug target interactions on a single living cell. An approach by optical microspectroscopy, *Proc. SPIE-Int. Soc. Opt. Eng.*, **1403**, 695–707.

62. Manfait, M., Morjani, H., and Nabiev, I. (1992). Molecular events on simple living cancer cells as studied by spectrofluorometry and micro-SERS Raman spectroscopy, *J. Cell. Pharmaco.*, **3**, 120–125.

63. Nabiev, I., Morjani, H., and Manfait, M. (1991). Selective analysis of antitumor drug interaction with living cancer cells as probed by surface-enhanced Raman spectroscopy, *Eur. Biophys. J.*, **19**, 311.

64. Kneipp, K., Haka, A. S., Kneipp, H., Badizadegan, K., Yoshizawa, N., Boone, C., Shafer-Peltier, K. E., Motz, J. T. Dasari, R. R., and Feld, M. S. (2002). Surface-enhanced Raman spectroscopy in single living cells using gold nanoparticles, *Appl. Spectrosc.*, **56**, 150–154.

65. Kneipp, J., Kneipp, H., McLaughlin, M., Brown, D., and Kneipp, K. (2006). In vivo molecular probing of cellular compartments with gold nanoparticles and nanoaggregates, *Nano Lett.*, **6**, 2225–2231.

66. AshaRani, P. V., Mun, L. K. G., Hande, P. M., and Valiyaveettil, S. (2009). Cytotoxicity and genotoxicity of silver nanoparticles in human cells, *ACS Nano*, **3**(2), 279–290.

67. Karataş, O. F., Sezgin, E., Aydin, O., and Culha, M. (2009). Interaction of gold nanoparticles with mitochondria, *Colloids Surf. B*, **71**(2), 315–318.

68. Shamsaie, A., Jonczyk, M., Sturgis, J., Robinson, P. J., and Irudayaraj, J. (2007). Intracellularly grown gold nanoparticles as potential surface-enhanced Raman scattering probes, *J. Biomed. Opt.*, **12**(2), 020502.

69. Stokes, D. L., Chi, Z., and Vo-Dinh, T. (2004). Surface-enhanced Raman scattering-inducing nanoprobe for spectrochemical analysis, *Appl. Spectrosc.*, **58**, 292.

70. Hankus, M. E., Li, H., Gibson, G. J., and Cullum, B. M. (2006). Surface enhanced Raman scattering (SERS)-based nanoprobe for high resolution, non-scanning chemical imaging, *Anal. Chem.*, **78**, 7535–7546.

71. Lee, S., Kim, S., Choo, J., Shin, S. Y., Lee, Y. H., Choi, H. Y., Ha, S., Kang, K., and Oh, C. H. (2007). Biological imaging of HEK293 cells expressing PLCgamma 1 using surface-enhanced Raman microscopy, *Anal. Chem.*, **79**, 916–922.

72. Wang, Y., Seebald, J. L., Szeto, D. P., and Irudayaraj, J. (2010). Biocompatibility and biodistribution of surface-enhanced Raman scattering nanoprobes in zebrafish embryos: in vivo and multiplex imaging, *ACS Nano*, **4**(7), 4039–4053.

73. Porter, M. D., Driskell, J. D., Kwarta, K. M., Lipert, R. J., Neill, J. D., and Ridpath, J. F. (2006). Detection of viruses: atomic force microscopy and surface enhanced Raman spectroscopy, *Dev. Biol. (Basel).*, **126**, 31–39.
74. Shanmukh, S., Jones, L., Driskell, J., Zhao, Y. P., Dluhy, R., and Tripp, R. A. (2006). Rapid and sensitive detection of respiratory virus molecular signatures using a silver nanorod array SERS substrate, *Nano Letts.*, **6**, 2630–2636.
75. Holt, R. E., and Cotton, T. M. (1989). Surface-enhanced resonance Raman and electrochemical investigation of glucose oxidase catalysis at a silver electrode, *J. Am. Chem. Soc.*, **111**, 2815–2821.
76. Efrima, S., and Bronk, B. V. (1998). Silver colloids impregnating or coating bacteria, *J. Phys. Chem. B*, **102**, 5947–5950.
77. Jarvis, R. M., and Goodacre, R. (2004). Rapid discrimination of bacteria using surface enhanced Raman spectroscopy, *Anal. Chem.*, **76**, 40–47.
78. Jarvis, R. M., Brooker, A., and Goodacre, R. (2004). SERS for bacterial characterisation using a scanning electron microscope with a Raman spectroscopy interface, *Anal. Chem.*, **76**, 5198–5202.
79. Zeiri, L., Bronk, B. V., Shabtai, Y., Eichler, J., and Efrima, S. (2004). Surface-enhanced Raman spectroscopy as a tool for probing specific biochemical components in bacteria, *Appl. Spectrosc.*, **58**, 33–40.
80. Zeiri, L., and Efrima, S. (2005). Surface-enhanced Raman spectroscopy of bacteria: the effect of excitation wavelength and chemical modification of the colloidal milieu, *J. Raman Spectrosc.*, **36**, 667–675.
81. Sengupta, A., Laucks, M. L., and Davis, E. J. (2005). Surface-enhanced Raman *spectroscopy* of bacteria and pollen, *Appl. Spectrosc.*, **59**, 1016–1023.
82. Sengupta, A., Mujacic, M., and Davis, E. J. (2006). Detection of bacteria by surface enhanced Raman spectroscopy, *Anal. Bioanal. Chem.*, **386**, 1379–1386.
83. Laucks, M. L., Sengupta, A., Junge, K., Davis, E. J., and Swanson, B. D. (2005). Comparison of psychrophilic arctic sea-ice bacteria and common mesophylic bacteria using surface-enhanced Raman spectroscopy, *Appl. Spectrosc.*, **59**, 1222.
84. Premasiri, W. R., Moir, D. T., Klempner, M. S., Krieger, N., Jones II, G., and Ziegler, L. D. (2005). Characterization of the surface enhanced Raman scattering (SERS) of bacteria, *J. Phys. Chem. B*, **109**, 312–320.
85. Kahraman, M., Yazici, M. M., Sahin, F., Bayrak, O. F., and Culha, M. (2007). Reproducible surface-enhanced Raman scattering spectra of bacteria on aggregated silver nanoparticles, *Appl. Spectrosc.*, **61**, 479–485.
86. Kahraman, M., Yazici, M. M., Sahin, F., and Culha, M. (2008). Convective-assembly of bacteria for surface-enhanced Raman scattering, *Langmuir*, **3**, 894–901.
87. Sayin, I., Kahraman, M., Sahin, F., Yurdakul, D., and Culha, M. (2009). Characterization of yeast species using surface-enhanced Raman scattering, *Appl. Spectrosc.*, **63**(11), 1276–1282.
88. Cam, D., Keseroglu, K., Kahraman, M., Sahin, F., and Culha, M. (2010). Multiplex identification of bacteria in bacterial mixtures with surface-enhanced Raman scattering, *J. Raman Spectrosc.*, **41**(5), 484–489.
89. Deng, H., He, Q., Xu, Z., Wang, X., and Sheng, R. (1993). The study of turnip mosaic virus coat protein by surface enhanced Raman spectroscopy, *Spectrochim. Acta, Part A*, **49A**, 1709–1714.
90. Bao, P. D., Huang, T. Q., Liu, X. M., and Wu, T. Q. (2001). Surface-enhanced *Raman spectroscopy* of insect nuclear polyhedrosis virus, *J. Raman Spectrosc.*, **32**, 227–230.

91. Shanmukh, S., Jones, L., Driskell, J., Zhao, Y., Dluhy, R., and Tripp, R. A. (2006). Rapid and sensitive detection of respiratory virus molecular signatures using a silver nanorod array SERS substrate, *Nano Lett.*, **6**(11), 2630–2636.

92. Xu, W., Xu, S., Ji, X., Song, B., Yuan, H., Ma, L., and Bai, Y. (2005). Preparation of gold colloid monolayer by immunological identification, *Colloids Surf., B*, **40**(3–4), 169–172.

93. Crow, P., Molckovsky, A., Stone, N., Uff, J., Wilson, B., and Wongkeesong, L.-M. (2005). Assessment of fiberoptic near-infrared Raman spectroscopy for diagnosis of bladder and prostate cancer, *Urology*, **65**, 1126–1130.

94. Frankt, C. J., McCreey, R. I., and Redd, D. C. B. (1995). Raman spectroscopy of normal and diseased human breast tissues, *Anal. Chem.*, **67**, 777–783.

95. Caspers, P. J., Lucassen, G. W., Wolthuis, R., Bruining, H. A., and Puppels, G. J. (1998). In vitro and in vivo Raman spectroscopy of human skin, *Biospectroscopy*, **4**, 31–39.

96. Stone, N., Kendall, C., Shepherd, N., Crow, P., and Barr, H. (2002). Near-infrared Raman spectroscopy for the classification of epithelial pre-cancers and cancers, *J. Raman Spectrosc.*, **33**(7), 564 - 573.

97. Amharref, N., Beljebbar, A., Dukic, S., Venteo, L., Schneider, L., Pluot, M., and Manfait, M. (2007). Discriminating healthy from tumor and necrosis tissue in rat brain tissue samples by Raman spectral imaging, *Biochim. Biophys. Acta – Biomembr.*, **1768**(10), 2605–2615.

98. Andrade, P., Bitar, R., Yassoyama, K., Martinho, H., Santo, A., Bruno, P., and Martin, A. (2007). Study of normal colorectal tissue by FT-Raman spectroscopy, *Anal. Bioanal. Chem.*, **387**(5), 1643–1648.

99. Pinzaru, S. C., Andronie, L. M., Domsa, I., Cozar, O., and Astilean, S. (2008). Bridging biomolecules with nanoparticles: surface-enhanced Raman scattering from colon carcinoma and normal tissue, *J. Raman Spectrosc.*, **39**, 331–334.

100. Aydın, Ö., Kahraman, M., Kiliç, E., and Çulha, M. (2009). Surface-enhanced Raman scattering of rat tissues, *Appl. Spectrosc.*, **63**(6).

101. Aydin, Ö., Altas, M., Kahraman, M., Bayrak, O. F., and Culha, M. (2009). Differentiation of healthy brain tissue and tumors using surface-enhanced Raman scattering, *Appl. Spectrosc.*, **63**(10), 1095–1100.

Chapter 5

Diagnostic and Therapeutic Applications of Carbon Nanotubes

Balaji Panchapakesan,[a] Thomas Burkhead,[a] Ben King,[a] Peng Xu,[a] James Loomis,[a] and Eric Wickstrom[b]

[a]*Small Systems Laboratory, Department of Mechanical Engineering, University of Louisville, 247 Belknap Research Building, Louisville, KY 40292, USA*

[b]*Department of Biochemistry and Molecular Biology, Kimmel Cancer Center, Thomas Jefferson University, Philadelphia, PA 19107, USA*

b0panc01@louisville.edu

Nanotechnology encompasses the creation and utilization of materials at the level of atoms, molecules, and supramolecular structures. Nanomaterials already have an impact in each and every sphere of our lives, from cosmetics to cancer research. One such material that has received wide attention in nanomaterials research is carbon nanotubes, made of atoms of carbon arranged in the form of a cylinder. Accessing the electronic properties of carbon nanotubes renders their application in biological sensing for targeting receptors in cells for *ex vivo* diagnosis. Constructed as a transistor, nanotubes can act as probes for detecting molecular overexpressions in cancer cells. Similarly, the small size, hollow nature, and unique optical absorption properties of nanotubes make them quite useful for therapeutic applications in the selective destruction of cancer cells. Functionalized nanotubes attached to molecular targets in cells can trigger thermal ablation using light that can be used for the surgical and selective destruction of cancer cells.

5.1 Introduction

Nanotechnology encompasses the manufacturing and use of materials, devices, and systems at the level of atoms, molecules, and supramolecular structures. As a measure,

Biomedical Nanosensors
Edited by Joseph Irudayaraj

ISBN 978-981-4303-03-3 (Hardcover), 978-981-4303-04-0 (eBook)
www.panstanford.com

1 nm = 1 × 10^{-9} m, which is about 100,000 times smaller than the diameter of the human hair. At nanometer-length scales, new physical properties emerge in these materials and, therefore, new techniques are required to make them. What makes nanotechnology interesting is that they neither behave like macroscopic systems nor behave like atoms. Size constraints often produce qualitatively new behavior at the nanometer scale. For example, when the size of the nanomaterial reaches the characteristics length scales such as Debye length, it can result in qualitatively new modes of electric current and heat transport in these materials. Similarly, mechanical properties change dramatically as the grain size approaches nanometer-length scales. The use of noble metals, such as gold, silver, and platinum, as nanoparticles lends their applications in biological imaging, sensing, and selective destruction of cancer cells. The size and unique physical properties of materials at the nanometer scale are explored for fabricating sensors to detect cancer cells and contrast agents for enhanced imaging capabilities and targeted therapeutics for treatment.

5.2 Nanotechnology for Cancer

In general, nanotechnology for cancer consists of three main areas: (1) nanodetectors for sensing proteins and cancer cells, (2) nanoparticle or nanovector formulations for high-contrast imaging, and (3) nanotechnology-based drug delivery and therapeutic formulations. Figure 5.1 illustrates some examples of cancer-related nanotechnology and how it will be applied in blood vessels. Biologic nanosensors, fabricated using techniques borrowed from the integrated circuits industry, are emerging for potential use in detecting specific tumor markers. Nanosensors are precisely constructed using the top-down approach. In this process, nanoscale materials are precisely patterned on silicon substrates using lithography. UV and electron beams are focused on specific substrates to sculpt features on the nanoscale using patterned masks. Biologic entities such as antibodies, DNA, and proteins can be patterned on top of nanomaterials such as nanowires, nanoparticles, and carbon nanotubes. Examples of cancer-related nanotechnology include nanovectors [1, 2] such as liposomes for breast cancer therapy [3], nanoparticle contrast agents for high-contrast imaging of brain tumors [4], paramagnetic nanoparticles for imaging of clinically occult lymph node metastases in patients [5], nanomechanical cantilevers [6], nanowires [7], and the recent nanotube sensors for the detection of breast cancer cells [8].

The effective early detection of precancerous lesions remains an elusive goal. Clinical cancer-imaging technologies do not possess sufficient spatial resolution for early detection based on lesion anatomy. To identify malignancies based on their molecular expression profiles, all imaging technologies require contrast agents conjugated to molecular recognition and a targeting agent such as an antibody. Much of the interest in using nanoparticles for cancer detection originated with the use of semiconductor nanocrystals (quantum dots of cadmium selenide) as a tool for laboratory diagnostics of cancer [9]. These quantum dots change their optical properties with their size. When linked to an antibody or a molecule capable of binding to a substrate of interest, quantum dots act like beacons that light up when binding occurs.

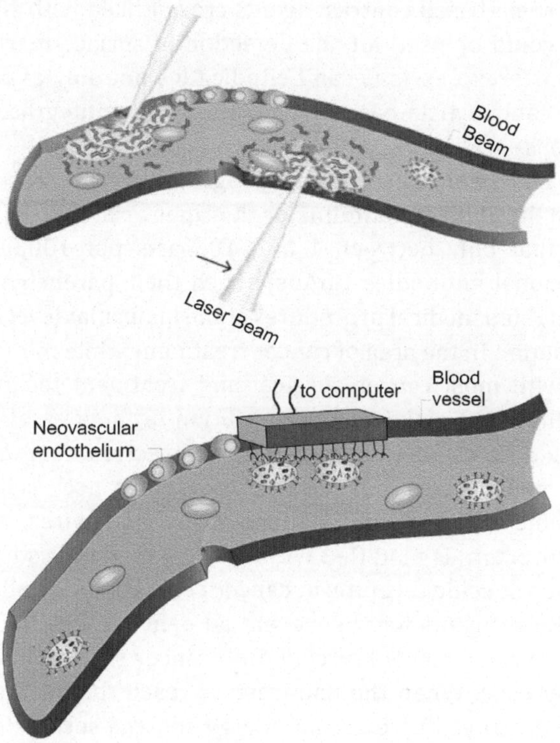

Figure 5.1 Examples of cancer-related nanotechnology. (a) Nanotube-based therapy; (b) nanotube diagnostic device-monitoring cells and proteins in blood. (Inspired from ref. 1.)

These dots have the potential for creating assays capable of detecting multiple substances simultaneously, such as insulin-like growth factor 1 receptor (IGF1R), human endothelial receptor 2 (Her2), actin, microfibril protein, and nuclear antigens [10]. Nanoparticle technologies are under development and testing as candidate multifunctional molecularly or physically targeted contrast agents for all clinical-imaging modalities, with the objectives of detecting smaller and earlier-stage cancer tumors, identifying molecular expressions of neoplasms and their microenvironment, and providing improved anatomic definition for lesions. Iron-based superparamagnetic nanoparticles have been demonstrated for high-contrast MRI of cancer. These nanoparticles are small, and their presence *in vivo* in small quantities is relatively nontoxic. For example, lymphotropic paramagnetic nanoparticles allow the MRI of clinically occult lymph nodes in patients who have prostate cancer that is not detectable using conventional technologies [5]. Dextran-coated paramagnetic iron oxide nanoparticles were shown to perform better than gadolinium-based MRI contrast agents for imaging brain tumors [11]. Dendrimers are fractal polymers that have been used as nanocarriers of gadolinium to image the lymphatic drainage of breast cancer in mice [12]. Cross-linked iron oxide nanoparticles as MRI contrast agents that were conjugated with annexin-V have been shown to recognize apoptotic cells [13]. Sustained angiogenesis is an important marker for use in early detection of cancer. Several groups have imaged angiogenesis with MRI in animal models of various nanoparticle formulations [14–18]. All these studies clearly indicate that nanoparticle-based contrast agents cross-linked with signal-amplifying and targeting molecules could be used for the detection of surface marker overexpressions *in vivo*. Such marker overexpressions can be indicative not only of cancer but also of its progression. Future nanoparticle-based formulations may integrate molecular imaging with treatment in steps.

Although there has been outstanding progress in the area of cancer biology, inadequacies exist in the ability to administer therapeutic moieties with high selectivity. It is worth noting that only between 1 and 10 parts per 100,000 of intravenously administered monoclonal antibodies (mAbs) reach their parenchymal targets [1]. The ability to carry out targeted medical procedures at the molecular level could bring about an unprecedented revolution in the area of cancer treatment while minimizing the collateral damage compared with most cancer surgical and treatment techniques in use today. Nanotechnology promises to make such a revolution in the area of cancer therapeutics using targeted drug-delivery methods called nanovectors that are in development today. So, what are nanovectors and how do they kill cancer cells? Nanovectors are multifunctional organic and inorganic nanoparticles, nanowires, and nanotubes. The surface of these nanovectors is modified with biologic materials such as antibodies that can be used to target a specific receptor in cancer cells. Further, hollow nanovectors can be filled with anticancer agents for their eventual delivery after reaching their targets. These nanovectors are then injected directly into cancer sites or into blood, where they target specific cancer cells. When the nanovectors reach their targets, they deliver the drug payload or are destroyed by external energy sources such as optics or magnetics, which enable these nanovectors to heat *in vivo* in a localized way. Such localized thermal treatments can kill cancer cells in parts of the body considered inaccessible using traditional surgical techniques. Although these ideas may sound like science fiction, they are being aggressively pursued in the laboratory setting today.

Two different approaches have been used — the first consists of therapeutic agents that are loaded inside a hollow nanoparticle with biologic surface modifiers, and the second consists of nanotubes that are functionalized with cancer-specific protein and heated using external sources to kill cancer cells in a localized environment. This type of multicomponent cell-killing strategy using nanoparticles and nanotubes can improve therapeutic efficacies and selectivity. Injectable nanovectors form a class of nanotechnology approaches that are of high interest for cancer treatment. For a nanovector to be highly effective, it should be selectively directed against cell clusters and able to kill malignant cancer cells and cells that are in early stages of transformation, without affecting patient quality of life. Further, therapeutic formulations of nanovectors should include proteins that can overcome biologic barriers that prevent the nanovectors from reaching their targets. Many different types of nanovectors have been proposed, including liposomes that use overexpressions of fenestrations in cancer neovasculature to increase drug concentration at tumor sites [3]. Liposomes are one of the most widely used nanovectors for different types of drug delivery modalities in the fight against cancer. There are hundreds to thousands of nanovectors that has been reported in the literature. However, only a fraction of them has been applied to cancer in diagnostics, imaging, and therapeutics. One such material that has been very popular among research groups is the use of carbon nanotubes made up of tiny atoms of carbon in the form of a cylinder. The following sections will review some of the fundamental properties of carbon nanotubes and their potential application in the area of cancer nanotechnology.

5.3 Carbon Nanotubes

The discoveries of carbon nanotubes can be historically dated back to the early 1970s and 1980s, when carbon fibers were grown through the thermal decomposition of hydrocarbons at high temperatures in the presence of transition metal catalyst nanoparticles [19–21]. However, the real burst of research about carbon nanotubes did not come until the observation of multiwall carbon nanotubes by Ijima in 1991 using high-resolution transmission electron microscopy (HRTEM) [22]. Ijima's hallmark work heralded the entry of many scientists into the field of carbon nanotubes, who were stimulated at first by the remarkable one-dimensional quantum effects predicted for their electronic properties and subsequently by the promise that the remarkable structure and properties of carbon nanotubes might give rise to some unique applications [23]. In the following two years, single-wall carbon nanotubes (SWCNTs) were also experimentally characterized by Ijima and Bethune [24–27]. Single-wall carbon nanotubes can be thought of as rolling up a single graphene sheet composed of only one layer of carbon atoms to form a hollow cylinder along the tube axis, while multiwall carbon nanotubes can be envisioned by rolling up more layers of graphene sheets to form concentric cylinders, with each one wrapped around the other layers. The graphene layers are coaxially arranged around the central axis of the nanotube, and the spacing between graphene layers is kept constant of 0.339 nm [28]. Single-wall and multiwall carbon nanotubes are intimately related to each other in structure and in physical properties, as single-wall nanotubes become multi wallnanotube by adding more shells around the primary tubes and each shell within the multiwall nanotube can be viewed as a single-wall nanotube with different diameters. Single-wall tubes are generally smaller in diameter (0.3–10 nm). However, multiwall tubes can be as big as tens to hundreds of nanometers in diameter. While the dimensions of carbon nanotubes are in nanometer scale in radial direction, the length of nanotubes along the tube axis can reach in millimeter or centimeter scale, thus resulting in extremely high aspect ratios of 10^5–10^6. This offers us the opportunity of studying the intriguing physical properties at nanometer-length scales, where quantum mechanical effects dominate in these one-dimensional systems.

Another important feature in the carbon nanotube atomic structure that is vital in deciding various physical properties is the arrangement of the carbon hexagons around the tube surfaces, in other words, the "helicity" of the honeycomb lattice of graphene layer with respect to the tube axis [22, 29]. When a layer of graphene is rolled up to form a nanotube, the edges of the atomic layer that carry unbonded carbon atoms must join perfectly with each other in order to form a seamless cylinder. Thus, the way the edges of graphene layer pair up result in specific directions of rolling and only from those directions can a tubular structure be formed. These specific directions can be characterized by the crystallographic orientations of the surface hexagons, which are noted by helicity indices [30].

As shown in Fig. 5.2, carbon nanotubes are formed by rolling the graphene sheet in the direction of $\overrightarrow{OA}$, which is expressed by a chiral vector:

$$C_h = na_1 + ma_2 \tag{5.1}$$

a_1 and a_2 are two crystallographically equivalent lattice vectors on a two-dimensional graphene sheet, and m and n are a pair of integer indices. A chiral angle θ is also defined in this figure, which is the angle between the chiral vector C_h and the zigzag direction ($\theta = 0$) parallel to the direction of unit vector of a_1:

$$\theta = tg^{-1}[\sqrt{3}n/(2m+n)] \tag{5.2}$$

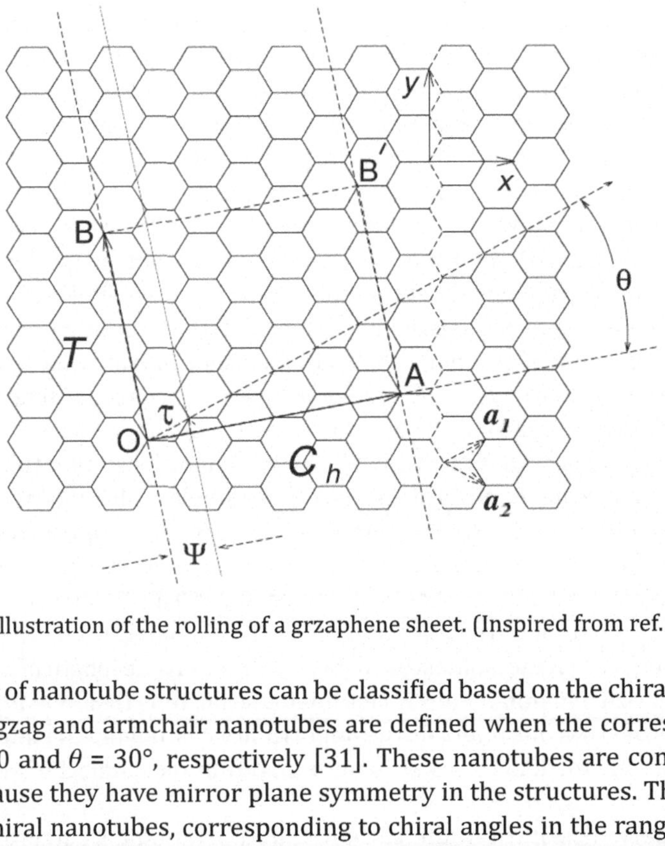

Figure 5.2 An illustration of the rolling of a grzaphene sheet. (Inspired from ref. [23])

Three types of nanotube structures can be classified based on the chiral vector and the chiral angle. Zigzag and armchair nanotubes are defined when the corresponding chiral angles are $\theta = 0$ and $\theta = 30°$, respectively [31]. These nanotubes are considered achiral nanotubes because they have mirror plane symmetry in the structures. The other type of nanotubes is chiral nanotubes, corresponding to chiral angles in the range $0 < \theta < 30°$. If the helicity indices (n, m) are used to label these nanotubes, armchair nanotubes are those corresponding to $n = m$. When $m = 0$, a zigzag nanotube is formed. Three-dimensional models of the two types of achiral nanotubes are shown in Fig. 5.3. Chiral nanotubes correspond to helicity indices n and m in other permutations. In practice, most carbon nanotubes do not form achiral structures with high symmetry. Instead, they like to form chiral nanotubes, whose helicities can be individually addressed experimentally using electron microscopes and electron diffraction techniques [32, 33]. For multiwall nanotubes, constant interlayer separation of continuous tubes puts an additional constraint on the tube structures, requiring each tube layer to have its own helicity.

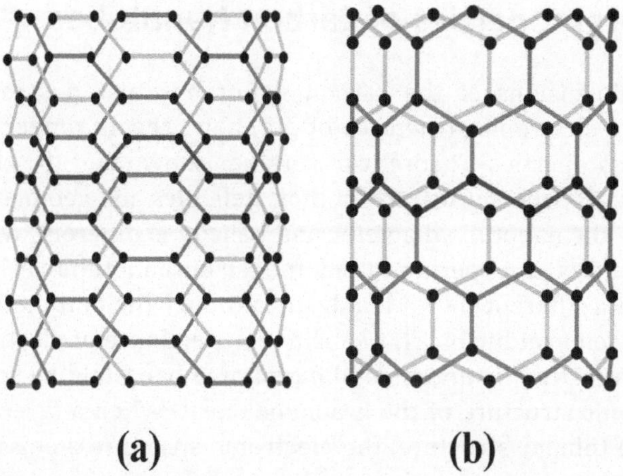

Figure 5.3 The models of (a) an armchair nanotube and (b) a zigzag nanotube. (Inspired from ref. [23])

However, multiwall nanotubes generally show a smaller number of helicities than the number of wall layers, indicating that some walls in multiwall tubes share the same helicity [30, 34]. Because a wide variety of helicity index permutations exists, many types of nanotubes with different helicities can be formed in real nanotube samples, which further lead to a wide range of nanotube geometries and, thus, a wide range of properties of nanotubes. More or less, most of the current applications of carbon nanotubes, especially in micro- and macroscales utilize the carbon nanotubes with properties averaged over the entire samples. The precise controlling of the helicities of carbon nanotubes within a narrow distribution through the nanotube growth is one of the critical issues to develop nanotubes with uniform properties [35–37].

Further, the diameters of carbon nanotubes are also related to their helicities and can be expressed as:

$$d_t = \sqrt{3}a_{c-c}(m^2 + mn + n^2)^{1/2}/\pi = C_h/\pi \tag{5.3}$$

In Eq. (5.2), C_h is the length of the chiral vector and a_{c-c} is the carbon bond length (1.42 Å). Generally, single-wall nanotubes remain straight and relatively defect free for most of the tube length before reaching the ends of the tubes, where cones and polyhedral cap structures occur due to the presence of pentagonal defects [30]. Further, smaller nanotubes (diameters < 2 nm) tend to have fewer structural defects and are structurally perfect than bigger nanotubes (diameter > 2 nm). As has been experimentally observed [38, 39], multiwall nanotubes generally have more defect populations along the tube axis due to their larger tubular dimensions. The presence of structural defects in both single-wall and multiwall tubes has profound impacts on their physical properties such as electrical conductivity, thermal conductivity, mechanical strength, optical absorption, and emission properties [23]. Through proper thermal treatments, most of the defects can be removed to leave relatively perfect carbon nanotubes [40, 41].

5.4 Electronic Properties of Carbon Nanotubes

The nanometer dimensions of the carbon nanotubes that possess strong quantum confinement to the electronic structures of graphene sheets render nanotubes unique electron transport properties. Theoretical work has shown that the electrical properties of carbon nanotube depend sensitively on their helicities and geometrical arrangements [23, 42–45]. When the nanotube diameter and helicity is different, which is the case for different helicity indices (n, m) permutations, the carbon nanotube can be either metallic or semiconducting with different energy bandgaps, although their intimate relative-graphene is a zero-bandgap semiconductor. This sensitive dependence of electronic properties on geometrical structures can be understood based on a band-folding scheme that owes to the unique electronic structure of the graphene sheet. When a layer of graphene sheet is rolled to form a tubular structure, the electronic structure along the circumferential direction of the tube must become quantized due to periodic boundary conditions. At the same time, the wave vector along the tube axis direction still remain continuous [46]. The quantization of wave vectors results in different electronic behaviors for different nanotubes (n, m). General rules can be stated as follows: (n, m) tubes, which are so-called armchair nanotubes, are metallic; (n, m) tubes where $n - m = 3j$ (j is a nonzero integer) are also metallic tubes; nanotubes with all other (n, m) permutations are semiconductors with an appreciable large bandgaps [23]. Statistically, 1/3 of a randomly produced single-wall nanotube sample is metallic nanotubes and 2/3 is semiconducting nanotubes. In many applications of carbon nanotubes, it is preferred that only one type of nanotubes is used instead of a mixture. For examples, carbon nanotube–based field effect devices, which are vital for nanoelectronics and sensing applications, rely on the capability of conductance modulation within semiconducting tubes [47–52], while metallic tubes are preferred for use in interconnections in nanoelectronics due to their high electronic conductance [53, 54]. Thus, selective enrichment of certain type of nanotubes in a nanotube product is a critical issue in many nanotube-related applications. It was reported that semiconducting nanotubes can be preferentially produced at the ratio of > 85% through plasma-enhanced chemical vapor deposition (PECVD) processes [55]. Other strategies involve the separation of two types of nanotubes [56–58] and the selective removal of certain types of nanotubes from the ensembles [59, 60]. An important property of semiconducting nanotubes is that the bandgap energy, which is essential in determining the electronic and optical properties of carbon nanotubes, is inversely proportional to the tube diameters [33]. The energy bandgap can be numerically expressed as:

$$E_g \sim 0.9/d_t(eV) \tag{5.4}$$

where d_t is the diameter of the nanotube as shown in Eq. 5.3. The bandgap tunability through diameters is an important merit of semiconducting nanotubes. The density of electronic carriers as well as the absorption bandgap of nanotubes can be potentially tuned by the tube diameters to match certain electronic and optical applications.

As stated before, one of the most important applications of semiconducting nanotubes is to construct field effect devices. Many experimental studies have been conducted on nanotube field effect transistors [47, 48, 50–52, 61–63]. In a typical nanotube transistor, the modulation of gate voltage induces large channel conductivity modulation with the on/off ratio approaching 10^6. Such field effect modulation would be sufficient for many applications in nanoelectronics. For example, prototype nanotube–based logic gate circuits have already been demonstrated [64–69]. Generally, as-produced carbon nanotubes behave like P-type semiconductor. P-type doping may be induced by a metal electrode due to the high work function of the metals [23]. The processing environment may also result in chemical and gas absorption (especially oxygen) for acceptors doping [23, 70]. An additional doping process such as alkaline metal doping can be utilized to change the nanotubes into an N-type semiconductor [71].

The unique electronic transport properties of carbon nanotubes also bring high electrical conducting capabilities of carbon nanotubes, exceeding the best of any metal of the same size, with charge carrier mobility reaching 100,000 cm^2/V-s [72] and current carrying capacity reaching 10^9 A/cm^2 [73]. Ballistic electron transport was observed in both defect-free multiwall and single-wall carbon nanotubes [74, 75]. Superconductivity of carbon nanotubes was also reported [76]. Another important aspect of electron transport of carbon nanotubes is the effects of defect-related structures on carbon nanotubes. A good example is that the introduction of pentagon-heptagon pair defects into the hexagonal network of a single nanotube can change the helicity of the tube and fundamentally alter its electronic structure [42, 77–82]. Based on such an idea, the defective carbon nanotube can be configured into functional nano devices to behave as nanoscale metal-semiconductor Schottky barriers, semiconductor heterojunctions, or metal-metal junctions with novel properties [23]. Further, extrinsic defects introduced onto nanotubes, such as gas molecules and chemical groups attached to the surface through covalent or noncovalent bonding, will have significant impacts on the electron transport of carbon nanotubes, thus opening a completely new area of electronic-sensing applications based on carbon nanotubes [70, 83–88].

5.5 Optical Properties of Carbon Nanotubes

The optical properties of carbon nanotubes are closely related to their electronic structures, which are dependent on the nanotube diameter and helicity. As stated before, the structure of a carbon nanotube can be specified by the chiral vector (Eq. 5.1). The electronic transport properties are determined by the helicity that $n - m = 3j$ (j is a nonzero integer) defines metallic carbon nanotubes, while others are semiconducting ones, affecting the corresponding optical responses in carbon nanotubes. The one-dimensional density of states (DOS) is given by the energy dispersion of carbon nanotubes, which can be obtained by zone folding of two-dimensional energy dispersion relations of graphite due to the tubular nature of the nanotubes. The periodic boundary conditions applied to carbon nanotubes quantize the wave vector along the circumferential direction of the tube into N discrete values, where N is number of hexagons of the graphite honeycomb lattice within the nanotube unit cell. At the same time, wave vectors along the tube axis direction remain continuous. Thus, the energy dispersion relations of carbon nanotubes can only adopt certain one-dimensional parallel line traces in the two-dimensional energy dispersion relations of graphite, whose positions near the K points in the Brillouin zone are determined by the chiral indices of the nanotubes. When the energy dispersion relations as a function of wave vector κ in nanotubes becomes flat in the Brillouin zone, the DOS becomes large due to the inverse dependence of the DOS on the first derivatives of energy dispersion in nanotubes. The one-dimensional van Hove singularities (vHs) in the DOS, which are proportional to $(E^2 - E_0^2)^{-1/2}$ at both the energy minima and maxima ($\pm E_0$) of the dispersion relations for carbon nanotubes, are very important for determining many physical properties of carbon nanotubes, such as optical absorption [89, 90] and resonant Raman spectroscopy [23, 91–93]. Calculations from the energy dispersion relations reveal the energy differences (bandgap) of metallic and semiconducting nanotubes between the highest valence band singularity and the lowest conduction band singularity in the one-dimensional electronic DOS, which are expressed as $E_{11}^M(d_t)$ and $E_{11}^S(d_t)$, respectively:

$$E_{11}^M(d_t) = 6a_{c\text{-}c}\gamma_0/d_t \quad \text{and} \quad E_{11}^S(d_t) = 2a_{c-c}\gamma_0/d_t \tag{5.5}$$

where d_t is the diameter of the nanotube and $a_{c\text{-}c}$ is the carbon lattice constant. The double subscripts in the expressions denote the corresponding orders of the valence π and conduction π^* energy bands, which are symmetrically located with respect to the Fermi energy. E_{nm}, meaning the energy bandgap between the nth valence band and n^{th} conduction band, could lead to optical transitions between these bands. For carbon nanotube samples mixed with metallic and semiconducting nanotubes with similar diameters, optical transitions determined by the energy gaps from both types of nanotubes may follow the order $E_{11}^S(d_t)$, $2E_{11}^S(d_t)$, $E_{11}^M(d_t)$, $4E_{11}^S(d_t)\ldots$, which starts from the lowest energy levels [32, 94–96]. An important so-called trigonal warping effect due to the distortion of the energy contour from circular patterns away from K points induces the splitting of vHs in metallic nanotubes [97], and thus affects electronic and optical transportation properties of the nanotubes. Using photothermal deflection spectroscopy [94], three absorption peaks were identified with transition energies located at 0.68, 1.2, and 1.7 eV, which correspond to the first and second optical transitions in semiconducting nanotubes and, subsequently, the first optical transition in metallic nanotubes. Further, carbon nanotubes also exhibit diameter-dependent optical absorption properties, shifting their absorption spectra for larger diameters. Further, the absorption properties of carbon nanotubes can also be modified by doping processes [90]. It was reported that the intensity of the absorption peaks decreased, especially for lower-energy absorption peaks, with the increase of dopant concentration. At a high level of doping, new doping-induced absorption peaks appeared, which may be due to conduction-to-conduction intersubband transitions for N-type nanotubes and valence-to-valence intersubband transitions for P-type nanotubes [90].

Newly discovered optical properties of carbon nanotubes such as strong optothermal effects [98–100] and optoelastic effects [101], although not fully understood yet, raise intensive research interests in discovering the underlying physical mechanisms and present important implications of many potential optical-related applications such as optical actuators. The strong optothermal effects were discovered through a simple experimental work of shining camera flash on fluffy carbon nanotube samples in air. It was found out that carbon nanotubes experience significant oxidation in the presence of oxygen or experience structural reconstruction in inert environments [100]. However, for nanotubes to oxidize, the temperature must rise above 600°C while at least 1,500°C is needed for nanotube structure reformation, showing the strong localized thermal effects of carbon nanotube under light exposure [102–104]. Efficient light absorption and high thermal confinement of nanotube structures were employed to explain the optothermal effects. These optothermal effects from nanotubes have already shown applications such as carbon nanotube segmentation [98] and thermal agents for cancer therapeutics [105, 106].

5.6 Biological Applications of Carbon Nanotubes

For applications of carbon nanotubes in the biological and biomedical area, the lack of solubility of carbon nanotubes has been a major technical barrier. Recently, there have been methods that have been demonstrated for chemically modifying carbon nanotubes, thereby rendering them soluble in water and paving the way for the manipulation of carbon nanotubes in physiological environments. Proteins, carbohydrates, DNA, and even antibodies have been functionalized on the surface of carbon nanotubes.

The biomodification of carbon nanotubes can be divided into three categories: (1) covalent attachment, (2) noncovalent attachment, and (3) the hybrid approach using an anchor molecule that is attached covalently to the biomolecule of interest. The most common method for the covalent functionalization of carbon nanotubes involves reactions with carboxylic acid (–COOH) residues on carbon nanotubes. This is done by oxidizing the carbon nanotubes with strong acids in defective sites of carbon nanotubes rather than sidewalls. For sidewall functionalization, arylation with diazonium salts are generally employed [107–111]. It should be remembered that depending on the application, the covalent functionalization of carbon nanotubes can alter the electron transport properties. The reaction with carboxylic acids can produce excessive oxidation, which, in turn, can alter the structure and properties of carbon nanotubes. Further, the electrocatalytic properties of carbon nanotubes can be altered by covalent functionalization. On the other hand, sidewall functionalization of carbon nanotube is less damaging and also does not alter the electron transport as well electrocatalytic properties of carbon nanotubes [112]. Recently, methods for attaching DNA to oxidized single-walled carbon nanotubes either in organic solvent or aqueous solutions has been demonstrated [113]. The sites of the DNA attachment to the nanotubes were verified by using gold nanoparticles modified with DNA of complementary sequence to the DNA-modified nanotubes and imaging with transmission electron microscopy.

The aromatic (π-electrons) hydrophobic surface of carbon nanotubes makes them ideal partners for noncovalent functionalization of biomolecules. For example, the transport of DNA inside a multiwall carbon nanotube is achieved using noncovalent functionalization methods. These methods are useful for attaching a range of biomolecules, including heavily doped DNA, polysaccharides, amylase, proteins, polyethylene glycol, and antibodies [114–118]. The major advantage of noncovalent functionalization is that the carbon nanotubes are not altered in any significant way. This could play a vital role in designing nanotube-based devices for biomolecular interactions on the surface of carbon nanotubes. Overall, most of the biofunctionalization of carbon nanotubes is aimed at either designing nanotube-based *in vitro* devices for detecting proteins and antibodies against autoimmune diseases or for the application of carbon nanotubes *in vivo* for localized therapy. One of the major stumbling blocks for the *in vivo* application of carbon nanotubes is the lack of sufficient data on the toxicity of carbon nanotubes.

5.7 Bioelectronic Devices Based on Carbon Nanotubes

Carbon nanotube–based electronics and bioelectronics have been the focus of many research articles in recent years. This is due to the fact that all the atoms in a single-wall carbon nanotube are surface atoms. Therefore, any change in surface characteristics such as structure, molecular adsorption and desorption, chemical bonding and stress, or pressure on the surface can dramatically impact the electronic properties of carbon nanotubes. These electronic properties can be monitored at the single-electron level using a transistor configuration. Since single nanotubes are almost one-dimensional conductors, the current flow in these systems are also in one dimension and most of the current in these systems are on the surface. This makes them highly sensitive to external perturbations [119]. This also means that sensors based on carbon nanotubes can possess extraordinary sensitivity compared to their macroscale and microscale counterparts. Single nanotube–based electronic elements as sensors could potentially be used in gas, liquid, aqueous, and vacuum environments for measuring pressure, stress, concentration of analytes, radiation, and trace elements in liquids. When biological entities such as proteins, macromolecules, antibodies, receptors, and even cells come in contact with the nanotube, they can change the flow of the current due to charge transfer [120–144]. Further, when biological entities undergo unbinding or binding events, these events can influence the current flow in a single nanotube that can be monitored accurately using a transistor configuration [143]. Binding events between proteins can lead to a change in enthalpy or entropy or free energy. Bound on the surface of the carbon nanotubes, these changes can cause the current flow to alter dramatically and, therefore, sophisticated sensors based on carbon nanotubes can be envisioned to monitor these changes rapidly. Further, nonspecific binding can be avoided using a polymer coating on the surface of the nanotube.

There has been significant research interest in the area of bioelectronic devices based on carbon nanotubes. Recent reports have shown that carbon nanotube transistors could be used for the detection of streptavidin-biotin-binding events [121, 141, 143–147]. The binding events changed the conductance that is monitored using a transistor device. Nonspecific binding was overcome by modifying the nanotubes with polyethylene oxide chains. Similar to this study, in another report, a polyethylene glycol (PEG) layer was used to avoid nonspecific binding on carbon nanotubes. These studies have shown the potential of using nanotube-based bioelectronic elements as highly sensitive devices for the detection of proteins. Nanotube-based transistors have also been used to transduce enzymatic reactions [121]. The attachment of glucose oxidase to the sidewall of the nanotube led to a change in device conductance. Large reversible changes in conductance were observed with changes in pH. Recently airbrushed single-wall carbon nanotube transistors have been used to detect clinically important receptors such as IGF1R and Her2 in breast cancer cells [8]. Single-nanotube field-effect transistor (FET) arrays, functionalized with IGF1R-specific and Her2-specific antibodies, exhibit selective sensing of live, intact MCF7 and BT474 human breast cancer cells. It was postulated that the free-energy change due to multiple simultaneous cell-antibody binding events exerted stress along the nanotube surface, decreasing its electrical conductivity due to an increase in the bandgap. Because the free-energy change upon cell-antibody binding, the stress exerted on the nanotube, and the change in conductivity are specific to a specific antigen-antibody interaction, these properties might be used as a fingerprint for the molecular sensing of circulating cancer cells. The field of bioelectronics based on carbon nanotubes continues to grow as new types of device designs, new functionalization strategies, and new signal analysis are being explored. The goal of bioelectronic sensors based on carbon nanotube research should gear toward making hundreds and thousands of arrays with functionalized carbon nanotube elements for the detection of proteins that could be prognostic indicators of a wide variety of autoimmune diseases in biological fluids such as urine or saliva or blood within a few minutes. This would make a significant impact in *ex vivo* monitoring of clinically relevant diseases. However, the fabrication of such a device itself is challenging and brings in many different disciplines such as carbon nanotube growth techniques, Raman spectroscopy, scanning and transmission electron microscopy, biological functionalization methods, cell culture, electronic circuits, and finally signal analysis.

5.8 Applications of Nanotube Transistors for Detecting Circulating Cancer Cells

The identification and quantitation of biological molecules to generate a complex molecular profile is required for the diagnosis, monitoring, and prognostic evaluation of complex diseases such as cancer [148]. Despite outstanding progress in the area of cancer biology, significant challenges remain in translating biological knowledge of cancer surface markers into clinically relevant devices that could be used as diagnostic or monitoring tools for cancer management. Developing high-throughput diagnostic cell and tissue analysis for disease detection has remained a challenge, however. Techniques for multiplexed analysis of extracted proteins for disease monitoring can be divided into four different categories: (1) radioactive, and fluorescent reporting of antigen-antibody binding [149, 150], time-of-flight mass spectroscopy [151, 152], (3) electrophoretic separation and antigen-antibody binding [153, 154], and (4) the detection of changes in surface mechanical and electrical properties due to antigen-antibody binding [138, 148]. Combining the knowledge of cancer biology, molecular targeting, and nanotechnology, however, might enable the design of portable or handheld devices for the diagnosis, monitoring, and treatment of cancer. Circulating cancer cells often express characteristic cell surface markers. Human MCF7 ER+ breast cancer cells exhibit high expression of the insulin-like growth factor 1 receptor (IGF1R), while BT474 ER– cancer cells have lower expression of IGF1R but high expression of human endothelial receptor 2 (Her2) [149–152]. The experiments resulting in molecular targeting and sensing of circulating cancer cells consist of the preparation of single-wall carbon nanotubes device arrays, antibody functionalization of single-wall carbon nanotubes, characterization of carbon nanotube transistors, incubation of nanotube-antibody hybrids with cells, and finally the monitoring of the conductance of nanotube transistors in different nanotube-antibody hybrids and cells [8].

Figure 5.4 is an optical image of a single-wall carbon nanotube field-effect transistor (SWCNT FET) chip array consisting of 20 pairs of electrodes with a single carbon nanotube between each pair. The nanotubes were anchored between the pairs of electrodes by airbrushing. Figure 5.5 is the scanning electron microscope (SEM) image of a nanotube anchored between pairs of electrodes. The spacing between the electrodes is 1 μm, and the nanotube diameter is ~10 nm from the inset in Fig. 5.5. Figure 5.6 presents the effects of increasing antibody concentration on the nanotube FET. The device behaves like a P-type semiconducting FET due to the depletion of current at all antibody concentrations tested. The negatively charged antibodies deplete the current with an increasing source-to-drain voltage. There is a strong decrease in the conductance of the channel with an increasing antibody concentration showing the modulation of transfer characteristics. Figure 5.7 shows the set of experiments conducted on cancer cells. It can be seen that the cells expressing characteristic surface receptors underwent greater change in conductance compared to the cells that did not overexpress the receptor. The binding of antibodies to their corresponding antigens creates a change in free energy that, in turn, can generate stress on a surface. This has been used to create bioassays of prostate-specific antigens using deflections of a microcantilever [155]. The results show that cancer cell binding to cognate antibodies on a nanotube device alters the electron transport properties of the nanotube. Since the nanotubes are not free to deflect and constrained between electrodes, one is left with the explanation that the nanotubes undergo strain that changes the electron transport properties.

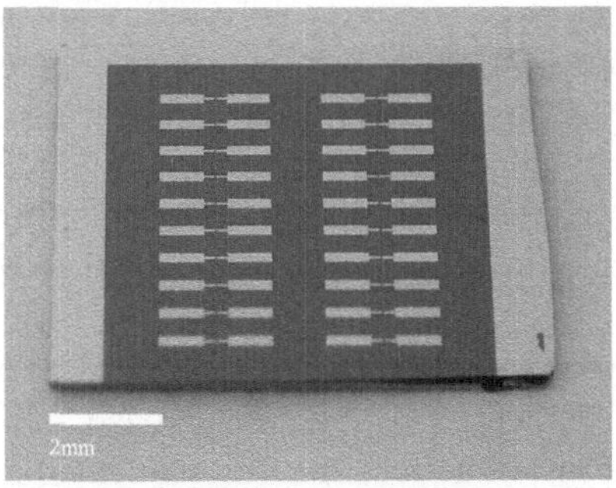

Figure 5.4 A 20-element array for the multicomponent detection of surface receptors in cancer cells. (Reprinted with permission from ref. 8.)

Figure 5.5 An SEM image of a small bundle of nanotubes between electrodes; inset is the higher magnification of the SEM image. (Reprinted with permission from ref. 8.)

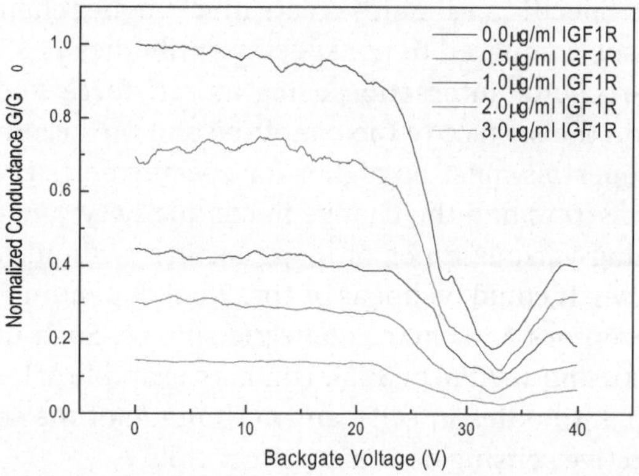

Figure 5.6 An increasing concentration of antibodies on a nanotube FET. (Reprinted with permission from ref. 8.) See also Color Insert.

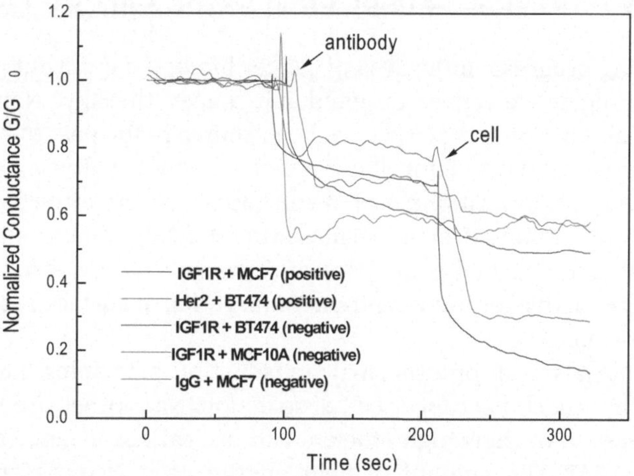

Figure 5.7 Experiments on cancer cells showing decreasing conductance change when antibodies and cancer cells are applied to the device. Antibodies and cells were dropped at about 100th second and 220th second. Two specific pairs of antibodies and cells showed conductance reduction and three nonspecific pairs had little impact. (Reprinted with permission from ref. 8.) See also Color Insert.

This hypothesis is supported by the theory of electron transport in single-wall carbon nanotubes that predicted that strain can have a large effect on the band structure of a nanotube [156], which, in turn, has an influence on its electron transport properties.

It was postulated that the binding of cancer cells simultaneously to multiple antibodies anchored to a nanotube creates a surface strain that changes the bandgap and the conductivity of the nanotube. The putative increase in the bandgap due to the cell-binding effect would result in a decrease in conductivity. It has been seen in recent reports that for all semiconducting nanotubes, the application of strain $< 0.1\%$ leads to an increase in the bandgap, thereby decreasing the conductivity of the device [155]. The results support this hypothesis, implying that the strain originates on the surface of the nanotube due to biomolecular interactions that are specific and, therefore, can be used as a "fingerprint" for molecular recognition. Specific antibodies create much higher change in free energy upon binding to their antigens compared to nonspecific antibodies [157]. This has also shown to be true for ligand-receptor interaction using atomic force microscope (AFM) [158]. Therefore, the stress on the surface of the nanotube and the accompanying change in the electron transport properties must be higher for specific pairs compared to nonspecific pairs [8]. The origin of stress and the change in conductivity can, therefore, be used as a "fingerprint" for sensing various types of biomolecules without labeling. The nature of stress remains unknown. It could be linear or torsional depending on how the antibodies interact with their receptors and their geometric shapes. Such understanding requires detailed investigations using an AFM or STM. This is an example of how molecular targeting, receptor binding with antibodies in cells, and carbon nanotube transistor configuration can lead to highly selective sensing of breast cancer cells.

5.9 Integrated Molecular Targeting and Selective Photothermal Therapy of Breast Cancer Cells Using Carbon Nanotubes

Despite outstanding progress in the area of cancer biology, significant challenges remain in administering highly selective, targeted anticancer therapy. Nanoparticles [159], nanoshells [160], and nanotubes [161] have been shown in the past to be quite applicable for cancer imaging and therapy. Subcellular nanostructures exhibiting advanced physical properties for biological applications can have an important impact on the introduction and delivery of DNA, proteins, and drug molecules into the living cells. In addition, the unique electronic, optical, and thermal properties and the subcellular size of nanostructures makes them very attractive for multicomponent targeting of surface receptors to enhance the efficacy of anticancer agents.

Nanoscale materials with optothermal transduction principles have been shown to be an interesting approach for cancer cell destruction. Nanotubes are highly suitable for this technique because of their high efficiency in optical absorbance and photothermal cell destruction [164]. The amount of light energy used is quite small compared to competing techniques as the optical absorbance of nanotubes is high in the 700–1,100 nm near-infrared (NIR) window. Therefore, multicomponent targeting with photothermal cell destruction might lead to the development of cancer therapeutics, microsurgery, and cell repair machineries. SWCNTs are unique nanomaterials with remarkable optical properties. The thermal energy generated during optical absorption in carbon nanotubes has been used to created nanobombs to destroy breast cancer cells [162]. Nanobombs may be highly effective in killing cancer cells that are malignant because the temperatures attained in a localized way are much higher than with other nanovectors in the thermal ablation of cancer cells. Although the temperatures attained in nanobombs are high, they are localized to a nanoscopic area that may not produce any pain in the patients when administered *in vivo*. Further, the complete destruction of nanotubes inside the body may make them ideal for handling toxicity problems that may be associated with nanotubes and nanoparticles. Further, nanobombs can kill cancer cells at specific places by internalization of the nanobombs within the cancer cells. The internalization of the nanobombs, followed by triggering of the nanobombs inside the cells, can cause cell death and can also sever the biological signaling pathways that promote cancer remission. Nanobombs can also find possible applications in blowing up clots and cholesterol in blood vessels and can be used as a nanosurgical tool for the removal of microscopic lesions in parts of the brains that are often inaccessible, using traditional surgical tools.

While nanoshells have been targeted to the Her2 surface receptor of cancer cells in the past [163, 164], recent reports have shown the integrated targeting of both IGF1R and Her2 surface receptors and photothermal cell destruction using SWCNTs [161]. Multicomponent targeting of more than one surface receptor can lead to higher efficacies in therapeutics. IGF1R and Her2 are both relevant surface markers that are overexpressed in a wide variety of cancer cells compared to normal cells and, therefore, have great potential for sensing, imaging, and therapy of cancer. Circulating cancer cells often express characteristic cell surface markers. Human MCF7 ER+ breast cancer cells exhibit high expression of the IGF1R, while BT474 ER– cancer cells have lower expression of IGF1R but high expression of Her2 [165]. The experiments resulting in molecular targeting and photothermal selective cell destruction consist of the preparation of SWCNTs, antibody functionalization of SWCNTs, evaluation of the toxicity of SWCNTs, incubation of nanotube-antibody hybrids with cells, and finally dosing of the suspension with NIR light for a few minutes. These methods are detailed elsewhere [161].

In general, molecular targeting is achieved by the functionalization of IGF1R and Her2-specific antibodies on top of SWCNTs as shown in Fig. 5.8. Since each carbon atom is a potential binding site for the antibody, a higher concentration of antibodies on top of SWCNTs can lead to selective molecular targeting. The SWCNTs are about 1.4 nm in diameter and the typical length varied from 500 nm to 1 μm. The shorter SWCNTs tend to be straighter with fewer curves, which makes endocytosis easier. In general, below the persistence length, the SWCNTs tend to act like rigid rods, which leads to better internalization of SWCNTs. The characterization of the presence of antibodies on SWCNTs is done using transmission electron microscopy as shown in Fig. 5.9. Potentially each group of carbon atoms equal to the cross-sectional area of the constant region of a mAb constitutes a binding site in the SWCNT that can act as a biological carrier to transport macromolecules inside the cells. Past AFM studies have shown nanotube-antibody hybrids to be 5–8 nm high. From transmission electron microscope (TEM) images presented here in Fig. 5.9, it can be observed that each one of the antibody aggregates varies between 66–100 nm. Hence it is expected that about 10–25 antibody molecules per site has been aggregated. Since the nanotubes conjugated with antibodies were rinsed three times before imaging, it is expected that all the antibody aggregates are covalently bonded to the carbon nanotubes. This shows that a higher number of antibody molecules per site can be attached to SWCNTs. Cancer cells incubated with a fluorescent specific-antibody-SWCNT complex were imaged by confocal microscopy in Fig. 5.10. Images were obtained at the same spot but using different filters. SWCNTs functionalized with Alexa Fluor 488 appeared green inside the cell and on the edges of the cell membrane while Alexa Fluor 555 delivered red color to the antibodies and observed mostly inside the cells. The less intense green fluorescence was caused by the limited binding effect of the dye and SWCNTs. The curvature of the SWCNTs also made them conjugate on the cell membrane, which was confirmed by the bright-green edges of the cells. Staining of cells with internalized SWCNTs with Trypan blue did not indicate cell death after incubation. No difference in the internalization was observed for IGF1R and Her2-specific antibodies that were functionalized on the surface of the SWCNTs. This process is also continuous and with long periods (>30 minutes) of time, the SWCNT get highly internalized into the cells due to the binding of the antibodies to their corresponding receptors. Nonspecific antibodies did not show such internalization but were attached to the circumference of the cell clusters. In recent studies, control studies using confocal microscopy indicated that when nanotubes were terminated with PEG molecules to prevent binding with biomolecules and treated with IGF1R-specific antibodies that were tagged with Alexa Fluor 555, and incubated with MCF7 cells, they showed no green fluorescence on the cells, illustrating no binding of SWCNTs on the membrane and no internalization as shown in Fig. 5.11a and 5.11b. Similarly, in a second control experiment, the cells were only incubated with SWCNTs tagged with Alexa Fluor 488; the absence of green fluorescence demonstrated that nanotubes by themselves do not get internalized within this period of time. Finally, SWCNTs tagged with Alexa 488 and IGF1R-specific antibodies tagged with Alexa 555 dye were mixed and the suspension was then incubated with MCF7 cancer cells for 30 minutes; the resulting images showed clear green fluorescence of nanotubes both on top of the cells as shown in Fig. 5.11c and Fig. 5.11d. These results clearly indicate that nanotubes are attached through the interaction of specific antibodies tagged to the nanotube surfaces with their corresponding antigens.

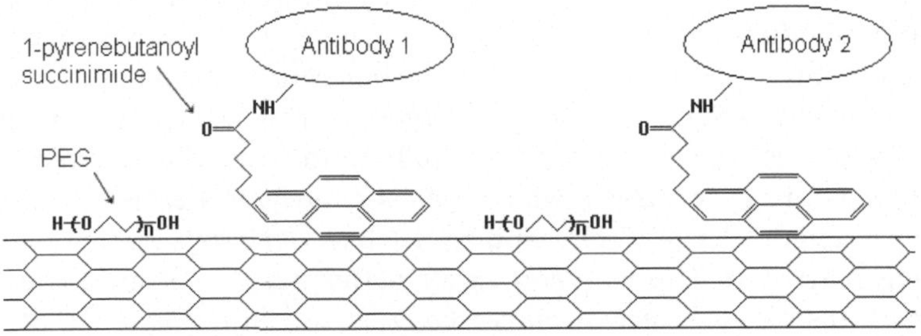

Figure 5.8 A functionalization schematic representation of SWCNTs with mAbs. 1-pyrenebutanoyl succinimide was adsorbed on the nanotube through π-π stacking. The succinimidyl ester can react with available lysine sidechain amines on antibodies to form amide bonds. PEG acts as an insulator to prevent potential biofouling. (Reprinted with permission from ref. 161.)

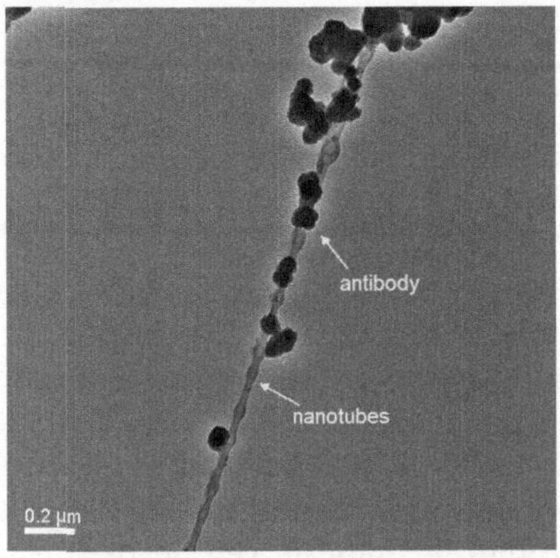

Figure 5.9 Transmission electron micrograph of the functionalization of a mAb aggregates specific to IGF1 receptor in breast cancer cells. Each site consists of 10–25 antibody molecules. (Reprinted with permission from ref. 161.)

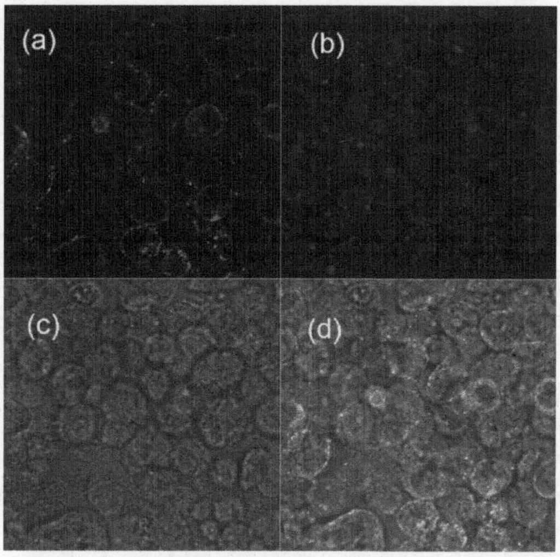

Figure 5.10 Confocal microscopy of specific antibody-facilitated SWCNT internalization of MCF7 cancerous cells. SWCNTs and IGF1R antibodies were tagged with Alexa Fluor 488 and Alexa Fluor 555 dye respectively. The green (a) and red (b) are fluorescent SWCNTs and antibodies, respectively. By overlapping (a), (b), and the optical image (c), facilitated SWCNT endocytosis could be confirmed by the image (d). (Reprinted with permission from ref. 161.) See also Color Insert.

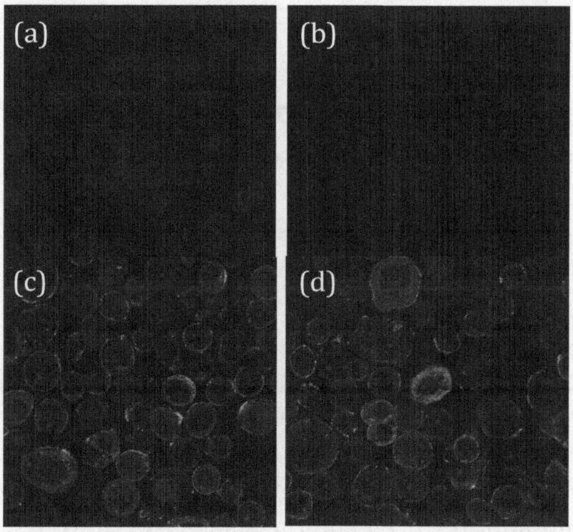

Figure 5.11 (a) The control group, SWCNT conjugated with Alexa Fluor 488, was completely terminated with PEG to prevent the binding with biomolecules (no 1-pyrenebutanoyl succinimide was adsorbed on the sidewall of the SWCNT). Such conjugate was mixed with complex Alexa Fluor 555 IGF1R and kept at room temperature for 30 minutes. MCF7 cancer cells were again incubated with in such a solution for 30 minutes. After which, the cells were triple rinsed and resuspended to remove excessive fluorescent dyes. Under this configuration, almost no green signal could be detected on or inside the cells. (b) In this control experiment, only SWCNT and Alexa Fluor 488 were incubated with MCF7 cells for 30 minutes and confocal images showed diminished green color. (c) and (d) are repeated experiments of the internalization of SWCNT. (Reprinted with permission from ref. [161].) See also Color Insert.

Finally, selective cell killing is achieved due to the high optical absorbance of SWCNT to infrared photons. Targeting surface receptors and NIR dosing, one can achieve a high degree of efficacy in the destruction of cancer cells. While past studies have shown the utility of nanotubes to transport DNA and act as NIR agents to kill cells [166], recent studies have shown that multicomponent targeting of more than one surface receptor (IGF1R and Her2) using SWCNT and simultaneous photothermal therapy due to their high optical absorbance can improve therapeutic efficacies. The energy used in the destruction of cancer cells was estimated to be ~200 nW per cell, too low to create any damage for the normal cells compared to past nanotechnology-based cell-killing techniques. Further, it was shown that nonspecific-antibody-tagged SWCNT when incubated with cells produced none or very less collateral damage. These results in principle show that some molecular targeting can be achieved using specific antibodies and the collateral damage can be minimized using NIR dosing of nanoparticles. By using nanostructures as biological transport carriers to target more than one surface marker, one can increase the efficiency of therapeutics that can lead to higher selectivity. Multicomponent targeting and internalization followed by photothermal cell killing can, therefore, improve the efficiency and selectivity of therapeutics. A wide variety of proteins and molecules capable of overcoming biophysical barriers can be functionalized on the surface of the nanotubes as transport carriers. Nanotubes act as nanoscale thermal sources for material processing at the nanoscale and also for changing the kinetics of reactions of proteins inside cells. These could become building blocks of noninvasive future cell repair machines using nanotechnology and optics.

5.10 Conclusions and Future Applications of Carbon Nanotubes for Biomedical Nanotechnology

Nanotechnology has shown the way for revolution in cancer research using molecular targeting and minimally invasive techniques. Nanotechnology can be used in three different areas of cancer research. These are (1) molecular imaging, (2) *ex vivo* diagnostics, and (3) *in vivo* therapeutics. However, there are tremendous challenges facing nanotechnologists in the future. First, although nanotechnology-based sensors work exceptionally well in the laboratory, detecting proteins *in vivo* using nanotechnology-based sensors is still a challenge. Molecular detection *in vivo* involves the use of sensors that have to be implanted inside the body to relay sensed information to external electronics or the use of nanosensors in combination with techniques such as MRI or CT scan. Despite years of research, the most important issue facing these tiny sensors is biofouling, or nonspecific adsorption of serum proteins on the sensing surface. This adsorption results in a rapid loss of sensitivity of the sensor. A challenge for researchers in this community may be to make self-cleaning sensors. As the field of nanotechnology grows, nanobiorobots that could potentially remove microscopic tumors in parts of the body that may be considered inaccessible with traditional surgery may become a reality. For such robots to become a reality, scientists from different disciplines such as physics, chemistry, biology, and electrical and mechanical engineering have to be involved in a common mission. Second, most of the nanovectors such as quantum dots that are used in high-contrast imaging are toxic and cannot be applied to the body. Although injectable nanovectors such as nanoparticles may not be toxic, they pose some risk when left in the body. Over long periods, nanoparticles may aggregate and potentially block arteries and veins or even the kidneys, thereby creating a host of new problems that have yet not been explored. For any nanovector to be successful in the clinic, it has to be completely destroyed or biodegradable *in vivo*. In that case, the use of biodegradable polymeric nanoparticles or naturally occurring materials as nanoparticles might give better biodistribution in the body. Nanovectors might also trigger immunogenic response. For example, antibodies specific to fullerenes have been described, and dendrimers and protein–dendrimer conjugates have shown strong immunogenic responses in these studies [167, 168]. Therefore, counter measures must be developed to suppress such reactions. Third, there are numerous biophysical barriers such as increased osmotic pressures in malignant cancer tissues that make drug delivery ineffective [169]. Due to these increased pressures within the cells, the diffusion of therapeutic agents and nanovectors may become problematic. Hence, new solutions such as surface modifiers are necessary for overcoming biophysical barriers. Another solution may be the development of designer drugs to reduce the osmotic pressure in cells and tissues. One creative approach may be the use of multiple agents and approaches within a single nanovector. For example, using chemotherapeutic drugs and radiologic drugs encapsulated within nanoparticles may be highly successful in overcoming the barriers. Although hundreds of nanovectors have been proposed in the literature, to be successful, several techniques may have to be combined in to a single nanovector "cocktail" to improve targeting efficacy, to overcome biologic barriers, and to kill cancer cells. In all these techniques, engineering principles are being increasingly applied to solve a medical-related problem, thereby giving birth to the field of "nano-oncology" for cancer treatment. These techniques can be customized for killing different types of cancer, and, therefore, nanovectors can serve as a generic platform for various types of cancer treatments.

Carbon nanotubes have emerged as a potential material for the creation of *ex vivo* diagnostic devices for detecting molecular overexpression in cells and detection of very small levels of proteins in blood. Further, the optical properties of carbon nanotubes have been successfully utilized for the selective killing of cancer cells. Despite this success, challenges remain. Growth of nanotubes directly onto prepatterned devices at specific locations and imparting biological functionality to arrays of devices are still a challenge and are just starting to be addressed. Second, the toxicity of nanotubes has to be still addressed for its application *in vivo*. Third, processes for the precise cutting of nanotubes into very small lengths (<500 nm) needs to developed further for *in vivo* therapeutics. Fourth, the use of carbon nanotube thin films instead of a network of nanotubes or bundles needs to be investigated. Thin films of carbon nanotubes can be processed and patterned very easily using batch fabrication techniques. Arrays of such devices could in the future emerge as potential contenders for cancer research.

Acknowledgments

This publication was made possible in part by the National Science Foundation CAREER award, grant no. ECCS-0853066; and NIH 1R15CA156322

References

1. M. Ferrari, "Cancer nanotechnology: Opportunities and challenges," *Nature Reviews Cancer*, vol. 5, pp. 161–171, Mar 2005.
2. M. Ferrari, "Nanovector therapeutics," *Current Opinion in Chemical Biology*, vol. 9, pp. 343–346, Aug 2005.
3. J. W. Park, "Liposome-based drug delivery in breast cancer treatment," *Breast Cancer Research*, vol. 4, pp. 93–97, 2002.
4. M. F. Kircher, U. Mahmood, R. S. King, R. Weissleder, and L. Josephson, "A multimodal nanoparticle for preoperative magnetic resonance imaging and intraoperative optical brain tumor delineation," *Cancer Research*, vol. 63, pp. 8122–8125, Dec 1 2003.
5. M. Harisinghani, J. Barentsz, and R. Weissleder, "Use of MRI to detect lymph-node metastases in prostate cancer - Reply," *New England Journal of Medicine*, vol. 349, pp. 1186–1186, Sep 18 2003.
6. A. Majumdar, "Bioassays based on molecular nanomechanics," *Disease Markers*, vol. 18, pp. 167–174, 2002.
7. Y. Cui, Q. Q. Wei, H. K. Park, and C. M. Lieber, "Nanowire nanosensors for highly sensitive and selective detection of biological and chemical species," *Science*, vol. 293, pp. 1289–1292, Aug 17 2001.
8. N. Shao, E. Wickstrom, and B. Panchapakesan, "Nanotube-antibody biosensor arrays for the detection of circulating breast cancer cells," *Nanotechnology*, vol. 19, Nov 19 2008.
9. B. Dubertret, P. Skourides, D. J. Norris, V. Noireaux, A. H. Brivanlou, and A. Libchaber, "In vivo imaging of quantum dots encapsulated in phospholipid micelles," *Science*, vol. 298, pp. 1759–1762, Nov 29 2002.
10. X. Y. Wu, H. J. Liu, J. Q. Liu, K. N. Haley, J. A. Treadway, J. P. Larson, N. F. Ge, F. Peale, and M. P. Bruchez, "Immunofluorescent labeling of cancer marker Her2 and other cellular targets with semiconductor quantum dots," *Nature Biotechnology*, vol. 21, pp. 41–46, Jan 2003.
11. H. Kobayashi, S. Kawamoto, Y. Sakai, P. L. Choyke, R. A. Star, M. W. Brechbiel, N. Sato, Y. Tagaya, J. C. Morris, and T. A. Waldmann, "Lymphatic drainage imaging of breast cancer in mice by micro-magnetic resonance lymphangiography using a nano-size paramagnetic contrast agent," *Journal of the National Cancer Institute*, vol. 96, pp. 703–708, May 5 2004.
12. E. A. Schellenberger, A. Bogdanov, A. Petrovsky, V. Ntziachristos, R. Weissleder, and L. Josephson, "Optical imaging of apoptosis as a biomarker of tumor response to chemotherapy," *Neoplasia*, vol. 5, pp. 187–192, May–Jun 2003.
13. P. M. Winter, A. M. Morawski, S. Caruthers, T. Harris, J. S. Allen, H. Y. Zhang, R. W. Fuhrhop, E. K. Lacy, T. A. Williams, G. M. Lanza, and S. A. Wickline, "Specific molecular imaging of vasa vasorum in early atherosclerosis with alpha(nu)beta(3)-integrin targeted nanoparticles," *Circulation*, vol. 108, pp. 168–168, Oct 28 2003.
14. S. A. Anderson, R. K. Rader, W. F. Westlin, C. Null, D. Jackson, C. M. Lanza, S. A. Wickline, and J. J. Kotyk, "Magnetic resonance contrast enhancement of neovasculature with alpha(v)beta(3)-targeted nanoparticles," *Magnetic Resonance in Medicine*, vol. 44, pp. 433–439, Sep 2000.
15. M. Krause, K. K. Kwong, J. Xiong, E. S. Gragoudas, and L. H. Y. Young, "MRI of blood volume and cellular uptake of superparamagnetic iron in an animal model of choroidal melanoma," *Ophthalmic Research*, vol. 34, pp. 241–250, Jul–Aug 2002.

16. C. H. Dodd, H. C. Hsu, W. J. Chu, P. G. Yang, H. G. Zhang, J. D. Mountz, K. Zinn, J. Forder, L. Josephson, R. Weissleder, J. M. Mountz, and J. D. Mountz, "Normal T-cell response and in vivo magnetic resonance imaging of T cells loaded with HIV transactivator-peptide-derived superparamagnetic nanoparticles," *Journal of Immunological Methods*, vol. 256, pp. 89–105, Oct 1 2001.
17. S. J. Lee, J. R. Jeong, S. C. Shin, J. C. Kim, Y. H. Chang, Y. M. Chang, and J. D. Kim, "Nanoparticles of magnetic ferric oxides encapsulated with poly(D,L latide-co-glycolide) and their applications to magnetic resonance imaging contrast agent," *Journal of Magnetism and Magnetic Materials*, vol. 272, pp. 2432–2433, May 2004.
18. H. Kobayashi, S. Kawamoto, M. W. Brechbiel, M. Bernardo, N. Sato, T. A. Waldmann, Y. Tagaya, and P. L. Choyke, "Detection of lymph node involvement in hematologic malignancies using micromagnetic resonance lymphangiography with a gadolinum-labeled dendrimer nanoparticle," *Neoplasia*, vol. 7, pp. 984–991, Nov 2005.
19. A. Oberlin, M. Endo, and T. Koyama, "Filamentous growth of carbon through benzene decomposition," *Journal of Crystal Growth*, vol. 32, p. 335, 1976.
20. A. Oberlin, M. Endo, and T. Koyama, "High resolution electron microscope observations of graphitized carbon fibers," *Carbon*, vol. 14, p. 133, 1976.
21. S. Iijima, "Direct observation of the tetrahedral bonding in graphitized carbon black by high resolution electron microscopy," *Journal of Crystal Growth*, vol. 50, p. 675, 1980.
22. S. Iijima, "Helical microtubules of graphitic carbon," *Nature*, vol. 354, pp. 56–58, 1991.
23. M. S. Dresselhaus, G. Dresselhaus, and P. Avouris, *Carbon Nanotubes Synthesis, Structure, Properties, and Applications*, 1 ed. vol. 80. Berlin: Springer, 2001.
24. S. Iijima and T. Ichihashi, "Single-shell carbon nanotubes of 1-nm diameter," *Nature*, vol. 363, pp. 603–615, 1993.
25. M. S. Dresselhaus, G. Dresselhaus, and R. Saito, "Carbon fibers based on C_{60} and their symmetry," *Physical Review B*, vol. 45, p. 6234, 1992.
26. P. M. Ajayan and S. Lijima, "Smallest carbon nanotube," *Nature*, vol. 358, p. 23, 1992.
27. D. S. Bethune, C. H. Klang, M. S. de Vries, G. Gorman, R. Savoy, J. Vazquez, and R. Beyers, "Cobalt-catalysed growth of carbon nanotubes with single-atomic-layer walls," *Nature*, vol. 363, p. 605, 1993.
28. J. C. Charlier and J. P. Michenaud, "Energetics of multilayered carbon tubules," *Physical Review Letters*, vol. 70, p. 1858, 1993.
29. X. F. Zhang, X. B. Zhang, G. Van Tendeloo, S. Amelinckx, M. Op de Beeck, and J. Van Landuyt, "Carbon nano-tubes; their formation process and observation by electron microscopy," *Journal of Crystal Growth*, vol. 130, p. 368, 1993.
30. P. M. Ajayan and T. W. Ebbesen, "Nanometre-size tubes of carbon," *Reports on Progress in Physics*, vol. 60, p. 1025, 1997.
31. C. T. White, D. H. Robertson, and J. W. Mintmire, "Helical and rotational symmetries of nanoscale graphitic tubules," *Physical Review B*, vol. 47, p. 5485, 1993.
32. M. Y. Sfeir, T. Beetz, F. Wang, L. Huang, X. M. H. Huang, M. Huang, J. Hone, S. O'Brien, J. A. Misewich, T. F. Heinz, L. Wu, Y. Zhu, and L. E. Brus, "Optical Spectroscopy of Individual Single-Walled Carbon Nanotubes of Defined Chiral Structure," *Science*, vol. 312, p. 554, 2006.
33. T. W. Odom, J.-L. Huang, P. Kim, and C. M. Lieber, "Atomic structure and electronic properties of single-walled carbon nanotubes," *Nature*, vol. 391, p. 62, 1998.

34. S. V. Ahir and E. M. Terentjev, "Polymers Containing Carbon Nanotubes: Active Composite Materials," in *Polymeric Nanostructures and Their Applications*. vol. 1, H. S. Nalwa, Ed., ed Valencia: American Scientific Publishers, 2006, p. 1.
35. X. Blase, J. C. Charlier, A. De Vita, R. Car, P. Redlich, M. Terrones, W. K. Hsu, H. Terrones, D. L. Carroll, and P. M. Ajayan, "Boron-Mediated Growth of Long Helicity-Selected Carbon Nanotubes," *Physical Review Letters*, vol. 83, p. 5078, 1999.
36. T. Mauricio, "Controlling Nanotube Chirality and Crystallinity by Doping," *Small*, vol. 1, pp. 1032–1034, 2005.
37. J. G. Wiltshire, L.-J. Li, L. M. Herz, R. J. Nicholas, M. Glerup, J.-L. Sauvajol, and A. N. Khlobystov, "Chirality-dependent boron-mediated growth of nitrogen-doped single-walled carbon nanotubes," *Physical Review B*, vol. 72, p. 205431, 2005.
38. M. Endo, K. Takeuchi, K. Kobori, K. Takahashi, H. W. Kroto, and A. Sarkar, "Pyrolytic carbon nanotubes from vapor-grown carbon fibers," *Carbon*, vol. 33, p. 873, 1995.
39. C.-H. Kiang, W. A. Goddard, R. Beyers, and D. S. Bethune, "Carbon nanotubes with single-layer walls," *Carbon*, vol. 33, p. 903, 1995.
40. M. T. Martínez, M. A. Callejas, A. M. Benito, M. Cochet, T. Seeger, A. Ansón, J. Schreiber, C. Gordon, C. Marhic, O. Chauvet, and W. K. Maser, "Modifications of single-wall carbon nanotubes upon oxidative purification treatments," *Nanotechnology*, vol. 14, p. 691, 2003.
41. Z. Osvath, G. Vertesy, L. Tapaszto, F. Weber, Z. E. Horvath, J. Gyulai, and L. P. Biro, "Atomically resolved STM images of carbon nanotube defects produced by Ar^+. irradiation," *Physical Review B*, vol. 72, p. 045429, 2005.
42. L. Chico, V. H. Crespi, L. X. Benedict, S. G. Louie, and M. L. Cohen, "Pure Carbon Nanoscale Devices: Nanotube Heterojunctions," *Physical Review Letters*, vol. 76, p. 971, 1996.
43. J. W. Mintmire, B. I. Dunlap, and C. T. White, "Are fullerene tubules metallic?," *Physical Review Letters*, vol. 68, p. 631, 1992.
44. X. Blase, L. X. Benedict, E. L. Shirley, and S. G. Louie, "Hybridization effects and metallicity in small radius carbon nanotubes," *Physical Review Letters*, vol. 72, p. 1878, 1994.
45. R. Saito, M. Fujita, G. Dresselhaus, and M. S. Dresselhaus, "Electronic structure of chiral graphene tubules," *Applied Physics Letters*, vol. 60, p. 2204, 1992.
46. A. Jorio, R. Saito, T. Hertel, R.B. Weisman, G. Dresselhaus, and M. S. Dresselhaus, "Carbon Nanotube Photophysics," *MRS Bull*, vol. 29, p. 276, 2004.
47. R. Martel, T. Schmidt, H. R. Shea, T. Hertel, and A. Ph, "Single- and multi-wall carbon nanotube field-effect transistors," *Applied Physics Letters*, vol. 73, pp. 2447–2449, 1998.
48. S. J. Tans, A. R. M. Verschueren, and C. Dekker, "Room-temperature transistor based on a single carbon nanotube," *Nature*, vol. 393, p. 49, 1998.
49. J. A. Misewich, R. Martel, P. Avouris, J. C. Tsang, S. Heinze, and J. Tersoff, "Electrically Induced Optical Emission from a Carbon Nanotube FET," *Science*, vol. 300, pp. 783–786, May 2, 2003 2003.
50. K. Keren, R. S. Berman, E. Buchstab, U. Sivan, and E. Braun, "DNA-Templated Carbon Nanotube Field-Effect Transistor," *Science*, vol. 302, pp. 1380–1382, November 21, 2003.
51. S. Heinze, J. Tersoff, R. Martel, V. Derycke, J. Appenzeller, and P. Avouris, "Carbon Nanotubes as Schottky Barrier Transistors," *Physical Review Letters*, vol. 89, p. 106801, 2002.
52. A. Javey, J. Guo, Q. Wang, M. Lundstrom, and H. Dai, "Ballistic carbon nanotube field-effect transistors," *Nature*, vol. 424, p. 654, 2003.

53. Y. Homma, T. Yamashita, Y. Kobayashi, and T. Ogino, "Interconnection of nanostructures using carbon nanotubes," *Physica B: Condensed Matter*, vol. 323, p. 122, 2002.

54. A. Naeemi and J. D. Meindl, "Impact of electron-phonon scattering on the performance of carbon nanotube interconnects for GSI," *Electron Device Letters, IEEE*, vol. 26, pp. 476–478, 2005.

55. Y. Li, D. Mann, M. Rolandi, W. Kim, A. Ural, S. Hung, A. Javey, J. Cao, D. Wang, E. Yenilmez, Q. Wang, J. F. Gibbons, Y. Nishi, and H. Dai, "Preferential Growth of Semiconducting Single-Walled Carbon Nanotubes by a Plasma Enhanced CVD Method," *Nano Lett.*, vol. 4, pp. 317–321, 2004.

56. R. Krupke, F. Hennrich, H. v. Lohneysen, and M. M. Kappes, "Separation of Metallic from Semiconducting Single-Walled Carbon Nanotubes," *Science*, vol. 301, p. 344, 2003.

57. M. Zheng, A. Jagota, E. Semke, B. A. Diner, R. S. Mclean, S. R. Lustig, R. E. Richardson, and N. G. Tassi, "DNA-assisted dispersion and separation of carbon nanotubes," *Nature Materials*, vol. 2, pp. 338–342, 2003.

58. M. Zheng, A. Jagota, M. S. Strano, A. P. Santos, P. Barone, S. G. Chou, B. A. Diner, M. S. Dresselhaus, R. S. McLean, G. B. Onoa, G. G. Samsonidze, E. D. Semke, M. Usrey, and D. J. Walls, "Structure-Based Carbon Nanotube Sorting by Sequence-Dependent DNA Assembly," *Science*, vol. 302, p. 1545, 2003.

59. A. Hassanien, M. Tokumoto, P. Umek, D. Vrbanic, M. Mozetic, D. Mihailovic, P. Venturini, and S. Pejovnik, "Selective etching of metallic single-wall carbon nanotubes with hydrogen plasma," *Nanotechnology*, vol. 16, p. 278, 2005.

60. G. Zhang, P. Qi, X. Wang, Y. Lu, X. Li, R. Tu, S. Bangsaruntip, D. Mann, L. Zhang, and H. Dai, "Selective Etching of Metallic Carbon Nanotubes by Gas-Phase Reaction," *Science*, vol. 314, p. 974, 2006.

61. S. Li, Z. Yu, S. F. Yen, W. C. Tang, and P. J. Burke, "Carbon Nanotube Transistor Operation at 2.6 GHz," *Nano Letters*, vol. 4, p. 753, 2004.

62. H. W. C. Postma, T. Teepen, Z. Yao, M. Grifoni, and C. Dekker, "Carbon Nanotube Single-Electron Transistors at Room Temperature," *Science*, vol. 293, pp. 76–79, July 6, 2001.

63. R. D. Antonov and A. T. Johnson, "Subband Population in a Single-Wall Carbon Nanotube Diode," *Physical Review Letters*, vol. 83, p. 3274, 1999.

64. A. Bachtold, P. Hadley, T. Nakanishi, and C. Dekker, "Logic Circuits with Carbon Nanotube Transistors," *Science*, vol. 294, pp. 1317–1320, November 9, 2001.

65. T. Rueckes, K. Kim, E. Joselevich, G. Y. Tseng, C.-L. Cheung, and C. M. Lieber, "Carbon Nanotube-Based Nonvolatile Random Access Memory for Molecular Computing," *Science*, vol. 289, p. 94, 2000.

66. P. G. Collins, M. S. Arnold, and P. Avouris, "Engineering Carbon Nanotubes and Nanotube Circuits Using Electrical Breakdown," *Science*, vol. 292, p. 706, 2001.

67. Z. Chen, J. Appenzeller, Y.-M. Lin, J. Sippel-Oakley, A. G. Rinzler, J. Tang, S. J. Wind, P. M. Solomon, and P. Avouris, "An Integrated Logic Circuit Assembled on a Single Carbon Nanotube," *Science*, vol. 311, p. 1735, 2006.

68. H. T. Soh, C. F. Quate, A. F. Morpurgo, C. M. Marcus, J. Kong, and H. Dai, "Integrated nanotube circuits: Controlled growth and ohmic contacting of single-walled carbon nanotubes," *Applied Physics Letters*, vol. 75, p. 627, 1999.

69. V. Derycke, R. Martel, J. Appenzeller, and P. Avouris, "Carbon Nanotube Inter- and Intramolecular Logic Gates," *Nano Letters*, vol. 1, p. 453, 2001.

70. P. G. Collins, K. Bradley, M. Ishigami, and A. Zettl, "Extreme Oxygen Sensitivity of Electronic Properties of Carbon Nanotubes," *Science*, vol. 287, pp. 1801–1804, March 10, 2000.
71. K. Jing, Z. Chongwu, Y. Erhan, and D. Hongjie, "Alkaline metal-doped n-type semiconducting nanotubes as quantum dots," *Applied Physics Letters*, vol. 77, pp. 3977–3979, 2000.
72. T. Durkop, B. M. Kim, and M. S. Fuhrer, "Properties and applications of high-mobility semiconducting nanotubes," *Journal of Physics: Condensed Matter*, vol. 16, p. R553, 2004.
73. B. Q. Wei, R. Vajtai, and P. M. Ajayan, "Reliability and current carrying capacity of carbon nanotubes," *Applied Physics Letters*, vol. 79, p. 1172, 2001.
74. W. A. D. Heer, "Nanotubes and the Pursuit of Applications," *MRS Bull*, vol. 29, p. 281, 2004.
75. S. Frank, P. Poncharal, Z. L. Wang, and W. A. Heer, "Carbon Nanotube Quantum Resistors," *Science*, vol. 280, p. 1744, 1998.
76. M. Terrones, "SCIENCE AND TECHNOLOGY OF THE TWENTY-FIRST CENTURY: Synthesis, Properties, and Applications of Carbon Nanotubes," *Annual Review of Materials Research*, vol. 33, p. 419, 2003.
77. P. Lambin, A. Fonseca, J. P. Vigneron, J. B. Nagy, and A. A. Lucas, "Structural and electronic properties of bent carbon nanotubes," *Chemical Physics Letters*, vol. 245, p. 85, 1995.
78. R. Saito, G. Dresselhaus, and M. S. Dresselhaus, "Tunneling conductance of connected carbon nanotubes," *Physical Review B*, vol. 53, p. 2044, 1996.
79. L. Chico, L. X. Benedict, S. G. Louie, and M. L. Cohen, "Quantum conductance of carbon nanotubes with defects," *Physical Review B*, vol. 54, p. 2600, 1996.
80. B. I. Dunlap, "Relating carbon tubules," *Physical Review B*, vol. 49, p. 5643, 1994.
81. J. C. Charlier, T. W. Ebbesen, and P. Lambin, "Structural and electronic properties of pentagon-heptagon pair defects in carbon nanotubes," *Physical Review B*, vol. 53, p. 11108, 1996.
82. T. W. Ebbesen and T. Takada, "Topological and SP3 defect structures in nanotubes," *Carbon*, vol. 33, p. 973, 1995.
83. W. Joseph, "Carbon-Nanotube Based Electrochemical Biosensors: A Review," *Electroanalysis*, vol. 17, pp. 7–14, 2005.
84. Y. Lin, S. Taylor, H. Li, K. A. S. Fernando, L. Qu, W. Wang, L. Gu, B. Zhou, and Y.-P. Sun, "Advances toward bioapplications of carbon nanotubes," *Journal of Materials Chemistry*, vol. 14, pp. 527–541, 2004.
85. R. J. Chen, S. Bangsaruntip, K. A. Drouvalakis, N. Wong Shi Kam, M. Shim, Y. Li, W. Kim, P. J. Utz, and H. Dai, "Noncovalent functionalization of carbon nanotubes for highly specific electronic biosensors," *PNAS*, vol. 100, pp. 4984–4989, April 29, 2003.
86. J. Kong, N. R. Franklin, C. Zhou, M. G. Chapline, S. Peng, K. Cho, and H. Dai, "Nanotube Molecular Wires as Chemical Sensors," *Science*, vol. 287, p. 622, 2000.
87. J. Kong, M. G. Chapline, and H. Dai, "Functionalized Carbon Nanotubes for Molecular Hydrogen Sensors," *Advanced Materials*, vol. 13, pp. 1384–1386, 2001.
88. M. W. C. M. Punit Kohli, "Nanotube Membrane Based Biosensors," *Electroanalysis*, vol. 16, pp. 9–18, 2004.
89. H. Ajiki and T. Ando, "Aharonov-Bohm effect in carbon nanotubes," *Physica B: Condensed Matter*, vol. 201, p. 349, 1994.
90. S. Kazaoui, N. Minami, R. Jacquemin, H. Kataura, and Y. Achiba, "Amphoteric doping of single-wall carbon-nanotube thin films as probed by optical absorption spectroscopy," *Physical Review B*, vol. 60, p. 13339, 1999.

91. A. M. Rao, E. Richter, S. Bandow, B. Chase, P. C. Eklund, K. A. Williams, S. Fang, K. R. Subbaswamy, M. Menon, A. Thess, R. E. Smalley, G. Dresselhaus, and M. S. Dresselhaus, "Diameter-Selective Raman Scattering from Vibrational Modes in Carbon Nanotubes," *Science*, vol. 275, p. 187, 1997.
92. A. Kasuya, Y. Sasaki, Y. Saito, K. Tohji, and Y. Nishina, "Evidence for Size-Dependent Discrete Dispersions in Single-Wall Nanotubes," *Physical Review Letters*, vol. 78, p. 4434, 1997.
93. M. A. Pimenta, A. Marucci, S. A. Empedocles, M. G. Bawendi, E. B. Hanlon, A. M. Rao, P. C. Eklund, R. E. Smalley, G. Dresselhaus, and M. S. Dresselhaus, "Raman modes of metallic carbon nanotubes," *Physical Review B*, vol. 58, p. R16016, 1998.
94. H. Kataura, Y. Kumazawa, Y. Maniwa, I. Umezu, S. Suzuki, Y. Ohtsuka, and Y. Achiba, "Optical properties of single-wall carbon nanotubes," *Synthetic Metals*, vol. 103, p. 2555, 1999.
95. S. M. Bachilo, M. S. Strano, C. Kittrell, R. H. Hauge, R. E. Smalley, and R. B. Weisman, "Structure-Assigned Optical Spectra of Single-Walled Carbon Nanotubes," *Science*, vol. 298, pp. 2361–2366, December 20, 2002.
96. M. Ichida, S. Mizuno, Y. Tani, Y. Saito, and A. Nakamura, "Exciton Effects of Optical Transitions in Single-Wall Carbon Nanotubes," *Journal of the Physical Society of Japan*, vol. 68, pp. 3131–3133, 1999.
97. R. Saito, G. Dresselhaus, and M. S. Dresselhaus, "Trigonal warping effect of carbon nanotubes," *Physical Review B*, vol. 61, p. 2981, 2000.
98. D.-H. Lien, H.-F. Kuo, and W.-K. Hsu, "Segmentation of single-walled carbon nanotubes by camera flash," *Applied Physics Letters*, vol. 88, p. 093113, 2006.
99. J. Smits, B. Wincheski, M. Namkung, R. Crooks, and R. Louie, "Response of Fe powder, purified and as-produced HiPco single-walled carbon nanotubes to flash exposure," *Materials Science and Engineering A*, vol. 358, p. 384, 2003.
100. P. M. Ajayan, M. Terrones, A. de la Guardia, V. Huc, N. Grobert, B. Q. Wei, H. Lezec, G. Ramanath, and T. W. Ebbesen, "Nanotubes in a Flash–Ignition and Reconstruction," *Science*, vol. 296, p. 705, April 26, 2002.
101. Y. Zhang and S. Iijima, "Elastic Response of Carbon Nanotube Bundles to Visible Light," *Physical Review Letters*, vol. 82, pp. 3472–3475, 1999.
102. S. C. Tsang, P. J. F. Harris, and M. L. H. Green, "Thinning and opening of carbon nanotubes by oxidation using carbon dioxide," *Nature*, vol. 362, p. 520, 1993.
103. P. M. Ajayan, T. W. Ebbesen, T. Ichihashi, S. Iijima, K. Tanigaki, and H. Hiura, "Opening carbon nanotubes with oxygen and implications for filling," *Nature*, vol. 362, p. 522, 1993.
104. P. Nikolaev, A. Thess, A. G. Rinzler, D. T. Colbert, and R. E. Smalley, "Diameter doubling of single-wall nanotubes," *Chemical Physics Letters*, vol. 266, p. 422, 1997.
105. N. W. Shi Kam, M. O'Connell, J. A. Wisdom, and H. Dai, "Carbon nanotubes as multifunctional biological transporters and near-infrared agents for selective cancer cell destruction," *PNAS*, vol. 102, pp. 11600–11605, August 16, 2005.
106. B. Panchapakesan, S. Lu, K. Sivakumar, K. Teker, G. Cesarone, and E. Wickstrom, "Single-Wall Carbon Nanotube Nanobomb Agents for Killing Breast Cancer Cells," *NanoBiotechnology*, vol. 1, pp. 133–140, 2005.
107. A. Hirsch, "Functionalization of single-walled carbon nanotubes," *Angewandte Chemie-International Edition*, vol. 41, pp. 1853–1859, 2002.

108. S. Niyogi, M. A. Hamon, H. Hu, B. Zhao, P. Bhowmik, R. Sen, M. E. Itkis, and R. C. Haddon, "Chemistry of single-walled carbon nanotubes," *Accounts of Chemical Research*, vol. 35, pp. 1105–1113, Dec 2002.
109. Y. P. Sun, K. F. Fu, Y. Lin, and W. J. Huang, "Functionalized carbon nanotubes: Properties and applications," *Accounts of Chemical Research*, vol. 35, pp. 1096–1104, Dec 2002.
110. X. Lu, F. Tian, Y. B. Feng, X. Xu, N. Q. Wang, and Q. N. Zhang, "Sidewall oxidation and complexation of carbon nanotubes by base-catalyzed cycloaddition of transition metal oxide: A theoretical prediction," *Nano Letters*, vol. 2, pp. 1325–1327, Nov 2002.
111. X. Lu, F. Tian, N. Q. Wang, and Q. N. Zhang, "Organic functionalization of the sidewalls of carbon nanotubes by Diels-Alder reactions: A theoretical prediction," *Organic Letters*, vol. 4, pp. 4313–4315, Nov 28 2002.
112. C. V. Nguyen, L. Delzeit, A. M. Cassell, J. Li, J. Han, and M. Meyyappan, "Preparation of nucleic acid functionalized carbon nanotube Arrays," *Nano Letters*, vol. 2, pp. 1079–1081, Oct 2002.
113. W. R. Yang, M. J. Moghaddam, S. Taylor, B. Bojarski, L. Wieczorek, J. Herrmann, and M. J. McCall, "Single-walled carbon nanotubes with DNA recognition," *Chemical Physics Letters*, vol. 443, pp. 169–172, Aug 6 2007.
114. S. C. Tsang, Z. J. Guo, Y. K. Chen, M. L. H. Green, H. A. O. Hill, T. W. Hambley, and P. J. Sadler, "Immobilization of platinated and iodinated oligonucleotides on carbon nanotubes," *Angewandte Chemie-International Edition in English*, vol. 36, pp. 2198–2200, Nov 3 1997.
115. N. Nakashima, Y. Tomonari, and H. Murakami, "Water-soluble single-walled carbon nanotubes via noncovalent sidewall-functionalization with a pyrene-carrying ammonium ion," *Chemistry Letters*, pp. 638–639, Jun 5 2002.
116. J. Chen, H. Y. Liu, W. A. Weimer, M. D. Halls, D. H. Waldeck, and G. C. Walker, "Noncovalent engineering of carbon nanotube surfaces by rigid, functional conjugated polymers," *Journal of the American Chemical Society*, vol. 124, pp. 9034–9035, Aug 7 2002.
117. A. Star, Y. Liu, K. Grant, L. Ridvan, J. F. Stoddart, D. W. Steuerman, M. R. Diehl, A. Boukai, and J. R. Heath, "Noncovalent side-wall functionalization of single-walled carbon nanotubes," *Macromolecules*, vol. 36, pp. 553–560, Feb 11 2003.
118. R. J. Chen, S. Bangsaruntip, K. A. Drouvalakis, N. W. S. Kam, M. Shim, Y. M. Li, W. Kim, P. J. Utz, and H. J. Dai, "Noncovalent functionalization of carbon nanotubes for highly specific electronic biosensors," *Proceedings of the National Academy of Sciences of the United States of America*, vol. 100, pp. 4984–4989, Apr 29 2003.
119. P. G. Collins, A. Zettl, H. Bando, A. Thess, and R. E. Smalley, "Nanotube nanodevice," *Science*, vol. 278, pp. 100–103, Oct 3 1997.
120. L. P. Jones, S. Stefansson, M. S. Kim, and S. N. Ahn, "Comparison of Radioimmuno and Carbon Nanotube Field-Effect Transistor Assays for Measuring Insulin-Like Growth Factor-1 in a Preclinical Model of Human Breast Cancer," *Journal of Nanobiotechnology*, vol. 9, Sep 2 2011.
121. P. A. Hu, J. Zhang, Z. Z. Wen, and C. Zhang, "Network single-walled carbon nanotube biosensors for fast and highly sensitive detection of proteins," *Nanotechnology*, vol. 22, Aug 19 2011.
122. J. W. Ko, J. M. Woo, J. Ahn, J. H. Cheon, J. H. Lim, S. H. Kim, H. Chun, E. Kim, and Y. J. Park, "Multi-Order Dynamic Range DNA Sensor Using a Gold Decorated SWCNT Random Network," *Acs Nano*, vol. 5, pp. 4365–4372, Jun 2011.
123. J. P. Kim, S. Hong, and S. J. Sim, "Apta-Biosensors for Nonlabeled Real Time Detection of Human IgE Based on Carbon Nanotube Field Effect Transistors," *Journal of Nanoscience and Nanotechnology*, vol. 11, pp. 4182–4187, May 2011.

124. K. I. Chen, B. R. Li, and Y. T. Chen, "Silicon nanowire field-effect transistor-based biosensors for biomedical diagnosis and cellular recording investigation," *Nano Today*, vol. 6, pp. 131–154, Apr 2011.

125. T. H. Kim, H. S. Song, H. J. Jin, S. H. Lee, S. Namgung, U. K. Kim, T. H. Park, and S. Hong, ""Bioelectronic super-taster" device based on taste receptor-carbon nanotube hybrid structures," *Lab on a Chip*, vol. 11, pp. 2262–2267, 2011.

126. P. A. Hu, J. Zhang, L. Li, Z. L. Wang, W. O'Neill, and P. Estrela, "Carbon Nanostructure-Based Field-Effect Transistors for Label-Free Chemical/Biological Sensors," *Sensors*, vol. 10, pp. 5133–5159, May 2010.

127. M. Abe, K. Murata, and K. Matsumoto, "Dependence of sensitivity of biosensor for carbon nanotube field-effect transistor with top-gate structures," *Journal of Applied Physics*, vol. 107, Apr 15 2010.

128. S. M. Khamis, R. R. Johnson, Z. T. Luo, and A. T. C. Johnson, "Homo-DNA functionalized carbon nanotube chemical sensors," *Journal of Physics and Chemistry of Solids*, vol. 71, pp. 476–479, Apr 2010.

129. Y. X. Huang, P. V. Palkar, L. J. Li, H. Zhang, and P. Chen, "Integrating carbon nanotubes and lipid bilayer for biosensing," *Biosensors & Bioelectronics*, vol. 25, pp. 1834–1837, Mar 15 2010.

130. S. Mao, G. H. Lu, K. H. Yu, and J. H. Chen, "Specific biosensing using carbon nanotubes functionalized with gold nanoparticle-antibody conjugates," *Carbon*, vol. 48, pp. 479–486, Feb 2010.

131. S. Kim, T. G. Kim, H. R. Byon, H. J. Shin, C. Ban, and H. C. Choi, "Recognition of Single Mismatched DNA Using MutS-Immobilized Carbon Nanotube Field Effect Transistor Devices," *Journal of Physical Chemistry B*, vol. 113, pp. 12164–12168, Sep 10 2009.

132. J. P. Kim, B. Y. Lee, J. Lee, S. Hong, and S. J. Sim, "Enhancement of sensitivity and specificity by surface modification of carbon nanotubes in diagnosis of prostate cancer based on carbon nanotube field effect transistors," *Biosensors & Bioelectronics*, vol. 24, pp. 3372–3378, Jul 15 2009.

133. R. A. Villamizar, A. Maroto, and F. X. Rius, "Improved detection of Candida albicans with carbon nanotube field-effect transistors," *Sensors and Actuators B-Chemical*, vol. 136, pp. 451–457, Mar 2 2009.

134. H. R. Byon, S. Kim, and H. C. Choi, "Label-Free Biomolecular Detection Using Carbon Nanotube Field Effect Transistors," *Nano*, vol. 3, pp. 415–431, Dec 2008.

135. E. Stern, A. Vacic, and M. A. Reed, "Semiconducting Nanowire Field-Effect Transistor Biomolecular Sensors," *Ieee Transactions on Electron Devices*, vol. 55, pp. 3119–3130, Nov 2008.

136. J. P. Kim, B. Y. Lee, S. Hong, and S. J. Sim, "Ultrasensitive carbon nanotube-based biosensors using antibody-binding fragments," *Analytical Biochemistry*, vol. 381, pp. 193–198, Oct 15 2008.

137. R. A. Villamizar, A. Maroto, F. X. Rius, I. Inza, and M. J. Figueras, "Fast detection of Salmonella Infantis with carbon nanotube field effect transistors," *Biosensors & Bioelectronics*, vol. 24, pp. 279–283, Oct 15 2008.

138. H. Yoon, J. H. Kim, N. Lee, B. G. Kim, and J. Jang, "A novel sensor platform based on aptamer-conjugated polypyrrole nanotubes for label-free electrochemical protein detection," *Chembiochem*, vol. 9, pp. 634–641, Mar 3 2008.

139. C. C. Cid, J. Riu, A. Maroto, and F. X. Rius, "Carbon nanotube field effect transistors for the fast and selective detection of human immunoglobulin G," *Analyst*, vol. 133, pp. 1005–1008, 2008.

140. M. Abe, K. Murata, A. Kojima, Y. Ifuku, M. Shimizu, T. Ataka, and K. Matsumoto, "Quantitative detection of protein using a top-gate carbon nanotube field effect transistor," *Journal of Physical Chemistry C*, vol. 111, pp. 8667–8670, Jun 21 2007.

141. X. J. Zhou, J. M. Moran-Mirabal, H. G. Craighead, and P. L. McEuen, "Supported lipid bilayer/carbon nanotube hybrids," *Nature Nanotechnology*, vol. 2, pp. 185–190, Mar 2007.

142. K. Maehashi, T. Katsura, K. Kerman, Y. Takamura, K. Matsumoto, and E. Tamiya, "Label-free protein biosensor based on aptamer-modified carbon nanotube field-effect transistors," *Analytical Chemistry*, vol. 79, pp. 782–787, Jan 15 2007.

143. G. Gruner, "Carbon nanotube transistors for biosensing applications," *Analytical and Bioanalytical Chemistry*, vol. 384, pp. 322–335, Jan 2006.

144. K. Bradley, M. Briman, A. Star, and G. Gruner, "Charge transfer from adsorbed proteins," *Nano Letters*, vol. 4, pp. 253–256, Feb 2004.

145. P. Hu, A. Fasoli, J. Park, Y. Choi, P. Estrela, S. L. Maeng, W. I. Milne, and A. C. Ferrari, "Self-assembled nanotube field-effect transistors for label-free protein biosensors," *Journal of Applied Physics*, vol. 104, Oct 1 2008.

146. M. Nikiforov, "Nanotube FET detects streptavidin," *Mrs Bulletin*, vol. 28, pp. 473–473, Jul 2003.

147. A. Star, J. C. P. Gabriel, K. Bradley, and G. Gruner, "Electronic detection of specific protein binding using nanotube FET devices," *Nano Letters*, vol. 3, pp. 459–463, Apr 2003.

148. G. H. Wu, R. H. Datar, K. M. Hansen, T. Thundat, R. J. Cote, and A. Majumdar, "Bioassay of prostate-specific antigen (PSA) using microcantilevers," *Nature Biotechnology*, vol. 19, pp. 856–860, Sep 2001.

149. J. P. Whelan, A. W. Kusterbeck, G. A. Wemhoff, R. Bredehorst, and F. S. Ligler, "Continuous-Flow Immunosensor for Detection of Explosives," *Analytical Chemistry*, vol. 65, pp. 3561–3565, Dec 15 1993.

150. P. J. Devine, N. A. Anis, J. Wright, S. Kim, A. T. Eldefrawi, and M. E. Eldefrawi, "A Fiberoptic Cocaine Biosensor," *Analytical Biochemistry*, vol. 227, pp. 216–224, May 1 1995.

151. H. Zhu, J. F. Klemic, S. Chang, P. Bertone, A. Casamayor, K. G. Klemic, D. Smith, M. Gerstein, M. A. Reed, and M. Snyder, "Analysis of yeast protein kinases using protein chips," *Nature Genetics*, vol. 26, pp. 283–289, Nov 2000.

152. H. Davies, L. Lomas, and B. Austen, "Profiling of amyloid beta peptide variants using SELDI Protein Chip (R) arrays," *Biotechniques*, vol. 27, pp. 1258–1261, Dec 1999.

153. N. H. H. Heegaard and R. T. Kennedy, "Identification, quantitation, and characterization of biomolecules by capillary electrophoretic analysis of binding interactions," *Electrophoresis*, vol. 20, pp. 3122–3133, Oct 1999.

154. W. Mullett, E. P. C. Lai, and J. M. Yeung, "Immunoassay of fumonisins by a surface plasmon resonance biosensor," *Analytical Biochemistry*, vol. 258, pp. 161–167, May 1 1998.

155. M. F. Hagan, A. Majumdar, and A. K. Chakraborty, "Nanomechanical motion generated by biomolecular binding.," *Abstracts of Papers of the American Chemical Society*, vol. 222, pp. U354–U354, Aug 2001.

156. H. Maki, T. Sato, and K. Ishibashi, "Direct observation of the deformation and the band gap change from an individual single-walled carbon nanotube under uniaxial strain," *Nano Letters*, vol. 7, pp. 890–895, Apr 2007.

157. G. H. Wu, H. F. Ji, K. Hansen, T. Thundat, R. Datar, R. Cote, M. F. Hagan, A. K. Chakraborty, and A. Majumdar, "Origin of nanomechanical cantilever motion generated from biomolecular interactions," *Proceedings of the National Academy of Sciences of the United States of America*, vol. 98, pp. 1560–1564, Feb 13 2001.

158. V. T. Moy, E. L. Florin, and H. E. Gaub, "Intermolecular Forces and Energies between Ligands and Receptors," *Science*, vol. 266, pp. 257–259, Oct 14 1994.

159. L. Brannon-Peppas and J. O. Blanchette, "Nanoparticle and targeted systems for cancer therapy," *Advanced Drug Delivery Reviews*, vol. 56, pp. 1649–1659, Sep 22 2004.

160. L. R. Hirsch, R. J. Stafford, J. A. Bankson, S. R. Sershen, B. Rivera, R. E. Price, J. D. Hazle, N. J. Halas, and J. L. West, "Nanoshell-mediated near-infrared thermal therapy of tumors under magnetic resonance guidance," *Proceedings of the National Academy of Sciences of the United States of America*, vol. 100, pp. 13549–13554, Nov 11 2003.

161. N. Shao, S. Lu, E. Wickstrom, and B. Panchapakesan, "Integrated molecular targeting of IGF1R and HER2 surface receptors and destruction of breast cancer cells using single wall carbon nanotubes," *Nanotechnology*, vol. 18, Aug 8 2007.

162. B. Panchapakesan, S. Lu, K. Sivakumar, K. Teker, G. Cesarone, and E. Wickstrom, "Single wall carbon nanotube nanobomb agents for killing breast cancer cells," *Nanobiotechnology*, vol. 1, pp. 133–140, 2005.

163. L. R. Bickford, G. Agollah, R. Drezek, and T. K. Yu, "Silica-gold nanoshells as potential intraoperative molecular probes for HER2-overexpression in ex vivo breast tissue using near-infrared reflectance confocal microscopy," *Breast Cancer Research and Treatment*, vol. 120, pp. 547–555, Apr 2010.

164. C. Loo, A. Lowery, N. J. Halas, J. West, and R. Drezek, "Immunotargeted nanoshells for integrated cancer imaging and therapy," *Nano Letters*, vol. 5, pp. 709–711, Apr 2005.

165. D. Sachdev and D. Yee, "The IGF system and breast cancer," *Endocrine-Related Cancer*, vol. 8, pp. 197–209, Sep 2001.

166. N. W. S. Kam, M. O'Connell, J. A. Wisdom, and H. J. Dai, "Carbon nanotubes as multifunctional biological transporters and near-infrared agents for selective cancer cell destruction," *Proceedings of the National Academy of Sciences of the United States of America*, vol. 102, pp. 11600–11605, Aug 16 2005.

167. O. D. Hendrickson, N. S. Fedyunina, A. A. Martianov, A. V. Zherdev, and B. B. Dzantiev, "Production of anti-fullerene C(60) polyclonal antibodies and study of their interaction with a conjugated form of fullerene," *Journal of Nanoparticle Research*, vol. 13, pp. 3713–3719, Sep 2011.

168. S. R. Wilson, "Fullerene nanotools from biology: Anti-C60 antibodies and their characterization by X-ray crystallography.," *Abstracts of Papers of the American Chemical Society*, vol. 220, pp. U433–U433, Aug 20 2000.

169. M. Sarntinoranont, F. Rooney, and M. Ferrari, "Interstitial stress and fluid pressure within a growing tumor," *Annals of Biomedical Engineering*, vol. 31, pp. 327–335, Mar 2003.

Chapter 6

Graphene for Biosensing Applications

Romaneh Jalilian,[a,b] Luis A. Jauregui,[a,c] Kyuwan Lee,[a,d,e] Yong P. Chen,[a,b,c] and Joseph Irudayaraj[a,d,e]

[a]Birck Nanotechnology Center, [b]Department of Physics, [c]School of Electrical and Computer Engineering, [d]School of Agricultural and Biological Engineering, and [e]Bindley Bioscience Center, Purdue University, West Lafayette, IN 47907, USA

yongchen@purdue.edu, josephi@purdue.edu

6.1 Introduction

Graphene (a single layer of graphite, although the term is also used to collectively refer to few layers of graphene) is the two-dimensional form of carbon with planar sp^2 bonding. It is also the building block for many carbon nanomaterials, including carbon nanotubes. Since the discovery of electrically isolated graphene in 2004 [1], this novel material has received tremendous interest due to a large number of unique and exceptional properties [2–5]. Many of these properties are highly tunable and sensitively dependent on various parameters. This, along with the fact that graphene is robust and typically exposed to the environment with an extraordinary specific surface area (2620 m^2/g), makes graphene and graphene-based nanomaterials promising as exquisite sensor materials. This review will discuss the fabrication methods, physical/chemical properties, and potential applications of graphene in biosensing, and current status of such research. Work in this area has been scarce but has gained considerable momentum and is expected to grow rapidly in the near future as researchers from diverse disciplines (including biomedical sciences) start to appreciate and utilize the unique properties of graphene for a wide variety of bionanotechnology-related applications.

Biomedical Nanosensors
Edited by Joseph Irudayaraj

ISBN 978-981-4303-03-3 (Hardcover), 978-981-4303-04-0 (eBook)
www.panstanford.com

After a review of current methods in the synthesis of graphene materials and fabrication of graphene devices in Section 6.2, we will review the major physical and chemical properties of graphene in Section 6.3 as a foundation to understanding and developing schemes of sensor operation. Some of the most interesting properties especially relevant to employing graphene in sensors (particularly biosensors) with remarkable functionality and performance are highlighted here:

(1) Exposed to environment with robust properties: Graphene can also behave well in liquid and graphene field-effect transistors (FETs) operating in liquids have been demonstrated [6].
(2) High surface area: 2620 m^2/g for both sides of graphene (e.g., suspended graphene) and 1310 m^2/g for one-side (e.g., supported on a substrate). This also means that a graphene-based composite material can have a huge surface area. This property has also been used to demonstrate, for example, an "ultracapacitor" based on graphene [7].
(3) Graphene (being a sp^2-bonded carbon network) can easily be functionalized to interact with many chemical and biological agents, with the power tool set available from organic chemistry [4, 8].
(4) Excellent electrical properties [9], especially high electrical conductivity and carrier mobility, and ultralow electronic noise: Moreover, many electronic properties of graphene are high *tunable*, including the bandgap, carrier type and density (field effect), and conductivity. Such exceptional and tunable (which can make it sensitive to external perturbations) electronic properties of graphene are appealing for graphene sensors based on conductivity measurement. For example, at room temperature, a graphene sensor that detects the adsorption or desorption of single gaseous molecules has been demonstrated [10] based on the extraordinary sensitivity of graphene to local electrical perturbation.
(5) Unique optical properties: Graphene is largely transparent (despite being highly conductive) [11, 12], with many distinctive optical properties and spectroscopic signatures (e.g., Raman [13]). These properties also often affected by external perturbations and may be utilized in optical-based sensing. Moreover, graphene has been shown to be able to strongly influence the optical/spectroscopic properties of objects (e.g., molecules) adsorbed on it [14, 15], opening many opportunities for sensing.

Other appealing properties of graphene as a sensor material include high thermal conductivity [16], excellent mechanical properties [17, 18], and relative ease of planar patterning using standard micro/nano fabrication (offering outstanding potentials to develop miniaturized sensors with high spatial resolution).

In Section 6.4 of this review, we will discuss how graphene may be affected by (or affect) external objects or environmental "stimuli" for sensing applications (including particularly, but not limited to, biosensing). From a fundamental point of view, graphene can be thought of both as a giant carbon molecule and a two-dimensional solid-state system possessing both types of properties while not completely described by either point of view alone. This places graphene in a unique position to interact with other molecular or nanoscale objects (e.g., biomolecules) and create responses in a measurable way, making graphene an exceptionally appealing sensor material. For example, recent experiments have shown that the electrical property of graphene can be affected by biological objects on its surface [19] and graphene can serve as a platform to detect biomolecules and interactions (e.g., binding/unbinding) optically [14].

The last section (Section 6.5) will give some closing remarks and present some promising future directions in the study of graphene and its employment for biosensing applications.

6.2 Fabrication of Graphene-Based Materials and Devices

6.2.1 Graphene Materials Fabrication

The most common method to fabricate graphene is by mechanical exfoliation [1, 20] from graphite onto a substrate (e.g., silicon wafer). This straightforward method, which typically gives graphene flakes with relatively small size (<100 μm), is best suited to make single devices to perform fundamental studies and proof-of-concept experiments. An extension of this method by nanoimprinting [21] promises to make arrays of many graphene devices over a large area. Graphene can also be synthesized on a variety of substrates, including silicon carbide (SiC) by thermal decomposition [22] and transition metals by chemical vapor deposition (CVD) or surface segregation [23, 24]. These synthetic methods are particularly interesting for growing graphene on a large scale for practical applications. In addition, a large number of "graphene-based" materials can be synthesized by various methods from (often chemically functionalized) graphene sheets and their mixture with other materials (such as in a composite) to fulfill certain functions.

Micromechanical exfoliation (cleavage) of bulk graphite was the first technique that allowed the deposition of (electrically) isolated graphene on a substrate. This method is simply repeated by peeling off graphite with adhesive tapes until the thinnest flakes are found and transferred to a substrate by the physical contact of tape and the substrate [20], as demonstrated in Fig. 6.1.

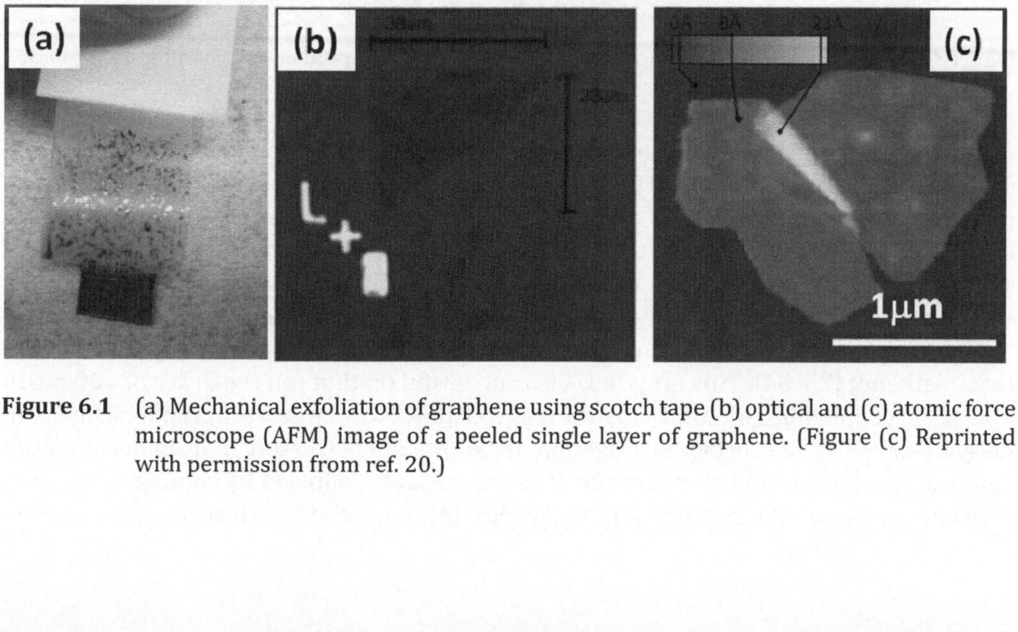

Figure 6.1 (a) Mechanical exfoliation of graphene using scotch tape (b) optical and (c) atomic force microscope (AFM) image of a peeled single layer of graphene. (Figure (c) Reprinted with permission from ref. 20.)

Several more sophisticated exfoliation methods have been developed. Electrostatic deposition of loosely bonded graphene flakes from highly oriented pyrolytic graphite (HOPG) is another alternative method for extracting graphene from graphite [25]. Liang *et al.* demonstrated graphene exfoliation using pillars on a stamp and then used the transfer printing method to place the exfoliated islands of graphene to a device-active area on a substrate [21]. Also, incorporating the electrostatic deposition and stamp technique, Liang reported electrostatic exfoliation of nanometer to micrometer-sized few-layer graphene (FLG) from patterned HOPG onto substrates [26]. Many methods have also been developed to perform chemical and/or solution-assisted exfoliation of graphite to extract graphene or chemically modified graphene (CMG) [27, 28].

Another technique for fabricating graphene is by the sublimation of Si from the SiC surface to produce epitaxial graphene sheets on SiC (a semiconductor) [22, 29] (Fig. 6.2). Prior to growth, substrates are typically etched in hydrogen. The temperature required for the growth of graphene film ranges from 1,350–1,650°C. At such high temperatures, Si atoms leave the surface, and graphitization occurs. The quality and thickness of graphene layers grown depend on many parameters, including which surface of SiC is used. More recently, it has become possible to grow large-area monolayer graphene epitaxially on SiC [30].

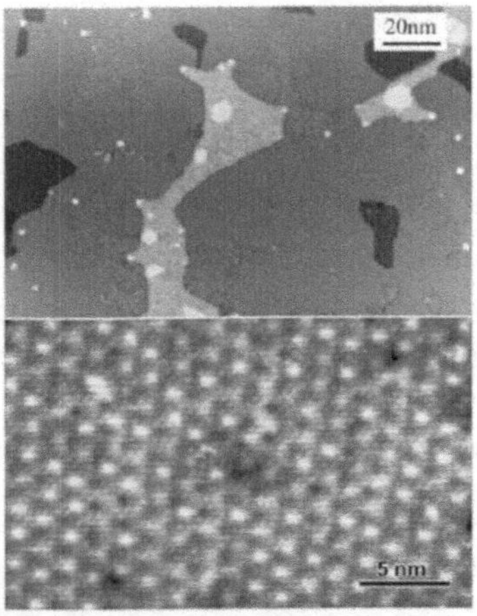

Figure 6.2 Scanning tunneling microscope (STM) images of epitaxial graphene growth on SiC. (Reprinted with permission from ref. 29).

Another method to synthesize graphene is by CVD growth on metal surfaces. This is a decades-old technique [24, 31]. Recently, it has been demonstrated that such growth performed on copper foils [23, 32] is particularly successful in producing large-size monolayer graphene (reaching 30 inches in size [33]) that can be transferred to arbitrary substrates (including flexible substrates). The graphene grows as film directly on the copper surface by a surface-catalyzed CVD process, at temperatures up to 1,000°C and low or ambient pressure, by the extraction of carbon from the decomposition of hydrocarbons (e.g., methane) [23, 34]. This growth is also successful on thin (on the order of 100 nm in thickness) copper films on solid substrates as well [35]. Also, large-area graphene films can be synthesized by the surface segregation of carbon atoms (dissolved from decomposed hydrocarbon gases) from Ni under the ambient pressure followed by cooling down from high temperatures. The hydrocarbon molecules decompose at the Ni surface, and carbon atoms diffuse into the metal. Using appropriate cooling rates, carbon atoms can segregate at the surface to form graphene layers [36]. Large-size transferrable graphene films (most often containing a mixture of monolayer and multilayer) have been grown by such a method on Ni foils [36, 37] and Ni thin films [38, 39]. Various examples of metal-based CVD growth of graphene are shown in Fig. 6.3.

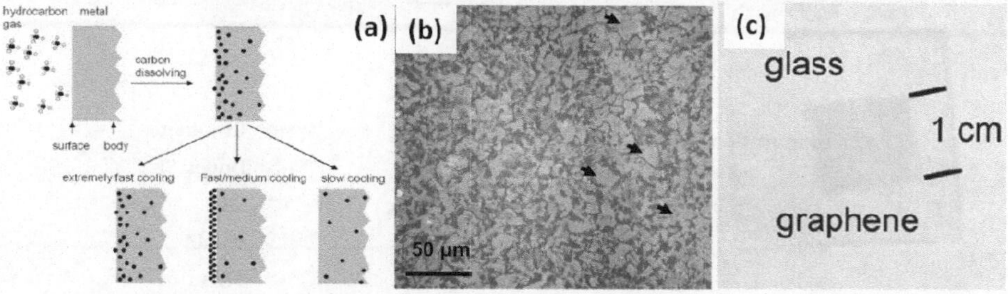

Figure 6.3 (a) A schematic representation of hydrocarbon decomposition, carbon deposition (on surface), dissolution (into metal), and various segregation/diffusion behavior upon cooling; (b) and optical image of CVD graphene grown using Ni film consisting of one to two layers of graphene (the black arrows); (c) monolayer CVD graphene film grown on Cu and then transferred onto a glass plate. (Fig. 6.3a, b, and c reprinted with permission from ref. 36, 39, and 23, respectively.) See also Color Insert.

Another route to synthesize graphene is by the chemical reduction of graphene oxide (GO) [40, 41] as graphite oxide (synthesized by oxidizing graphite with acids) is water soluble and much easier to exfoliate [8]. GO is hydrophilic (in contrast to graphene, which is hydrophobic), allowing stable aqueous dispersion in water. It is convenient to exfoliate GO into nano-sized single layers, which is the promising mass production of both GO and graphene. Using such a method, one often obtains partially oxidized graphene or graphene with a chemical functional group attached. Various methods are being also developed to exfoliate graphite in solutions with appropriate chemical properties [8, 27] to obtain graphene or chemically functionalized graphene.

In addition to graphene thin films (single or few atomic layers), graphene-based conducting composites [42] have been successfully demonstrated, where a relatively large amount (several percent of volume fraction) of graphene flakes is dispersed in a polymer (e.g., polysterene) matrix to make a macroscopic, solid form of composite (which is conductive due to a graphene network) with excellent material and electronic properties and a large surface area, which may be particularly interesting for sensing applications.

6.2.2 Graphene Device Fabrication

We briefly describe here the typical lithography fabrication process of a graphene device with electrical connection. Graphene is placed (by exfoliation or other ways of transfer) on a degenerately doped Si substrate with an oxide overlayer (Fig. 6.4). The doped Si substrate can be used as a back gate to tune the carrier type and concentration in graphene by an electric field effect (see Section 6.3). The electric contact leads on graphene are fabricated by e-beam and/or optical lithography. The fabrication steps are presented in Fig. 6.4 (a–e). Also, by etching the SiO_2 underneath graphene after the fabrication of the graphene sensor, it is possible to make suspended graphene. The change of graphene electrical property integrated into a device can be measured due to its chemical, physical, and mechanical responses to changes in the surrounding environment.

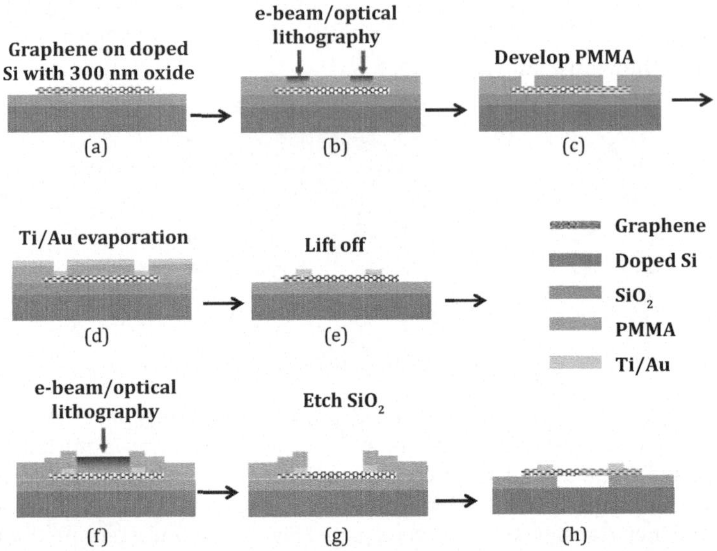

Figure 6.4 (a) Graphene deposition, (b) e-beam/optical lithography, (c) developing exposed regions on e-beam/photo resist (poly[methyl methacrylate], or PMMA, is an example of e-beam resist), (d) metal evaporation, and (e) lift off. To suspend the graphene in the device, (f) second e-beam optical lithography opening a window over graphene, (g) wet etching of SiO_2 by hydrofluoric acid, and (h) remove the extra photo/e-beam resist and dry the sample by a critical point dryer. See also Color Insert.

6.3 Basic Properties of Graphene

Graphene is formed by sp^2 carbon-carbon bonds in a honeycomb shape. The unit cell is formed by two carbon atoms (referred to as atom A and atom B). Figure 6.5 shows the representation of a graphene sheet and its respective unit cell.

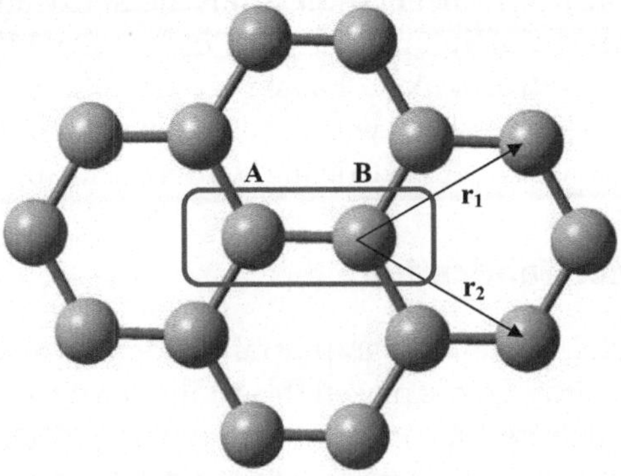

Figure 6.5 A portion of a graphene sheet; each light-blue ball is a carbon atom, and the sticks are the representation of the sp^2-bonds between carbon atoms. One cite in a unit cell of graphene is marked by a red hallmark. See also Color Insert.

6.3.1 Electrical Properties

As an electronic material, graphene has the highest mobility at room temperature among all known materials. For this reason, graphene has been considered as one of the most promising material for "post-silicon" (beyond Moore's law) electronics in semiconductor industry and much work has started to explore the applications of graphene in high-speed, high-performance, and low-power electronic devices.

The electronic band structure of graphene can be constructed using the tight-binding method; the size of this matrix will depend on the number of valence orbitals. Carbon has six electrons in total. There are four electrons in the valence band, and four basis functions ($2s$, $2p_x$, $2p_y$, and $2p_z$) will be needed to represent each carbon atom. However, the unit cell is formed by two carbon atoms; therefore, eight basis functions will be necessary per unit cell [43]. It is interesting to point out that the levels related to the orbitals $2s$, $2p_x$ and $2p_y$ are far separated from the Fermi energy compared to the levels related to the orbitals $2p_z$. Therefore, the conduction is mostly related to the levels related to the $2p_z$ orbitals, simplifying the number of basis functions to just two needed per unit cell. Solving the Hamiltonian describing electrons moving in the graphene lattice, a linear relation between momentum ($p = \hbar k$) and electron energy (E) is obtained:

$$E = v_F p = \hbar v_F k \text{ , where } v_F = 10^6 \text{ m/s is called the Fermi velocity.}$$

One of the most striking differences between this result from graphene compared to other semiconductors is that here, E is just linear dependent with k, leading to a constant velocity, called Fermi velocity, with an effective mass equal to zero. This Fermi velocity does not depend on energy or momentum, while in other semiconductors, the Fermi velocity depends on the root square of energy and there is a finite effective mass for electrons. The constant velocity plays a similar role as the speed of light in usual relativistic massless particles as photons. This unique energy-momentum dispersion of charge carriers and a nonzero Berry's phase [44] related to the lattice structure (A and B atoms being symmetric) have important implications for the electronic transport of carbon materials, including graphene and single-walled carbon nanotubes (SWCNT, which are rolled-up graphene sheets).

Experiments have shown mean free paths for graphene and SWCNT as long as ~10 μm at low temperatures (20 K) and ~1 μm at room temperature [45], showing that ballistic devices working at room temperature can be made for microscale devices. This ballistic behavior at room temperature is a direct consequence of the nonzero Berry's phase happening in graphene that protects electrons from being scattered.

Both the type and density of charge carriers in graphene can be tuned by an electric field (provided by, for example, a gate voltage). Varying the carrier density tunes the conductivity of graphene, resulting in a field effect device [1]. The typical dependence of resistivity on the gate voltage for graphene devices has a sharp peak on the order of kilo-ohms, while the resistivity drops far away from this peak to hundreds of ohms (Fig. 6.6). The gate voltage induces a surface charge density, which shifts the position of the Fermi level in the graphene. This electric field can transform the graphene channel into either completely n-type (electron) or completely p-type (hole) conductor and a mixed state where both electrons and holes are present; this behavior is called an ambipolar electric field effect. A positive (negative) back-gate voltage induces electrons (holes) in concentrations proportional to the back-gate voltage value [1].

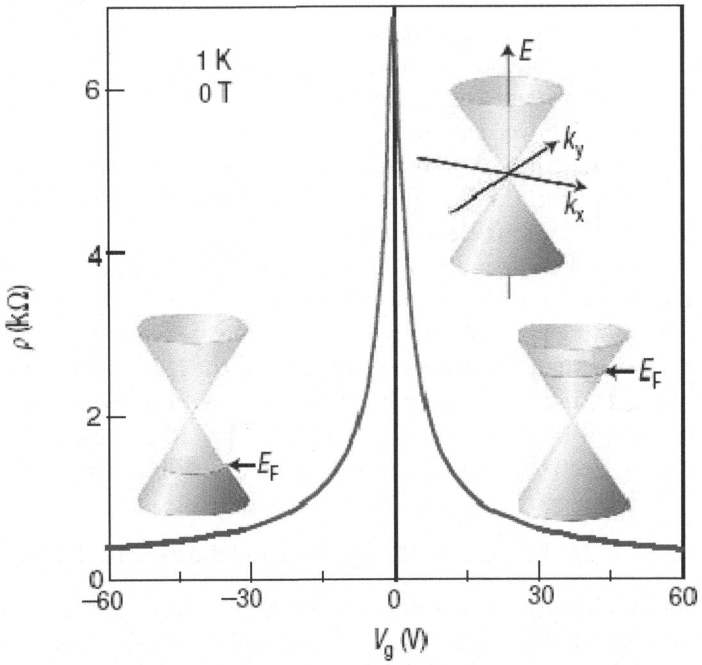

Figure 6.6 The ambipolar effect in graphene devices. The Fermi level changes when the back gate is applied from a negative to a positive value. (Reprinted with permission from ref. 5.)

The sharp peak in Fig. 6.6 is called the Dirac point. An important observation in graphene is that the conductivity of graphene at zero charge density (Dirac point) doesn't vanish; instead, it approaches a value on the similar order of the quantum conductance $4e^2/h$ [46]. Near the Dirac point, the resistivity (inverse of conductivity) is very sensitive to even a small and *local* change in the carrier density [10], or a local electrostatic potential. This can be a key mechanism for novel schemes of using graphene to detect charges or electric fields in biological systems.

Moreover, often the position of the neutrality point (Dirac point) as well as the carrier mobility of graphene is dependent on external doping, which in term can depend on the local environment of graphene [47, 48]. Chen *et al.* [47] doped graphene by potassium atoms under ultrahigh vacuum; the minimum of conductivity depended on the charged-impurity density, and the carrier mobility decreases with an increase of charge impurity density; but these are reversible by heating the sample to remove the potassium atoms. Schedin *et al.* [48] doped graphene with NO_2, H_2O, I, NH_3, CO, and ethanol, showing that the first three gases acted as acceptors while the last three compounds acted as donors.

Electronic noise is an important issue for sensor operation. Graphene is a highly conductive material with a zero electronic bandgap (metallic conductivity), leading to a low Johnson noise even at low charge densities (near the Dirac point) and at room temperatures. These properties make graphene an exceptional electronic low-noise material [10]. Coupled with graphene high sensitivity to a change in charge density or an applied external field near the charge neutrality point (Dirac point) enables such a high signal-to-noise ratio that makes graphene sensors capable of detecting change in the local charge concentration by less than 1 electron [10]. Based on such a principle, a room-temperature graphene sensor that detects adsorption or desorption of single molecules has been recently demonstrated [10]. Graphene typically has a small amount of crystal defects [20, 46, 48, 49], which can lead to a low level of pink (1/*f*) noise. Two sheets of graphene stacked together (as in graphite), known as bilayer graphene, have also been studied by Lin *et al.* [50] who found that bilayer graphene has in particular a lower noise figure than single-layer graphene, even lower than the Johnson noise limit. Figure 6.7 shows that a bilayer graphene device has a smoother resistance-gate voltage response compared to the single-layer device and bilayer graphene resistivity around the Dirac point is smaller than in a single layer. These observations lead us to think and demonstrate that bilayer graphene devices are less noisy than single-layer graphene ones. The power spectral density was measured on several devices with different geometries; the noise figure versus resistance over length is depicted in Fig. 6.7. As shown in Fig. 6.7, the noise characteristic for a single-layer device is constant for different resistances. However, bilayer devices possess lower noise than single-layer ones, with a decreasing noise characteristic when the resistance is increased.

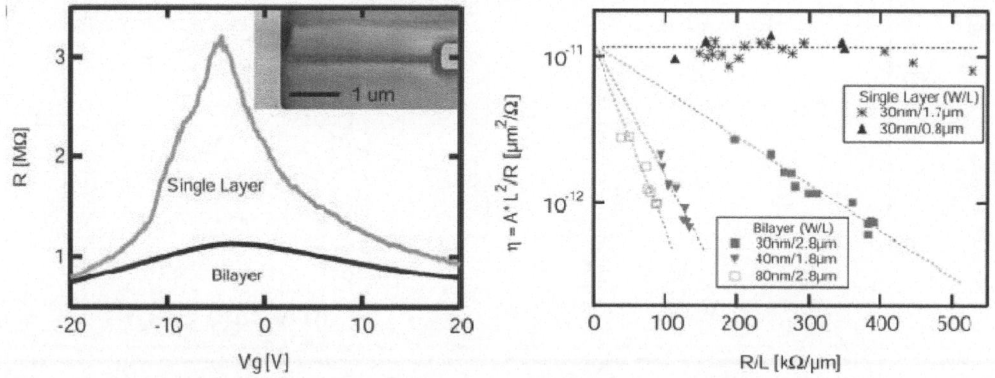

(a) (b)

Figure 6.7 (a) Back-gate voltage sweep for single- and double-layer graphene devices with the same geometrical dimensions, showing a smoother signal for bilayer graphene compared to single-layer graphene; (b) noise figure versus resistance/length for single- and double-layer graphene devices. (Reprinted with permission from ref. 50.)

The electronic properties of graphene as noise or carrier mobility are highly dependent on the interaction with the substrate, commonly SiO_2. This dependence can be shown by removing the substrate. Bolotin *et al.* [51] and Du *et al.* [52] found that after suspending a single layer graphene sheet (Fig. 6.8) by removing the substrate, the carrier mobility of graphene increases from values ranging between 2,000 $cm^2\,V^{-1}s^{-1}$ and 25,000 $cm^2\,V^{-1}s^{-1}$ to values reaching 60,000–230,000 $cm^2\,V^{-1}s^{-1}$ after current-induced annealing.

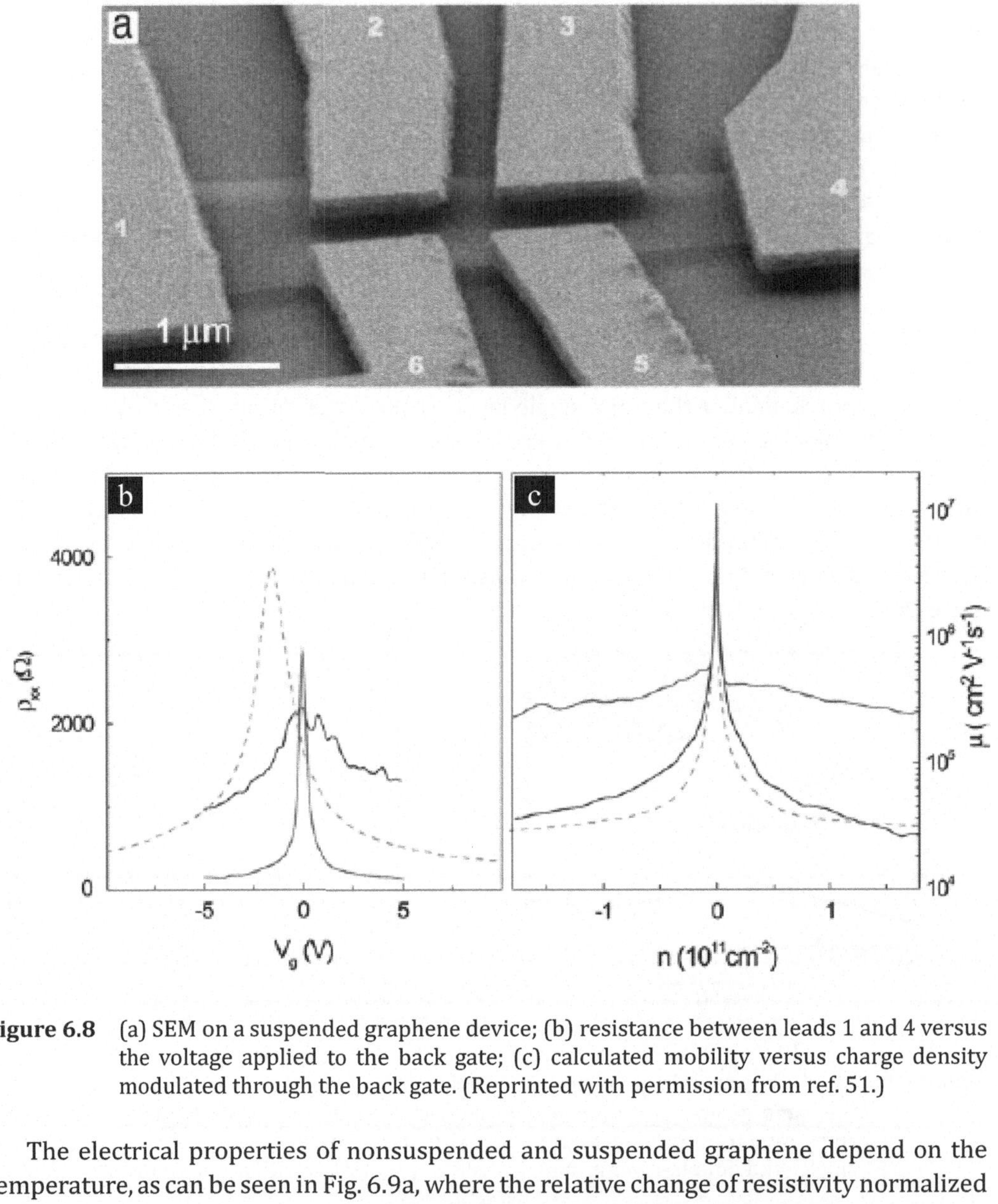

Figure 6.8 (a) SEM on a suspended graphene device; (b) resistance between leads 1 and 4 versus the voltage applied to the back gate; (c) calculated mobility versus charge density modulated through the back gate. (Reprinted with permission from ref. 51.)

The electrical properties of nonsuspended and suspended graphene depend on the temperature, as can be seen in Fig. 6.9a, where the relative change of resistivity normalized by the room temperature value ($\rho(T)/\rho(300\text{ K})$) is plotted for two representative carrier densities $n = 0$ (Dirac point) and $n = 3 \times 10^{12}\text{ cm}^{-2}$ (high density). A metallic behavior (monotonic reduction of $\rho(T)/\rho(300\text{ K})$ for a decreasing temperature) is observed, interpreted by that the residual electron-phonon scattering is part of the scattering mechanism in the high-density limit of conduction. However, for $V_g - V_{\text{Dirac}} = 0$ (Dirac point), the resistivity increases with a decreasing temperature, indicating nonmetallic behavior of the carriers at the Dirac point [53] for this moderate-mobility sample (unsuspended). For high-mobility suspended [51, 54] or unsuspended [55] graphene devices, metallic behavior can be found at all densities. As shown in Fig. 6.9b, the temperature dependence is also sample dependent, suggesting that the transport at the charge neutrality point (Dirac point) is dominated by extrinsic scattering [56].

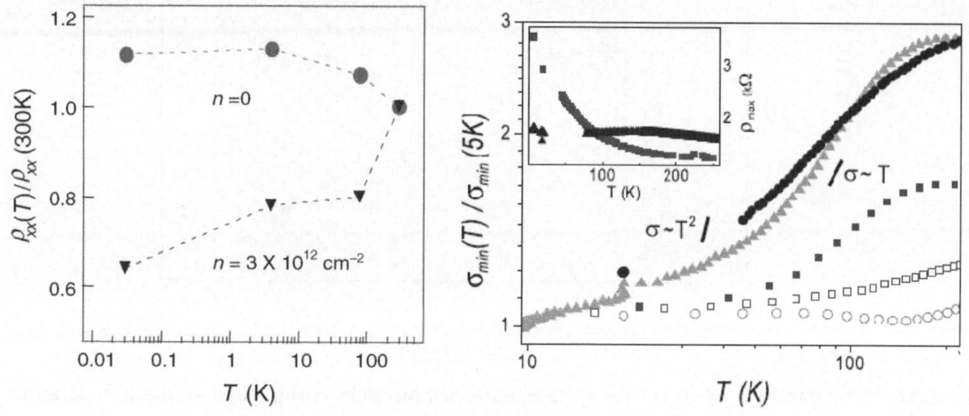

(a) (b)

Figure 6.9 (a) $\rho(T)/\rho(300\text{ K})$ at two different densities, Dirac point (n = 0) and high density (n = 3 × 10^{12}), for a moderate mobility (nonsuspended) graphene device [57]. (b) $\sigma_{min}(T)/\sigma_{min}(5\text{ K})$ at the Dirac point for high-mobility (suspended) graphene devices, as a function of T for three devices before (S1. , S2 ○) and after (S1 ■, S2 ●, S3 ▲) annealing. (Reprinted with permission from ref. 54.) See also Color Insert.

Intrinsically, graphene is a zero bandgap semiconductor (semimetal), which means graphene is always in the "ON" state. For many sensors or transistor applications, we need to switch from the "ON" to the "OFF" state. Many researchers have been working on opening a bandgap in graphene. And a bandgap can be opened by various procedures or mechanisms: geometric confinement/patterning (e.g., nanoribbons [28, 58] or antidotes [59]), electric field in bilayer graphene [60], interaction with substrate [61], or controlled oxidation [62]. Most of the bandgap reached so far ranges from 0 to 0.4 eV.

Graphene nanoribbons (GNR) are strips of graphene that are obtained by geometric confinement/patterning. These strips can be armchair or zigzag, depending on their edge structure. Calculations show zigzag GNRs are always metallic; in contrast, armchair GNRs are semiconducting, with a bandgap inversely proportional to its width [63]. Han *et al.* [64] and Chen *et al.* [65] studied how an apparent bandgap is opened in graphene nanoribbons with rough edges for different ribbon widths as it can be visualized in Fig. 6.10, where the measured bandgap increases for a decrease in the width of the nanoribbons. Han *et al.* reported a 300 meV bandgap induced for ~15 nm wide graphene nanoribbons. Pedersen *et al.* modeled a periodic array of holes (antidote lattice) in graphene to show that such an artificial nanomaterial can be controlled to create a bandgap with a very pronounced optical-absorption edge [59].

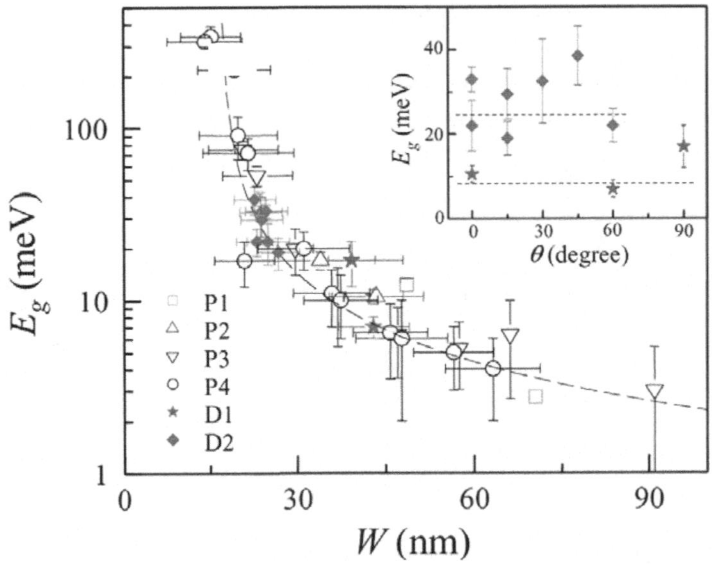

Figure 6.10 The bandgap opened in graphene nanoribbons versus the width of the nanoribbons. (Reprinted with permission from ref. 58.)

An electric field can open a bandgap in bilayer graphene devices. Oostinga *et al.* [66] used a double-gate (back-gate and top-gate electrodes) bilayer graphene FET device, which induces a chemical potential difference between each of the graphene layers and opens a bandgap as can be seen in Fig. 6.11. They could open a small bandgap close to 10 meV.

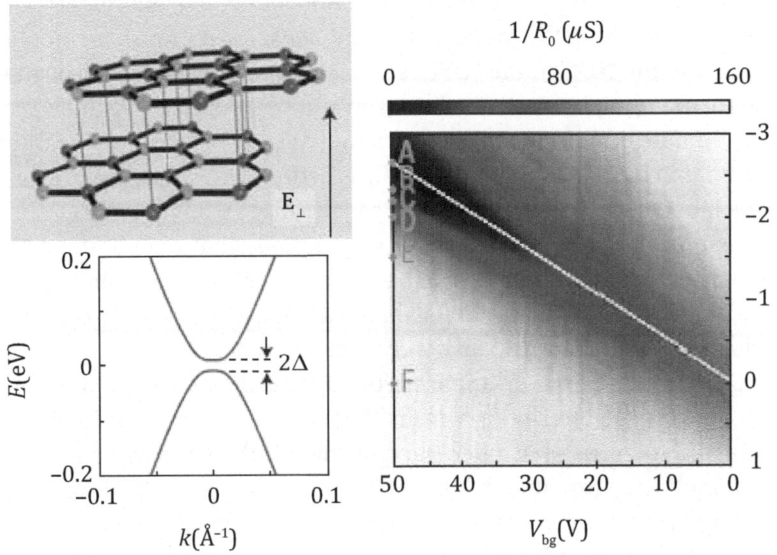

Figure 6.11 A bandgap opening in bilayer graphene by a perpendicular electrical field applied using a top-gated device. (Reprinted with permission from ref. 60.)

Another way to create a bandgap in graphene was shown by Zhou *et al.* [67] on epitaxially grown graphene on SiC grown. A bandgap of ~0.26 eV was measured on these samples. The proposed origin of this bandgap opening is the breaking of A-B site symmetry due to the graphene-substrate interaction.

6.3.2 Optical Properties

We will briefly review the optical properties of graphene, an understanding of which is critical for developing sensor devices. First, understanding the optical visibility of graphene using its optical contrast is essential for imaging, as well as for interrogating the number of layers. Graphene has much stronger visibility in reflection than transmission. Especially, when the bare substrate thickness is appropriate to achieve the resonant light transmission, the visibility of graphene reaches the optimum conditions [68]. It is because the interaction of light with electrons in graphene renders graphene relatively opaque in spite of its monolayer structure [69]. In other words, while the transparent substrate is transmitting the light, graphene reflects a certain amount of light and this makes graphene visible under a normal microscope. Therefore, it is necessary to consider the type of substrate that will be used to acquire a clear image of graphene.

As a direct result of Fresnel's theory, the light intensity reflected from graphene can be written as [70]:

$$I(n_1)=\left|\frac{r_1e^{i(\phi_1+\phi_2)}+r_2e^{-i(\phi_1-\phi_2)}+r_3e^{-i(\phi_1+\phi_2)}+r_1r_2r_3e^{i(\phi_1-\phi_2)}}{r_1r_2e^{-i(\phi_1-\phi_2)}+r_2r_3e^{i(\phi_1-\phi_2)}+r_3r_1e^{-i(\phi_1+\phi_2)}+e^{i(\phi_1+\phi_2)}}\right|^2 \tag{6.1}$$

where $r_i = (n_{i-1} - n_i)/(n_{i-1} + n_i)$ is the index of refraction, I is the intensity of light, n is the refractive index, and $\phi = 2\pi nd/\lambda$ is the phase shift, respectively.

By definition, the contrast C is described as

$$C=\frac{I(n_1=1)-I(n_1)}{I(n_1=1)} \tag{6.2}$$

For optimum visibility, the wavelength of monochromatic light used should meet the resonance condition of the substrate combining Eq. (6.1) and (6.2), and the bilayer gives a better visibility than the monolayer in the visual frequency range of the light source.

While light reflection provides better visibility of graphene, transmission is also very interesting when graphene is to be used in many optical devices such as liquid crystal displays or transparent electrodes. This is because graphene has low resistivity, high transparency, and chemical stability compared to other metal oxides [72]. Naturally, the transmission of graphene doesn't depend on the substrate thickness. Even suspended single-layer graphene can be solely identified by transmission due to graphene's unique electronic structure. Light is interacting with relativistic electrons (with linear energy-momentum dispersion) in graphene, and the opacity of graphene is determined only by the fine structure constant $\alpha = 2\pi e^2/hc \sim 1/137$. This is a universal characteristic of quantum electrodynamics of two-dimensional Dirac electrons regardless of material parameters. This is why graphene absorbs a sizable amount of light ($\pi\alpha$ = 2.3%) in spite of its single-atomic-layer structure. The experimental result agrees with the above theoretical expectation from two-dimensional Dirac electrons calculated with Fermi's golden rule.

This optical property of graphene is related to its electrical property showing a universal dynamic conductivity over visible frequencies due to the behavior of ideal Dirac electrons [12] (Fig. 6.13). This can be directly used not only for counting the number of layers but also for optical device replacing other metal oxides because graphene is an excellent transparent and conductive material with good chemical stability. For example, Blake *et al.* demonstrated that it can replace indium tin oxide (ITO) of liquid crystal display (LCD), removing many current problems such as the image-sticking problem [72].

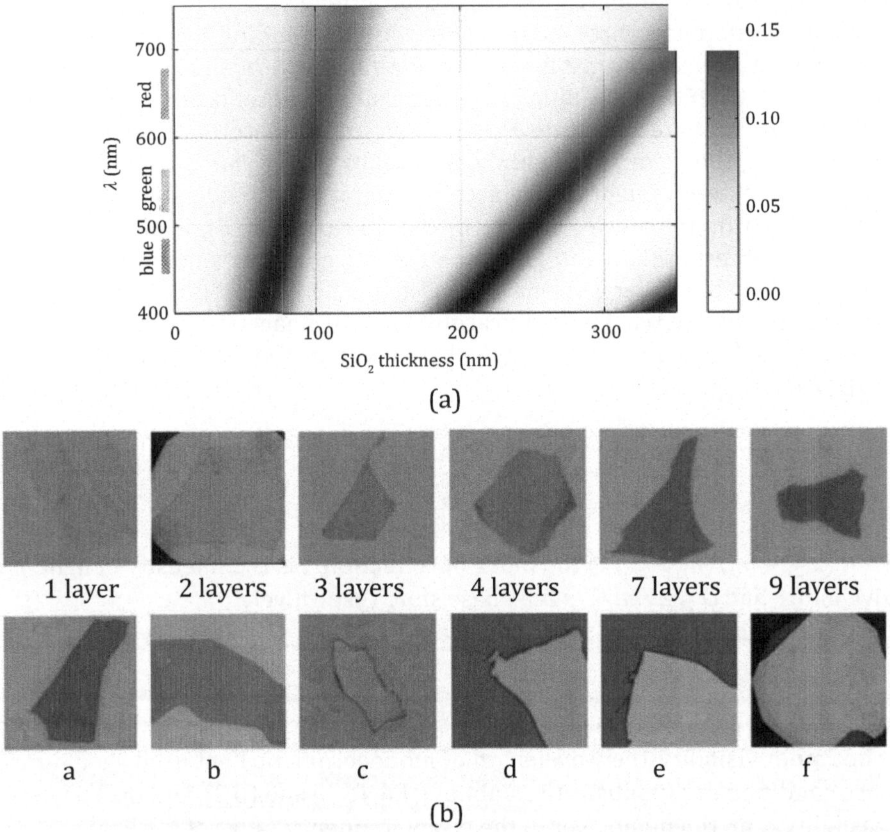

Figure 6.12 (a) The contrast plotted according to the equation; (b) the contrast shown in the optical images of graphene sheets depending on the number of layers. Pictures a–f are the images of graphene thicker than 10 layers. The number of layers increases from a to f. (Figures (a) and (b) reprinted with permission from ref. 70 and 71, respectively.) See also Color Insert.

While the natural transmission of white light through graphene doesn't depend on the material parameters, graphene also has nonlinear optical properties, allowing different transmission rates when laser pulses of different energy densities are used. It is because the light-electron interaction causes the optical limiting effect of graphene, and this also depends on the environmental refractive index [73]. Wang *et al.* described how thermal heating and ionization of graphene increases the light scattering of graphene. A nonlinear optical reaction brings a better visibility to graphene as it increases the scattering of the incident light. The more intense laser beam is attenuated by graphene, while the weaker ambient light is highly transmitted.

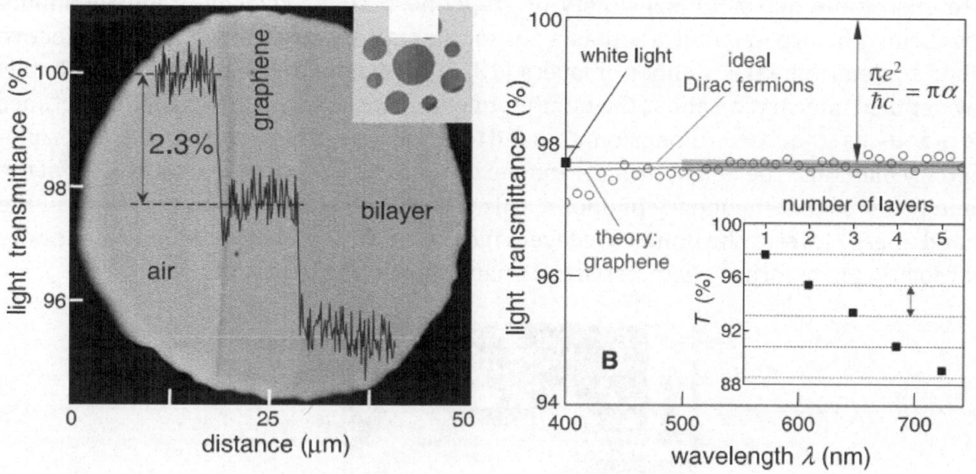

(a) (b)

Figure 6.13 (a) An aperture of 50 μm in size was partially covered by a single and bilayer of graphene. The blue line profile indicates the light transmission along the yellow line only dependent on the number of layers; inset — the original sample preparation. (b) A comparison of experimental data of transmission of single-layer graphene (circles) to theoretical expectations of graphene (green line) and ideal Dirac fermion (red line); inset — transmission depending on the number of layers. (Reprinted with permission from ref. 12.) See also Color Insert.

These nonlinear optical properties of graphene originate from nonequilibrium distributions of photoexcited carriers, of which energy is smaller than the optical phonon energy in the low temperature region [74]. This is why the photoresponse of graphene at a long wavelength excitation is strong and its nonlinear response threshold is low. Nonlinear optical properties of graphene can be characterized by the Z-scan method, and the optical-limiting studies on graphene showed that graphene has a nonlinear absorption and scattering with the use of 532 nm and 1064 nm laser pulses (Fig. 6.14).

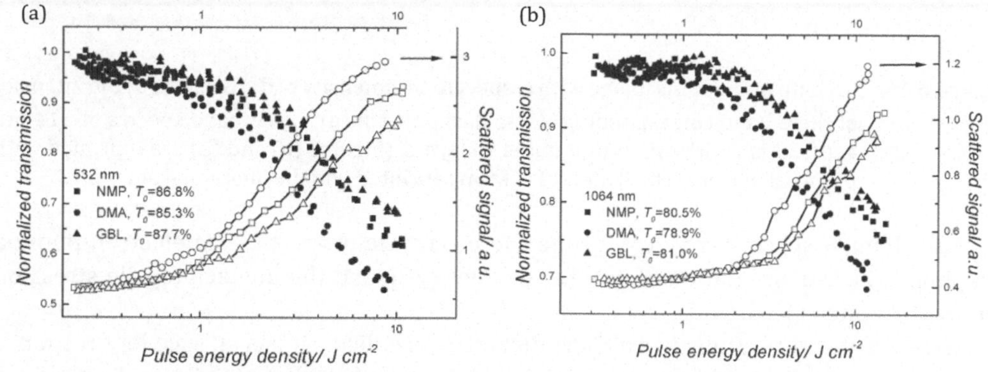

Figure 6.14 Graphene shows nonlinear transmission and scattering in various solvents for (a) 532 nm and (b) 1,064 nm excitations. (Reprinted with permission from ref. 73.)

To ensure the maximum sensitivity of graphene, a single layer of graphene should be confirmed before its application as a sensor. Raman spectroscopy is a very effective method to determine the number of layers [13, 75, 76]. Raman peaks at G′ and G bands show detailed information about the number of graphene layers. The G′ band (sometimes referred to as the "two-dimensional" band) represents the second order of zone-boundary phonons. The neighboring graphene layers interfere with the energy level of the second order of zone-boundary phonons. This results in the splitting and the shift of the G′-band energy level as the number of layers increases. The split and shifted G′-bands are superpositioned to form a characteristic Raman G′ peak (Fig. 6.15).

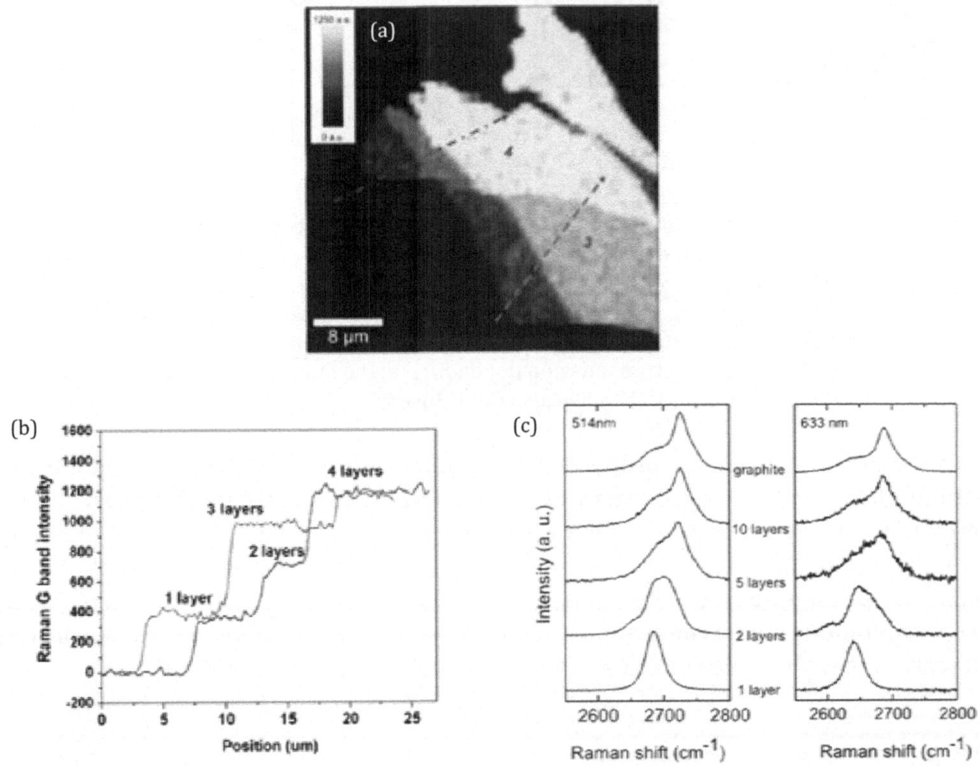

Figure 6.15 (a) Raman-mapping image with respect to the intensity of the G band; (b) the intensity of the G band corresponding to the dash lines in (a); (c) G′ band spectra at 514 nm and 633 nm with different number of layers. (Figures (a) and (b) are reprinted with permission from ref. 72 and (c) is from reprinted with permission from ref. 13.)

Also, Raman spectroscopy can probe the electronic properties via electron-phonon coupling [77], the orientation of graphene edges [78], and the uniaxial tensile stress on strained graphene [79].

Since graphene naturally has a short distance periodicity, it has an excellent response to high-frequency light, which enhances the anisotropy of conductivity. Since the bandgap and photon-induced conduction is suppressed along the zigzag direction of graphene, high-frequency absorption depends on the polarization of the light. Here, graphene acts like a polarizer [80]. Since this phenomenon depends on the geometry of graphene, it is also sensitive to temperature and physical tension on graphene.

6.3.3 Other Properties

Since electrical and/or optical signals are most commonly measured in sensor operations, we have focused on electrical and optical properties related to graphene. Graphene, in fact, possesses many other extraordinary properties, ranging from chemical [8, 81] and thermal [16] to mechanical. Such properties of graphene could also be relevant for sensing operation or used as sensing mechanisms. Below, we discuss the mechanical properties of graphene and its possible applications in biosensing.

Graphite is composed of a stack of two-dimensional graphene sheets held by van der Waals forces. Lee *et al.* [82] measured the elastic properties and breakdown strength of suspended graphene using an AFM. They found a breaking strength of 42 Nm^{-1} with a corresponding Young's modulus of E = 1.0 TPa. Previous experiments on bulk graphite found a Young's modulus of 1.02 ± 0.03 TPa, while suspended multilayer graphene yields a Young's modulus of 0.5TPa [83]. Therefore, graphene is a very strong and rigid material. These exceptional mechanical properties can be exploited in sensors based on nanoelectromechanical resonators (Fig. 6.16) [17, 84, 85], where high-frequency resonance could be achieved. For any nanoelectromechanical system (NEMS), the resonant frequency depends on the speed of sound on the material $v = \sqrt{E/\rho}$, where E is Young's modulus and ρ is the material density. Compared to Si or gallium arsenide (GaAs), carbon presents the highest value of speed of sound [84] due to a very high value of Young's modulus with a low material density.

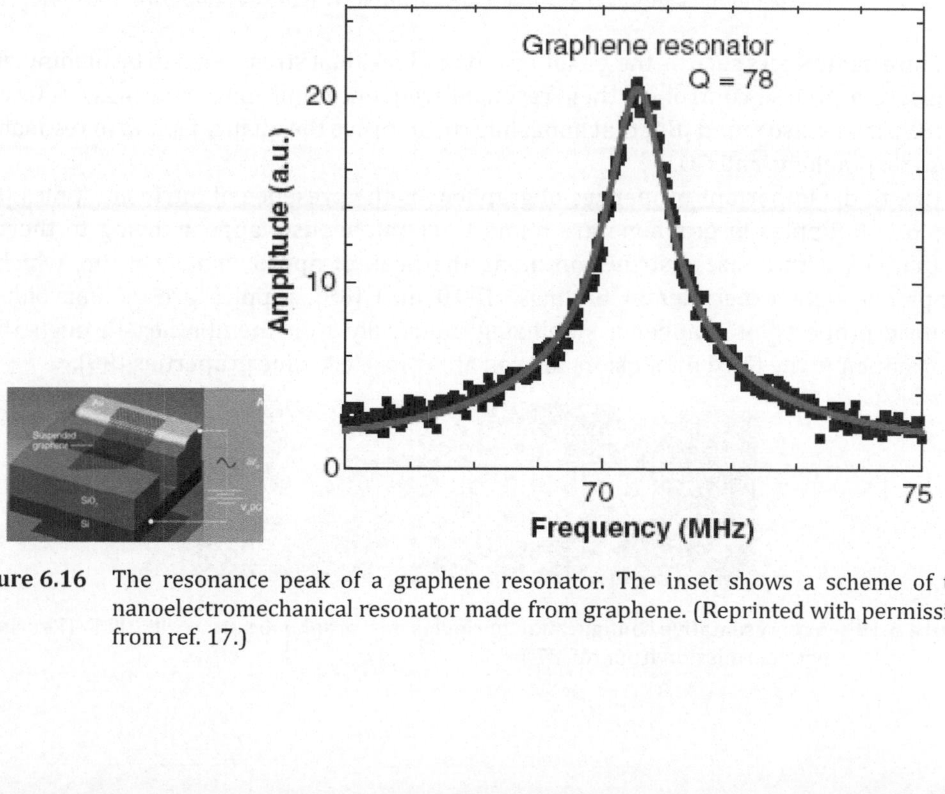

Figure 6.16 The resonance peak of a graphene resonator. The inset shows a scheme of the nanoelectromechanical resonator made from graphene. (Reprinted with permission from ref. 17.)

When a single layer of suspended graphene is used in such a nanoelectromechanical system (NEMS) resonator, a high force sensitivity below 1 fN per root hertz is demonstrated [17]. This sensitivity is suitable for the use of graphene as mass, force, and charge (electrostatic force) sensors. For a resonator, the minimum mass to be detected is:

$$\delta M \sim 2m_{\text{eff}}/Q \times 10^{-\text{DR}/20}$$

Graphene is an excellent choice for mass detector because of its low effective mass. Graphene dynamic range (DR) is sample dependent and is defined as the decibel ratio between the amplitude of onset of nonlinearity to the noise floor. In the supporting material of ref. 17, DR was found to be 60 dB, giving a room temperature mass sensitivity of 450 zeptograms. Such resonators can in principle be used to weigh biomolecules, cells, and nanoparticles in fluids as it was done in ref. 86 by using suspended silicon resonators with microchannels, but the sensitivity can be improved by using graphene (Fig. 6.17).

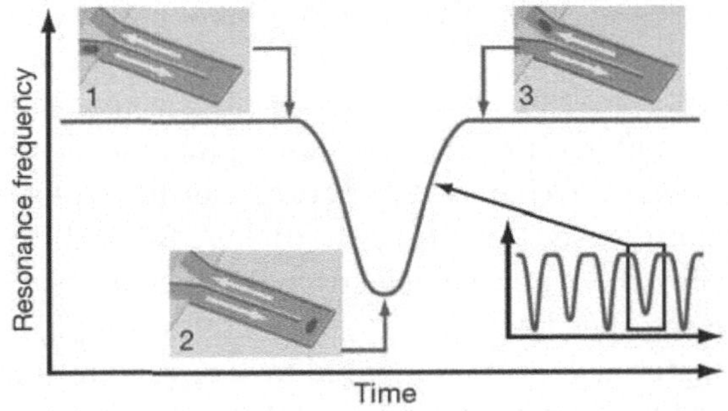

Figure 6.17 Time-dependent resonant frequency while a biomolecule passes through a silicon microchannel cantilever resonator. (Reprinted with permission from ref. 86.)

More parameters such as the quality factor and residual stress need to be manipulated to ensure a precise control on their resonant frequency and minimum mass detection [17, 85]. It was also found [84] that annealing can improve the quality factors in resonators based on graphene and GO.

One of the important properties of graphene is the presence of intrinsic ripples [87] (Fig. 6.18). Ripples in graphene are found to spontaneously appear owing to thermal fluctuations with a size distribution along the peak of approximately 8 nm, which is compatible with experimental findings (5–10 nm) [88]. Ripples are an unavoidable intrinsic property of graphene, stabilizing atomically thin membranes through their deformation in the third dimension and can affect its electronic properties [89].

Figure 6.18 A representative configuration of ripples on a graphene surface at 300 K. (Reprinted with permission from ref. 87.)

6.4 Biosensing Applications of Graphene

In this section, we review a number of recent experiments that demonstrate the potential of graphene for sensing applications. Most of these experiments are relevant for biosensors and a few prototypes have been developed.

6.4.1 Electrical Sensing

6.4.1.1 Gas sensing

As mentioned earlier, Schedin *et al.* [10] reported the capability of individual gas molecule detection by graphene FET. The principle of this device is based on changes in the electrical conductivity due to the adsorption of gas molecules on the surface of graphene. The adsorbed molecules change the local carrier concentration acting as donors or acceptors. The unique electronic properties of graphene maximize the sensing effect. Graphene is highly conductive, with a low Johnson noise in the transport measurement. In this experiment, the graphene device was mounted in an evacuated chamber and exposed to a selected chemical diluted in pure helium or nitrogen at atmospheric pressure. The tested chemicals are NO_2, NH_3, H_2O, and CO in concentrations of 1 part per million (ppm). A clear change was detected immediately after letting the chemicals in, followed by saturation (Fig. 6.19).

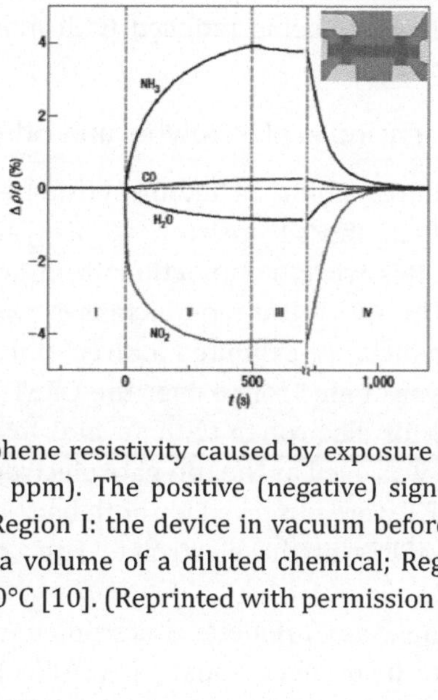

Figure 6.19 Changes in graphene resistivity caused by exposure to various gases (NH_3, CO, H_2O, NO_2) diluted (1 ppm). The positive (negative) sign of changes is due to electron (hole) doping. Region I: the device in vacuum before its exposure to gases; Region II: exposure to a volume of a diluted chemical; Region III: evacuation; Region IV: annealing at 150°C [10]. (Reprinted with permission from ref. 48.)

6.4.1.2 Metal doping

Chen *et al.* [90] reported on the change of electronic characteristics of graphene when potassium atoms are deposited onto its surface in vacuum. The mobility of graphene decreases after more potassium doping. The conductivity versus gate voltage shows the minimum ($V_{g,\,min}$) conductivity shifts to more negative gate voltages for more potassium exposure. Moreover, the researchers tried to dope graphene and changed the position of the Dirac point by the deposition of potassium atoms on a graphene device as shown in Fig. 6.20.

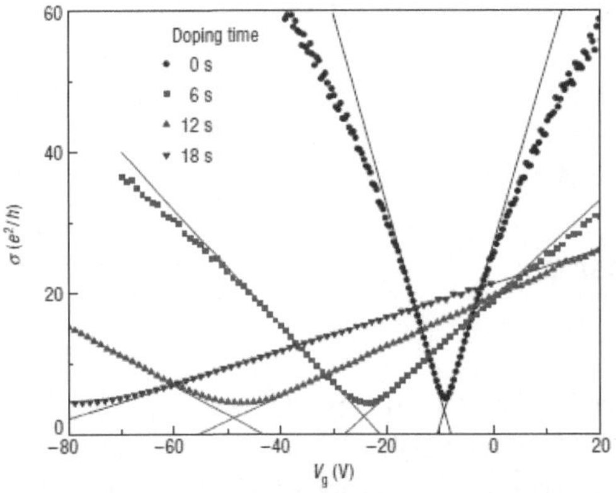

Figure 6.20 Potassium doping of graphene. The conductivity versus gate voltage (V_g) curves for the pristine sample and three different doping concentrations at T = 20 K in UHV. The crossing of the lines defines the points of the residual conductivity and the gate voltage at minimum conductivity ($V_{g,\,min}$) for each data set. (Reprinted with permission from ref. 47.)

The field effect characteristic is turning more n-type, and the transconductance (proportional to mobility), which can be extracted from the slope of conductance with respect to the back-gate voltage, is being reduced with more deposition of potassium atoms on the graphene device.

6.4.1.3 Chemical and biosensing in electrolytes and other liquids

Graphene shows a high sensing capability in liquid environment. Ohno *et al.* [91] reported electrolyte-gated graphene field-effect transistors (GFETs) for electrically detecting pH and protein adsorptions. In this work, nonfunctionalized graphene was used as a FET channel in an electrolyte. The GFET transconductances was higher in the electrolyte than in vacuum, and the conductance exhibited a direct linear increase with electrolyte pH. The electrical-double layer of electrolyte over the GFET acts as a top-gate insulator with a thickness of 1–5 nm for electrolyte with several millimolars concentration. An Ag/AgCl reference electrode was used as the top-gate electrode. Ambipolar electric field-effect characteristics of a GFET were observed for both back-gate and top-gate operation. Transfer characteristics of GFETs in the electrolyte (e.g., pH 5.8) indicate their high potentials for use in (Fig. 6.21).

Another example of graphene as a pH meter was studied with a few layers of graphene assembled as a solution-gate field-effect transistor (SGFET) with an electrolyte acting as the dietetic [92]. The modulation of the graphene (SGFET channel) conductance was achieved by applying a gate potential from an electrode placed on top of the channel, across the electrolyte. In this experiment one to two or three to four layers epitaxially grown graphene was used to demonstrate pH sensing using cyclic voltammetry (CV) and frequency-dependent impedance spectroscopy of graphene. Frequency-dependent impedance spectroscopy in the potential range from −1.0 V to +1.0 V at frequencies ranging from 5 Hz to 120 Hz shows peaks and a pH-sensitive maxima (Fig. 6.22).

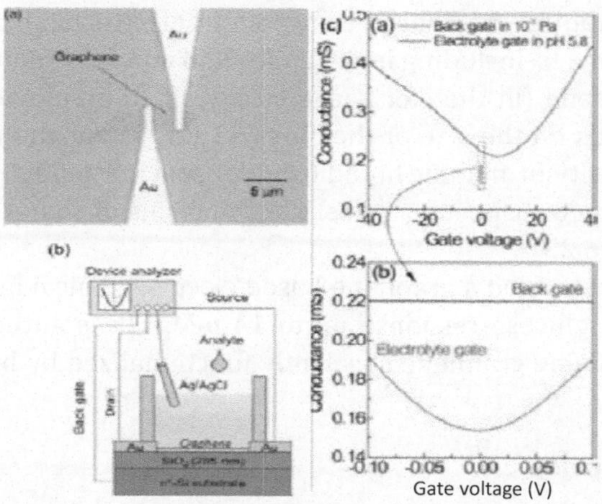

Figure 6.21 GFET-based sensors in an electrolyte. (Reprinted with permission from ref. 91.)

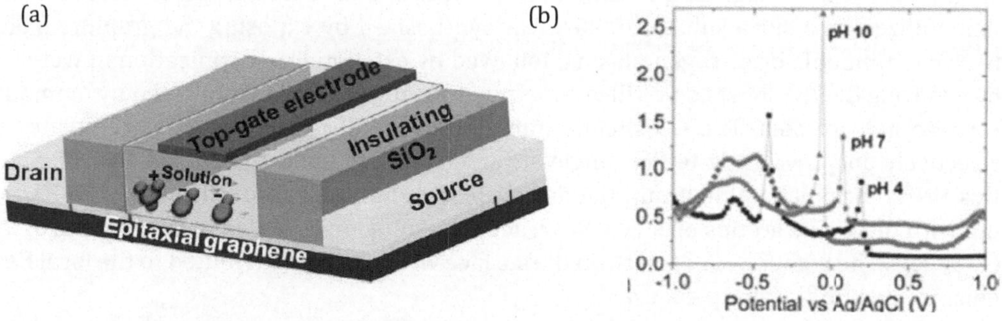

Figure 6.22 (a) A solution-gated epitaxial graphene pH sensor; (b) interfacial capacitance of graphene from impedance measurements. Shown is the pH dependence of interfacial capacitance peak potential (collected at a fixed frequency of 11 Hz). (Reprinted with permission from ref. 92.)

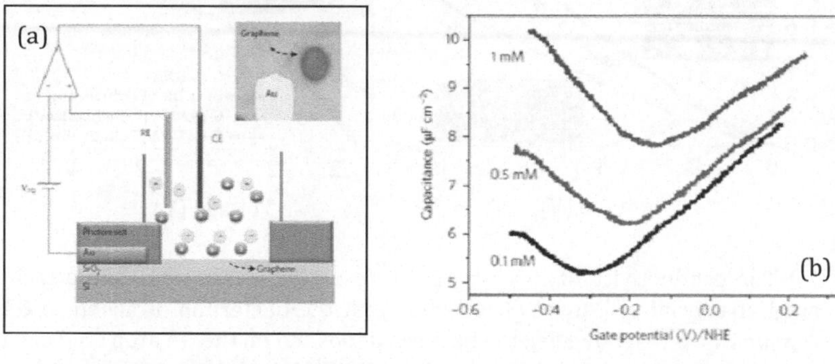

Figure 6.23 (a) A schematic representation of the quantum capacitance measurement setup in which a graphene sheet on a silicon/SiO_2 substrate is connected to a gold electrode; (b) the capacitance of graphene measured in a NaF aqueous solution at different ionic concentrations (0.1 mM, 0.5 mM, and 1 mM, from bottom to top). (Reprinted with permission from ref. 93.)

Chen *et al.* have studied the charge transport in electrochemically gated graphene under different electrolytes, including ionic liquids and aqueous solutions. A double layer is formed above graphene (in the electrochemically gated graphene FET). The double-layer thickness depends on the size of the ions (~1nm). Graphene is able to determine the mobile carrier density in an ionic liquid using a capacitor model, and Chen *et al.* have interpreted the electron transport characteristics in terms of charged impurity–induced scattering [6, 93] (Fig. 6.23).

Shan *et al.* [94] constructed a graphene-based electrochemical biosensor for glucose, which achieved linear glucose response up to 14 mM. These authors also studied the properties of water-soluble graphene covalently functionalized by biocompatible poly-l-lysine [95].

6.4.1.4 Biosensor device

CMG, with a two-dimensional nanostructure and tunable surface chemistry, can interface strongly with the biological systems [19] (Fig. 6.24). The CMGs and their biohybrids can be synthesized using GO, graphene amine (GA), or plasma-modified graphene amine (PGA) immobilized on a silica substrate. PGA was synthesized by exposing the graphite flakes to either ammonia or nitrogen plasma, followed by exfoliation via sonication in water or by exposing the GO sheets on a silica substrate to hydrogen plasma followed by ammonia (or nitrogen) plasma. The GO sheets immobilized on silica were used as templates to selectively and covalently tether single-stranded DNA to build the G-DNA hybrid. Since this DNA's terminal amine group bonds covalently with the carboxylic group on GO and not with the amine groups on silica, the reaction resulted in selective DNA tethering of the GO sheets. The higher DNA density on the surface wrinkles was attributed to the local field enhancement at the sharp edges.

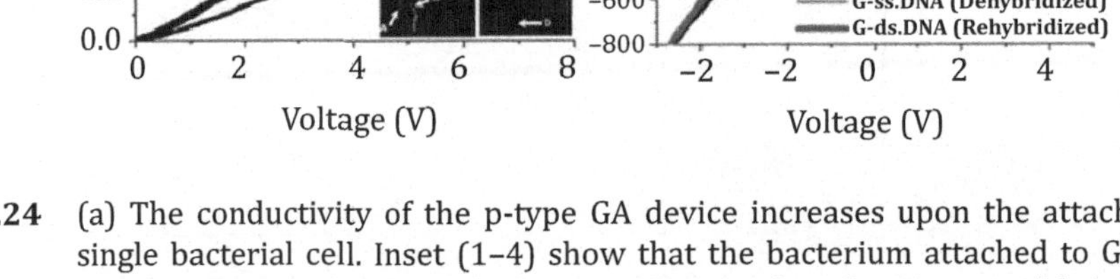

Figure 6.24 (a) The conductivity of the p-type GA device increases upon the attachment of a single bacterial cell. Inset (1–4) show that the bacterium attached to GA on silica remains alive (A); however, the bacteria deposited on the GA atop gold electrodes are dead (D) after electrical measurements. (b) DNA transistor: ss-DNA tethering on GO increases the conductivity of the device. Hybridization and dehybridization of DNA on the G-DNA device result in the reversible increase and restoration of conductivity. Inset shows a G-DNA (ds) sheet with wrinkles and folds clearly visible. (Reprinted with permission from ref. 19.) See also Color Insert.

Graphene-based glucose biosensors have been demonstrated, for example, based on the immobilization of glucose oxidase in platinum nanoparticles/graphene/chitosan nanocomposite film [96] or a oxidase-graphene-chitosan-modified electrode for direct electrochemistry and glucose sensing [97]. Electrochemical sensing and a biosensing platform based on chemically reduced GO have been reported [98]. Understanding the electrochemistry of graphene and related materials have becoming increasingly important for biosensing applications [99, 100].

6.4.2 Optical Sensing

Optical measurements of graphene and/or chemicals bonded to it are useful to confirm a successful modification of graphene. Applications of chemically modified graphene have been studied very intensely because modified graphene is used for many different experiments to detect various molecules. Among them, GO is oxygenated bearing hydroxyl, epoxide, and carbonyl functional groups, and it is very attractive to many researchers for its special properties[42]. Other additional functionalization can be easily equipped, and this strongly implies its possible use in biosensing and drug delivery. Infrared (IR) spectroscopy reveals the existence of –OH, C=O, and C=C bonding of GO [101, 102]. By this method, the further modification of GO, such as GO–PEG for spectral imaging and drug delivery, is verified [103] (Fig. 6.25). This method is especially useful to check the successful chemical modification of graphene for specific detection, applicable for biosensing and drug delivery.

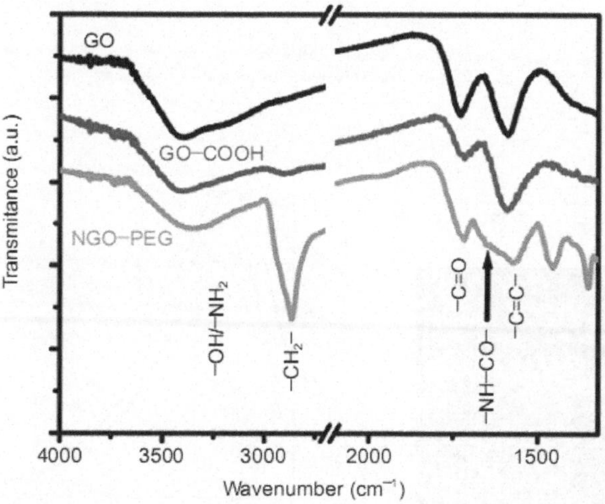

Figure 6.25 IR spectra of chemical bondings on graphene. GO, GO–COOH, and NGO–PEG peaks are seen. (Reprinted with permission from ref. 103.)

Other optical methods are also utilized to study optical properties of graphene and molecules adsorbed to it. UV-Vis-NIR spectroscopy detects the optical absorption peak at 230 nm originated from the π-plasmon of carbon [104]. The more modification to graphene results in a broader absorption bandwidth; GO–COOH and GO–PEG showed much broader bands than GO. Also, the fluorescence of GO excited in the visible range reveals a peak at

570 nm when it is excited by the 400 nm laser, while GO–PEG has a fluorescence peak at 520 nm. Photoluminescence in the IR range also shows the difference between GO and GO–PEG. This is useful in bioimaging because it is almost free from cellular autofluorescence. Recently, it was also shown that graphene-based materials can effectively quench the fluorescence (Fig. 6.26) of other molecules (such as dye molecules) nearby [105, 106] (Fig. 6.26a, b). Lu *et al.* [14] have used GO to sense DNA and proteins based on GO's fluorescence quenching (Fig. 6.26c, d). They demonstrated that the quenched fluorescence can be recovered upon unbinding to as ssDNA from GO caused by the binding of the ssDNA to a target (complementary DNA). Experiments suggesting the biostability of DNA and its protection from cleavage by DNase I (Tang *et al.* 2010]) and DNA-mediated self-assembly of graphene was also demonstrated (Patil *et al.* 2009]. Multicolor fluorescence-based DNA analysis by labeling DNA probes with different fluorophores was later suggested as an improvement (He *et al.* 2010). Molecular beacon–based approaches were also suggested by Lu *et al.* (2010a, b) to detect single-base DNA mismatch *in vitro* and surviving mRNA in cells. Fluorescence resonance energy transfer graphene sensors using quantum dots was demonstrated to detect target DNA (Dong *et al.* 2010) and thrombin (Chang *et al.* 2010). DNA–GO constructs uptake have also been demonstrated in living cells (Wang *et al.* 2010), including drug delivery (Zhang *et al.* 2010). It is clear that graphene could be integrated with fluorescing molecules for both *in vitro* and *in vivo* sensing.

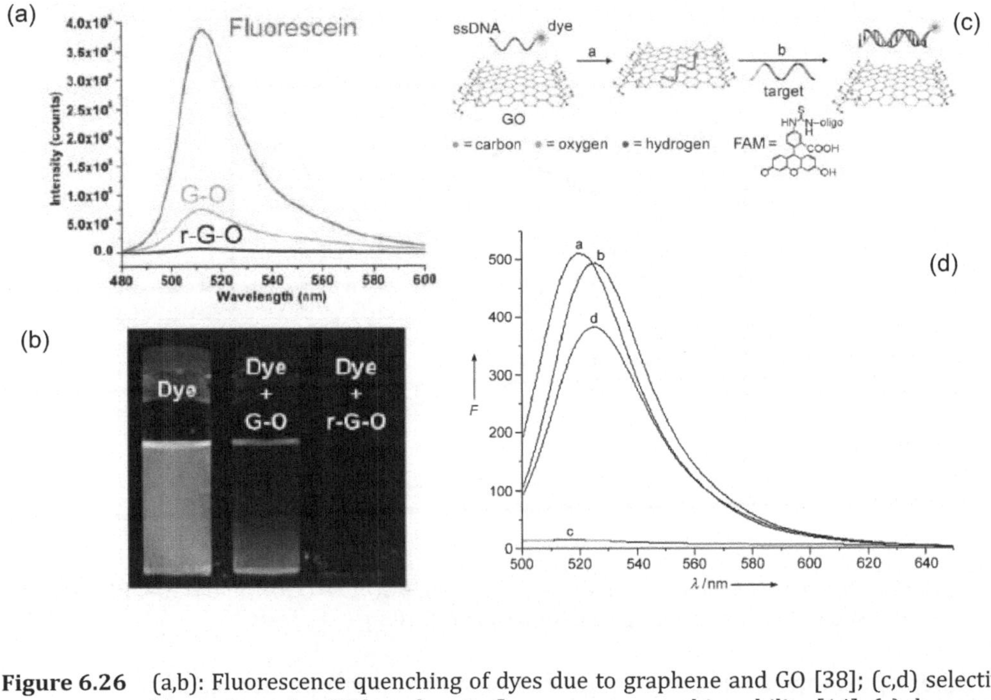

Figure 6.26 (a,b): Fluorescence quenching of dyes due to graphene and GO [38]; (c,d) selective biosensing using GO based on its fluorescence quenching ability [14]; (c) the sensor operation principle; (d) fluorescence emission spectra for a. P1 in Tris-HCl buffer, b. P1 + (300 nM) HIV1, c. P1 + GO, and d. P1 + GO + (300 nM) HIV1 [14]. (Fig. 6.26 a,b and c,d reprinted with permission from ref. 38 and 14, respectively.) See also Color Insert.

Raman spectroscopy is also one of the most useful detection techniques. Raman spectroscopy probes the phonon vibration mode of graphene and electron-phonon interaction. Raman is an excellent tool to monitor the quality of graphene, detecting the charged impurity doping in graphene and defects in graphene [13, 107] and can be sensitive to chemical modifications of graphene [81]. In addition to measuring graphene, one can also probe the Raman spectra of the molecules adsorbed on graphene [15].

Doping and charge impurities in graphene can strongly affect its Raman spectrum. With the existence of doping/charge impurities, the *G* peak is upshifted and its bandwidth is stiffened [77]. *G* peak is upshifted because of Kohn anomaly [108]. Also, the *G* peak is stiffened because phonon decay into electron-hole pairs is blocked by Pauli exclusion. The two-dimensional peak also slightly shifts and, more importantly, its peak intensity strongly varies with the charge impurities. The shift of two-dimensional peak is such that hole-doping induces an upshift and the electron-doping induces a downshift. Lastly, charged impurity doesn't generate the *D* peak while the defects and edges of graphene do [107]. Using this method with Raman spectroscopy, the qualities of graphene and its doping can be monitored [109, 110], with potential applications in biosensing [111].

Not only the Raman spectrum of graphene can be affected by molecules adsorbed on graphene, there are possibilities that graphene may also enhance and affect the Raman spectra of the adsorbed molecules for direct sensing of those molecules through their molecular Raman fingerprints. For example, it has been shown that graphene can enhance (close to 20-fold demonstrated so far) the Raman signals (which would have been unobservable without graphene, e.g., if the molecules are on a Si wafer) of various dye molecules adsorbed on graphene (Fig. 6.27) [15]. The enhancement has been interpreted as due to a chemical mechanism involving the charge transfer between the molecules and graphene. Since the electronic properties of graphene are highly tunable by chemical or electrical (e.g., a gate) manipulations, it would be interesting to explore how such tunability (which, in turn, may influence the charge transfer between graphene and adsorbed molecules) can affect the Raman enhancement and Raman fingerprints of the adsorbed molecules, potentially opening new venues for selective Raman biosensing.

It is worthwhile to point out that in many ways, graphene can be applied for sensing and biosensing as other carbon materials, especially carbon nanotubes, have been. Carbon nanotubes and graphene are both suitable candidate materials for electronic, photonic, and optoelectronic nano-devices and the intense studies on carbon nanotubes, and their sensing applications can often serve as inspirations and/or informative comparison points for applications of graphene. For example, similar to graphene, Raman spectroscopy has been used as a tool to study charge transfer in doped carbon nanotubes [112] and how the direction of carbon nanotube crystals determines its electronic properties [113]. It is also shown that the defects or local environment of SWCNTs can cause different fluorescence spectra emission energy and bandwidth of the nanotubes without fluctuation [114]. The Raman signals of the carbon nanotubes are also sensitive to various chemical modifications, such as oxidization [115]. A carbon nanotube has been used as a Raman label [116] because it generates a strong Raman signal with multiple modifications without photobleaching or quenching. Proteins with the concentration as low as 1 fM could be detected and quantified by this method. Also, gene or drug delivery due to the multifunctionality of carbon nanotubes is a promising application because graphene protects the gene/drug attached to it until it is released [117, 118]. Moreover, the electronic sensing by carbon nanotubes with special chemical modification is also demonstrated [119]. Most applications listed above can be directly adapted to graphene. Furthermore, the dangling bonds at the edges of a graphene nanoribbon may be chemically functionalized to perform selective biological sensing. The two-dimensional features of graphene can facilitate the application of graphene as a powerful biosensing tool.

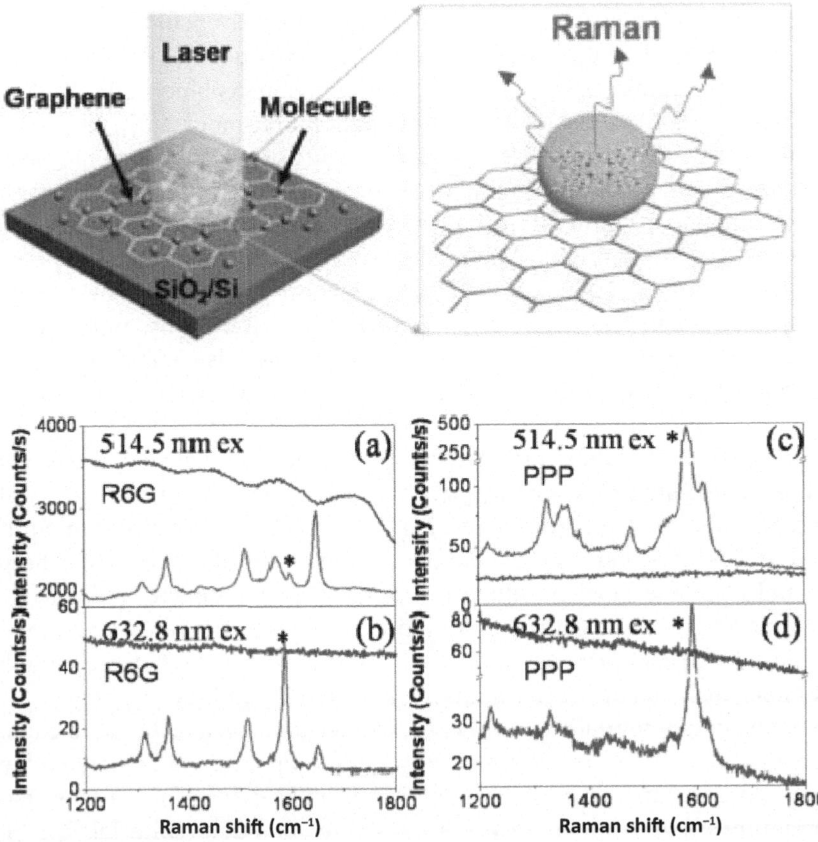

Figure 6.27 The enhancement of the Raman signals of dye molecules on graphene. (Reprinted with permission from ref. 15.)

In addition to sensing experiments, fundamental studies (both experimentally and theoretically) on the interaction of biological molecules, such as DNA/nucleotides, with graphene have been performed [120, 121]. Density functional theory–based methods can be used to model how organic molecules interact with graphene to change its electronic properties [122]. Catalyst-free growth, orientation, and biosensing properties of multilayer graphene nanoflake films with sharp-edged planes have been studied [123]. GO has been found to exhibit peroxidase catalytic activity that could catalyze 3,3,5,5-tetramethylbenzidine (TMB) oxidation with H_2O_2 for glucose detection (Song *et al.* 2010). Liu *et al.* (2010) decorated GO with protein or DNA to fabricate nanoparticle assemblies on GO as a strategy to synthesize new hybrid materials. We expect that such biomimetic materials could have a variety of functionalities in biomedical sensing.

6.5 Closing Remarks and Future Work

While emphasizing the nature of graphene as a material, the primary focus of this article was on harnessing its potential as a sensor in biological systems. Graphene in biosensing is only at the initial stages. We expect to see an increased thrust in the development of biomimetic graphene materials to address specific cellular functions. Applications *in vivo* will obviously require the development of multifunctional graphene materials that are not cytotoxic while simultaneously enhancing the detection and functional characteristics of the materials, including photothermal therapy. The unparalleled use of graphene in gas sensing will be further advanced to detect volatiles as well as other biothreat agents using integrated portable devices for detecting gaseous compounds in ultralow quantities. While constructing sensor-related applications using graphene and graphene biocomposites, surface fouling and sample presentation steps need to be given careful attention.

References

1. Novoselov, K. S., Geim, A. K., Morozov, S. V., Jiang, D., Zhang, Y., Dubonos, S. V., Grigorieva, I. V., and Firsov, A. A. (2004). Electric field effect in atomically thin carbon films, *Science*, **306**(5696), 666–669.
2. Geim, A. K., and Kim, P. (2008). Carbon wonderland, *Sci. Am.*, **298**(4), 90.
3. Geim, A. K. (2009). Graphene: status and prospects, *Science*, **324**(5934), 1530–1534.
4. Rao, C. N. R., Sood, A. K., Subrahmanyam, K. S., and Govindaraj, A. (2009). Graphene: the new two-dimensional nanomaterial, *Angew. Chem., Int. Ed.*, **48**(42), 7752–7777.
5. Geim, A. K., and Novoselov, K. S. (2007). The rise of graphene, *Nat. Mater.*, **6**(3), 183–191.
6. Chen, F., Qing, Q., Xia, J., Li, J., and Tao, N. (2009). Electrochemical gate-controlled charge transport in graphene in ionic liquid and aqueous solution, *J. Am. Chem. Soc.*, **131**(29), 9908–9909.
7. Stoller, M. D., Park, S., Zhu, Y., An, J., and Ruoff, R. S. (2008). Graphene-based ultracapacitors, *Nano Lett.*, **8**(10), 3498–3502.
8. Park, S., and Ruoff, R. S. (2009). Chemical methods for the production of graphenes, *Nat. Nano*, **4**(4), 217–224.
9. Neto, A. H. C., Guinea, F., Peres, N. M. R., Novoselov, K. S., and Geim, A. K. (2009). The electronic properties of graphene. *Rev. Mod. Phys.*, **81**, 109.
10. Schedin, F., Geim, A. K., Morozov, S. V., Hill, E. W., Blake, P., Katsnelson, M. I., and Novoselov, K. S. (2007). Detection of individual gas molecules adsorbed on graphene, *Nat. Mater.*, **6**(9), 652–655.
11. Li, X., Zhu, Y., Cai, W., Borysiak, M., Han, B., Chen, D., Piner, R. D., Colombo, L., and Ruoff, R. S. (2009). Transfer of large-area graphene films for high-performance transparent conductive electrodes, *Nano Lett.*, **9**(12), 4359–4363.

12. Nair, R. R., Blake, P., Grigorenko, A. N., Novoselov, K. S., Booth, T. J., Stauber, T., Peres, N. M. R., and Geim, A. K. (2008). Fine structure constant defines visual transparency of graphene, *Science*, **320**(5881), 1308.

13. Ferrari, A. C., Meyer, J. C., Scardaci, V., Casiraghi, C., Lazzeri, M., Mauri, F., Piscanec, S., Jiang, D., Novoselov, K. S., Roth, S., and Geim, A. K. (2006). Raman spectrum of graphene and graphene layers, *Phys. Rev. Lett.*, **97**(18), 187401–187404.

14. Chun-Hua, L., Huang-Hao, Y., Chun-Ling, Z., Xi, C., and Guo-Nan, C. (2009). A graphene platform for sensing biomolecules. *Angew. Chem., Int. Ed.*, **48**(26), 4785–4787.

15. Ling, X., Xie, L., Fang, Y., Xu, H., Zhang, H., Kong, J., Dresselhaus, M. S., Zhang, J., and Liu, Z. (2009). Can graphene be used as a substrate for Raman enhancement? *Nano Lett.*, **10**(2), 553–561.

16. Balandin, A. A., Ghosh, S., Bao, W., Calizo, I., Teweldebrhan, D., Miao, F., and Lau, C. N. (2008). Superior thermal conductivity of single-layer graphene, *Nano Lett.*, **8**(3), 902–907.

17. Bunch, J. S., van der Zande, A. M., Verbridge, S. S., Frank, I. W., Tanenbaum, D. M., Parpia, J. M., Craighead, H. G., and McEuen, P. L. (2007). Electromechanical resonators from graphene sheets, *Science*, **315**(5811), 490–493.

18. Lee, C., Wei, X., Kysar, J. W., and Hone, J. (2008). Measurement of the elastic properties and intrinsic strength of monolayer graphene, *Science*, **321**(5887), 385–388.

19. Mohanty, N., and Berry, V. (2008). Graphene-based single-bacterium resolution biodevice and DNA transistor: interfacing graphene derivatives with nanoscale and microscale biocomponents, *Nano Lett.*, **8**(12), 4469–4476.

20. Novoselov, K. S., Jiang, D., Schedin, F., Booth, T. J., Khotkevich, V. V., Morozov, S. V., and Geim, A. K. (2005). Two-dimensional atomic crystals, *Proc. Natl. Acad. Sci. USA*, **102**(30), 10451–10453.

21. Liang, X., Fu, Z., and Chou, S. Y. (2007). Graphene transistors fabricated via transfer-printing in device active-areas on large wafer, *Nano Lett.*, **7**(12), 3840–3844.

22. Berger, C., Song, Z., Li, X., Wu, X., Brown, N., Naud, C., Mayou, D., Li, T., Hass, J., Marchenkov, A. N., Conrad, E. H., First, P. N., and de Heer, W. A. (2006). Electronic confinement and coherence in patterned epitaxial graphene, *Science*, **312**(5777), 1191–1196.

23. Li, X., Cai, W., An, J., Kim, S., Nah, J., Yang, D., Piner, R., Velamakanni, A., Jung, I., Tutuc, E., Banerjee, S. K., Colombo, L., and Ruoff, R. S. (2009). Large-area synthesis of high-quality and uniform graphene films on copper foils, *Science*, **324**(5932), 1312–1314.

24. Wintterlin, J., and Bocquet, M. L. (2009). Graphene on metal surfaces, *Surf. Sci.*, **603**(10–12), 1841–1852.

25. Anton, N. S., Mehdi, M. Y., Romaneh, J., Ouseph, P. J., Cohn, R. W., and Sumanasekera, G. U. (2007). Electrostatic deposition of graphene, *Nanotechnology*, (13), 135301.

26. Liang, X., Chang, A. S. P., Zhang, Y., Harteneck, B. D., Choo, H., Olynick, D. L., and Cabrini, S. (2009). Electrostatic force assisted exfoliation of prepatterned few-layer graphenes into device sites. *Nano Lett.*, **9**(1), 467–472.

27. Hernandez, Y., Nicolosi, V., Lotya, M., Blighe, F. M., Sun, Z., De, S., McGovern, I. T., Holland, B., Byrne, M., Gun'Ko, Y. K., Boland, J. J., Niraj, P., Duesberg, G., Krishnamurthy, S., Goodhue, R., Hutchison, J., Scardaci, V., Ferrari, A. C., and Coleman, J. N. (2008). High-yield production of graphene by liquid-phase exfoliation of graphite, *Nat. Nano*, **3**(9), 563–568.

28. Li, X., Wang, X., Zhang, L., Lee, S., and Dai, H. (2008). Chemically derived, ultrasmooth graphene nanoribbon semiconductors, *Science*, **319**(5867), 1229–1232.

29. de Heer, W. A., Berger, C., Wu, X., First, P. N., Conrad, E. H., Li, X., Li, T., Sprinkle, M., Hass, J., Sadowski, M. L., Potemski, M., and Martinez, G. (2007). Epitaxial graphene, *Solid State Commun.*, **143**(1–2), 92–100.

30. Emtsev, K. V., Bostwick, A., Horn, K., Jobst, J., Kellogg, G. L., Ley, L., McChesney, J. L., Ohta, T., Reshanov, S. A., Rohrl, J., Rotenberg, E., Schmid, A. K., Waldmann, D., Weber, H. B., and Seyller, T. (2009). Towards wafer-size graphene layers by atmospheric pressure graphitization of silicon carbide, *Nat. Mater.*, **8**(3), 203–207.

31. Chuhei, O., and Ayato, N. (1997). Ultra-thin epitaxial films of graphite and hexagonal boron nitride on solid surfaces, *J. Phys.: Condens. Matter*, (1), 1.

32. Cao, H., Yu, Q., Jauregui, L. A., Tian, J., Wu, W., Liu, Z., Jalilian, R., Benjamin, D. K., Jiang, Z., and Bao, J. (2009). Wafer-scale graphene synthesized by chemical vapor deposition at ambient pressure, *Appl. Phys. Lett.*, **96**, 122106.

33. Bae, S., Kim, H. K., Xu, X., Balakrishnan, J., Lei, T., Song, Y. I., Kim, Y. J., Ozyilmaz, B., Ahn, J. H., and Hong, B. H. (2009). 30-inch roll-based production of high-quality graphene films for flexible transparent electrodes, *Nat. Nanotechnol.*, **5**, 574–578.

34. Yu, Q., Lian, J., Siriponglert, S., Li, H., Chen, Y., and Pei, S.-S. (2008). Graphene synthesis by surface segregation on Ni and Cu, *Arxiv preprint arXiv:0804.1778v1.*

35. Levendorf, M. P., Ruiz-Vargas, C. S., Garg, S., and Park, J. (2009). Transfer-free batch fabrication of single layer graphene transistors, *Nano Lett.*, **9**(12), 4479–4483.

36. Yu, Q., Lian, J., Siriponglert, S., Li, H., Chen, Y. P., and Pei, S.-S. (2008). Graphene segregated on Ni surfaces and transferred to insulators, *Appl. Phys. Lett.*, **93**(11), 113103–113105.

37. Cao, H., Yu, Q., Pandey, D., Zemlianov, D., Colby, R., Childres, I., Drachev, V., Stach, E., Lian, J., and Li, H. (2009). Large scale graphene films synthesized on metals and transferred to insulators for electronic applications, *Adv. Mater.*, **107**(4), 1–15.

38. Kim, K. S., Zhao, Y., Jang, H., Lee, S. Y., Kim, J. M., Kim, K. S., Ahn, J.-H., Kim, P., Choi, J.-Y., and Hong, B. H. (2009). Large-scale pattern growth of graphene films for stretchable transparent electrodes, *Nature*, **457**(7230), 706–710.

39. Reina, A., Jia, X., Ho, J., Nezich, D., Son, H., Bulovic, V., Dresselhaus, M. S., and Kong, J. (2009). Layer area, few-layer graphene films on arbitrary substrates by chemical vapor deposition, *Nano Lett.*, **9**(8), 3087–3087.

40. Gilje, S., Han, S., Wang, M., Wang, K. L., and Kaner, R. B. (2007). A chemical route to graphene for device applications, *Nano Lett.*, **7**(11), 3394–3398.

41. Gomez-Navarro, C., Weitz, R. T., Bittner, A. M., Scolari, M., Mews, A., Burghard, M., and Kern, K. (2007). Electronic transport properties of individual chemically reduced graphene oxide sheets, *Nano Lett.*, **7**(11), 3499–3503.

42. Stankovich, S., Dikin, D. A., Dommett, G. H. B., Kohlhaas, K. M., Zimney, E. J., Stach, E. A., Piner, R. D., Nguyen, S. T., and Ruoff, R. S. (2006). Graphene-based composite materials, *Nature*, **442**(7100), 282–286.

43. Datta, S. (2005). *Quantum Transport: Atom to Transistor*, 2nd edn, Cambridge University Press, Cambridge, UK.

44. Zhang, Y. B., Tan, Y. W., Stormer, H. L., and Kim, P. (2005). Experimental observation of the quantum Hall effect and Berry's phase in graphene, *Nature*, **438**(7065), 201–204.

45. Purewal, M. S., Hong, B. H., Ravi, A., Chandra, B., Hone, J., and Kim, P. (2007). Scaling of resistance and electron mean free path of single-walled carbon nanotubes, *Phys. Rev. Lett.*, **98**(18), 4.

46. Geim, A. K., and Novoselov, K. S. (2007). The rise of graphene. *Nat. Mater.*, **6**(3), 183–191.

47. Chen, J. H., Jang, C., Adam, S., Fuhrer, M. S., Williams, E. D., and Ishigami, M. (2008). Charged-impurity scattering in graphene, *Nat. Phys.*, **4**(5), 377–381.

48. Schedin, F., Geim, A. K., Morozov, S. V., Hill, E. W., Blake, P., Katsnelson, M. I., and Novoselov, K. S. (2007). Detection of individual gas molecules adsorbed on graphene, *Nat. Mater.*, **6**(9), 652–655.

49. Novoselov, K. S., Geim, A. K., Morozov, S. V., Jiang, D., Katsnelson, M. I., Grigorieva, I. V., Dubonos, S. V., and Firsov, A. A. (2005). Two-dimensional gas of massless Dirac fermions in graphene, *Nature*, **438**(7065), 197–200.

50. Lin, Y.-M., and Avouris, P. (2008). Strong suppression of electrical noise in bilayer graphene nanodevices. *Nano Lett.*, **8**(8), 2119–2125.

51. Bolotin, K. I., Sikes, K. J., Jiang, Z., Klima, M., Fudenberg, G., Hone, J., Kim, P., and Stormer, H. L. (2008). Ultrahigh electron mobility in suspended graphene, *Solid State Commun.*, **146**(9–10), 351–355.

52. Du, X., Skachko, I., Barker, A., and Andrei, E. Y. (2008). Approaching ballistic transport in suspended graphene, *Nat. Nanotechnol.*, **3**(8), 491–495.

53. Tan, Y. W., Zhang, Y., Stormer, H. L., and Kim, P. (2007). Temperature dependent electron transport in graphene, *Eur. Phys. J.*, **148**, 15–18.

54. Bolotin, K. I., Sikes, K. J., Hone, J., Stormer, H. L., and Kim, P. (2008). Temperature-dependent transport in suspended graphene, *Phys. Rev. Lett.*, **101**(9), 096802.

55. Chen, J.-H., Cullen, W. G., Jang, C., Fuhrer, M. S., and Williams, E. D. (2009). Defect scattering in graphene, *Phys. Rev. Lett.*, **102**(23), 236805.

56. Bolotin, K. I., Sikes, K. J., Hone, J., Stormer, H. L., and Kim, P. (2008). Temperature-dependent transport in suspended graphene, *Phys. Rev. Lett.*, **101**(9), 4.

57. Tan, Y. W., Zhang, Y., Stormer, H. L., and Kim, P. (2007). Temperature dependent electron transport in graphene, *Eur. Phys. J. Spec. Top.*, **148**(1), 15–18.

58. Han, M. Y., Özyilmaz, B., Zhang, Y., and Kim, P. (2007). Energy band-gap engineering of graphene nanoribbons, *Phys. Rev. Lett.*, **98**(20), 206805.

59. Pedersen, T. G., Flindt, C., Pedersen, J., Jauho, A.-P., Mortensen, N. A., and Pedersen, K. (2008). Optical properties of graphene antidot lattices, *Phys. Rev. B: Condens. Matter Mater. Phys.*, **77**(24), 245431–245436.

60. Oostinga, J. B., Heersche, H. B., Liu, X., Morpurgo, A. F., and Vandersypen, L. M. K. (2008). Gate-induced insulating state in bilayer graphene devices, *Nat. Mater.*, **7**(2), 151–157.

61. Zhou, S. Y., Gweon, G. H., Fedorov, A. V., First, P. N., de Heer, W. A., Lee, D. H., Guinea, F., Castro Neto, A. H., and Lanzara, A. (2007). Substrate-induced bandgap opening in epitaxial graphene, *Nat. Mater.*, **6**(10), 770–775.

62. Wu, X., Sprinkle, M., Li, X., Ming, F., Berger, C., and de Heer, W. A. (2008). Epitaxial-graphene/graphene-oxide junction: an essential step towards epitaxial graphene electronics, *Phys. Rev. Lett.*, **101**(2), 026801.

63. Barone, V., Hod, O., and Scuseria, G. E. (2006). Electronic structure and stability of semiconducting graphene nanoribbons, *Nano Lett.*, **6**(12), 2748–2754.

64. Han, M. Y., Ozyilmaz, B., Zhang, Y. B., and Kim, P. (2007). Energy band-gap engineering of graphene nanoribbons, *Phys. Rev. Lett.*, **98**(20), 4.

65. Chen, Z. H., Lin, Y. M., Rooks, M. J., and Avouris, P. (2007). Graphene nano-ribbon electronics, *Physica E*, **40**(2), 228–232.

66. Oostinga, J. B., Heersche, H. B., Liu, X. L., Morpurgo, A. F., and Vandersypen, L. M. K. (2008). Gate-induced insulating state in bilayer graphene devices, *Nat. Mater.*, **7**(2), 151–157.

67. Zhou, S. Y., Gweon, G. H., Fedorov, A. V., First, P. N., De Heer, W. A., Lee, D. H., Guinea, F., Neto, A. H. C., and Lanzara, A. (2007). Substrate-induced bandgap opening in epitaxial graphene, *Nat. Mater.*, **6**(10), 770–775.

68. Abergel, D. S. L., Russell, A., and Fal'ko, V. I. (2007). Visibility of graphene flakes on a dielectric substrate, *Appl. Phys. Lett.*, **91**(6), 063125–063127.

69. Avouris, P., Chen, Z., and Perebeinos, V. (2007). Carbon-based electronics, *Nat. Nano*, **2**(10), 605–615.

70. Blake, P., Hill, E. W., Neto, A. H. C., Novoselov, K. S., Jiang, D., Yang, R., Booth, T. J., and Geim, A. K. (2007). Making graphene visible, *Appl. Phys. Lett.*, **91**(6), 063124–063126.

71. Ni, Z. H., Wang, H. M., Kasim, J., Fan, H. M., Yu, T., Wu, Y. H., Feng, Y. P., and Shen, Z. X. (2007). Graphene thickness determination using reflection and contrast spectroscopy, *Nano Lett.*, **7**(9), 2758–2763.

72. Blake, P., Brimicombe, P. D., Nair, R. R., Booth, T. J., Jiang, D., Schedin, F., Ponomarenko, L. A., Morozov, S. V., Gleeson, H. F., Hill, E. W., Geim, A. K., and Novoselov, K. S. (2008). Graphene-based liquid crystal device, *Nano Lett.*, **8**(6), 1704–1708.

73. Jun, W., Yenny, H., Mustafa, L., Jonathan, N. C., and Werner, J. B. (2009). Broadband nonlinear optical response of graphene dispersions, *Adv. Mater.*, **21**(23), 2430–2435.

74. Vasko, F. T., and Ryzhii, V. (2008). Photoconductivity of intrinsic graphene, *Phys. Rev. B: Condens. Matter Mater. Phys.*, **77**(19), 195433–195440.

75. Graf, D., Molitor, F., Ensslin, K., Stampfer, C., Jungen, A., Hierold, C., and Wirtz, L. (2007). Spatially resolved Raman spectroscopy of single- and few-layer graphene, *Nano Lett.*, **7**(2), 238–242.

76. Gupta, A., Chen, G., Joshi, P., and Tadigadapa, S. (2006). Eklund, Raman scattering from high-frequency phonons in supported n-graphene layer films, *Nano Lett.*, **6**(12), 2667–2673.

77. Yan, J., Zhang, Y., Kim, P., and Pinczuk, A. (2007). Electric field effect tuning of electron-phonon coupling in graphene, *Phys. Rev. Lett.*, **98**(16), 166802.

78. Casiraghi, C., Hartschuh, A., Qian, H., Piscanec, S., Georgi, C., Fasoli, A., Novoselov, K. S., Basko, D. M., and Ferrari, A. C. (2009). Raman spectroscopy of graphene edges, *Nano Lett.*, **9**(4), 1433–1441.

79. Huang, M., Yan, H., Chen, C., Song, D., Heinz, T. F., and Hone, J. (2009). Phonon softening and crystallographic orientation of strained graphene studied by Raman spectroscopy, *Proc. Natl. Acad. Sci. USA*, **106**(18), 7304–7308.

80. Zhang, C., Chen, L., and Ma, Z. (2008). Orientation dependence of the optical spectra in graphene at high frequencies, *Phys. Rev. B: Condens. Matter Mater. Phys.*, **77**(24), 241402–4.

81. Elias, D. C., Nair, R. R., Mohiuddin, T. M. G., Morozov, S. V., Blake, P., Halsall, M. P., Ferrari, A. C., Boukhvalov, D. W., Katsnelson, M. I., Geim, A. K., and Novoselov, K. S. (2009). Control of graphene's properties by reversible hydrogenation: evidence for graphane, *Science*, **323**(5914), 610–613.

82. Lee, C., Wei, X. D., Kysar, J. W., and Hone, J. (2008). Measurement of the elastic properties and intrinsic strength of monolayer graphene, *Science*, **321**(5887), 385–388.

83. Frank, I. W., Tanenbaum, D. M., Van der Zande, A. M., and McEuen, P. L. (2007). Mechanical properties of suspended graphene sheets, *J. Vac. Sci. Technol., B: Microelectron. Nanometer Struct.*, **6**, 2558–2561.

84. Robinson, J. T., Zalalutdinov, M., Baldwin, J. W., Snow, E. S., Wei, Z. Q., Sheehan, P., and Houston, B. H. (2008). Wafer-scale reduced graphene oxide films for nanomechanical devices, *Nano Lett.*, **8**(10), 3441–3445.

85. Shivaraman, S., Barton, R. A., Yu, X., Alden, J., Herman, L., Chandrashekhar, M. V. S., Park, J., McEuen, P. L., Parpia, J. M., Craighead, H. G., and Spencer, M. G. (2009). Free-standing epitaxial graphene, *Nano Lett.*, **9**(9), 3100–3105.

86. Burg, T. P., Godin, M., Knudsen, S. M., Shen, W., Carlson, G., Foster, J. S., Babcock, K., and Manalis, S. R. (2007). Weighing of biomolecules, single cells and single nanoparticles in fluid, *Nature*, **446**(7139), 1066–1069.

87. Fasolino, A., Los, J. H., and Katsnelson, M. I. (2007). Intrinsic ripples in graphene, *Nat. Mater.*, **6**(11), 858–861.

88. Meyer, J. C., Geim, A. K., Katsnelson, M. I., Novoselov, K. S., Booth, T. J., and Roth, S. (2007). The structure of suspended graphene sheets, *Nature*, **446**(7131), 60–63.

89. Morozov, S. V., Novoselov, K. S., Katsnelson, M. I., Schedin, F., Elias, D. C., Jaszczak, J. A., and Geim, A. K. (2008). Giant intrinsic carrier mobilities in graphene and its bilayer, *Phys. Rev. Lett.*, **100**(1), 016602–016605.

90. Chen, J.-H., Jang, C., Xiao, S., Ishigami, M., and Fuhrer, M. S. (2008). Intrinsic and extrinsic performance limits of graphene devices on SiO2, *Nat. Nano*, **3**(4), 206–209.

91. Ohno, Y., Maehashi, K., Yamashiro, Y., and Matsumoto, K. (2009). Electrolyte-gated graphene field-effect transistors for detecting pH and protein adsorption, *Nano Lett.*, **9**(9), 3318–3322.

92. Ang, P. K., Chen, W., Wee, A. T. S., and Loh, K. P. (2008). Solution-gated epitaxial graphene as pH sensor, *J. Am. Chem. Soc.*, **130**(44), 14392–14393.

93. Xia, J., Chen, F., Li, J., and Tao, N. (2009). Measurement of the quantum capacitance of graphene, *Nat Nano*, **4**, 505–509.

94. Shan, C., Yang, H., Song, J., Han, D., Ivaska, A., and Niu, L. (2009). Direct electrochemistry of glucose oxidase and biosensing for glucose based on graphene, *Anal. Chem.*, **81**(6), 2378–2382.

95. Shan, C., Yang, H., Han, D., Zhang, Q., Ivaska, A., and Niu, L. (2009). Water-soluble graphene covalently functionalized by biocompatible poly-l-lysine, *Langmuir*, **25**(20), 12030–12033.

96. Wu, H., Wang, J., Kang, X., Wang, C., Wang, D., Liu, J., Aksay, I. A., and Lin, Y. (2009). Glucose biosensor based on immobilization of glucose oxidase in platinum nanoparticles/graphene/chitosan nanocomposite film, *Talanta*, **80**(1), 403–406.

97. Kang, X., Wang, J., Wu, H., Aksay, I. A., Liu, J., and Lin, Y. (2009). Glucose oxidase-graphene-chitosan modified electrode for direct electrochemistry and glucose sensing, *Biosens. Bioelectron.*, **25**(4), 901–905.

98. Zhou, M., Zhai, Y., and Dong, S. (2009). Electrochemical sensing and biosensing platform based on chemically reduced graphene oxide, *Anal. Chem.*, **81**(14), 5603–5613.

99. Alwarappan, S., Erdem, A., Liu, C., and Li, C.-Z. (2009). Probing the Electrochemical Properties of Graphene Nanosheets for Biosensing Applications. *The Journal of Physical Chemistry C*, **113**(20), 8853–8857.

100. Martin, P. (2009). Electrochemistry of graphene: new horizons for sensing and energy storage, *Chem. Rec.*, **9**(4), 211–223.

101. Hontoria-Lucas, C., López-Peinado, A. J., López-González, J. d. D., Rojas-Cervantes, M. L., and Martín-Aranda, R. M. (1995). Study of oxygen-containing groups in a series of graphite oxides: physical and chemical characterization. *Carbon*, **33**(11), 1585–1592.

102. Szabó, T., Berkesi, O., and Dékány, I. (2005). DRIFT study of deuterium-exchanged graphite oxide, *Carbon*, **43**(15), 3186–3189.

103. Sun, X., Liu, Z., Welsher, K., Robinson, J. T., Goodwin, A., Zaric, S., and Dai, H. (2008). Nano-graphene oxide for cellular imaging and drug delivery, *Nano Res.*, **1**(3), 203–212.

104. Attal, S., Thiruvengadathan, R., and Regev, O. (2006). Determination of the concentration of single-walled carbon nanotubes in aqueous dispersions using UV-visible absorption spectroscopy, *Anal. Chem.*, **78**(23), 8098–8104.

105. Kim, J., Cote, L. J., Kim, F., and Huang, J. (2009). Visualizing graphene based sheets by fluorescence quenching microscopy, *J. Am. Chem. Soc.*, **132**(1), 260–267.

106. Xie, L., Ling, X., Fang, Y., Zhang, J., and Liu, Z. (2009). Graphene as a substrate to suppress fluorescence in resonance Raman spectroscopy, *J. Am. Chem. Soc.*, **131**(29), 9890–9891.

107. Casiraghi, C., Pisana, S., Novoselov, K. S., Geim, A. K., and Ferrari, A. C. (2007). Raman fingerprint of charged impurities in graphene, *Appl. Phys. Lett.*, **91**(23), 233108–233110.

108. Pisana, S., Lazzeri, M., Casiraghi, C., Novoselov, K. S., Geim, A. K., Ferrari, A. C., and Mauri, F. (2007). Breakdown of the adiabatic born-Oppenheimer approximation in graphene, *Nat. Mater.*, **6**(3), 198–201.

109. Das, A., Pisana, S., Chakraborty, B., Piscanec, S., Saha, S. K., Waghmare, U. V., Novoselov, K. S., Krishnamurthy, H. R., Geim, A. K., Ferrari, A. C., and Sood, A. K. (2008). Monitoring dopants by Raman scattering in an electrochemically top-gated graphene transistor, *Nat. Nano*, **3**(4), 210–215.

110. Xiaochen, D., Dongliang, F., Wenjing, F., Yumeng, S., Peng, C., and Lain-Jong, L. (2009). Doping single-layer graphene with aromatic molecules, *Small*, **5**(12), 1422–1426.

111. Das, B., Voggu, R., Rout, C. S., and Rao, C. N. R. (2008). Changes in the electronic structure and properties of graphene induced by molecular charge-transfer, *Chem. Commun.*, (41), 5155–5157.

112. Rao, A. M., Eklund, P. C., Bandow, S., Thess, A., and Smalley, R. E. (1997). Evidence for charge transfer in doped carbon nanotube bundles from Raman scattering, *Nature*, **388**(6639), 257–259.

113. Jorio, A., Saito, R., Hafner, J. H., Lieber, C. M., Hunter, M., McClure, T., Dresselhaus, G., and Dresselhaus, M. S. (2001). Structural (n, m) determination of isolated single-wall carbon nanotubes by resonant Raman scattering, *Phys. Rev. Lett.*, **86**(6), 1118.

114. Hartschuh, A., Pedrosa, H. N., Novotny, L., and Krauss, T. D. (2003). Simultaneous fluorescence and Raman scattering from single carbon nanotubes, *Science*, **301**(5638), 1354–1356.

115. Christian, S., Carola, M., and Claus, M. S. (2008). Oxidation induced shifts of Raman modes of carbon nanotubes, *Phys. Status Solidi B*, **245**(10), 2205–2208.

116. Chen, Z., Tabakman, S. M., Goodwin, A. P., Kattah, M. G., Daranciang, D., Wang, X., Zhang, G., Li, X., Liu, Z., Utz, P. J., Jiang, K., Fan, S., and Dai, H. (2008). Protein microarrays with carbon nanotubes as multicolor Raman labels, *Nat. Biotech.*, **26**(11), 1285–1292.

117. Prato, M., Kostarelos, K., and Bianco, A. (2008). Functionalized carbon nanotubes in drug design and discovery, *Acc. Chem. Res.*, **41**(1), 60–68.

118. Wu, Y., Phillips, J. A., Liu, H., Yang, R., and Tan, W. (2008). Carbon nanotubes protect DNA strands during cellular delivery, *ACS Nano*, **2**(10), 2023–2028.

119. Martinez, M. T., Tseng, Y.-C., Ormategui, N., Loinaz, I., Eritja, R., and Bokor, J. (2009). Label-free DNA biosensors based on functionalized carbon nanotube field effect transistors, *Nano Lett.*, **9**, 530–536.

120. Gowtham, S., Scheicher, R. H., Ahuja, R., Pandey, R., and Karna, S. P. (2007). Physisorption of nucleobases on graphene: density-functional calculations, *Phys. Rev. B: Condens. Matter Mater. Phys.*, **76**(3), 033401–033404.

121. Neenu, V., Umesha, M., Achutharao, G., Anindya, D., Prabal, K. M., Ajay, K. S., and Rao, C. N. R. (2009). Binding of DNA nucleobases and nucleosides with graphene, *Chem. Phys. Chem.*, **10**(1), 206–210.

122. Lu, Y. H., Chen, W., Feng, Y. P., and He, P. M. (2009). Tuning the electronic structure of graphene by an organic molecule, *J. Phys. Chem. B*, **113**(1), 2–5.

123. Nai Gui, S., Pagona, P., Martin, M., Ming, C., Artemis, S., Alessandro, P., Sarnjeet, S. D., and Helder, M. (2008). Catalyst-free efficient growth, orientation and biosensing properties of multilayer graphene nanoflake films with sharp edge planes, *Adv. Funct. Mater.*, **18**(21), 3506–3514.

Chapter 7

Single-Molecule Fluorescence Spectroscopy Techniques for Biomedicine

Jose M. Moran-Mirabal,[a,b,c] Harold G. Craighead,[b] and Larry P. Walker[c]

[a]*Department of Chemistry and Chemical Biology, McMaster University, Hamilton, Ontario, Canada*
[b]*School of Applied and Engineering Physics, Cornell University, Ithaca, NY, USA*
[c]*Department of Biological and Environmental Engineering, Cornell University, Ithaca, NY, USA*
moran-mirabal@mcmaster.ca

Over the past decades, the development of highly sensitive detectors and the synthesis of fluorophores with large cross sections and quantum yields have enabled the detection of single fluorescent molecules. This has allowed life scientists to probe the dynamic properties of single molecules (SMs) in solution, immobilized on solid surfaces, and even inside cells. Measurements of biomolecular properties have thus been obtained with higher precision than ever before. In addition, the observation of SMs has evidenced behaviors otherwise obscured by ensemble averaging. The tradeoffs in performing single-molecule fluorescence spectroscopy (SMFS) are that observation times are limited by photobleaching of the fluorophores and large datasets are required to ensure statistical significance. This chapter explores the fundamentals, experimental implementation, and observables obtained through SMFS, and their potential biomedical applications. An introduction containing the fundamentals of fluorescence techniques is presented, followed by a description of typical experimental setups, and an overview of the observables that can be accessed through different SMFS techniques. The design and implementation of micro- and nanostructured devices for SMFS is also reviewed. The chapter ends with a discussion of the areas of opportunity for SM biomedical studies.

Biomedical Nanosensors
Edited by Joseph Irudayaraj

ISBN 978-981-4303-03-3 (Hardcover), 978-981-4303-04-0 (eBook)
www.panstanford.com

7.1 Single-Molecule Fluorescence Spectroscopy Fundamentals

7.1.1 Why Single-Molecule Fluorescence Spectroscopy?

Since the pioneering work of Thomas Hirschfeld in the mid-1970s, involving labeling of immunoglobulins with ~100 fluorophores and their successful detection [1], the field of SM spectroscopy has seen an exponential expansion. It seems that not a day goes by without a new study introducing novel SM techniques, improving on them, or utilizing them to gain insight into biomolecular function. In the early days, the main challenges in SM detection were related to the discrimination of signal above the background, reduction of the excitation volume, and the temporal resolution required for SM displacement tracking. Initial studies conducted by Moerner and Orrit demonstrated the ability to detect SMs immobilized on a crystal at cryogenic temperatures, where excitation and emission bands are more sharply defined and cross sections much larger than at room temperature [2, 3]. A giant leap was made in the mid-1990s, with developments in scanning optical microscopy and improvements in detection devices (such as charge-coupled devices, or CCDs) that led to the successful demonstration of SM detection at room temperature via near-field spectroscopy, confocal microscopy, wide-field epi-illumination, and total internal reflection fluorescence microscopy (TIRF) [4–7]. Improvements in detection, removal of background fluorescence, and reduction of the excitation volume allowed the study of more complex biological systems. In this way, it was possible to visualize the turnover of ATPs during myosin movement [8], the rotational motion of F1-ATPase [9], the sliding of kinesin over microtubules [10], and the diffusion of molecules in lipid membranes [11], to name a few examples. After these initial seminal studies, a myriad of applications and novel SM-detection techniques have followed, but the question remains: what can SM techniques provide that other ensemble methods cannot?

The ability to observe individual molecules and record their association, displacement, conformation, or biological activity attracts scientists because it accesses more directly the properties and characteristics of the biomolecules of interest than ensemble techniques. Ensemble techniques are clearly valuable because they provide the average value of the experimental observable measured. On the other hand, the measurement of individual molecules allows the construction of a histogram for the discrete values of the experimental observable and allows the extraction of more information than just the observable mean [12]. Thus, a fundamental advantage offered by SM techniques is the possibility to detect the heterogeneous nature of biomolecules, where subpopulations can be identified, characterized, and sorted according to their behavior. This leads to the detection of rare populations that are otherwise averaged out, such as intermediates in biochemical reactions and intermediate conformations in folding pathways. SM experimentation also allows the study of fluctuating dynamics within a system in equilibrium, without the requirement of any type of synchronization. In this way, it has been possible to observe two fundamental types of disorder in biomolecular populations — static and dynamic disorder. Static disorder refers to the presence of biomolecular subpopulations with different properties that remain constant over time. An example of this type of behavior has been reported by Xie and Lu [13–15], who studied the oxidation of a cholesterol derivative by monitoring the fluorescence from the oxidized state of flavin adinine dinucleotide (FAD) in the catalytic site of cholesterol oxidase (Fig. 7.1). In these studies, the loss of fluorescence by the reduction of FAD worked as a molecular reporter on the turnover rate of cholesterol oxidase. It was observed that the turnover from individual molecules had different rates for the limiting step of activation (Fig. 7.1). On the other hand, dynamic disorder is the type of variability in molecular properties that a SM presents over time. For the same system described above, the length of adjacent turnovers for a SM was compared to the length of its turnovers separated by a given number of intermediate turnovers. It was observed that the rate for individual events depended strongly on previous turnover history, with short turnover times followed by short turnover times and long ones followed more often by long turnover times. This memory effect could not have been observed without the possibility of accessing SM data. Thus, SM techniques complement the information obtained from ensemble measurements by allowing the access to information about the heterogeneity of individual molecules derived from transient states or the presence of subpopulations with different behaviors.

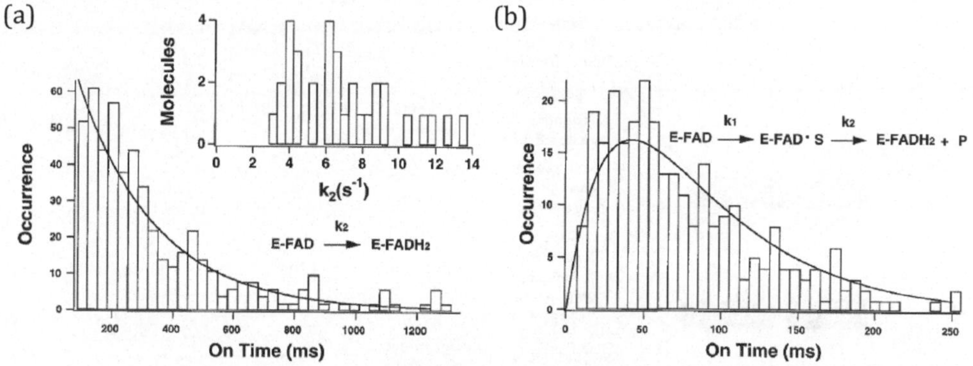

Figure 7.1 Experiments demonstrating the existence of static and dynamic disorder in SM data. (a) The histogram shows the distribution of on-times in a SM turnover history (k_2 being the rate-limiting step of the conversion of a cholesterol derivative). The inset shows the distribution of k_2 derived from 33 molecules in the same sample, showing static disorder. (b) The histogram shows the on-times for a SM turnover history taken with cholesterol as the substrate. The fit presented is the convolution of two exponentials, exemplifying dynamic disorder. (Reprinted with permission from ref. 15.)

An additional advantage of SM methods is the high spatiotemporal resolution that can be achieved. It must be recognized that no single technique is able to yield every piece of information from the system studied, nor can a single technique measure all experimental observables with the highest accuracy possible. The spatiotemporal scales for biomolecular properties accessible via SMFS methods are dictated by the choice of reporter molecules, experimental setup, and detector type. In some cases, the experimental setups for different techniques are not compatible. Thus, it is of primordial importance to recognize the information that is sought from the experiment and choose the experimental approach accordingly. The time resolution for SMFS can span from the nanosecond range, where ultrafast dynamics such as conformational changes can be monitored, to minutes, where molecular trajectories can be followed for as long as the fluorescence reporter lasts (Fig. 7.2). The time resolution required depends on the type of observation volume and detector that is used. On the other hand, the spatial resolution for SMFS techniques ranges from the subnanometer range, where conformational changes can be monitored by techniques sensitive to the local fluorophore environment, to the micron scale, where molecular trajectories for SMs can be followed by single-particle or single-fluorophore tracking (Fig. 7.2). The spatial resolution is also determined by the type of excitation volume used and the optics employed to image the excitation volume. In the following sections, the experimental implementations for SMFS will be explored and the ties to the accessible experimental observables identified for each one of the techniques discussed.

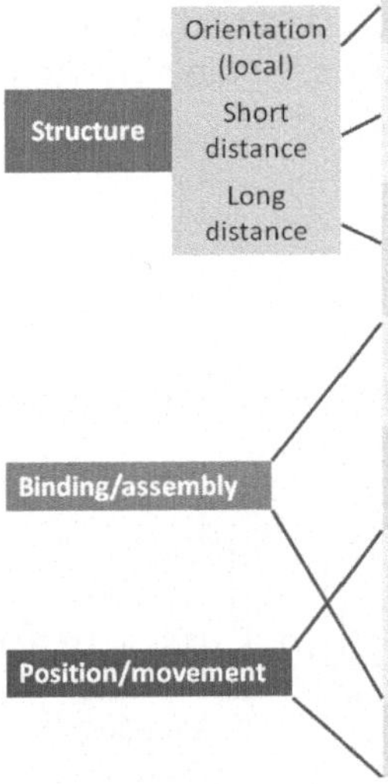

Technique	Observable	Resolution of observable	Time resolution
Polarization	Polarization or anisotropy	> 5	ms
Quenching, energy transfer, switch	Intensity, lifetime	< 30Å	ms
Fluorescence Resonance Energy Transfer	Intensity, lifetime	30-100Å	µs
Photon counting Histogram	Photon output, coincidence	ns	ns
Fluorescence Correlation Spectroscopy	Correlation function	ns	ns
Super Resolution Techniques (STED, PALM, STORM)	Point Spread Function	> 1nm	> 10ms
SPT, SFT, FIONA	Point spread function	> 1nm	> ms

Figure 7.2 The classification of SMFS techniques according to their spatiotemporal resolution and the observables that can be obtained from them. (Adapted from ref. 16.)

7.1.2 Fluorescence Principles and Properties

Fluorescence has its origins in the spontaneous emission of a photon from a molecule where the absorption of an incident photon causes an electron to transition to an excited singlet state. Because the excited electron is spin-paired to an electron in the ground orbital, transition from the excited to the ground state is permitted and occurs rapidly via photon emission [17]. Fluorescence in the visible wavelength range is typically observed from molecules with multiple aromatic rings, such as commercially available dyes from the fluorescein, rhodamine, or cyanine families. The Jablonski diagram (Fig. 7.3) is a schematic representation of the processes involved in the absorption and emission of light, as well as the intermediate states that affect the emission rate and quantum yield [17]. Fluorescence emission typically occurs in four steps following photon absorption: first, an electron transitions from a singlet ground state to an excited state ($\sim 10^{-15}$ s) where the fluorophore can exist in a number of vibrational energy levels; second, the molecule relaxes ($\sim 10^{-12}$ s) to the lowest vibrational excited state, a process termed internal conversion; third, after a characteristic time in the excited state (fluorescence lifetime, ~10^{-8} s), the molecule typically transitions to a higher vibrational level of the ground state via radiative (photon emission) or nonradiative decay (e.g., quenching or intersystem crossing); finally, the molecule rapidly (10^{-12} s) relaxes to the vibrational level of the ground state in thermal equilibrium with the environment [17, 18]. Because energy is lost to vibrations during the excitation-deexcitation cycle, emitted photons have lower energies and the emission spectrum of a fluorophore is shifted toward longer wavelengths, a phenomenon known as "Stokes shift."

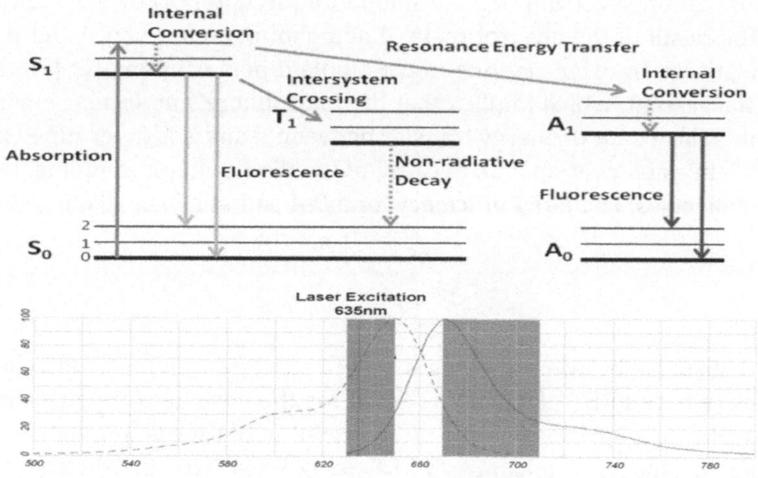

Figure 7.3 Processes involved in fluorescence absorption and emission. A Jablonski diagram representing the transitions from ground to excited states and the various relaxations and decays possible.

The rate of radiative decay from the excited state, and consequently the number photons emitted from a SM, can be reduced by a series of processes involving the molecular environment, excitation intensity, and fluorophore stability. At low excitation intensities, the photon output of the fluorophore is dominated by the absorption process and the output increases linearly with excitation power [18], whereas at high excitation intensities, the singlet excited state is saturated and the fluorescence intensity depends directly on the fluorescence lifetime of the fluorophore. Additionally, at high excitation intensities, the probability of intersystem crossing is higher and can lead to triplet state saturation (dark states) with long-lived lifetimes that can significantly reduce the fluorophore photon output [19, 20]. Thus, for high emission rates, it is preferable to have fluorophores with short-lived excited states, that is, short fluorescence lifetimes, and low intersystem crossing probabilities, which can be ameliorated by using triplet state quenchers. A second factor that can limit the fluorophore photon output is fluorescence quenching, which can occur by a number of mechanisms, such as collisions, proximity to metallic surfaces, molecular associations, and the presence of ions in solution. In all of these cases, the fluorophore is not changed but loses the energy from the excited state due to vibrational relaxations and energy or electron transfer. Examples of collision quenchers are oxygen, halogens, amines, and electron-deficient molecules [17]. Oxygen species and hydroxyl radicals also play a role in the reduction of the photon output via degradation reactions. Thus, solution degassing and the inclusion of oxygen scavenging and reducing agents, triplet state quenchers, and oxidation/reduction systems are frequently added to buffers used in SMFS to remove dissolved oxygen, reduce photo-degradation, and enhance emission [18, 21–23].

Förster resonance energy transfer (FRET) is another process that reduces the photon output but that has been turned into a SM tool and successfully used in the measurement of molecular distances. FRET occurs when the emission spectrum of a fluorophore (donor, *D*) and the absorption spectrum of a chromophore (acceptor, *A*) overlap [24]. However, FRET is not the result of the absorption by *A* of a photon emitted by *D*, but the result of non radiative energy transfer via long-range dipole-dipole coupling [25]. In addition, *A* need not be fluorescent, which implies that the fluorophore's molecular environment is very important. The extent of energy transfer between *D* and *A* is determined both by the excitation and emission overlap and, because of the dipole-dipole coupling, the distance between the molecules. The FRET efficiency for a *D-A* pair at a fixed distance is:

$$E = \frac{R_0^6}{R_0^6 + r^6} \tag{7.1}$$

where r is the distance between the donor and the acceptor and R_0 is the characteristic Förster radius (where FRET efficiency is 50%) for the specific *D-A* pair. Typical values for R_0 are 20–100 Å, which makes this effect useful in SM measurements of intra- and intermolecular distances. To maximize FRET assay sensitivity, the distances measured through *D-A* quenching should be $0.7R_0 < r < 1.5R_0$ [19]. This technique has been used in a number of SM applications, some of which will be described later, and a number of excellent reviews on the topic exist [25–29].

Fluorescence anisotropy is another property that can be exploited to study SM properties and interactions. Fluorophores have greater absorption efficiency for light that is polarized parallel to their transition dipole moments [17]. Thus, upon illumination, fluorophores with the correct orientation are more likely to be excited, creating selectivity for certain fluorophore orientations. Furthermore, emission is also preferentially oriented toward a fixed molecular axis. In systems freely diffusing in solution, the rotational molecular motion has characteristic times (~50 ps) much faster than the fluorescence lifetime, allowing emission to be randomized. However, in systems where the fluorophore orientation is preserved for times comparable to or longer than the fluorophore's lifetime (~10 ns), fluorescence emission will be preferentially polarized along the molecule's emission axis. Anisotropy measurements can thus report on changes in the environment of the biomolecules of interest, such as viscosity changes in lipid membranes or changes in the rotational diffusion induced by molecular aggregation. A system's anisotropy (ρ) is measured as a ratio of the emission intensities parallel ($I_{\parallel}$) and perpendicular ($I_{\perp}$) to the polarization of the excitation source [17]:

$$\rho = \frac{I_{\parallel} - I_{\perp}}{I_{\parallel} + 2I_{\perp}} \tag{7.2}$$

7.1.3 Dye Characteristics and Background Considerations for Successful SMFS

The detection of SMs via fluorescence spectroscopy relies on the capture of thousands of photons arising from many excitation-deexcitation cycles of the fluorophore attached to the molecule of interest. Thus, the intrinsic characteristics of the fluorophore, such as excitation-emission spectra, brightness, quantum yield, fluorescence lifetime, cross-section, saturation, and photostability, play a fundamental role in the experimental implementation of SMFS [17, 19]. In addition to this, the elimination of the background signal arising from elastic Rayleigh and Raman scattering from the excitation source and optics, solvent and glass impurities, and detector noise is essential to provide good contrast for the signal observed from the SMs [30, 31]. In this way, the successful detection of SMs and their study rely heavily on the optimization of the signal-to-noise ratio (SNR) by maximizing the photon output while minimizing spurious signal or background.

Potentially, the most important properties when choosing the right fluorophore for the application in mind are the fluorophore's intrinsic characteristics. The absorption and emission spectra of the fluorophore must be chosen to match available excitation sources and excitation and emission optics. Because background can be generated by the inelastic scattering of the excitation light, it is important to use high-quality filters that only allow narrow excitation and emission wavelength bands through but that are matched to the fluorophore's spectra to maximize excitation and detection efficiency. The excitation and emission wavelengths of the fluorophore will also have an impact on the amount of intrinsic background fluorescence for the system of study. For most biological materials, autofluorescence is higher in the blue-green spectrum (400–550 nm) and lower in the orange-near-infrared spectrum (550–700 nm), making dyes with emission in the latter range preferable. The absorption and emission spectra are also key in the selection of fluorophores to be used in FRET studies, where the absorption band of the acceptor dye must be matched with the emission band of the donor [25].

A second dye characteristic that is important to maximize the rate of photon emission is the absorption cross section (σ_p), which can be interpreted as the effective molecular area capable of absorbing a single photon from the excitation source [30]. The probability that an incident photon will bump the fluorophore into its excited state can be calculated as the ratio of σ_p to the cross-sectional area of the focused excitation source. Thus, choosing a fluorophore with large σ_p ensures that excitation is efficient. The cross section for a fluorophore can be easily calculated from its extinction coefficient ε and Avogadro's number (N_A), by the relation [32]:

$$\sigma_p - \frac{2303\varepsilon}{N_A} \tag{7.3}$$

The other process involved in determining the number of emitted photons per molecule is the fluorescence quantum yield (Q), which is defined as the probability of photon emission per photon absorbed [30]. The quantum yield is calculated using the rates involved in depopulating the excited state, both by the emission of a photon (fluorescence decay rate Γ) and the rates for the nonradiative decay processes (Σk_{nr}) such as intersystem crossing, internal conversion, or resonance energy transfer [17]:

$$Q = \frac{\Gamma}{\Gamma + \sum_{i} k_{nr}} \tag{7.4}$$

The quantum yield for fluorophores for SMFS should be as high as possible, to ensure the efficient deexcitation via radiative decay, and minimize losses to non radiative decay processes. Fluorophores with rigid structures usually have higher quantum yields because their structures do not undergo bending or distortion motions in the excited state, thus favoring the emission of a photon and minimizing the non radiative decay [30]. A typical measure of the rate of photon emission for a given fluorophore is its intrinsic "brightness," which is defined as the product of the extinction coefficient and the quantum yield [19]. Because the extinction coefficient is proportional to the fluorophore's cross section, this number reports directly on the efficiency of excitation photon absorption and radiative decay. Some commercially available fluorophores have large cross sections and quantum yields that approach unity, making them excellent candidates for SM work (Table 7.1).

Another intrinsic characteristic of the fluorophore that can have a direct impact on the number of photons emitted per molecule per unit time, especially at high excitation intensities, is the fluorescence lifetime (τ_F). This parameter is defined as the average time that the fluorophore spends in its excited state prior to radiative decay, and typical values range from 5 ns to 10 ns. The fluorescence lifetime is calculated as the inverse of the sum of all radiative and nonradiative decay rates, or Γ/Q [17]. The number of photons emitted from the excitation-deexcitation cycles of a fluorophore increases as the excitation power is increased as long as it does not reach saturation levels. Upon saturation of the singlet excited state, the effective absorption cross section decreases and the emission becomes limited by the fluorescence lifetime [30]. Because of this effect, fluorophores with short fluorescence lifetimes are preferred for SMFS. It must be noted that at or above saturation levels, the signal intensity from the fluorophore does not increase while the background does, ultimately hindering the SNR.

Table 7.1 Intrinsic properties of natural and synthetic fluorophores

Fluorophore	Absorbance maximum (nm)	Emission maximum (nm)	Extinction coefficient ($cm^{-1} M^{-1}$)	Quantum yield	Brightness	Laser line (nm)	Reference
L-Tyrosine	275	303	1,405	0.13	422		[17]
Tryptophan	280	348	5,579	0.14	781		[17]
POPOP	300	380–450	47,000	0.97	45,590		[17]
Quinine Sulfate	350	425	5,700	0.58	3306	350	[17]
Green	395	509	30,000	0.8	24,000	457	[19]
Fluorescent Protein (GFP)	470		7,000		5,600		
Fluorescein	490	520	67,000	0.5–0.9	33,500–60,300	488	[19]
Rhodamine6G	488	530	27,000	0.98		488	[33]
Alexa 488	494	520	80,000	0.5–0.9	40,000–72,000	488	[19]
RhodamineB	514	560	39,000	0.31	12,090	514	[33]
Tetramethyl Rhodamine	554	585	85,000	0.2–0.5	17,000–42,500	514, 532	[19]
Cy3	550	585	150,000	0.2	75,000	514, 532	[19]
Cresyl Violet	580	620	59,800	0.53	31,694	568	[33]
Texas Red	596	620	85,000	0.5	42,500	514,532	[19]
Cy5	650	670	250,000	0.15	37,500	633, 635, 647	[19]

Additional fluorophore characteristics worth considering for SMFS are the propensity of the fluorophore to undergo intersystem crossing from the excited state, photoisomerization, and photodegradation. These are phenomena that can reduce the photon output by introducing long-lived "dark" states where the molecules does not emit photons. However, these effects can be minimized by the appropriate selection of buffers and additives that enhance the rate of fluorescence emission. The most common cause for the reduction of the photon output is photodegradation of the fluorophore, also termed "photobleaching." This term encompasses all photophysical effects that irreversibly change the fluorophore to a state where absorption and emission of photons is no longer possible. Photooxidation seems to be the most common photophysical effect resulting in fluorophore photodegradation [34]. Thus, oxygen species are typically removed from solution by solution degassing, flooding with inert gasses [35], and the addition of oxygen scavenging agents, such as ascorbic acid [22, 23, 36] or enzymatic oxygen scavenging systems [37–39]. However, oxygen is a good triplet state quencher, and its removal can lead to an increase in the triplet state lifetime, also reducing the photon output and even promoting fluorophore destruction via triplet state intermediates. To alleviate this effect, triplet state quenchers can also be added to the buffer solution and include Trolox, *n*-propyl-gallate, *β*-Mercaptoethanol, and *L*-Glutathione, among others [23]. It must be noted that not all fluorophores react in the same fashion to these additives and they must be tested individually for the specific effects induced in a particular fluorophore. Recently, a reduction/oxidation system (ROXS) has been introduced as a general approach to eliminate all photophysical effects simultaneously and thus increase the rate of photon emission and the fluorophore lifetime [22]. The ROXS system relies on the removal of reactive oxygen species, and quenching of triplet state as well as on the production of charge-separated molecular states by electron transfer. The combination of oxidizing and reducing agents is used to quickly depopulate the triplet state via electron transfer and recover the ionized intermediates through an additional redox reaction. The ROXS system has been shown to improve the photon output and prolong the lifetime of various fluorophores and shows great promise as a universal system to be used in SMFS [21, 22].

As a final note, we remark upon the importance of the reduction of background noise, which is essential to SMFS. Noise from Raman and Rayleigh scattering of the excitation light by the objective and sample can be alleviated by the appropriate choice of high-quality filters that only allow the desired wavelengths to reach the detector. Additional sources of noise include the intrinsic fluorescence from immersion oils, which can be reduced by picking ultralow-fluorescence oils; dust and impurities in buffer solutions, which can be removed by filtering and using the highest-quality reagents available; and impurities on the surface of the optical window used for experimental observation, which can be drastically reduced by using high-quality glass surfaces (e.g., borosilicate, quartz, or fused silica) with low surface roughness and appropriate cleaning protocols that eliminate organic contaminants. The combination of all these measures, together with the selection of an appropriate fluorophore, ensures that the maximum number of photons is detected against a minimum background signal, providing adequate contrast and thus higher accuracy in SM detection. Typical measures of the signal intensity against background noise are the SNR and the signal-to-background ratio (SBR). The latter of these is the simplest to calculate and only informs of the brightness of the molecule and the quality of the sample preparation and is calculated as

$$\mathrm{SBP} = \frac{\text{Signal Intensity}}{\text{Background Intensity}} \tag{7.5}$$

Basché and collaborators presented an equation that takes into account multiple factors intrinsic to the fluorophore and the detection system to quantify the SNR [40]:

$$\mathrm{SNR} = \frac{DQ\left(\frac{\sigma_{\mathrm{p}}}{A}\right)\left(\frac{P_0}{h\nu}\right)T}{\sqrt{\left(\frac{DQ\sigma_{\mathrm{p}}P_0T}{Ah\nu}\right) + C_{\mathrm{B}}P_0T + N_{\mathrm{d}}T}} \tag{7.6}$$

where Q is the fluorescence quantum yield, σ_{p} the absorption cross section, T the detection or integration interval, A the cross-sectional area of excitation, $\left(\frac{P_0}{h\nu}\right)$ the incident photon flux per second, N_{D} the detector dark count, C_{B} the background count, and D the detection efficiency. While this equation is a complete description of the emission-detection system, it is possible that not all parameters within the equation are known or can be measured. Thus, a simpler empirical form of measuring the SNR is by taking the signal above the background intensity and dividing it by the standard deviation of background fluctuations [41]. This is the ultimate measurement of the quality of our SM signal and the possibility of detecting it in our experimental system.

7.2 Experimental Implementations of SMFS Techniques

Multiple configurations exist for the implementation of SM detection and SMFS, all of which aim at reducing the background fluorescence from the specimen and achieving the highest SNR possible. These experimental configurations consist in the most general form of three basic elements: excitation source, optics for excitation and collection, and one or multiple highly sensitive detectors. The experimental implementation for SMFS can be divided broadly into two main groups, those using wide-field detection and those exploiting point-detection. The choice of experimental configuration depends on the observable that is to be measured and the temporal and spatial resolution desired. As mentioned before, no single technique can yield all the experimental information from a given system and careful selection of the SM apparatus to be used is crucial for the successful measurement of the biomolecular properties of interest.

7.2.1 Widefield Techniques

Widefield techniques allow the simultaneous visualization of multiple SMs by the illumination of a large field of view. The fluorescence emitted from a sample is collected through a high numerical aperture objective and imaged onto a high sensitivity two-dimensional detector array (e.g., an intensified charge-coupled device [ICCD] or an electron-multiplying charge-coupled device [EMCCD] camera). The temporal resolution of these methods depends on the frame-transfer rate of the detector and the exposure times needed to collect enough photons to produce SM images. This limits the temporal resolution to the millisecond regime. The spatial localization accuracy for widefield techniques is dictated by the magnification, the background from the sample image, and the number of photons collected per pixel. Although 10–30 nm accuracy is typical, nanometer accuracy is possible and 1.5 nm localization accuracy has been reported in the study of molecular motors [42].

7.2.1.1 Epifluorescence microscopy

The simplest way to observe SMs is to use widefield epifluorescence microscopy. This method was first used to study individual molecular motors in the mid-1990s [8, 43]. As the name of the technique implies, the excitation source illuminates a large sample area. In its implementation, an arc lamp or collimated laser is focused on the back aperture of a high numerical aperture objective (Fig. 7.4a), which results in large excitation areas with diameters that range from tens to hundreds of microns [30]. The emitted fluorescence is then collected through the objective and passed though a dichroic mirror and an emission filter that select the wavelength of the emitted fluorescence and reject background noise from the laser, reflections, scattering, and autofluorescence from the sample and optics. Because the goal is to collect as many photons emitted from the sample as possible, high-magnification large-numerical-aperture objectives are employed (typically numerical aperture [NA] 1.2 for water immersion objectives and 1.45–1.65 for oil immersion objectives). The selection of excitation and emission filters also affects the amount of background noise and should be selected to closely match the fluorescent dye characteristics. The signal that remains after the emission filter is then collected by a high-sensitivity two-dimensional detector. Several types of CCD cameras exist with high sensitivity, capable of imaging SMs and even single fluorophores, and millisecond time resolution. The most commonly used nowadays in SM imaging is the EMCCD. In this type of detector, charges stored in the CCD are multiplied before they are converted to a readout voltage, which allows high readout speeds with reduced readout noise. A drawback of the widefield technique is its poor axial discrimination, which can introduce significant background fluorescence form out-of-focus planes. Nevertheless, in thin samples such as solid-supported lipid bilayers or cell lamellipodia [11, 44–46], and monolayers of immobilized biomolecules [8, 43, 47–49], widefield epifluorescence can be a simple and inexpensive technique to conduct SM experiments.

7.2.1.2 Total internal reflection fluorescence microscopy

A second widefield technique commonly used for observing the localization and dynamics of multiple molecules simultaneously is TIRF microscopy. In TIRF, the incidence of a laser beam upon the interface between a high (n_1) and a low (n_2) index of refraction material (usually glass-water interface) at angles larger than the critical angle θ_c causes the laser beam to undergo TIR. The critical angle is calculated using the formula:

$$\sin\theta_c = \frac{n_2}{n_1} \tag{7.7}$$

While under TIR no light is propagated through the interface, the electromagnetic field at the interface does not drop suddenly to zero. Instead, the field decays exponentially within the lower index of refraction material with a characteristic penetration depth (d) calculated as

$$d = \frac{\lambda_0}{4\pi\left(n_1^2 \sin^2\theta - n_2^2\right)^{\frac{1}{2}}} \tag{7.8}$$

where λ_0 is the wavelength of the incident light, n_1 and n_2 are the higher and lower indexes of refraction of the materials at the interface, and θ is the incident angle. Values for d typically range between 50 nm and 200 nm and are only limited by the incident angles accessible through the illumination geometry chosen. This evanescent field at the interface can excite fluorescence emission from fluorophores located near the interface. Because of the selectivity of the evanescent field, the excitation volume is large in the transverse dimensions but highly confined in the axial dimension. This greatly reduces the background noise contributed by the bulk of the sample and allows imaging of SMs close to the glass-water interface with a high SNR. SM detection through TIRF was pioneered by Hirschfeld through the detection of labeled immunoglobulins [1]. Since these initial studies, TIRF has been used to image SMs and even single fluorophores immobilized on solid surfaces [50, 51], attached to molecular motors [8, 10, 42, 52, 53] anchored on model lipid membranes or cellular membranes [54], and even freely diffusing species.

Two experimental implementations have been described for TIRF microscopy — prism [8] and objective [55–57] based. In prism-based TIRF (Fig. 7.4b), a prism is coupled to a glass slide or coverslip via an index-matching medium (e.g., glycerol or oil). Then, TIR is achieved as a laser beam is coupled through the prism at an angle larger than the critical angle. The fluorescence emitted from the illuminated sample can be imaged through the microscope into a high-sensitivity detector, such as an EMCCD. As the name suggests, in through-the-objective-type TIRF, the laser beam is coupled to the glass-water interface through the objective (Fig. 7.4c). With the use of a moving mirror or lens, the position of the laser beam can be shifted toward the edge of the back aperture of the objective, causing the light to impinge on the interface at an angle. Provided a high-numerical-aperture objective is used (NA 1.4 or higher), the incidence will occur at supercritical angles, establishing TIR. The fluorescence excited from the sample can then be imaged through the same objective into a high-sensitivity detector. Key advantages of through-the-objective TIRF are that the use of high-magnification, high-NA objectives allows improved lateral resolution, that the top of the sample can be open and easily accessible, and that switching between widefield epifluorescence and TIRF can be achieved easily by changing the position of the incident beam [16]. Several excellent reviews of the implementation of TIRF and the comparison between the experimental configurations have been published [55, 56, 58–62].

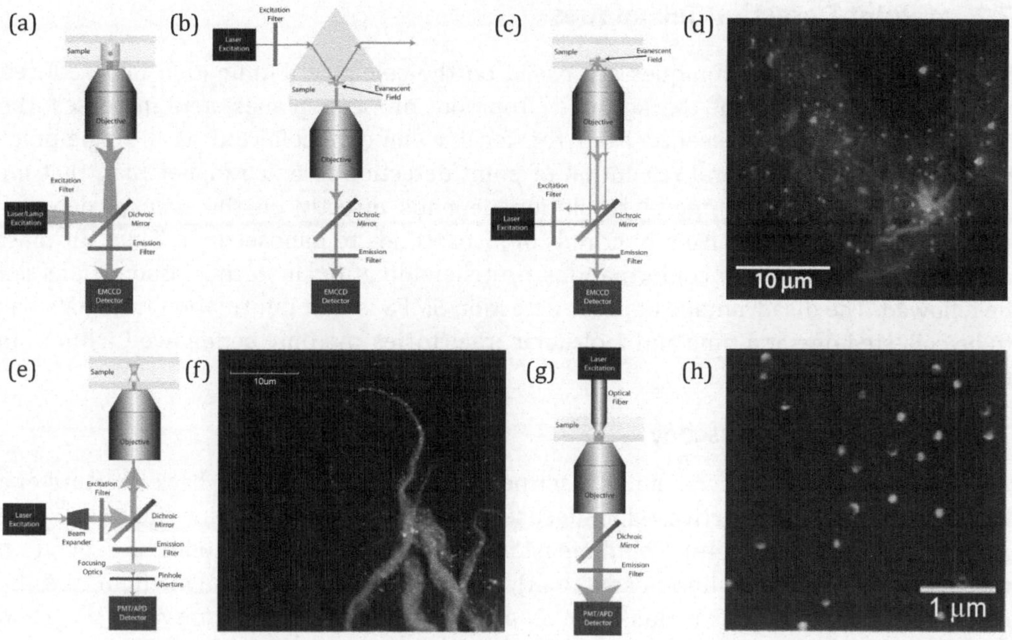

Figure 7.4 Experimental implementations for SMFS. (a) Widefield epifluorescence is obtained by focusing a high-intensity excitation source on the back aperture of the objective to obtain a wide illumination field. The fluorescence emitted from the sample is collected through the objective. Images are obtained with a high-sensitivity camera. (b) and (c) The experimental implementation of prism and objective-based TIRF microscopy. In both setups, illumination is provided by a laser source that impinges on a glass-water interface at supercritical angles and undergoes TIR. Coupling of the laser beam to the sample is achieved either via a prism or an objective. The evanescent field excites fluorescence from the sample. In both implementations, the emitted light is collected through the objective and imaged onto a high-sensitivity camera. (d) A sample image produced by objective-based TIRF. Single fluorescently labeled cellulases (red) are imaged bound on fluorescently labeled cellulose fibers (green). (e) Confocal microscopy is implemented by expanding a laser beam to fill the back aperture of the objective, producing a diffraction-limited illumination volume. Fluorescence is collected by the objective and focused onto a conjugate plane, where an aperture is positioned to reject out-of-focus light. Light is then coupled into a high-sensitivity detector (e.g., a photomultiplier tube [PMT] or an avalanche photodiode [APD]). Reconstruction of the image is achieved through stage or sample scanning. (f) A sample image obtained through confocal microscopy. The sample is identical to that from (d). (g) Near-field scanning optical microscopy (NSOM) is implemented by coupling a laser beam through a fiber that ends in a subwavelength aperture in a metallic coating. The aperture is brought into close proximity of the sample so that the light propagating through the aperture cannot diffract and produces a subdiffraction near-field excitation volume. The excited fluorescence is collected through an objective and imaged onto a high-sensitivity detector (e.g., PMT or APD). (h) A single-fluorophore image taken with NSOM. The anisotropy in the fluorophore signature corresponds to the immobilized fluorophore dipole. (Panel h reprinted with permission from ref. 4). See also Color Insert.

7.2.2 Point-Detection Techniques

Point-detection SMFS techniques are based on the selective illumination of a reduced volume, which limits the background from out-of-focus planes and increases the SNR. Because the fluorescence from the focal volume is collected as a single-point measurement, the temporal resolution of point detection is much higher than that for widefield techniques. Temporal resolution depends directly on the type of detector used but typically ranges from hundreds of picoseconds to nanoseconds. Thus, ultrafast changes such as molecular conformations, protein folding, and molecular associations can be followed. The disadvantage of point-detection SMFS is that information from SMs has to be collected one at a time and molecular trajectories can only be followed within the focal volume dimensions.

7.2.2.1 Confocal microscopy

In confocal microscopy, a collimated laser beam is used to overfill the back aperture of a high-NA microscope objective, which focuses the light into a diffraction-limited volume at the focal plane (Fig. 7.4e). The dimensions of the focal volume produced depend on the wavelength of the excitation source (λ), the refractive index of the medium, and the objective NA. If perfect diffraction-limited optics are considered, the transverse (r_{x-y}) and axial (r_z) radii of the focal volume can be calculated from the Rayleigh criterion as

$$r_{x-y} = \frac{0.61\lambda}{\text{NA}} \text{ and } r_z = \frac{2n\lambda}{\text{NA}^2} \tag{7.9}$$

where λ is the wavelength of the excitation light, NA is the numerical aperture, and n is the index of refraction of the medium. From this, the confocal excitation volume can be calculated as the volume of an ellipsoid: $V_c = \frac{4}{3}\pi r_x^2 - y^{r_z}$. A confocal volume calculated this way for a sample imaged in water with a 1.45 NA objective and 488 nm excitation is on the order of 0.1 fL. Such a small focal volume allows the excitation of SMs with a high SNR because it eliminates background fluorescence and collects emitted fluorescence with high efficiency [30]. The fluorescence emitted from the molecules in the focal volume is then collected through the objective, filtered through the dichroic mirror and emission filter, and focused onto a small pinhole aperture. This aperture is positioned at the conjugate focal plane (thus the name confocal microscopy) and serves to reject out-of-focus light. The confocal aperture must be matched to the objective used, and its diameter (typically 20–100 μm) should be larger than the transverse diameter of the focal volume multiplied by the objective's magnification [63–65]. Spatial filtering through the use of the confocal aperture improves the SNR of the fluorescence detected and is an essential component in confocal SM detection. The core of a multimode optical fiber can be used to collect the fluorescence light as an alternative to the confocal aperture — the advantage is ease of coupling to the detection but with added losses due to optical transmission.

After passing through the confocal aperture, the collected light is recorded by a sensitive photon-detection device, which converts incident photons into electrical signals. For low numbers of emitted photons, as is the case in SMFS, the detectors of choice tend to be PMTs or APDs, which amplify the signal produced by incident photons into detectable voltages. The principle of operation of PMTs relies on a single incident photon producing an electron from a photocathode via the photoelectric effect. The produced electron is then focused by an electrode toward the electron multiplier, where the process of secondary emission multiplies the incident electron. These electrons reach the anode, producing a sharp photocurrent burst that can be detected and identified with the arrival of a single photon. APDs are semiconductor devices where an incident photon results in the production of an electron-hole pair. Because of the reverse bias in the semiconductor material, the pair undergoes a rapid avalanche of impact ionizations, yielding an increasing number of electrons, which are detected as a digital waveform indicating the arrival of a photon at the detector. APDs are typically cooled to reduce the dark noise arising from stochastic impact ionizations and allow the observation of single-photon events. APDs operating in a high-gain regime are said to be operating in the Geiger or photon-counting mode.

The detected signal from the illuminated focal volume represents a single pixel of information. Thus, the confocal setup can be used in the stationary mode to detect emission from a localized SM or in the scanning mode, where an image is reconstructed pixel by pixel via scanning of the sample or laser beam. A stationary mode allows high temporal resolution, as the signal is constantly collected and recorded from the photon-counting device, but has no spatial resolution. The scanning implementation allows high spatial resolution (Fig. 7.4f), which can be extended to the axial dimension by the scanning of multiple planes (or z-stacking) but has a limited temporal resolution due to the dwell time for each collected pixel. The main advantage of the scanning confocal setup is the possibility of reconstructing the fluorescence arising from complex three-dimensional structures with a high SNR. Thus, confocal excitation has been used effectively in SM studies both in solution [66] and on solid surfaces [5, 30, 67, 68].

Focal volumes similar to those of confocal microscopy can be obtained through two-photon excitation (TPE) without the need for a pinhole aperture [69]. TPE is based on the almost simultaneous absorption of two long-wavelength photons by the fluorophore and its transition to an excited state, from which radiative decay can occur. Because TPE is a second-order nonlinear process, the excitation depends on the square of the laser intensity [70]. To achieve the photon densities required for efficient TPE, high-intensity, short-laser pulses are required, such as those obtained from mode-locked Ti:sapphire lasers. The effective volume for TPE is thus confined to the region where the laser is most tightly focused, reducing the photobleaching of out-of-focus sample planes and increasing the SNR. TPE microscopy can also be used in the stationary or scanning mode to study SMs.

7.2.2.2 Photon-counting histogram and fluorescence correlation spectroscopy

Two techniques that use a confocal excitation volume in stationary mode to extract molecular properties and dynamic information from SMs are the photon-counting histogram (PCH) and fluorescence correlation spectroscopy (FCS). Because of their frequent use in SMFS, they deserve additional discussion regarding the methods and how to analyze the information obtained. The techniques are complementary because they provide distinct information, with PCH providing information about the brightness and number of molecules in the focal volume and FCS providing information on the timescale of dynamic processes and the number of molecules present in the focal volume. The basis of both techniques is the detection of the burst of photons emitted by single-fluorescent molecules traversing a confocal detection volume with a high temporal resolution down to the nanosecond regime (Fig. 7.5). For this purpose, PMTs or APDs in the photon-counting mode are commonly used to count the collected photons as a function of time. Because the confocal excitation volume should be occupied by a SM, these experiments are conducted at subnanomolar concentrations [71]. A further constraint is that the observation time for any individual molecule is limited to a few milliseconds by the residence time of the molecule in the focal volume, which precludes the study of long dynamics of individual molecules [16].

The PCH is a representation of the measurements of many SM events that gives insight into ensemble properties while preserving the nature of the individual SM measurements. The technique is based on the Poissonian nature of photon-counting statistics [72]. A detailed theoretical treatment, including the contribution of shot noise and the point spread function (PSF) on the distribution of detected events, was elegantly presented by Gratton *et al.* for a Gaussian (confocal) and a Gaussian-Lorentzian (TPE) excitation volume [73]. In practice, the concentration of the fluorescent species is adjusted such that the probability of occupancy of the focal volume by more than one molecule is very low. The fluctuation in the number of particles within the excitation volume is then described by Poisson statistics [74] so the probability of finding N molecules within the excitation volume is given by:

$$P_N(N) = \frac{\overline{N}^N e^{-\overline{N}}}{N!} \tag{7.10}$$

where $\overline{N}$ is the average number of molecules within the excitation volume at the chosen concentration. Once a suitable concentration of molecules is chosen to ensure that only SM recordings are observed, fluorescence photons (or photon bursts) from the molecules traversing the focal volume are recorded with a high temporal resolution (Fig. 7.5). The temporal history of recorded SM events is called a photon time trace, which represents the raw data in both PCH and FCS. The data in the time trace has information at a much higher temporal resolution than typically needed for most SM measurements and is thus binned on a relevant timescale, which allows the observation of the molecular property of interest. In most PCH implementations, there is a need to separate SM events from background noise via a thresholding algorithm. Typical thresholding algorithms rely on the observation of a signal with a minimal SNR (e.g., 3–5), to consider it a true SM event. Once SM events have been recognized, the photon count (or brightness) for each event is calculated. The collection of SM events represents the population to be studied with the PCH technique. The SM events can then be binned according to their brightness and the resulting distribution plotted as a brightness-versus-frequency histogram (or PCH, [Fig. 7.5]). The PCH can be analyzed and fit to theoretical function forms in order to obtain information about the way the molecules are distributed according to their brightness. A detailed description of an algorithm to discriminate fluorescent species that have overlapping populations in the PCH has been recently described by Rye [75]. Through PCH analysis, the separation of biomolecules with different sizes (and thus fluorescence intensity) [76] and aggregation states [75] has been achieved. A further extension of the

PCH analysis can be performed if more than one fluorescence wavelength are recorded. In this case, the coincidence (or lack thereof) of photon bursts can be evaluated from individual molecules displaying both fluorescent tags or from the association of individual molecules containing spectrally different tags [77–79]. This allows the study of molecular interactions through PCH. An advantage of PCH worth mentioning is that, because the collection time for the number of molecules required to generate the histogram is typically slow, the temporal evolution of the populations contained within the sample can be followed in the minute-to-hour time range [75].

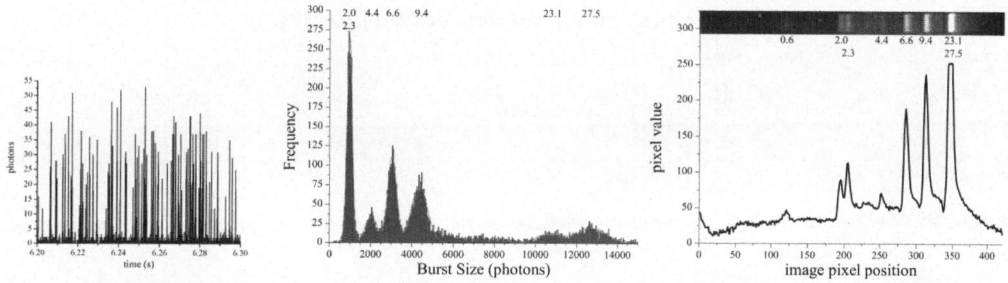

(a) (b) (c)

Figure 7.5 The fragment sizing of DNA molecules through SMFS. (a) A photon intensity trace taken from a focal volume overlaid on a microfluidic channel where single DNA fragments are flowed. (b) A PCH from the mixture of DNA fragments. Because the DNA was uniformly labeled with an intercalating dye, the fluorescence intensity can be correlated with the DNA size. The red line is a fit of multiple Gaussian distributions representing the individual fragments. (c) The intensity profile from an agarose gel separating the DNA mixture into the different fragments. (Reprinted with permission from ref. 76.) See also Color Insert.

FCS was first introduced in the 1970s by Elson, Magde, and Webb to study fluctuations arising due to thermal effects in a system in quasi-static equilibrium [80–83]. The principle behind FCS is that the fluctuations in the fluorescence intensity from molecules traversing a fixed focal volume can be analyzed to extract dynamic characteristics of the system, such as diffusion, conformational changes, enzymatic and chemical reactions, and molecular associations. Thus, FCS is sensitive to any change in molecular states that results in a fluctuation of the fluorescence intensity. Although FCS was described decades earlier, it wasn't until the mid-1990s, when Rigler implemented FCS using a confocal excitation volume, thus reducing the focal volume and significantly improving the SNR, that the technique became widespread [5, 66, 84]. Experimentally, the setup is identical to that used for PCH and the raw data acquired is the same photon-intensity trace. However, the data is binned appropriately in time and the analysis of the intensity trace in FCS is done through the normalized autocorrelation function defined as

$$G(\tau)=\frac{\left(\delta F(t)\cdot\delta F(t+\tau)\right)}{\left(F(t)\right)^2} \tag{7.11}$$

where $\delta F(t)$ represents the fluorescence intensity fluctuation and $F(t)$ is the fluorescence intensity at time t. $G(\tau)$ is essentially a measure of the self-similarity of the fluorescence signal after a lag time τ (Fig. 7.6) [85]. In practice, this curve is generated by autocorrelating the binned photon time trace. To obtain the experimental observables, the autocorrelation curve is fitted to a theoretical function specific for the shape of the excitation volume and the experimental conditions. The theoretical form of the autocorrelation function has been worked out for many of the common excitation volumes, such as confocal [66, 86], TPE [87], TIRF [17], as well as some unique setups where the excitation is confined by micro- or nanostructures [88–90]. Additional consideration in calculating the model autocorrelation function is also given to flow in the system, the reduced diffusion due to surface confinement or diffusion hindrance [91–94], the surface binding [61, 95], and the photophysical characteristics of the dye, such as triplet state, isomerization-induced dark states, blinking, and photodegradtion [20, 96].

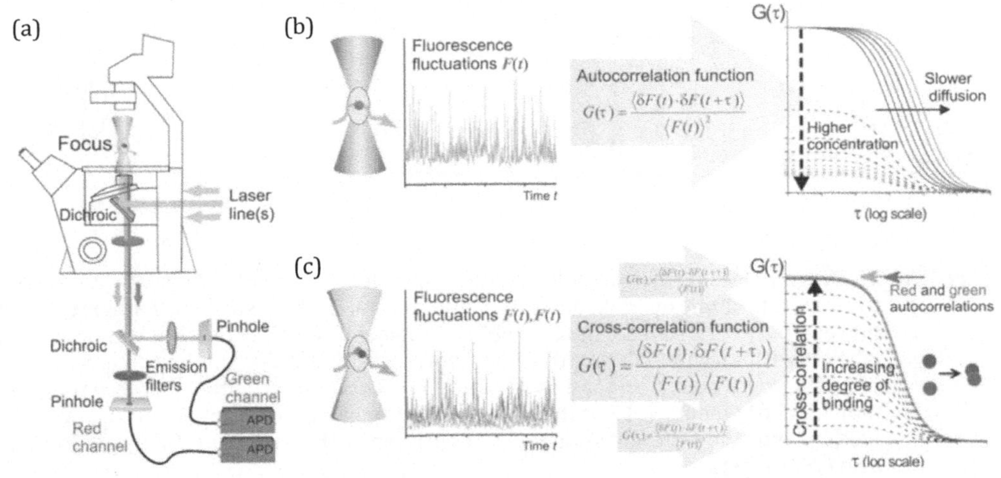

Figure 7.6 The principle of operation of FCS and fluorescence cross-correlation spectroscopy (FCCS). (a) An experimental setup for FCS and FCCS. (b) FCS utilizes the photon-intensity trace obtained from molecules passing through a focal volume excited with a single laser line. The fluctuations from the intensity trace are self-correlated to obtain the autocorrelation curve. From the autocorrelation curve, the diffusion behavior can be extracted from the half-decay time and concentration can be calculated from the autocorrelation curve intercept. (c) FCCS utilizes the photon-intensity trace obtained from molecules labeled with two spectrally distinct dyes, which diffuse through the focal volume excited with two independent laser lines. The fluctuations in each channel are correlated with those in the second channel, generating the cross-correlation curve. From the cross-correlation curve, the diffusive behavior can be ascertained from molecules tagged with the two fluorophores diffusing together. The amplitude of the curve determines the extent of colocalization of the two species. (Reprinted with permission from ref. 97). See also Color Insert.

As an example of how the autocorrelation curve is interpreted, consider a curve obtained from a system that only involves the motion of an SM traversing the focal volume (Fig. 7.6). Two main parameters can be extracted from the generated curve — the y-intercept ($G(0)$) and the half-decay value of the autocorrelation curve (τ_D). $G(0)$ relates inversely to the average number of molecules present in the focal volume ($1/N$), while the half decay represents the average residence time of the molecules within the focal volume.

As the number of molecules within the excitation volume increases, the fluctuations for individual molecules become smaller in comparison to the average intensity and $G(0)$ decreases. Also, if the molecules diffuse slower, the average residence time within the focal volume increases, shifting the half-decay point of the autocorrelation curve to the right. The residence time can then be related to the diffusion coefficient if the mean square displacement of a freely diffusing molecule is set equal to the width of the excitation volume. For an FCS experiment, this self-similarity operation is performed over the whole collection of molecules that traverse the focal volume. Thus, from FCS measurements, the concentration and diffusion coefficient can be easily calculated.

As described previously, FCS is sensitive to fluctuations in the fluorescence arising from changes in the molecular structure, diffusion, binding, and chemical reactions that alter the emission rate. However, FCS is relatively insensitive to molecular associations where both of the molecules interacting are of similar molecular weights [17]. To circumvent this limitation, FCCS was developed as an extension to FCS. In FCCS, two laser beams are aligned such that their focal volumes are co-localized. A similar result can be obtained with TPE by using a single laser beam that can be used to excite two spectrally distinct fluorophores. The fluorescence emitted from the focal volume is collected through dual-band optics, which allows the passage of both emission wavelengths. Later, the collected light is split into the two colors and the photons detected by independent APD or PMT detectors. The photon time traces recorded for the two different wavelengths are then correlated, generating the cross-correlation curve (Fig. 7.6). The theory of FCCS has been covered in detail and many excellent reviews of the technique exist in the literature [85, 97–102].

In FCCS, it is the coincidence in fluctuations from the two channels that is of primordial interest [85]. The experimental implementation requires that each one of the two ligands to be studied is labeled with a spectrally distinct dye (Fig. 7.6). Then, the coincidence of fluorescence fluctuations, or photon bursts, indicates the concomitant movement of the two ligands through the focal volume, demonstrating binding. The cross-correlation curve intercept is proportional to the fraction of SM events that show correlated fluctuations in both channels. Thus, as opposed to the autocorrelation function in FCS, as the concentration of bound molecules increases, the amplitude of the cross-correlation curve increases. Also, because the molecules traverse the focal volume together, the half-decay time corresponds to the average residence time for the molecular complex studied. It is important to note that for FCCS, it is necessary to work at low concentrations that ensure a low probability of two unbound molecules diffusing simultaneously through the focal volume. With this technique, it has been possible to study DNA hybridization [98], enzymatic cleavage [103], and protein oligomerization [104]. More recently, the technique has been extended to the study of dissociation of proteins in intracellular environments [105], and in combination with fluorescence lifetime measurements, it has been used to study the interaction of an antibody labeled with Alexa 488 and a GFP-labeled receptor. This latter application demonstrated the possibility of performing cross-correlation spectroscopy (CCS) studies with molecules that are excited by a single wavelength and have similar emission spectra but that can be separated based on additional information on the fluorophore's characteristics [106].

7.2.2.3 Near-field scanning optical microscopy

NSOM is another SM technique proposed by Betzig in the mid-1980s and further developed during the 1990s to break the limitation on resolution imposed by the diffraction limit [4, 107–109]. As opposed to the techniques presented previously, NSOM is a near-field technique that requires the excitation source to be placed in close proximity to the molecule to be imaged. The principle behind NSOM is that if the excitation light is passed through a subwavelength aperture, the light transmitted will produce in the near field an illumination volume with dimensions resembling those of the aperture (Fig. 7.4). Thus, if the excitation source is placed within 5 to 10 nm of the sample, the photons emitted from the aperture will not diffract and will produce an illumination volume with subdiffraction dimensions. As the distance from the aperture increases, the light diffracts and the confinement is lost. In the first NSOM implementations, to produce a suitable excitation source for visible light, a single-mode optical fiber was heated and elongated until it broke, producing a tip with a diameter much smaller than the original fiber [108]. This tip was then partially coated with a thin layer of aluminum to produce an opaque cladding and leave a subwavelength aperture at the tip. Combining the excitation through the optical fiber, with shear-force feedback loop, the source could be brought into close proximity to the sample (~1 nm) and excite fluorescence from a region of the sample smaller than 100 nm [18, 110]. The fluorescence was then collected in the far field through a microscope objective and recorded via a high-sensitivity photon detector.

The main advantage of NSOM is the reduced excitation volume that can reach subdiffraction resolution. If used in the scanning mode, a whole image can be reconstructed from areas microns in size. However, NSOM suffers from some important drawbacks that limit the usefulness of the technique, such as low power output [111], a cumbersome experimental implementation, and sample perturbation due to the proximity of the fiber [112]. Nevertheless, recent advances in the development of optical fibers with higher coupling efficiency, as well with the combination of the excitation source with force microscopy-type setups, have allowed the efficient acquisition of topographical and fluorescence imaging of SMs even on cell surfaces [109, 113–116]. NSOM research has grown in the past few years, and a number of reviews detail in depth some of the more recent advances achieved [117–119].

7.3 Observables Measured with SMFS Techniques

The techniques described in the previous section can be used to measure different experimental observables in a variety of biological systems. In this section, examples from the literature will be presented for the various experimental observables measured with different SMFS techniques. The examples presented here are a small subset of those published and are not intended to be an exhaustive review.

7.3.1 Polarization: Local Orientation of Molecules

The easiest way that one can imagine to measure the orientation of molecules would be to visualize them with sufficient resolution to determine their positions and orientations at all times. This can sometimes be achieved if a fluorophore with low rotational mobility is bound to the molecule of interest and polarized light is used to excite the fluorescence. Then, the parallel and perpendicular components of the resulting fluorescence can be analyzed to determine the orientation of the molecule of interest. The direct measurement of orientation is possible with large macromolecules that have an anisotropic form by taking successive recordings of their position. In one of the earliest papers demonstrating SM imaging of biomolecules at room temperature, Kinoshita's group recorded images of single F_1-ATPase molecules rotating as they converted ATP into ADP [9]. In this experiment, a fluorescently labeled actin filament was attached to the γ-subunit of the ATPase as a marker. The ATPase molecules were immobilized on a glass surface and imaged at 133 ms intervals using epifluorescence. They observed molecules in presence and absence of ATP and determined that rotational movement was unidirectional and ATP driven. This study helped confirm conjectures based on structural, biochemical, and spectroscopic studies but also demonstrated that the ATPase motor could move under high loads [9].

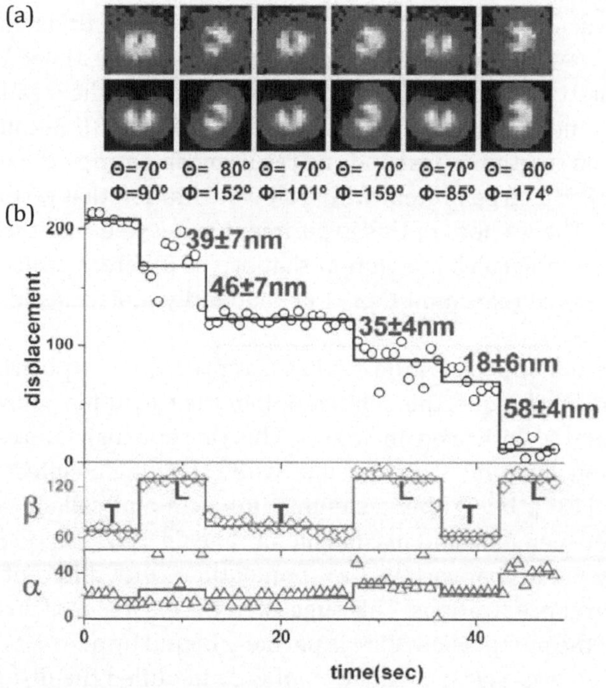

Figure 7.7 The determination of the displacement and orientation of myosin molecules stepping on actin filaments. (a) Defocused imaging of myosin molecules compared with modeled PSFs. It is shown that the acquired images correspond to a unique three-dimensional orientation of the dye. (b) Traces of the displacement and polarization orientation of an SM tracked. (Reprinted with permission from ref 12.)

More recently, the possibility of localizing molecules with high accuracy has allowed the use of polarized light to simultaneously detect the positions and orientations of individual myosin V molecules [120]. In this study, myosin was labeled with a bifunctional rhodamine molecule on one of its calmodulin (CaM) light chains and was imaged using polarized TIRF together with fluorescence imaging with one nanometer accuracy (FIONA) [42]. It was demonstrated that for every 74 nm displacement, two reorientational motions of the probe occurred and that there was additional flexibility of the myosin molecules to move around the actin filament. The same experimental system was also studied with a novel technique termed defocused orientation and position imaging [121, 122]. In this technique, the excitation beam is circularly polarized and the SM image is obtained in an out-of-focus plane. The resulting PSF reflects the position and three-dimensional orientation of the dipole moment of the fluorophore. Upon comparison of the images with model PSFs, the precise orientations of the labeled CaM light chain domain were obtained and a more complete description of the behavior of myosin V was achieved (Fig. 7.7).

While direct imaging is possible for some macromolecules where their positions and orientations change slowly, this is not possible for molecules in solution that change their orientations too quickly to be visualized at video frame rates. In this case, polarized confocal excitation is used in conjunction with high temporal resolution detection of the parallel and perpendicular components of the emitted fluorescence by two independent detectors. The intensities of these channels can then be used to determine the system's anisotropy. In this way, the rotational relaxation times can be estimated. In an example of such an implementation, Seidel and collaborators compared the steady state anisotropy of rhodamine 123 and yellow fluorescent protein (YFP) [123]. Because the rhodamine molecule is small, its rotation is fast and the resulting anisotropy is centered on zero. On the other hand, YFP is a large protein with a slower rotation that results in an estimated anisotropy of 0.34. The authors of this experiment suggested that this method can help distinguish between molecules in solution that are of different molecular weights (and have different rotational relaxation times) or to identify molecules in solution that have bound a target ligand.

A further extension of polarization studies is represented by polarization modulation spectroscopy. In this technique, the polarization of the excitation source is rotated while the emission is analyzed with fixed polarizers. This causes a high fluorescence modulation depth on orientationally immobile molecules while showing no effect on orientationally mobile molecules [124]. Using this technique, Johnson and collaborators have studied the interactions between CaM and plasma membrane Ca^{+2} ATPase (PMCA) [125–128]. In these studies, they found that various CaM populations with different mobilities existed at different calcium concentrations. This suggested the existence of an intermediate state in the activation of the pump where CaM is partially bound by one of its globular domains. A further conclusion was that the oxidation of CaM modified the distribution of binding events, making it unable to dissociate from PMCA.

7.3.2 Conformation and Dynamics of Molecules: Short Range Distances

Measurement of inter- and intramolecular distances has been one of the areas that have received most attention in SM experimentation. Two main approaches exist in the measurement of small distances with SMFS. The first is the use of FRET, where the nonradiative energy transfer efficiency between two fluorophores is used to measure distances in the range of 20 Å to 100 Å. The second relies on direct imaging and deconvolution of the image into the individual PSFs arising from single fluorophores. In this section, only the former will be explored while the latter will be discussed in Section 7.3.4. Many implementations of FRET have been described, and only a few of them are presented in this section. For more information, the reader is directed to a wealth of reviews that have been published on the theory and applications of FRET [25–27, 29, 113, 124, 129, 130].

The implementation of FRET requires that a donor (D) and an acceptor (A) dyes be localized on an individual molecule or on distinct molecules and come into close proximity as the result of conformational changes or molecular associations. In practice, the molecules of interest are illuminated with an excitation source matched with the donor absorption band. If the D-A pair is in close proximity, nonradiative energy transfer will occur, resulting in acceptor excitation and concomitant fluorescence emission. The fluorescence emitted for both D and A channels is recorded over time, and the FRET transfer efficiency (E) is calculated according to Eq. 7.1. Because E depends strongly on the separation between the D-A pair, ratiometric intensity measurements of fluorescence coming from D and A can be used as a "spectroscopic ruler" [24]. The experimental implementation of short-range distance measurement through FRET can be done in two modalities — using immobilized biomolecules and imaging them through one of the widefield techniques or scanning confocal microscopy, or probing molecules in solution with point-probe detection. A particular disadvantage of using immobilized biomolecules is that surface interactions can have an effect on the conformational changes and thus on the measured FRET efficiencies although cases have been reported where immobilization did not affect the outcome of FRET experiments [131]. A further consideration is that photophysical effects intrinsic to the chosen dyes can bias the measured FRET efficiency, especially when the fluorophores have low rotational mobility and their instantaneous dipole orientation affects E. Thus, fluorophores with high rotational mobility are preferred so that effects due to the fluorophores' orientations are minimized by the positional averaging at any given sample time.

The first demonstration of single-pair FRET (spFRET) was presented in 1996 by Ha and collaborators [132]. In this study, the authors used NSOM to excite fluorescence from D-A pairs linked by a short DNA molecule at the SM level. In addition to fluorescence intensities from D and A fluorophores, the authors also obtained emission spectra for each molecule assayed. They observed that the emission characteristics of the fluorophores had higher variability than expected from ensemble measurements, demonstrating the heterogeneity in the local environment surrounding each individual molecule. Confirmation that each two-color image corresponded to an SM containing only one D and one A was achieved by the photobleaching of one of the fluorophores. Following this experiment, Deniz and collaborators demonstrated the feasibility of measuring FRET efficiencies from molecules in solution [133]. By using multiple DNA constructs as rigid separators, they measured E as a function of distance and compared the resulting distributions to theoretical values (Fig. 7.8). Additionally, they used spFRET to monitor the denaturation of a DNA hairpin as a function of urea concentration. This study demonstrated the possibility of monitoring conformational changes in biomolecules through the use of spFRET.

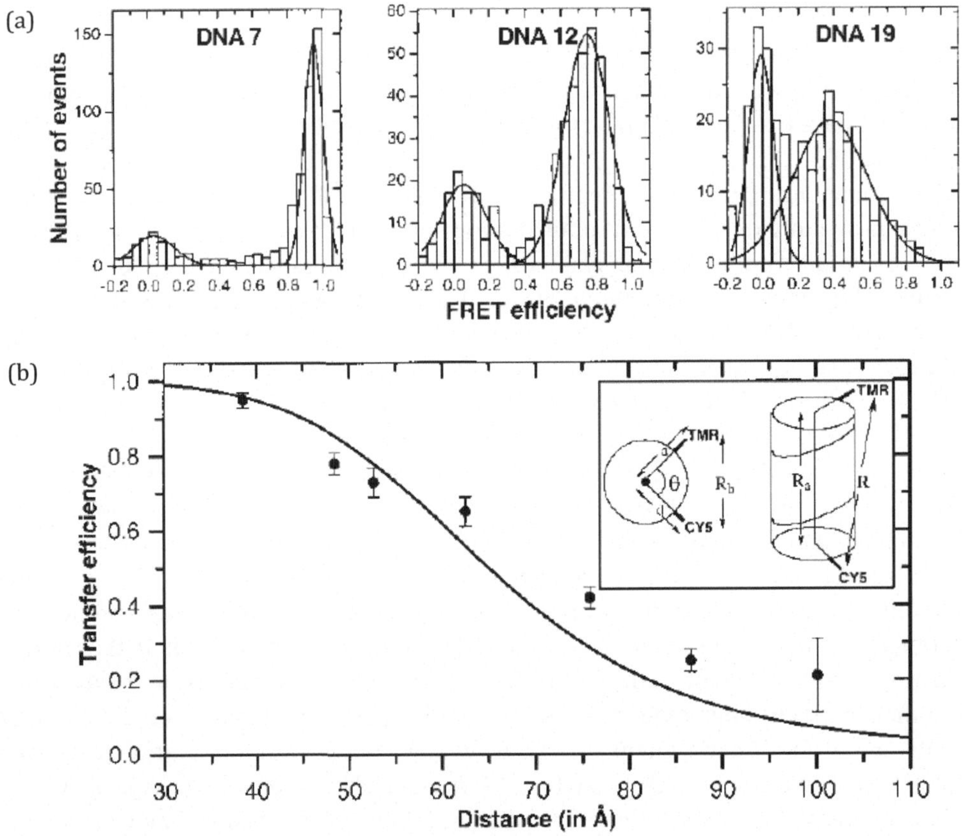

Figure 7.8 FRET efficiency assessed using DNA molecules of varying sizes as molecular rulers. (a) FRET histograms extracted from time traces for double-stranded DNAs of 7, 12, and 19 base pairs. Solid lines show Gaussian fits of the resulting distributions. (b) Mean FRET efficiencies extracted from FRET histograms plotted as a function of distance for the seven DNA constructs. (Reprinted with permission from ref. 133).

More recently, spFRET has been applied to the study of RNA catalysis and folding [131]. In this study, surface-immobilized ribozyme molecules were labeled with Cy5 as the acceptor dye while the substrate oligonucleotide was labeled with the donor dye Cy3. Upon binding of the oligonucleotide (or docking), the ribozyme undergoes a reversible folding transition that changes the *D-A* distance from 7 nm to 1–2 nm. By monitoring the FRET efficiency and residence times for each of the two conformations for a SM over multiple docking-undocking transitions, the authors were able to obtain transition rates for two previously observed states plus a novel intermediate that had not been observed through ensemble measurements. Another study that exploited spFRET to study conformational transitions was performed by Deniz and collaborators on α-synuclein, an intrinsically disordered protein linked to Parkinson's disease [134]. The authors probed membrane associations, conformational changes, and dynamics using a combination of spFRET and correlation methods. Using this methodology, they were able to reconstruct the folding energy landscapes for the protein and identified two main structural minima. The partition of the protein to either of the two structures was found to be dependent on the concentration of binding sites and the curvature of lipid membranes with which the protein associates. These examples of spFRET applications underscore the importance that this technique can have in elucidating structural information that would not be available by any other means, especially in systems that cannot be crystallized.

spFRET can yield information on the structure of molecules but is at a loss when the interactions between molecules need to be studied unless the *D-A* pair is located at the interface between the binding partners. Moreover, FRET techniques are limited when binding stoichiometry is an important observable in the experimental system. Recently, an extension of spFRET has been introduced that yields both information on FRET efficiency and additionally on donor-acceptor stoichiometry. This technique employs alternating laser excitation (ALEX) from two or more laser sources focused in a point-probe detection volume and records the resulting emission with PMT or APD detectors [135, 136]. Because *D* and *A* excitations are alternated, segments of the photon time trace contain fluorescence due to *D* or *A* excitation, allowing one to obtain fluorescence intensities for both fluorophores independent on the proximity of the probes. Thus, in addition to donor fluorescence emitted due to direct *D*-excitation, fluorescence signals emitted by the acceptor due to FRET are alternated with fluorescence signals emitted due to *A*-excitation [135]. Using a ratio of the total *A*-fluorescence observed to the total *D-A* combined fluorescence, a distance-independent figure (S) can be computed that directly reports on *D-A* stoichiometry. Two-dimensional E versus S plots where individual molecules are represented help discriminate between different aggregation and conformational states. The method was first presented and validated using DNA constructs of different lengths with *D* and *A* fluorophores attached to the ends of the oligonucleotide, with mixtures of these oligonucleotides, and with a protein-DNA-binding system where they were able to identify binding and oligomerization at the same time that FRET efficiency was monitored [136]. The authors of a recent review propose the use of ALEX to obtain structural and binding information from a given system for time frames spanning the millisecond to nanosecond range [135]. The contention is that utilizing the appropriate excitation and detection techniques, ALEX can be extended so that slow and fast dynamics can be observed. The same group has extended the implementation of ALEX to three excitation colors to monitor multiple distances and stoichiometric ratios [137]. While this allows access to a higher number of ratios, the experimental implementation and interpretation of the resulting data becomes more challenging, especially when cross talk between the different colors occurs.

7.3.3 Molecular Displacement: SM Tracking

Possibly the most intuitive of all experimental observables, molecular displacement has received much attention and has been pushed to the limit of molecular dimension accuracy, even under challenging imaging conditions within the cellular environment. The principle behind single-fluorophore tracking (SFT), single-molecule tracking (SMT), or single-particle tracking (SPT) for objects larger than SMs, such as beads, nanocrystals, and quantum dots, is the recording of a succession of fluorescence images with a high

SNR where single emitting entities can be identified. These entities are then localized with a subpixel resolution and tracked for a finite period of time through computational routines. In this section, the discussion will focus on SMT, where each molecule is labeled with a single fluorophore. A wealth of information regarding SPT is available, both for fluorescence and contrast methods, for the interested reader [138–149].

Experimentally, SM images are typically obtained with widefield fluorescence methods (epifluorescence or TIRF) and recorded via a high-sensitivity detector, such as an EMCCD. The images obtained are then processed in two basic steps — (1) the identification of SMs and discrimination from background and (2) the tracking of SMs across frames. Methods to discriminate fluorescence arising from SMs from background within an image include image filtering, convolution with model distributions, and thresholding based on the background level. The SNR for image analysis is defined as the ratio of the signal minus the mean background over the standard deviation of the background intensity. Thus, a threshold based on SNR levels is usually the easiest method used for the discrimination of SM signals. After the identification of pixels corresponding to fluorescent signal from SMs, it is necessary to localize the center of the molecules detected to within subpixel accuracy. A formula for the uncertainty of localization ($\sigma\mu$) achievable for a Gaussian-like PSF in an experimental system has been derived by Thompson and collaborators [150]:

$$\sigma_{\mu}^{2}=\frac{\sigma^{2}}{N}+\frac{a^{2}}{12N}+\frac{8\pi\sigma^{4}b^{2}}{a^{2}N^{2}} \tag{7.12}$$

where σ is the standard deviation of the PSF, N is the number of detected photons, a is the effective pixel size of the detector, and b is the background noise. A more generalized formula for the uncertainty of localization of emitters with non-Gaussian shape has been derived by Ober and collaborators [151]. In Eq. 7.12, the first term is the limit imposed by the statistical nature of photon detection and is the dominating factor when large numbers of photons are detected. The second term represents the effect of the finite size of the detector pixel. Finally, the third term represents the contribution of background noise to the accuracy of localization, and is the dominating factor at low photon-detection levels [150, 152]. From this formula, it can be seen that although the full width at half maximum of a diffraction limited PSF is on the order of $\lambda/2\text{NA}$, the localization of the fluorescent molecule can be determined with much higher precision by finding the center of the resulting PSF.

Several different methods for finding the center of the PSF for SMT have been utilized, including the calculation of the center of mass [153] and fitting of a two-dimensional Gaussian shell [144]. These methods have been compared quantitatively by Cheezum under conditions that mimic most SM experiments [154]. From the localization of objects with above and subresolution PSFs, it was determined that the two-dimensional Gaussian approach yielded the most reproducible and unbiased results, especially for conditions mimicking single-fluorophore localization. From Eq. 7.12, it can be seen that by reducing the background signal and increasing the number of photons collected, accuracy in the range of 1–10 nm should be possible. The increase in photon collection depends on the excitation power, the photophysical characteristics of the chosen fluorophore, and the molecular environment it encounters. On the other hand, the reduction in the background signal depends on the rejection of the scattered light by high-quality optical filters and the reduction of fluorescence from sample impurities by a reduction in the focal volume. Such an optimization of experimental parameters has been recently realized by Yildiz and collaborators, who demonstrated better than 1.5 nm localization accuracy [42].

The second step in SMT implementation is the tracking of the localized SM centers across the frames within the recorded experiment. Tracking an SM in a sequence of images requires an algorithm that can account for high molecular density, displacement, particle trajectory heterogeneity, fluorescence fluctuations, instantaneous appearance and disappearance of molecules, and merging and splitting of molecular trajectories [41]. Although some of these challenges can be mitigated by using systems with a low particle density, this condition prohibits the study of particle-particle interactions and yields a low number of data points per experiment and in some systems, the particle density cannot be tightly controlled, such as in the cellular environment. Tracking algorithms based on the center of mass, surface fitting, and image correlation have been compared quantitatively [154, 155]. Most tracking algorithms link molecules detected in each image based on a set of rules that assigns a cost to the each molecule-molecule link established between frames. These costs can be associated with a physical distance, PSF characteristics, or trajectory characteristics. The matrix representing the costs for all linkages between molecules can then be minimized, generating the most likely trajectories either locally (frame-to-frame, temporally greedy algorithms) or globally (all possible trajectories, global algorithms). The implementation of a robust algorithm based on the mathematical framework of linear assignment problem has been recently reported by Jaqaman and collaborators [41]. This algorithm allows the user to detect the particles within an image via a variety of methods (the center of mass, PFS fitting, etc.) and then links the positions of the detected molecules taking into account the costs described above using a global optimization. This algorithm is capable of handling high particle densities, appearance and disappearance of molecules, a low SNR, and particle trajectory merging and splitting, with the only caveat of requiring large computational power.

The initial SMT experiments yielded localization accuracy in the 10–50 nm range for molecules labeled with one or more fluorophores even within the cellular environment [44, 45, 144, 153, 156–158]. More recent developments in the implementation of SMT, such as a better detector, enhanced stability of translational stages with nanometer accuracy, and the use of highly stable fluorophores in the presence of additives that significantly reduce photodegradation, have yielded some impressive results where the accuracy of localization has improved to the 1–10 nm range. The technique of localizing the center of SMs and tracking them with high accuracy has been dubbed fluorescence imaging with one-nanometer accuracy (FIONA). This technique was first presented by the group led by Selvin in 2003 for the study of myosin V, a molecular motor that walks processively on actin filaments [42]. In this study, myosin molecules were tagged with a single Cy3-labeled CaM on one of the six binding sites localized on the light-chain domain. Thus, several populations of labeled myosins were obtained, with the fluorophore located at a distance x from the myosin stalk. The images of the moving molecules were obtained with 0.5 s acquisition times, which yielded 5,000–10,000 collected photons per image and a localization accuracy of 1.5–3 nm. The individual molecules observed could be followed continuously for ~50 s, which means that for bright molecules, a total of more than one million photons could be collected! With such a precise localization, the authors could observe single myosin steps of 37 + 2x and 37 − 2x nm, which strongly supported the notion that myosin V walks hand over hand and not in an inchworm fashion (Fig. 7.9). Using the FIONA technique, the movement of kinesin has also been studied with high-precision localization [53]. This study demonstrated that kinesin walks hand over hand with 17.4 nm steps. The same technique was later used for the study of myosin VI constructs, where the large step distribution was investigated along with the possibility of this type of myosin to function as a monomer or a dimer [52]. It must be noted that high-precision localization can also be used for SPT of molecules labeled with Q-dots [159] and for cellular vesicles labeled with thousands of GFPs [160].

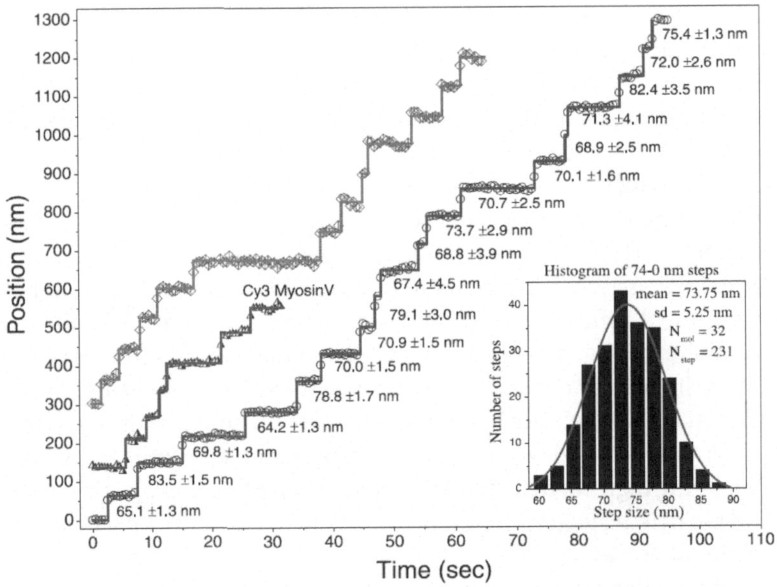

Figure 7.9 Stepping traces obtained through FIONA of three different myosin V molecules displaying 74 nm steps and the histogram (inset) of the distribution of 231 steps taken from 32 different myosin molecules. (Reprinted with permission from ref. 42.)

7.3.4 Diffusion, Binding, and Aggregation: Coincidence, Correlations

Interactions between molecules are another experimental observable studied through SMFS. As with many of the other observables, molecular interactions can be monitored by point-probe or widefield detection. The former allows monitoring of associations in solution, with high temporal resolution, but with the limit of measuring a single assembly at once and only in the time that it takes for it to diffuse through the focal volume. On the other hand, widefield imaging allows monitoring many molecular assemblies at a time with high localization accuracy but with the constraints of surface immobilization and/or a lower temporal resolution.

FCS is a technique ideally suited for binding and aggregation studies in solution or in lipid membranes because these events change the diffusion behavior of the species monitored. FCS reports the average residence time for molecules in the focal volume by measuring the self-similarity of a collection of SM photon bursts through the autocorrelation curve. Thus, processes such as aggregation, ligand binding, and hybridization, which change the effective hydrodynamic radius of the fluorescent species, can be monitored through changes in the autocorrelation curve. However, in order to easily measure distinct diffusion coefficients from molecules with different sizes, the hydrodynamic radius of the species to be distinguished must be different by at least a factor of 4 (which translates into a factor of 1.6 for diffusion coefficients) [161]. The first SM FCS diffusion measurements were conducted by Rigler using a confocal volume to detect rhodamine6G molecules in solution [5, 66]. In these studies, the theoretical basis was established for the analysis of the autocorrelation curves and the diffusion properties of the fluorophores were obtained. Since these early studies, advances in detectors, fluorescent dyes, and autocorrelation electronics have enabled the efficient detection of SMs with a high SNR via one or TPE.

An example of the application of FCS to study changes in diffusion is given in the work by Eigen, where the change in the diffusion of a fluorescently labeled primer was used to detect pathogenic DNA in small copy number [162, 163]. In this study, the DNA from the pathogen was incubated with a labeled primer and the change in the diffusion coefficient of the primer was monitored. The high sensitivity of FCS allowed the detection of as little as 10 target molecules in the presence of a large amount of nonspecific DNA without the need for polymerase chain reaction (PCR) amplification. This methodology was proposed as a simple and fast way to obtain diagnosis based on DNA sequence identification with low sample requirement. A second application of FCS was the study of DNA cleavage through a variety of restriction enzymes [164]. In this study, the degradation of fluorescently labeled DNA was characterized by a decrease in the autocorrelation curve, which corresponds to an increase in the number of fluorescent molecules, and a shift of the residence time to lower values, which corresponds to a decrease in the molecular sizes of the cut fragments. This approach was also used to detect point mutations in certain genes by analyzing the restriction fragments obtained [164]. The aggregation or polymerization of molecules in solution has also been studied through FCS. In a study by Bark *et al.*, the activation of fibrinogen by thrombin was followed through the formation of fibrin monomers and fibrils over time [165]. The aggregation of fibrin into oligomeric fibers causes a change in its diffusion coefficient that can be observed in the autocorrelation curve. Based on the observations on the distribution of aggregates over time, the authors proposed a model for the formation of fibrin networks. In a similar experiment, the polymerization of amyloid β-peptide (Aβ) was studied at various concentrations and in the presence of polymerization inhibitors [166]. It was found that Aβ aggregation is highly cooperative and that after the transition from monomers to dimers, large aggregates distribute into two well-defined populations.

The molecular diffusion of fluorescently labeled species and binding to membranes has also been studied through FCS. Monitoring of the diffusion of fluorescent species within a model or cellular membrane allows the elucidation of the fluid behavior of the membrane and can point toward hindered diffusion due to lipid and sterol composition. In this way, the diffusion properties of phase-separated membranes has been studied by measuring the diffusion coefficients for fluorescent lipid analogs that partition within each lipid phase [167, 168]. Furthermore, the diffusion coefficients of fluorescently labeled lipids embedded in membranes containing various concentrations of different phospholipids and sterols have been studied [169, 170]. FCS has also been used to study the diffusion characteristics of dyes and fluorescently labeled proteins within cellular compartments (such as the cytosol, [Fig. 7.10]) and the plasma membrane both by one- and two-photon excitation [171]. The study of diffusion in plasma membranes has shown that anomalous diffusion exists for certain membrane anchored proteins while other proteins and lipids exhibit correlation curves consistent with free diffusion in a two-dimensional plane [91, 171]. Thus, FCS allows the monitoring of SM diffusion along lipid membranes. The binding of a fluorescent molecule to membranes changes its diffusion behavior from the fast, free diffusion in solution to a slower diffusion along a two-dimensional surface. This change in diffusion behavior can be used to report on molecular associations and can be used to estimate binding kinetics. Thus, Rigler and collaborators used FCS to study the binding of proinsulin C-peptide to cell membranes [172]. In this study, fluorescently labeled C-peptide was incubated with several types of human cells at various concentrations, which allowed the reconstruction of binding curves and the estimation of the association constant. In addition, by using nonlabeled C-peptide, they were able to displace the bound fluorescent protein over time and obtain the dissociation constant. Multiple potential agonists were used to test their effect on binding, and it was concluded that C-peptide binds onto G-protein-coupled receptors. In a subsequent study, using FCS to probe binding, it was demonstrated that the binding of C-peptide to cellular targets is highly dependent on a single glutamic acid residue [173]. Other membrane-binding experiments that have been conducted by the group led by Rigler include the study of the binding of epidermal growth factor to its membrane receptors [174], the study of the binding of neuropeptide galanin to its membrane-bound receptor [175], and the interaction of ethanol/naltrexone at the μ-opioid receptor [176].

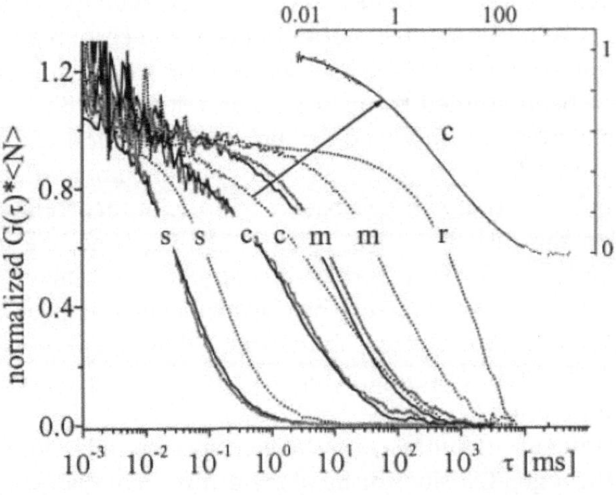

Figure 7.10 FCS autocorrelation curves of tetramethyl rhodamine dye in solution(s), in cell cytoplasm (c), and in the plasma membrane (m) and of IgE receptors labeled with Cy3 on the plasma membrane (r). Curves were taken with a confocal excitation volume (dotted lines for collection through a 100 μm pinhole and black lines for a 25 μm pinhole) and with TPE (gray lines). The inset demonstrates the data fitting with an anomalous diffusion model. (Reprinted with permission from ref. 171.)

Because of the limitation imposed by the need to have a large change in the hydrodynamic radius to detect a significant change in the diffusion coefficient, it is not possible to measure the binding of species with similar molecular weights by FCS. To overcome this limitation, dual-color fluorescence cross-correlation was established. In this technique, two independent excitation sources are combined in a single focal volume to excite fluorescence from two molecules labeled with different color fluorophores. The emitted fluorescence is collected through dual band optical filters and split into two independent high-sensitivity detectors (PMT or APD). The intensity trace for each color is recorded, and then the two traces are correlated to measure the coincidence of fluorescence bursts. If the two labeled molecules diffuse through the excitation volume independently, no correlation between the fluctuations of the two channels is observed. Yet, if the molecules diffuse together, coincidence in the intensity fluctuations of the two channels occurs and a correlation signal is observed. This concept was first demonstrated by monitoring the hybridization kinetics of two single-stranded DNA molecules [98]. The experiment involved using differently colored complementary ssDNA in solution and monitoring the intensity traces for both colors. As the ssDNAs hybridized the proportion of events showing coincident emission in both colors increased, giving an increased amplitude in the cross-correlation curve. In this way, the temporal course of hybridization could be followed and the kinetic parameters for the reaction obtained. FCCS has been further applied to the study of EcoRI enzyme kinetics in the degradation of dual-labeled double-stranded DNA [103]. In this case, the degradation of the doubly labeled DNA resulted in a decrease in the amplitude of the cross-correlation curve. By conducting the reaction at different substrate concentrations, the reaction rates could be extracted and fitted to Michaelis-Menten kinetics. This technique for measuring enzymatic catalysis was further expanded into a high-throughput method for screening and evolutionary biology, where the setup was optimized to yield rapid decisions on the activity of an isolate on a predetermined substrate [177]. This methodology has been proposed as a useful tool for selection in evolutionary biotechnology. FCCS has also been shown as a viable technique for studies within cellular environments. In an experiment involving cholera toxin with its alpha and beta subunits labeled with different fluorophores, the internalization of the toxin was followed within live cells [105]. This study found that the cholera toxin is internalized and remains intact until it reaches the Golgi apparatus, where its subunits are separated and cross correlation is lost. FCCS can be used in this way to study binding and aggregation, as well as dissociation or degradation, of proteins both in solution and within cellular environments. Furthermore, the technique applies to the three-dimensional motion of proteins in free diffusion or to proteins confined to diffusion on a surface. A recent study has further combined the concept of CCS with fluorescence lifetime imaging to enable CCS with fluorophores that absorb and emit in the same spectral regions but that are chemically different. In this study, the interaction of a GFP-labeled epidermal growth factor receptor (EGFR) with a neutralizing antibody labeled with Alexa 488 was studied in live HEK293 cells [106]. The two fluorescent moieties could be distinguished and their fate tracked independently by using their different fluorescent lifetimes. This study demonstrated an original combination of SMFS techniques to overcome the technical challenges imposed by the use of fluorophores with similar emission characteristics.

Since the first studies of subdiffraction co-localization of fluorescently labeled molecules were performed via NSOM with a localization accuracy of 100 nm [178], researchers have strived to improve accuracy and simplify experimental implementation. Co-localization of molecules through widefield fluorescence has been applied to study the binding and aggregation of molecules immobilized on surface structures or bound to a supporting lipid bilayer. The first experimental demonstration of SM colocalization through epifluorescence was presented by Schütz *et al.* by using two fluorophore-biotin conjugates bound to a single streptavidin molecule immobilized on a supported lipid bilayer [179]. By imaging the fluorescence emitted in the green and red channels and fitting the resulting emission profile to Gaussian surfaces, it was possible to colocalize the molecules with an accuracy of 40 nm. It was also noted that in some cases, when the fluorophores were close enough to each other, spFRET could be observed, unequivocally confirming the colocalization of molecules at a distance of at least 7 nm. Lacoste *et al.* later introduced the idea of superresolution co-localization of fluorescent particles to achieve sub-10 nm accuracy [180, 181]. Their experiment was performed using 40 nm spheres and nanocrystals that could be easily excited with a single wavelength in a confocal scanning setup, thus eliminating the need of multiple lasers to be aligned and scanned together to generate the final image. More recently, a couple of techniques were developed to co-localize identical single fluorophores based on continuous monitoring of the combined PSF before and after one of the fluorophores bleach. These techniques, developed independently and dubbed SHRImP [182] and NALMS [183], were developed to increase the accuracy of localization (to 10–200 nm) and to enable the detection at a higher temporal resolution than that achieved with scanning techniques. Both techniques were validated using double-stranded DNA as a molecular ruler. In 2005, an additional improvement was reported by the introduction of a technique called single-molecule high-resolution co-localization (or more colorfully, SHREC) [184]. This technique, an extension of FIONA to two colors, utilizes two alternating excitation sources to independently image two spectrally different fluorophores and then uses the Gaussian-fitting approach to obtain subpixel resolution. It must be noted that for this technique to achieve its maximum potential, the variations in position due to chromatic aberrations must be corrected. Churchman and collaborators tackled this issue by generating a correction map using fiduciary marks that could be stepped across the imaging field with nanometer precision [184]. Using this approach, they were able to obtain image registration to within 3.3 nm and single-fluorophore localization of better than 10 nm. The technique was validated using dsDNA as a molecular ruler and then further applied to observe the motion of myosin V. It was shown that for myosin labeled with different fluorophores in each of the light chains attached to the actin filament, alternating 72 nm steps could be observed, further confirming the hand-over-hand model for myosin motion (Fig. 7.11). SM colocalization with a high resolution in cells has proved more challenging, with few studies reported. The most recent details the colocalization of EGFP-cadherin with anti-EGFP labeled with Alexa Fluor 633, with a localization accuracy of 64–100 nm [185].

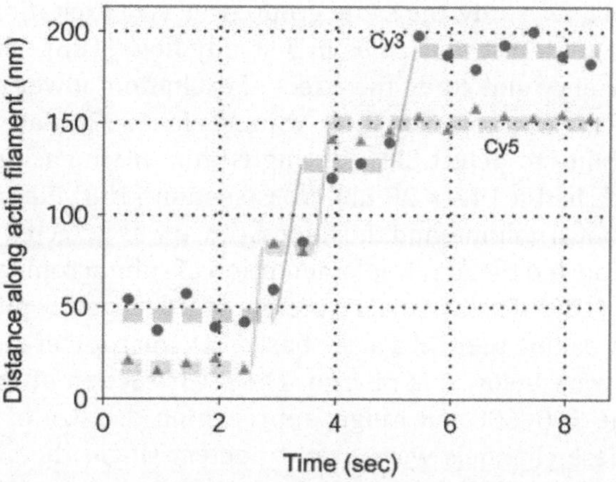

Figure 7.11 The application of SHREC to the study of myosin movement along an actin filament. The graph shows the time trace of the measured displacement for each of the two labels placed on the light chain of myosin V. CaMs labeled with spectrally distinct fluorophores Cy3 and Cy5 were exchanged on the myosin molecule. The precise localization of the fluorophores over time allows the observation of 72 nm steps taken by each of the labels, indicating that labeling occurred close to the motor domain. The alternating positions of the labels strongly support the hand-over-hand model. (Reprinted with permission from ref. 184).

7.4 Micro- and Nanostructured Surfaces for SMFS

As mentioned previously, success in SM detection is highly dependent on the reduction of background noise and the concomitant increase in the SNR. A way to decrease the fluorescent background is to reduce the focal volume through physical confinement via the use of micro- and nanofabricated structures in the form or channels, slits, and apertures. The use of low-fluorescence substrates for the fabrication of such devices allows their use for SMFS. Due to the reduction in the focal volume to subdiffraction sizes, nanostructured devices can be used to perform SMFS at high concentrations (up to micromolar concentrations), which allow the study of biomolecular function under physiologically relevant conditions. The use of nanostructured substrates for SMFS also allows the integration of fluorescence detection with lab-on-a-chip techniques. Thus, experimental conditions, such as potential, flow, concentration, and pH, can be controlled within these devices. Additionally, the size of the fabricated structures can influence the way molecules behave, inducing stretching, surface interactions, and diffusion anomalies.

7.4.1 Micro- and Nanofluidic Slits and Channels

The first descriptions of SM detection in micron-size fluidic channels were reported in the early 1990s with the use of electrophoresis capillaries [186, 187]. In a pioneering study, Winefordner *et al.* used a focused laser beam to fully illuminate the inside of a capillary with 11 μm inner diameter and excite fluorescence from a dilute solution of IR140 molecules [187]. In a later study, the same group used two excitation volumes separated by a known distance to detect molecules in a steady flow [186]. They aimed to achieve 100% detection efficiency and study the effect of excitation power on photodegradation of the molecules after the first excitation volume. In 1997, Mathis and collaborators demonstrated the ability to detect SMs flowing within micromachined silicon channels, the smallest of which had a 147 × 20 μm cross section [188]. Rhodamine6G molecules were excited in a confocal volume, and the fluorescence was detected by an APD. The same year, Lyon and Nie reported the first implementation of submicrometer-sized channels for the detection of SMs [189]. The channels were prepared by gently pulling electrophoresis capillaries without breaking them in a laser-based CO_2 micropipet puller until their cross-section dimensions were below one micron. The cross section of the resulting channels had diameters in the 500–600 nm range, approaching the size of a diffraction-limited confocal volume. These channels were used to detect single rhodamine molecules and T4 DNA fragments as they diffused through the focal volume with almost 100% detection efficiency. Electroosmotic flow was established by applying a voltage across the fluidic channel. Changing the voltage allowed control of the number of molecules crossing the excitation volume, with the tradeoff that the faster the molecules passed by, the lower was the number of photons detected per molecule. It was also observed that as T4 DNA fragments flowed by the focal volume, the number of detected photons had asymmetrical shapes, suggesting that the DNA was not completely stretched within the channel.

Micro- and nanofabrication through photolithography and etching became the common manufacturing processes for channels with controlled geometries in a wide variety of substrates, ranging from silicon to transparent glasses and, more recently, elastomeric compounds. The ease with which the geometry of fabricated devices could be tailored for the particular SM application allowed the establishment of novel analysis techniques. In this way, hydrodynamic focusing was achieved by introducing a tapered separation channel and a focusing sheath flow by cross channels [190]. Changes in the flow rate of the sheath stream allowed the separation stream to be constricted and increased the detection efficiency by matching the width of the focused stream to the confocal detection volume. This experimental setup was used to separate and detect single DNA molecules in the 100–1,000 bp range. DNA fragment sizing was also conducted by Foquet *et al.* in submicron size fluidic devices manufactured by a sacrificial-layer process [76]. Single DNA fragments in the 500–27,500 range were separated using a wide range of voltages to study the transition between diffusion and flow-dominated regimes (cf. Fig. 7.5). The authors also analyzed the SM data through photon-burst and FCS methodologies. In a later study, Foquet *et al.* refined the FCS analysis of SMs motion in nanofluidic channels, exploring the effects of channel confinement and describing the effects of electrokinetic flow on the FCS autocorrelation curve [88]. A different geometry of microfabricated device that has been used for long DNA separation was the entropic trap array, introduced by Han and Craighead [191]. This device presents a microfluidic channel with alternating shallow (75–100 nm) and deep (1.5–3 μm) regions through which DNA is driven by the application of a voltage bias (Fig. 7.12). Because the shallow regions were narrower than the hydrodynamic radius of the DNA to be separated, the alternating shallow regions acted as a sieve, effectively separating long DNAs in the 5,000–160,000 bp range. The principle behind this device is that the DNA molecules have to change their conformation from globular to elongated as they transition from deep to shallow regions, which has an entropy change associated with it. Thus, the molecules get transiently trapped before the shallow regions, with the trapping time and probability of going through the shallow region being dependent on the overall DNA length. A similar entropic recoil separation was presented by Cabodi *et al.* on the basis of DNA confinement by arrays of closely spaced nanopillars [192]. Entropic elongation and compression of DNA to study fundamental properties of DNA as a polymer chain have been recently conducted in nanofluidic channels with cross sections as small as 30 × 40 nm [193–196]. A different approach to biomolecular sorting based on the rectification of diffusion through two-dimensional arrays of microfabricated pillars (a technique known as "molecular ratchet") has been presented by the group led by Austin [197–201]. The principle behind this technique is to establish a flow of a solution containing a mixture of particles or molecules and pass this mixture of molecules through an array of microfabricated pillars with various spacing dimensions and induce the preferential movement of molecules according to size either by the asymmetric shape of the obstacles or by applying an asymmetrical pulsed electric field. In this way, large DNA fragments have been separated under continuous flow conditions, a technique not amenable to traditional gel-based separation techniques [199–201].

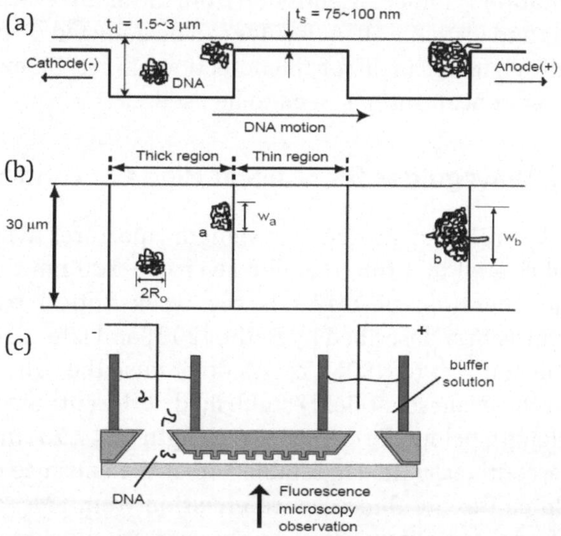

Figure 7.12 A schematic depiction of the principle behind the entropic separation of DNA molecules in nanofluidic devices. (a) A side view of the nanofluidic device showing the shallow and deep regions through which the DNA molecules are electrophoresed. Because the radius of gyration of the molecules is larger than the depth of the thin region, they get transiently trapped. (b) A top view of the device with molecules under an electric field probing the thin regions. If a molecule is trapped in a thin region, it will escape toward a deep region with a probability dependent on the length of the molecule, resulting in effective DNA separation. (c) A side view of the whole nanogfluidic device (not to scale). Reservoirs at both ends of the nanochannel are filled with a buffer solution. (Reprinted with permission from ref. 191.)

Submicron fluidic channels have also been applied to the identification of SM species within a mixture in real time. The ability to combine more than one excitation wavelength focused on a submicrometer fluidic channel allows the discrimination between molecules containing one or both fluorophores and their stoichiometry. In this way,

Stavis and collaborators determined the percentage of Alexa Fluor 488–labeled ssDNA functionalized with a biotin linker bound to streptavidin-coated quantum dots [79]. The intensity trace collected from each channel was analyzed to detect SM fluorescence bursts and calculate the number of SM events that presented the coincidence of fluorescence in both colors. In subsequent experiments, Stavis and collaborators demonstrated the possibility of differentiating between nucleic acid constructs containing different ratios of two spectrally distinct fluorophores [78] and the possibility of quantifying the amount of reacted amplicons in a PCR reaction [77].

The use of micro- and nanofluidic channels for SMFS presents several advantages over measurements performed in solution. First, the use of fluidic structures allows the direct control of flow conditions as well as control over the reaction environment of the molecules studied. Second, by using a structure with confined dimensions, scattering and background noise are greatly reduced, enhancing the SNR for SM events. Micro- and nanofabricated structures also present the benefit of controlled geometry, where secondary channels can be used to hydrodynamically alter the molecular environment or to deliver reactants of interest. Finally, because the channels can have dimensions below that of a confocal excitation volume, the emission from the SMs traversing the focal volume is more homogeneous and virtually all molecules can be detected. These advantages make micro- and nanofluidic channels ideal for the detection of rare SM events and well suited for assays where higher concentrations need to be used.

7.4.2 Zero-Mode Waveguides for SMFS at High Concentrations

Zero-mode waveguides (ZMWs) are subwavelength apertures with diameters in the 30–300 nm range fabricated in a thin metallic layer of ~100 nm thickness. The optical principle behind the operation of ZMWs is the weak optical transmission through subwavelength apertures first described by Bethe [202] and later used as the foundation for NSOM. However, in contrast to NSOM, ZMWs do not use the light transmitted through the nanoaperture but the evanescent field established at the entrance of it. As light from a laser source of wavelength below the critical wavelength ($\lambda = 1.7d$) impinges on a ZMW of diameter d, an evanescent decay field is established at the entrance of the nanoaperture. This evanescent field can excite fluorescence emission from SMs, creating an effective excitation volume of atto-zeptoliters. Because the excitation volume is greatly reduced, ZMWs allow SMFS at high concentrations (up to 100 μM). ZMWs have been fabricated through e-beam lithography [89, 203, 204] or focused-ion-beam milling [205] in aluminum, chrome, and gold films. Thus, large arrays of ZMWs can be produced on a single glass surface, allowing for highly parallelized SMFS. The theory and applications of ZMWs have been recently reviewed [206, 207].

Since the first implementation by Levene *et al.* [89], ZMWs have been applied to the study of SMs in a variety of biological systems, including DNA polymerization, aggregation kinetics, diffusion and binding to model membranes, and diffusion on cellular membranes. The first implementation of ZMWs was to detect the synthesis of a complementary DNA strand by mutant T7 DNA polymerases with SM resolution [89]. In this experiment, a polymerase was immobilized at the bottom of a ZMW and incubated with a mixture of M13 DNA, three unlabeled nucleotides (dATP, dGTP, and dTTP), and coumarin-labeled dCTP. As the labeled nucleotides were incorporated into the newly synthesized DNA strand, they became immobilized, resulting in a long fluorescence burst followed by bleaching. Each long burst of fluorescence corresponded to a single labeled nucleotide being incorporated. The authors observed that the polymerization of the chains typically stopped after 30 minutes, the time necessary to synthesize a full copy of the M13 DNA. This technique holds great promise for rapid DNA sequencing, and its development has been pursued by a newly established company (Pacific Biosciences). Recently, the targeted immobilization of DNA polymerases within ZMWs with selective passivation of the aluminum surface has been reported [208]. The successful incorporation of all-labeled nucleotides has also been demonstrated [209], paving the way for high-throughput sequencing of long DNA molecules with ZMWs. The first real-time sequencing data from continuous template-directed synthesis with fluorescently labeled deoxynucleotide triphosphates (dNTPs) obtained through the use of arrays of ZMWs has just been presented [210]. In this study, the authors were able to obtain continuous readouts of consecutive dNTP incorporations over thousands of base pairs. They have also shown that the accuracy for the consensus sequence generated through SM readouts with 15*x* coverage is greater than 99%. If the full potential of ZMW sequencing is realized, it would be possible to obtain a four order of magnitude improvement on the sequencing speed, with an array of 14,000 ZMWs producing a complete read of a diploid human genome per day [210].

(a) (b)

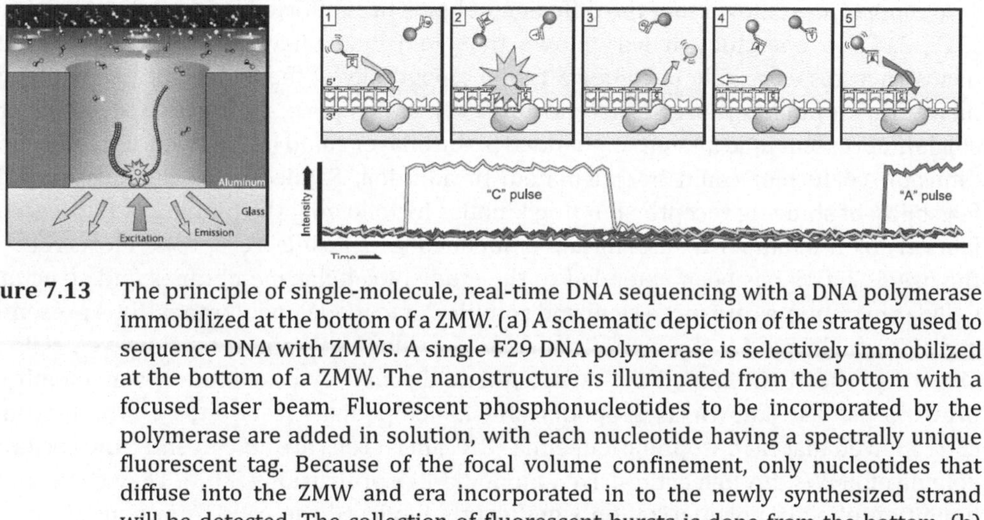

Figure 7.13 The principle of single-molecule, real-time DNA sequencing with a DNA polymerase immobilized at the bottom of a ZMW. (a) A schematic depiction of the strategy used to sequence DNA with ZMWs. A single F29 DNA polymerase is selectively immobilized at the bottom of a ZMW. The nanostructure is illuminated from the bottom with a focused laser beam. Fluorescent phosphonucleotides to be incorporated by the polymerase are added in solution, with each nucleotide having a spectrally unique fluorescent tag. Because of the focal volume confinement, only nucleotides that diffuse into the ZMW and era incorporated in to the newly synthesized strand will be detected. The collection of fluorescent bursts is done from the bottom. (b) A schematic depiction of the sequence readout from SM events. (1) A fluorescent phosphonucleotide diffuses into the ZMW containing the polymerase and the template strand and forms an association with its complementary base in the polymerase active site. (2) The turnover time of the polymerase causes an increased residence of the fluorophore in the ZMW, increasing the photon signature in the corresponding channel. (3) The phosphodiester bond formation results in the cleavage of the dye-linker construct, which diffuses out of the ZMW and no longer contributes to the detected signal. (4) The polymerase translocates to the next nucleotide position. (5) The cycle repeats. (Reprinted with permission from ref 210.)

ZMWs can also be applied to the study of biomolecular interactions in solution, such as binding, aggregation, or enzymatic catalysis. FCS and FCCS can be applied in combination with ZMWs to study the interaction between molecules labeled with one or two distinct fluorophores, with the added advantage of higher molar concentrations. In this way, the oligomerization of the CI repressor of bacteriophage λ has been studied [211]. Through FCS autocorrelation curves, the authors of the study have shown that at micromolar concentrations of the CI monomer, two fluorescent species with different diffusion coefficients exist corresponding to dimers and tetramers. In addition, the further decrease in the diffusion coefficient observed when the repressor oligomers were incubated with their target sequence within the nanostructures confirmed that the oligomers retained their natural activity. The implementation of FCCS with ZMWs highlights another advantage of these nanostructures, which is that they can be used to overcome the need for the precise alignment of multiple lasers in a single focal volume, as demonstrated by Wenger *et al.* [212]. In this experiment, the enzymatic cleavage of doubly labeled DNA molecules was followed in time through the monitoring of the concomitant decrease in the cross-correlation curve amplitude. These two implementations represent proof-of-principle experiments that show the potential of ZMWs for the study of binding, aggregation, and enzymatic catalysis at high molar ratios with SM sensitivity.

Finally, ZMWs have also been employed in the study of model and cellular membranes. One of the challenges of using ZMWs for membrane studies is the requirement that the bilayer conform to the topography of the nanostructure or at least span over the aperture. Two reports have shown that the diffusion of lipids in supported lipid bilayers is possible [213, 214]. In one study, it was shown that fluid lipid bilayers formed on top of the nanostructures were able to conform to the topography of the nanostructure while lipid bilayers in the liquid-ordered phase did not [213]. Both reports showed that the diffusion of single-fluorescent lipid analogs embedded in the bilayer could be followed and an accurate diffusion coefficient could be calculated. In addition, Samiee *et al.* demonstrated the feasibility of studying receptor-binding kinetics by following the binding of fluorescently labeled toxin subunits to membrane-embedded ganglioside receptors. More recently, the use of ZMWs has been extended to the study of cellular membranes and structures. Cellular membranes are more rigid and complex than model membranes, thus presenting a greater challenge for the conformal coating of the ZMW aperture. However, cells are able to probe their immediate environment through cytoskeleton-driven membrane protrusions, also known as filopodia and lamellipodia. In one study, Moran-Mirabal *et al.* showed that rat basophilic leukemia (RBL) mast cells are able to enter the excitation volume of ZMWs through actin-driven filopodia-like protrusions [203]. They showed that inhibition of actin polymerization significantly hindered the cells from penetrating the ZMWs. In addition, they showed that the polymerization and depolymerization of actin-EGFP constructs could be observed within ZMWs. Two studies conducted independently also demonstrated that the diffusion of fluorescently labeled lipid analogs could be followed within the nanostructures [203, 215]. Wenger and collaborators then studied the ultrafine membrane organization using ZMWs of different diameters. By studying a number of different molecules embedded in the cellular membrane and their diffusion within excitation volumes with cross sections in the 0.02–0.35 μm^2 range, they obtained diffusion laws that yielded estimates of the relevant size of the domains within which these molecules interact. These reports show the use of ZMWs for SMFS studies at high molar concentrations both on freely diffusing and membrane-embedded systems.

7.4.3 Solid-State Nanopores

Solid-state nanopores, apertures with diameters in the 1–30 nm range, have been successfully used to study the translocation of biomolecules using a combination of electrical detection, force measurements, and optical trapping [216–218]. However, few applications of the use of solid-state nanopores have been reported for optical studies and most of the reports to date have focused on large-scale fluorescence techniques, such as fluorescence recovery after photobleaching. Solid-state nanopores coupled with suitable detection methods could aid in the study of biomolecular properties with the high SNR provided by the isolation of the aperture, the high temporal resolution provided by single-point detection, and the high throughput through parallelization. Recently, Edel *et al.* showed the first implementation of solid-state nanopores with optical detection for SM translocation detection [219]. The authors monitored the translocation of fluorescently labeled λ DNA molecules through 300 nm solid-state nanopores fabricated on an aluminum-silicon-nitride membrane. Widefield and point-probe-type detection were used to study the translocation of DNA driven by an electric field. With an EMCCD camera, multiple events were recorded with a temporal resolution of 15 ms, where each event had a duration of ~360 ms. This length of translocation was significantly larger than that measured through ionic currents, which was attributed to the fact that the optical technique used records the DNA before and after complete translocation through the pore and to the length of the nanopore, which increases the entropic barrier. Point-probe detection of DNA translocation was achieved through the use of a confocal excitation volume focused at the opening of the nanopore. The use of an APD to detect the fluorescence increased the temporal resolution, and its sensitivity increased the SNR observed for each event. In this way, the authors detected SM translocation through nanopores at various driving voltages. This study presented a proof of the concept of the use of nanopores for SMFS and showed that these nanostructures can provide high SNR measurements of SMs with high throughput. These structures will undoubtedly be used in the future in a number of applications, including the detection of DNA hybridization, membrane pore formation, and biomolecular translocation through membrane-embedded pores. An advantage of nanopores that has not yet been demonstrated is the possibility to couple fluorescence observation with ionic current detection methods.

7.5 Potential Biomedical SMFS Applications

Throughout this chapter, the principles of SMFS have been presented, including the characteristics a dye must possess to qualify as a suitable fluorescence reporter that yields high SNR measurements, the typical experimental setups and techniques used in the collection of SM data, and a representative sample of the experimental observables that can be measured through SMFS. Some details and findings from hallmark studies have been discussed to demonstrate the way the SM techniques can provide insights into biomolecular mechanisms. Yet, the field of SMFS has not yet grown to its full potential and many applications are yet to be implemented, invented, or discovered. Most "-omic" fields are already implementing some of the techniques that form the foundation of SM measurements, although without the ultimate sensitivity or resolution allowed by SMFS [220].

The area that has received most attention for a direct SMFS application is genomic sequencing with high fidelity at high sequencing rates and low cost [89, 209, 210, 221–227]. The achievement of real-time sequencing of reads thousands of bp long has signaled the potential for sequencing complete genomes in a matter of days, with multiple-read coverage. The implementation of SMFS in combination with nanostructured surfaces has further enabled sequencing at higher concentrations. Though the sequencing of whole genomes has not been reported with such speed, it seems a matter of time before the first reports appear. The task of researchers is now to design experiments with a different frame of mind, where DNA sequencing is rapid and available on demand and current hybridization microarrays are replaced with SM detection of genes with low copy numbers in single cells. The availability of rapid sequencing opens the door for experimental testing of whole transcriptome sequencing for cells under unique environmental conditions or in distinct developmental stages. Additionally, the possibility of tracking single DNA molecules in cellular environments will allow following the fate of specific sequences targeted to disrupt transcription or silence mRNAs. Other properties of DNA and RNA molecules, such as their three-dimensional structure, will be undoubtedly studied through FRET, colocalization, and even translocation through solid-state nanopores. The use of these SMFS techniques in research will also be translated into the development of novel tools to expedite diagnostics and medical treatment. In particular, having rapid sequencing available can translate into new approaches to the medical treatment of hereditary and environment-induced diseases, although this could also have a strong repercussion on healthcare, insurance, and medical ethics.

Proteomic research can also benefit from SMFS techniques, in the elucidation of function, structure, binding, and interactions. As has been described earlier, the design of appropriately labeled proteins containing single FRET pairs can help investigate molecular conformations. This can further help elucidate the conformational changes incurred by a molecule during the binding of target molecules or catalysis. In addition, SMFS can aid in the study of protein aggregation kinetics, especially of proteins that accumulate within a specific tissue. The elucidation of the path of aggregation of proteins or protein derivatives under controlled conditions can help in the designing of methodologies to prevent diseases caused by protein agglutination in specific cell types or organ tissue. Inhibitors designed to reduce aggregation can also be tested on a small scale, with the best candidates screened in a parallel fashion through the use of focal volumes confined by nano- or microstructured devices. A challenge that will be presented to the application of SMFS to catalysis or protein activity is the development of ultrasensitive assays that allow the instantaneous quantification of products released by a single enzyme. In this regard, dual fluorescence will be needed with high accuracy of localization and temporal resolution to image-diffusing species produced by a localized SM. The possibility of using SM assays to determine activity can help elucidate the effect of inhibitory drugs or drug analogs and assess the induction of static and dynamic heterogeneity across biomolecular activity. In addition, protein microarrays could be further miniaturized to yield SM-binding detection in a highly parallelized fashion, with the capability of calculating SM association and dissociation rates.

SMFS studies on macromolecular structures have been discussed in the form of molecular tracking of motors along fibrils or tubules, the diffusion and binding of labeled molecules on model and cellular lipid membranes, or biomolecular assemblies. While this type of study will continue to help in the understanding of the basic structure and function of macromolecular structures and the molecules that interact with them, SMFS will eventually be taken to the extent that the disruption of the macromolecular structure will be studied with a high spatial accuracy. An example of one such application that can help in the production of bioenergy is the disruption of the cell wall by different types of proteins produced by bacterial and fungal pathogens. Similar studies could be conducted to understand the effect of proteins, lipids, and nucleic acids used in the evasion of immune defenses by pathogens. SMFS could also be used to detect binding events on low copy number receptors on cellular membranes, identifying early stages of cellular differentiation. In particular, the application of SMFS to the study of immune cells and the change in the composition of surface receptors as a function of environmental challenge or stimulus could yield further understanding on the pathways the acquired immune system takes to cope with diseases. In the same way, the interaction between drugs targeted to cell surface receptors and membrane channels or pores could be studied with high spatiotemporal accuracy.

The possibilities for the application of SMFS to biology and biomedical engineering are vast and only limited by the number of interesting experiments that can be imagined by the biomedical scientist. The detection devices used in SMs applications will undoubtedly improve over time, allowing increasingly sensitive detection even in challenging environments, such as the cytosol, and the elucidation of intrinsic cellular mechanisms. There will eventually be a confluence of technologies that allow us to bridge the gap between bulk measurements and SM events, yielding at the same time measurements of individual molecule characteristics and of population behavior.

References

1. Hirschfeld, T. (1976). Optical microscopic observation of single small molecules, *Appl. Opt.*, **15**(12), 2965–2966.
2. Moerner, W. E., and Kador, L. (1989). Optical-detection and spectroscopy of single molecules in a solid, *Phys. Rev. Lett.*, **62**(21), 2535–2538.
3. Orrit, M., and Bernard, J. (1990). Single pentacene molecules detected by fluorescence excitation in a para-terphenyl crystal, *Phys. Rev. Lett.*, **65**(21), 2716–2719.
4. Betzig, E., and Chichester, R. J. (1993). Single molecules observed by near-field scanning optical microscopy, *Science*, **262**(5138), 1422–1425.
5. Eigen, M., and Rigler, R. (1994). Sorting single molecules – application to diagnostics and evolutionary biotechnology, *Proc. Natl. Acad. Sci. USA*, **91**(13), 5740–5747.
6. Macklin, J. J., Trautman, J. K., Harris, T. D., and Brus, L. E. (1996). Imaging and time-resolved spectroscopy of single molecules at an interface, *Science*, **272**(5259), 255–258.

7. Trautman, J. K., and Macklin, J. J. (1996). Time-resolved spectroscopy of single molecules using near-field and far-field optics, *Chem. Phys.*, **205**(1–2), 221–229.

8. Funatsu, T., Harada, Y., Tokunaga, M., Saito, K., and Yanagida, T. (1995). Imaging of single fluorescent molecules and individual ATP turnovers by single myosin molecules in aqueous-solution, *Nature*, **374**(6522), 555–559.

9. Noji, H., Yasuda, R., Yoshida, M., and Kinosita, K., Jr. (1997). Direct observation of the rotation of F-1-ATPase, *Nature*, **386**(6622), 299–302.

10. Vale, R. D., Funatsu, T., Pierce, D. W., Romberg, L., Harada, Y., and Yanagida, T. (1996). Direct observation of single kinesin molecules moving along microtubules, *Nature*, **380**(6573), 451–453.

11. Schmidt, T., Schutz, G. J., Baumgartner, W., Gruber, H. J., and Schindler, H. (1995). Characterization of photophysics and mobility of single molecules in a fluid lipid-membrane, *J. Phys. Chem.*, **99**(49), 17662–17668.

12. Moerner, W. E., and Orrit, M. (1999). Illuminating single molecules in condensed matter, *Science*, **283**(5408), 1670–1676.

13. Lu, H. P., Xun, L. Y., and Xie, X. S. (1998). Single-molecule enzymatic dynamics, *Science*, **282**(5395), 1877–1882.

14. Lu, H. P., and Xie, X. S. (1997). Single-molecule spectral fluctuations at room temperature, *Nature*, **385**(6612), 143–146.

15. Xie, X. S., and Lu, H. P. (1999). Single-molecule enzymology, *J. Biol. Chem.*, **274**(23), 15967–15970.

16. Deniz, A. A., Mukhopadhyay, S., and Lemke, E. A. (2008). Single-molecule biophysics: at the interface of biology, physics and chemistry, *J. R. Soc. Interface*, **5**, 15–45.

17. Lakowicz, J. R. (1999). *Principles of Fluorescence Spectroscopy*, Kluwer Academic/Plenum, New York.

18. Nie, S., and Zare, R. N. (1997). Optical detection of single molecules, *Annu. Rev. Biophys. Biomol. Struct.*, **26**(1), 567–596.

19. Kapanidis, A. N., and Weiss, S. (2002). Fluorescent probes and bioconjugation chemistries for single-molecule fluorescence analysis of biomolecules, *J. Chem. Phys.*, **117**(24), 10953–10964.

20. Widengren, J., and Schwille, P. (2000). Characterization of photoinduced isomerization and back-isomerization of the cyanine dye Cy5 by fluorescence correlation spectroscopy, *J. Phys. Chem. A*, **104**(27), 6416–6428.

21. Cordes, T., Vogelsang, J., and Tinnefeld, P. (2009). On the mechanism of Trolox as antiblinking and antibleaching reagent, *J. Am. Chem. Soc.*, **131**(14), 5018–5019.

22. Vogelsang, J., Kasper, R., Steinhauer, C., Person, B., Heilemann, M., Sauer, M., and Tinnefeld, P. (2008). A reducing and oxidizing system minimizes photobleaching and blinking of fluorescent dyes, *Angew. Chem., Int. Ed.*, **47**(29), 5465–5469.

23. Rasnik, I., McKinney, S. A., and Ha, T. (2006). Nonblinking and longlasting single-molecule fluorescence imaging, *Nat. Methods*, **3**(11), 891–893.

24. Stryer, L., and Haugland, R. P. (1967). Energy transfer: a spectroscopic ruler, *Proc. Natl. Acad. Sci. USA*, **58**(2), 719–726.

25. Jares-Erijman, E. A., and Jovin, T. M. (2003). FRET imaging, *Nat. Biotechnol.*, **21**(11), 1387–395.

26. Clegg, R. M. (1992). Fluorescence resonance energy-transfer and nucleic-acids, *Methods Enzymol.*, **211**, 353–388.

27. Sekar, R. B., and Periasamy, A. (2003). Fluorescence resonance energy transfer (FRET) microscopy imaging of live cell protein localizations, *J. Cell Biol.*, **160**(5), 629–633.

28. Selvin, P. R. (2000). The renaissance of fluorescence resonance energy transfer, *Nat. Struct. Biol.*, **7**(9), 730–734.

29. Ha, T. (2001). Single-molecule fluorescence resonance energy transfer, *Methods*, **25**(1), 78–86.

30. Moerner, W. E., and Fromm, D. P. (2003). Methods of single-molecule fluorescence spectroscopy and microscopy, *Rev. Sci. Instrum.*, **74**(8), 3597–3619.

31. Weiss, S. (1999). Fluorescence spectroscopy of single biomolecules, *Science*, **283**(5408), 1676–1683.

32. Silbey, R. J., and Alberty, R. A. (2001). *Physical Chemistry*, John Wiley, New York.

33. Magde, D., Rojas, G. E., and Seybold, P. G. (1999). Solvent dependence of the fluorescence lifetimes of xanthene dyes, *Photochem. Photob l.*, **70**(5), 737–744.

34. Wilkinson, F., McGarvey, D. J., and Olea, A. F. (1994). Excited triplet-state interactions with molecular-oxygen – influence of charge-transfer on the bimolecular quenching rate constants and the yields of singlet oxygen [O*2(1.DELTA.g)] for substituted naphthalenes in various solvents, *J. Phys. Chem.*, **98**(14), 3762–3769.

35. Deschenes, L. A., and Bout, D. A. V. (2001). Molecular motions in polymer films near the glass transition: a single molecule study of rotational dynamics, *J. Phys. Chem. B*, **105**(48), 11978–11985.

36. Moran-Mirabal, J. N., Santhanam, N., Corgie, S. C., Craighead, H. G., and Walker, L. P. (2008). Immobilization of cellulose fibrils on solid substrates for cellulase-binding studies through quantitative fluorescence microscopy, *Biotechnol. Bioeng.*, **101**(6), 1129–1141.

37. Harada, Y., Sakurada, K., Aoki, T., Thomas, D. D., and Yanagida, T. (1990). Mechanochemical coupling in actomyosin energy transduction studied by invitro movement assay, *J. Mol. Biol.*, **216**(1), 49–68.

38. Yasuda, R., Noji, H., Kinosita, K., Jr., and Yoshida, M. (1998). F1-ATPase is a highly efficient molecular motor that rotates with discrete 120° steps, *Cell*, **93**(7), 1117–1124.

39. Iwana, A. H., Funatsu, T., Harada, Y., Tokunaga, M., Ohara, O., Morimoto, S., and Yanagida, T. (1997). Single molecular assay of individual ATP turnover by a myosin-GFP fusion protein expressed in vitro, *FEBS lett.*, **407**(2), 235–238.

40. Basche, T., Ambrose, W. P., and Moerner, W. E. (1992). Optical-spectra and kinetics of single impurity molecules in a polymer – spectral diffusion and persistent spectral hole burning, *J. Opt. Soc. Am. B*, **9**(5), 829–836.

41. Jaqaman, K., Loerke, D., Mettlen, M., Kuwata, H., Grinstein, S., Schmid, S. L., and Danuser, G. (2008). Robust single-particle tracking in live-cell time-lapse sequences, *Nat. Methods*, **5**(8), 695–702.

42. Yildiz, A., Forkey, J. N., McKinney, S. A., Ha, T., Goldman, Y. E., and Selvin, P. R. (2003). Myosin v walks hand-over-hand: single fluorophore imaging with 1.5-nm localization, *Science*, **300**(5628), 2061–2065.

43. Sase, I., Miyata, H., Corrie, J. E., Craik, J. S., and Kinosita, K., Jr. (1995). Real-time imaging of single fluorophores on moving actin with an epifluorescence microscope, *Biophys. J.*, **69**(2), 323–328.

44. Schutz, G. J., Kada, G., Pastushenko, V. P., and Schindler, H. (2000). Properties of lipid microdomains in a muscle cell membrane visualized by single molecule microscopy, *EMBO J.*, **19**(5), 892–901.
45. Schmidt, T., Schutz, G. J., Baumgartner, W., Gruber, H. J., and Schindler, H. (1996). Imaging of single molecule diffusion, *Proc. Natl. Acad. Sci. USA*, **93**(7), 2926–2929.
46. Douglass, A. D., and Vale, R. D. (2005). Single-molecule microscopy reveals plasma membrane microdomains created by protein-protein networks that exclude or trap signaling molecules in T cells, *Cell*, **121**(6), 937–950.
47. Lizardi, P. M., Huang, X., Zhu, Z., Bray-Ward, P., Thomas, D. C., and Ward, D. C. (1998). Mutation detection and single-molecule counting using isothermal rolling-circle amplification, *Nat. Genet.*, **19**(3), 225–232.
48. Bohmer, M., and Enderlein, J. (2003). Orientation imaging of single molecules by wide-field epifluorescence microscopy, *J. Opt. Soc. Am. B*, **20**(3), 554–559.
49. Neuweiler, H., Schulz, A., Böhmer, M., Enderlein, J., and Sauer, M. (2003). Measurement of submicrosecond intramolecular contact formation in peptides at the single-molecule level, *J. Am. Chem. Soc.*, **125**(18), 5324–5330.
50. Moerner, W. E., Peterman, E. J., Brasselet, S., Kummer, S., and Dickson, R. M. (1999). Optical methods for exploring dynamics of single copies of green fluorescent protein, *Cytometry*, **36**(3), 232–238.
51. Paige, M. F., Bjerneld, E. J., and Moerner, W. E. (2001). A comparison of through-the-objective total internal reflection microscopy and epifluorescence microscopy for single-molecule fluorescence imaging, *Single Mol.*, **2**(3), 191–201.
52. Yildiz, A., Park, H., Safer, D., Yang, Z., Chen, L. Q., Selvin, P. R., and Sweeney, H. L. (2004). Myosin VI steps via a hand-over-hand mechanism with its lever arm undergoing fluctuations when attached to actin, *J. Biol. Chem.*, **279**(36), 37223–37226.
53. Yildiz, A., Tomishige, M., Vale, R. D., and Selvin, P. R. (2004). Kinesin walks hand-over-hand, *Science*, **303**(5658), 676–678.
54. Sako, Y., Minoghchi, S., and Yanagida, T. (2000). Single-molecule imaging of EGFR signalling on the surface of living cells, *Nat. Cell Biol.*, **2**(3), 168–172.
55. Paige, M. F., Bjerneld, E. J., and Moerner, W. E. (2001). A comparison of through-the-objective total internal reflection microscopy and epifluorescence microscopy for single-molecule fluorescence imaging, *Single Mol.*, **2**(3), 191–201.
56. Axelrod, D. (2001). Selective imaging of surface fluorescence with very high aperture microscope objectives, *J. Biomed. Opt.*, **6**(1), 6–13.
57. Ambrose, W. P., Goodwin, P. M., and Nolan, J. P. (1999). Single-molecule detection with total internal reflection excitation: comparing signal-to-background and total signals in different geometries, *Cytometry*, **36**(3), 224–231.
58. Axelrod, D., Burghardt, T. P., and Thompson, N. L. (1984). Total internal-reflection fluorescence, *Annu. Rev. Biophys. Bioeng.*, **13**, 247–268.
59. Axelrod, D. (1981). Cell-substrate contacts illuminated by total internal-reflection fluorescence, *J. Cell Biol.*, **89**(1), 141–145.
60. Axelrod, D. (2001). Total internal reflection fluorescence microscopy in cell biology, *Traffic*, **2**(11), 764–774.

61. Thompson, N. L., Burghardt, T. P., and Axelrod, D. (1981). Measuring surface dynamics of biomolecules by total internal-reflection fluorescence with photobleaching recovery or correlation spectroscopy, *Biophys. J.*, **33**(3), 435–454.

62. Axelrod, D. (1989). Total internal-reflection fluorescence microscopy, *Methods Cell Biol.*, **30**, 245–270.

63. Kino, G. S., and Corle, T. R. (1989). Confocal scanning optical microscopy, *Phys. Today*, **42**(9), 55–62.

64. Xiao, G. Q., Corle, T. R., and Kino, G. S. (1988). Real-time confocal scanning optical microscope, *Appl. Phys. Lett.*, **53**(8), 716–718.

65. Corle, T. R., Chou, C. H., and Kino, G. S. (1986). Depth response of confocal optical microscopes, *Opt. Lett.*, **11**(12), 770–772.

66. Rigler, R., Mets, Ü., Widengren, J., and Kask, P. (1993). Fluorescence correlation spectroscopy with high count rate and low background: analysis of translational diffusion, *Eur. Biophys. J.*, **22**(3), 169–175.

67. Nie, S., Chiu, D. T., and Zare, R. N. (1994). Probing individual molecules with confocal fluorescence microscopy, *Science*, **266**(5187), 1018–1021.

68. Dickson, R. M., Norris, D. J., Tzeng, Y. L., and Moerner, W. E. (1996). Three-dimensional imaging of single molecules solvated in pores of poly(acrylamide) gels, *Science*, **274**(5289), 966–968.

69. Denk, W., Strickler, J. H., and Webb, W. W. (1990). Two-photon laser scanning fluorescence microscopy, *Science*, **248**(4951), 73–76.

70. Tinnefeld, P., and Sauer, M. (2005). Branching out of single-molecule fluorescence spectroscopy: challenges for chemistry and influence on biology, *Angew. Chem., Int. Ed.*, **44**, 2642–2671.

71. Deniz, A. A., Laurence, T. A., Dahan, M., Chemla, D. S., Schultz, P. G., and Weiss, S. (2001). Ratiometric single-molecule studies of freely diffusing biomolecules, *Annu. Rev. Phys. Chem.*, **52**(1), 233–253.

72. Mandel, L. (1958). Fluctuations of photon beams and their correlations, *Proc. Phys. Soc. London*, **72**(468), 1037–1048.

73. Chen, Y., Müller, J. D., So, P. T. C., and Gratton, E. (1999). The photon counting histogram in fluorescence fluctuation spectroscopy, *Biophys. J.*, **77**(1), 553–567.

74. Chandrasekhar, S. (1943). Stochastic problems in physics and astronomy, *Rev. Mod. Phys.*, **15**(1), 1.

75. Puchalla, J., Krantz, K., Austin, R., and Rye, H. (2008). Burst analysis spectroscopy: a versatile single-particle approach for studying distributions of protein aggregates and fluorescent assemblies, *Proc. Natl. Acad. Sci. USA*, **105**(38), 14400–14405.

76. Foquet, M., Korlach, J., Zipfel, W., Webb, W. W., and Craighead, H. G. (2002). DNA fragment sizing by single molecule detection in submicrometer-sized closed fluidic channels, *Anal. Chem.*, **74**(6), 1415–1422.

77. Stavis, S. M., Corgié, S. C., Cipriany, B. R., Craighead, H. G., and Walker, L. P. (2007). Single molecule analysis of bacterial polymerase chain reaction products in submicrometer fluidic channels, *Biomicrofluidics*, **1**(3), 034105–13.

78. Stavis, S. M., Edel, J. B., Li., Y., Samiee, K. T., Luo, D., and Craighead, H. G. (2005). Detection and identification of nucleic acid engineered fluorescent labels in submicrometre fluidic channels, *Nanotechnology*, **16**(7), S314–S323.

79. Stavis, S. M., Edel, J. B., Samiee, K. T., and Craighead, H. G. (2005). Single molecule studies of quantum dot conjugates in a submicrometer fluidic channel, *Lab Chip*, **5**(3), 337–343.

80. Magde, D., Webb, W. W., and Elson, E. L. (1978). Fluorescence correlation spectroscopy. III. Uniform translation and laminar-flow, *Biopolymers*, **17**(2), 361–376.

81. Elson, E. L., and Magde, D. (1974). Fluorescence correlation spectroscopy. I. Conceptual basis and theory, *Biopolymers*, **13**(1), 1–27.

82. Magde, D., Elson, E. L., and Webb, W. W. (1974). Fluorescence correlation spectroscopy. II. Experimental realization, *Biopolymers*, **13**(1), 29–61.

83. Magde, D., Webb, W. W., and Elson, E. (1972). Thermodynamic fluctuations in a reacting system – measurement by fluorescence correlation spectroscopy, *Phys. Rev. Lett.*, **29**(11), 705–708.

84. Edman, L., Mets, U., and Rigler, R. (1996). Conformational transitions monitored for single molecules in solution, *Proc. Natl. Acad. Sci. USA*, **93**(13), 6710–6715.

85. Haustein, E., and Schwille, P. (2004). Single-molecule spectroscopic methods, *Curr. Opin. Struct. Biol.*, **14**(5), 531–540.

86. Qian, H., and Elson, E. L. (1991). Analysis of confocal laser-microscope optics for 3-d fluorescence correlation spectroscopy, *Appl. Opt.*, **30**(10), 1185–1195.

87. Berland, K. M., So, P. T. C., and Gratton, E. (1995). 2-photon fluorescence correlation spectroscopy – method and application to the intracellular environment, *Biophys. J.*, **68**(2), 694–701.

88. Foquet, M., Korlach, J., Zipfel, W. R., Webb, W. W., and Craighead, H. G. (2004). Focal volume confinement by submicrometer-sized fluidic channels, *Anal. Chem.*, **76**(6), 1618–1626.

89. Levene, M. J., Korlach, J., Turner, S. W., Foquet, M., Craighead, H. G., and Webb, W. W. (2003). Zero-mode waveguides for single-molecule analysis at high concentrations, *Science*, **299**(5607), 682–686.

90. Wawrezinieck, L., Rigneault, H., Marguet, D., and Lenne, P. F. (2005). Fluorescence correlation spectroscopy diffusion laws to probe the submicron cell membrane organization, *Biophys. J.*, **89**(6), 4029–4042.

91. Schwille, P., Korlach, J., and Webb, W. W. (1999). Fluorescence correlation spectroscopy with single-molecule sensitivity on cell and model membranes, *Cytometry*, **36**(3), 176–182.

92. Wachsmuth, M., Waldeck, W., and Langowski, J. (2000). Anomalous diffusion of fluorescent probes inside living cell nuclei investigated by spatially-resolved fluorescence correlation spectroscopy, *J. Mol. Biol.*, **298**(4), 677–689.

93. Banks, D. S., and Fradin, C. (2005). Anomalous diffusion of proteins due to molecular crowding, *Biophys. J.*, **89**, 2960–2971.

94. Weiss, M., Hashimoto, H., and Nilsson, T. (2003). Anomalous protein diffusion in living cells as seen by fluorescence correlation spectroscopy, *Biophys. J.*, **84**(6), 4043–4052.

95. Thompson, N. L., and Axelrod, D. (1983). Immunoglobulin surface-binding kinetics studied by total internal-reflection with fluorescence correlation spectroscopy, *Biophys. J.*, **43**(1), 103–114.

96. Widengren, J., Mets, U., and Rigler, R. (1995). Fluorescence correlation spectroscopy of triplet-states in solution – a theoretical and experimental-study, *J. Phys. Chem.*, **99**(36), 13368–13379.

97. Bacia, K., Kim, S. A., and Schwille, P. Fluorescence cross-correlation spectroscopy in living cells, *Nat. Meth.*, **3**(2), 83–89.

98. Schwille, P., Meyer-Almes, F. J., and Rigler, R. (1997). Dual-color fluorescence cross-correlation spectroscopy for multicomponent diffusional analysis in solution, *Biophys. J.*, **72**(4), 1878–1886.

99. Rigler, R., Földes-Papp, Z., Meyer-Almes, F. J., Sammet, C., Völcker, M., and Schnetz, A. (1998). Fluorescence cross-correlation: a new concept for polymerase chain reaction, *J. Biotechnol.*, **63**(2), 97–109.

100. Webb, W. W. (2001). Fluorescence correlation spectroscopy: inception, biophysical experimentations, and prospectus, *Appl. Opt.*, **40**(24), 3969–3983.

101. Bacia, K., and Schwille, P. (2003). A dynamic view of cellular processes by in vivo fluorescence auto- and cross-correlation spectroscopy, *Methods*, **29**(1), 74–85.

102. Haustein, E., and Schwille, P. (2003). Ultrasensitive investigations of biological systems by fluorescence correlation spectroscopy, *Methods*, **29**(2), 153–166.

103. Kettling, U., Koltermann, A., Schwille, P., and Eigen, M. (1998). Real-time enzyme kinetics monitored by dual-color fluorescence cross-correlation spectroscopy, *Proc. Natl. Acad. Sci. USA*, **95**(4), 1416–1420.

104. Dittrich, P. S., Schafer, S. P., and Schwille, P. (2005). Characterization of the photoconversion on reaction of the fluorescent protein kaede on the single-molecule level, *Biophys. J.*, **89**(5), 3446–3455.

105. Bacia, K., Majoul, I. V., and Schwille, P. (2002). Probing the endocytic pathway in live cells using dual-color fluorescence cross-correlation analysis, *Biophys. J.*, **83**(2), 1184–1193.

106. Chen, J. J., and Irudayaraj, J. (2010). Fluorescence lifetime cross correlation spectroscopy resolves EGFR and antagonist interaction in live cells, *Anal. Chem.*, **82**(15), 6415–6421.

107. Betzig, E., and Trautman, J. K. (1992). Near-field optics – microscopy, spectroscopy, and surface modification beyond the diffraction limit, *Science*, **257**(5067), 189–195.

108. Betzig, E., Trautman, J. K., Harris, T. D., Weiner, J. S., and Kostelak, R. L. (1991). Breaking the diffraction barrier – optical microscopy on a nanometric scale, *Science*, **251**(5000), 1468–1470.

109. Trautman, J. K., Macklin, J. J., Brus, L. E., and Betzig, E. (1994). Near-field spectroscopy of single molecules at room-temperature, *Nature*, **369**(6475), 40–42.

110. Betzig, E., Finn, P. L., and Weiner, J. S. (1992). Combined shear force and near-field scanning optical microscopy, *Appl. Phys. Lett.*, **60**(20), 2484–2486.

111. Harris, T. D., Grober, R. D., Trautman, J. K., and Betzig, E. (1994). Superresolution imaging spectroscopy, *Appl. Spectrosc.*, **48**(1), A14–A21.

112. Ambrose, W. P., Goodwin, P. M., Keller, R. A., and Martin, J. C. (1994). Alterations of single molecule fluorescence lifetimes in near-field optical microscopy, *Science*, **265**(5170), 364–367.

113. Xie, X. S., and Dunn, R. C. (1994). Probing single molecule dynamics, *Science*, **265**(5170), 361–364.

114. Enderle, T., Ha, T., Chemla, D. S., and Weiss, S. (1998). Near-field fluorescence microscopy of cells, *Ultramicroscopy*, **7s1**(1–4), 303–309.

115. de Lange, F., Cambi, A., Huijbens, R., de Bakker, B., Rensen, W., Garcia-Parajo, M., van Hulst, N., and Figdor, C. G. (2001). Cell biology beyond the diffraction limit: near-field scanning optical microscopy, *J. Cell Sci.*, **114**(23), 4153–4160.

116. Koopman, M., de Bakker, B. I., Garcia-Parajo, M. F., and van Hulst, N. F. (2003). Shear force imaging of soft samples in liquid using a diving bell concept, *Appl. Phys. Lett.*, **83**(24), 5083–5085.

117. Barbara, P. F., Adams, D. M., and O'Connor, D. B. (1999). Characterization of organic thin film materials with near-field scanning optical microscopy (NSOM), *Annu. Rev. Mater. Sci.*, **29**(1), 433–469.

118. Hamers, R. J. (1996). Scanned probe microscopies in chemistry, *J. Phys. Chem.*, **100**(31), 13103–13120.

119. Bodnar, A., Nizsalóczki, E., Mocsár, G., Szalóki, N., Waldmann, T. A., Damjanovich, S., and Vámosi, G. (2008). A biophysical approach to IL-2 and IL-15 receptor function: localization, conformation and interactions. *Immunol. Lett.*, **116**(2), 117–125.

120. Syed, S., Snyder, G. E., Franzini-Armstrong, C., Selvin, P. R., and Goldman, Y. E. (2006). Adaptability of myosin V studied by simultaneous detection of position and orientation, *EMBO J.*, **25**(9), 1795–1803.

121. Toprak, E., Enderlein, J., Syed, S., McKinney, S. A., Petschek, R. G., Ha, T., Goldman, Y. E., and Selvin, P. R. (2006). Defocused orientation and position imaging (DOPI) of myosin V, **103**(17), 6495–6499.

122. Bartko, A. P., and Dickson, R. M. (1999). Imaging three-dimensional single molecule orientations, *J. Phys. Chem. B*, **103**(51), 11237–11241.

123. Schaffer, J., Volkmer, A., Eggeling, C., Subramaniam, V., Striker, G., and Seidel, C. A. M. (1999). Identification of single molecules in aqueous solution by time-resolved fluorescence anisotropy, *J. Phys. Chem. A*, **103**(3), 331–336.

124. Johnson, C. K., Osborn, K. D., Allen, M. W., and Slaughter, B. D. (2005). Single-molecule fluorescence spectroscopy: new probes of protein function and dynamics, *Physiology*, **20**, 10–14.

125. Osborn, K. D., Bartlett, R. K., Mandal, A., Zaidi, A., Urbauer, R. J., Urbauer, J. L., Galeva, N., Williams, T. D., and Johnson, C. K. (2004). Single-molecule dynamics reveal an altered conformation for the autoinhibitory domain of plasma membrane Ca^{2+}-ATPase bound to oxidatively modified calmodulin, *Biochemistry*, **43**(40), 12937–12944.

126. Slaughter, B. D., Allen, M. W., Unruh, J. R., Urbauer, R. J. B., and Johnson, C. K. (2004). Single-molecule resonance energy transfer and fluorescence correlation spectroscopy of calmodulin in solution, *J. Phys. Chem. B*, **108**(29), 10388–10397.

127. Allen, M. W., Urbauer, R. J. B., and Johnson, C. K. (2004). Single-molecule assays of calmodulin target binding detected with a calmodulin energy-transfer construct, *Anal. Chem.*, **76**(13), 3630–3637.

128. Allen, M. W., Urbauer, R. J. B., Zaidi, A., Williams, T. D., Urbauer, J. L., and Johnson, C. K. (2004). Fluorescence labeling, purification, and immobilization of a double cysteine mutant calmodulin fusion protein for single-molecule experiments, *Anal. Biochem.*, **325**(2), 273–284.

129. Hink, M. A., Bisseling, T., and Visser, A. (2002). Imaging protein-protein interactions in living cells, *Plant Mol. Biol.*, **50**(6), 871–883.

130. Hoppe, A., Christensen, K., and Swanson, J. A. (2002). Fluorescence resonance energy transfer-based stoichiometry in living cells, *Biophys. J.*, **83**(6), 3652–3664.

131. Zhuang, X., Bartley, L. E., Babcock, H. P., Russell, R., Ha, T., Herschlag, D., and Chu, S. (2000). A single-molecule study of RNA catalysis and folding, *Science*, **288**(5473), 2048–2051.

132. Ha, T., Enderle, T., Ogletree, D. F., Chemla, D. S., Selvin, P. R., and Weiss, S. (1996). Probing the interaction between two single molecules: fluorescence resonance energy transfer between a single donor and a single acceptor, *Proc. Natl. Acad. Sci. USA*, **93**(13), 6264–6268.

133. Deniz, A. A., Grunwell, J. R., Ha, T., Faulhaber, A. E., Chemla, D. S., Weiss, S., and Schultz, P. G. (1999). Single-pair fluorescence resonance energy transfer on freely diffusing molecules: observation of Förster distance dependence and subpopulations, *Proc. Natl. Acad. Sci. USA*, **6**(7), 3670–3675.

134. Ferreon, A. C. M., Gambin, Y., Lemke, E. A., and Deniz, A. A. (2009). Interplay of alpha-synuclein binding and conformational switching probed by single-molecule fluorescence, *Proc. Natl. Acad. Sci. USA*, **106**(14), 5645–5650.

135. Kapanidis, A. N., Laurence, T. A., Lee, N. K., Margeat, E., Kong, X., and Weiss, S. (2005). Alternating-laser excitation of single molecules, *Acc. Chem. Res.*, **38**(7), 523–533.

136. Kapanidis, A. N., Lee, N. K., Laurence, T. A., Doose, S., Margeat, E., and Weiss, S. (2004). Fluorescence-aided molecule sorting: analysis of structure and interactions by alternating-laser excitation of single molecules, *Proc. Natl. Acad. Sci. USA*, **101**(24), 8936–8941.

137. Lee, N. K., Kapanidis, A. N., Koh, H. R., Korlann, Y., Ho, S. O., Kim, Y., Gassman, N., Kim, S. K., and Weiss, (2007). Three-color alternating-laser excitation of single molecules: monitoring multiple interactions and distances, *Biophys. J.*, **92**(1), 303–312.

138. Saxton, M. J., and Jacobson, K. (1997). Single-particle tracking:applications to membrane dynamics, *Annu. Rev. Biophys. Biomol. Struct.*, **26**(1), 373–399.

139. Kusumi, A., Sako, Y., and Yamamoto, M. (1993). Confined lateral diffusion of membrane receptors as studied by single particle tracking (nanovid microscopy). Effects of calcium-induced differentiation in cultured epithelial cells. *Biophys. J.*, **65**(5), 2021–2040.

140. Pralle, A., Keller, P., Florin, E.-L., Simons, K., and Hörber, J. K. H. (2000). Sphingolipid-cholesterol rafts diffuse as small entities in the plasma membrane of mammalian cells, *J. Cell Biol.*, **148**(5), 997–1007.

141. Marshall, W. F., Straight, A., Marko, J. F., Swedlow, J., Dernburg, A., Belmont, A., Murray, A. W., Agard, D. A., and Sedat, J. W. (1997). Interphase chromosomes undergo constrained diffusional motion in living cells, *Curr. Biol.*, **7**(12), 930–939.

142. Hooper, N. M. (1999). Detergent-insoluble glycosphingolipid/cholesterol-rich membrane domains, lipid rafts and caveolae (review), *Mol. Memb . Biol.*, **16**(2), 145–156.

143. Jacobson, K., Sheets, E. D., and Simson, R. (1995). Revisiting the fluid mosaic model of membranes, *Science*, **268**(5216), 1441–1442.

144. Schutz, G. J., Schindler, H., and Schmidt, T. (1997). Single-molecule microscopy on model membranes reveals anomalous diffusion, *Biophys. J.*, **73**(2), 1073–1080.

145. Kusumi, A., Nakada, C., Ritchie, K., Murase, K., Suzuki, K., Murakoshi, H., Kasai, R. S., Kondo, J., and Fujiwara, T. (2005). Paradigm shift of the plasma membrane concept from the two-dimensional continuum fluid to the partitioned fluid: high-speed single-molecule tracking of membrane molecules, *Annu. Rev. Biophys. Biomol. Struct.*, **34**(1), 351.

146. Fu, A. H., Gu, W., Larabell, C., and Alivisatos, A. P. (2005). Semiconductor nanocrystals for biological imaging, *Curr. Opin. Neurobo l.*, **15**(5), 568–575.

147. Brandenburg, B., and Zhuang, X. W. (2007). Virus trafficking – learning from single-virus tracking, *Nat. Rev. Microbo l.*, **5**(3), 197–208.

148. Ritchie, K., and Kusumi, A. (2003). Single-particle tracking image microscopy, *Biophotonics, Pt. A*, **360**, 618–634.

149. Levi, V., and Gratton, E. (2007). Exploring dynamics in living cells by tracking single particles, *Cell Biochem. Biophys.*, **48**(1), 1–15.

150. Thompson, R. E., Larson, D. R., and Webb, W. W. (2002). Precise nanometer localization analysis for individual fluorescent probes, *Biophys. J.*, **82**(5), 2775–2783.

151. Ober, R. J., Ram, S., and Ward, E. S. (2004). Localization accuracy in single-molecule microscopy, *Biophys. J.*, **86**(2), 1185–1200.

152. Park, H., Toprak, E., and Selvin, P. R. (2007). Single-molecule fluorescence to study molecular motors, *Q. Rev. Biophys.*, **40**(1), 87–111.

153. Ghosh, R. N., and Webb, W. W. (1994). Automated detection and tracking of individual and clustered cell-surface low-density-lipoprotein receptor molecules, *Biophys. J.*, **66**(5), 1301–1318.

154. Cheezum, M. K., Walker, W. F., and Guilford, W. H. (2001). Quantitative comparison of algorithms for tracking single fluorescent particles, *Biophys. J.*, **81**(4), 2378–2388.

155. Carter, B. C., Shubeita, G. T., and Gross, S. P. (2005). Tracking single particles: a user-friendly quantitative evaluation, *Phys. Biol.*, **2**(1), 60–72.

156. Sonnleitner, A., Schutz, G. J., and Schmidt, T. (1999). Free Brownian motion of individual lipid molecules in biomembranes, *Biophys. J.*, **77**(5), 2638–2642.

157. Schutz, G. J., Trabesinger, W., and Schmidt, T. (1998). Direct observation of ligand colocalization on individual receptor molecules, *Biophys. J.*, **74**(5), 2223–2226.

158. Schutz, G. J., Sonnleitner, M., Hinterdorfer, P., and Schindler, H. (2000). Single molecule microscopy of biomembranes (review), *Mol. Memb . Biol.*, **17**(1), 17–29.

159. Reck-Peterson, S. L., Yildiz, A., Carter, A. P., Gennerich, A., Zhang, N., and Vale, R. D. (2006). Single-molecule analysis of dynein processivity and stepping behavior, **126**(2), 335–348.

160. Kural, C., Kim, H., Syed, S., Goshima, G., Gelfand, VI., and Selvin, P. R. (2005). Kinesin and dynein move a peroxisome in vivo: a tug-of-war or coordinated movement? *Science*, **308**(5727), 1469–1472.

161. Hess, S. T., Huang, S., Heikal, A. A., and Webb, W. W. (2002). Biological and chemical applications of fluorescence correlation spectroscopy: a review, *Biochemistry*, **41**(3), 697–705.

162. Walter, N. G., Schwille, P., and Eigen, M. (1996). Fluorescence correlation analysis of probe diffusion simplifies quantitative pathogen detection by PCR, *Proc. Natl. Acad. Sci. USA*, **93**(23), 12805–12810.

163. Oehlenschläger, F., Schwille, P., and Eigen, M. (1996). Detection of HIV-1 RNA by nucleic acid sequence-based amplification combined with fluorescence correlation spectroscopy, *Proc. Natl. Acad. Sci. USA*, **93**(23), 12811–12816.

164. Kinjo, M., Nishimura, G., Koyama, T., Mets, and Rigler, R. (1998). Single-molecule analysis of restriction DNA fragments using fluorescence correlation spectroscopy, *Anal. Biochem.*, **260**(2), 166–172.

165. Bark, N., Földes-Papp, Z., and Rigler, R. (1999). The incipient stage in thrombin-induced fibrin polymerization detected by FCS at the single molecule level, *Biochem. Biophys. Res. Commun.*, **260**(1), 35–41.

166. Tjernberg, L. O., Pramanik, A., Björling, S., Thyberg, P., Thyberg, J., Nordstedt, C., Berndt, K. D., Terenius, L., and Rigler, R. (1999). Amyloid beta-peptide polymerization studied using fluorescence correlation spectroscopy, *Chem. Biol.*, **6**(1), 53–62.

167. Scherfeld, D., Kahya, N., and Schwille, P. (2003). Lipid dynamics and domain formation in model membranes composed of ternary mixtures of unsaturated and saturated phosphatidylcholines and cholesterol, *Biophys. J.*, **85**(6), 3758–3768.

168. Korlach, J., Schwille, P., Webb, W. W., and Feigenson, G. W. (1999). Characterization of lipid bilayer phases by confocal microscopy and fluorescence correlation spectroscopy, *Proc. Natl. Acad. Sci. USA*, **96**(15), 8461–8466.

169. Bacia, K., Schwille, P., and Kurzchalia, T. (2005). Sterol structure determines the separation of phases and the curvature of the liquid-ordered phase in model membranes, *Proc. Natl. Acad. Sci. USA*, **102**(9), 3272–3277.

170. Kahya, N., and Schwille, P. (2006). How phospholipid-cholesterol interactions modulate lipid lateral diffusion, as revealed by fluorescence correlation spectroscopy, *J. Fluoresc.*, **16**(5), 671–678.

171. Schwille, P., Haupts, U., Maiti, S., and Webb, W. W. (1999). Molecular dynamics in living cells observed by fluorescence correlation spectroscopy with one- and two-photon excitation, *Biophys. J.*, **77**(4), 2251–2265.

172. Rigler, R., Pramanik, A., Jonasson, P., Kratz, G., Jansson, O. T., Nygren, P., Ståhl, S., Ekberg, K., Johansson, B., Uhlén, S., Uhlén, M., Jörnvall, H., and Wahren, (1999). J. Specific binding of proinsulin C-peptide to human cell membranes, *Proc. Natl. Acad. Sci. USA*, **96**(23), 13318–13323.

173. Pramanik, A., Ekberg, K., Zhong, Z., Shafqat, J., Henriksson, M., Jansson, O., Tibell, A., Tally, M., Wahren, J., Jörnvall, H., Rigler, R., and Johansson, J. (2001). C-peptide binding to human cell membranes: importance of Glu27, *Biochem. Biophys. Res. Comm.*, **284**(1), 94–98.

174. Pramanik, A., and Rigler, R. (2001). Ligand-receptor interactions in the membrane of cultured cells monitored by fluorescence correlation spectroscopy, *Biol. Chem.*, **382**(3), 371–378.

175. Pramanik, A., Olsson, M., Langel, Ü., Bartfai, T., and Rigler, R. (2001). Fluorescence correlation spectroscopy detects galanin receptor diversity on insulinoma cells, *Biochemistry*, **40**(36), 10839–10845.

176. Vukojevic, V., Ming, Y., D'Addario, C., Rigler, R., Johansson, B., and Terenius, L. (2008). Ethanol/naltrexone interactions at the mu-opioid receptor. CLSM/FCS study in live cells, *PLoS ONE*, **3**(12), e4008.

177. Koltermann, A., Kettling, U., Bieschke, J., Winkler, T., and Eigen, M. (1998). Rapid assay processing by integration of dual-color fluorescence cross-correlation spectroscopy: high throughput screening for enzyme activity, *Proc. Natl. Acad. Sci. USA*, **95**(4), 1421–1426.

178. Enderle, T., Ha, T., Ogletree, D. F., Chemla, D. S., Magowan, C., and Weiss, S. (1997). Membrane specific mapping and colocalization of malarial and host skeletal proteins in the *Plasmodium falciparum* infected erythrocyte by dual-color near-field scanning optical microscopy, *Proc. Natl. Acad. Sci. USA*, **94**(2), 520–525.

179. Schütz, G. J., Trabesinger, W., and Schmidt, T. (1998). Direct observation of ligand colocalization on individual receptor molecules, *Biophys. J.*, **74**(5), 2223–2226.

180. Lacoste, T. D., Michalet, X., Pinaud, F., Chemla, D. S., Alivisatos, A. P., and Weiss, S. (2000). Ultrahigh-resolution multicolor colocalization of single fluorescent probes, *Proc. Natl. Acad. Sci. USA*, **97**(17), 9461–9466.

181. Michalet, X., Lacoste, T. D., and Weiss, S. (2001). Ultrahigh-resolution colocalization of spectrally separable point-like fluorescent probes, *Methods*, **25**(1), 87–102.

182. Gordon, M. P., Ha, T., and Selvin, P. R. (2004). Single-molecule high-resolution imaging with photobleaching, *Proc. Natl. Acad. Sci. USA*, **101**(17), 6462–6465.

183. Qu, X., Wu, D., Mets, L., and Scherer, N. F. (2004). Nanometer-localized multiple single-molecule fluorescence microscopy, *Proc. Natl. Acad. Sci. USA*, **101**(s31), 11298–11303.

184. Churchman, L. S., Okten, Z., Rock, R. S., Dawson, J. F., and Spudich, J. A. (2005). Single molecule high-resolution colocalization of Cy3 and Cy5 attached to macromolecules measures intramolecular distances through time, *Proc. Natl. Acad. Sci. USA*, **102**(5), 1419–1423.

185. Koyama-Honda, I., Ritchie, K., Fujiwara, T., Lino, R., Murakoshi, H., Kasai, R. S., and Kusumi, A. (2005). Fluorescence imaging for monitoring the colocalization of two single molecules in living cells, *Biophys. J.*, **88**(3), 2126–2136.

186. Guenard, R. D., King, L. A., Smith, B. W., and Winefordner, J. D. (1997). Two-channel sequential single-molecule measurement, *Anal. Chem.*, **69**(13), 2426–2433.

187. Lee, Y.-H., Maus, R. G., Smith, B. W., and Winefordner, J. D. (1994). Laser-induced fluorescence detection of a single molecule in a capillary, *Anal. Chem.*, **66**(23), 4142–4149.

188. Mathis, H. P., Kalusche, G., Wagner, B., and McCaskill, J. S. (1997). Steps towards spatially resolved single molecule detection in solution, *Bioimaging*, **5**(3), 116–128.

189. Lyon, W. A., and Nie, S. (1997). Confinement and detection of single molecules in submicrometer channels, *Anal. Chem.*, **69**(16), 3400–3405.

190. Haab, B. B., and Mathies, R. A. (1999). Single-molecule detection of DNA separations in microfabricated capillary electrophoresis chips employing focused molecular streams, *Anal. Chem.*, **71**(22), 5137–5145.

191. Han, J., and Craighead, H. G. (2000). Separation of long DNA molecules in a microfabricated entropic trap array, *Science*, **288**(5468), 1026–1029.

192. Cabodi, M., Turner, S. W. P., and Craighead, H. G. (2002). Entropic recoil separation of long DNA molecules, *Anal. Chem.*, **74**(20), 5169–5174.

193. Reccius, C. H., Stavis, S. M., Mannion, J. T., Walker, L. P., and Craighead, H. G. (2008). Conformation, length, and speed measurements of electrodynamically stretched DNA in nanochannels, *Biophys. J.*, **95**(1), 273–286.

194. Levy, S. L., Mannion, J .T., Cheng, J., Reccius, C. H., and Craighead, H. G. (2008). Entropic unfolding of DNA molecules in nanofluidic channels, *Nano Lett.*, **8**(11), 3839–3844.

195. Mannion, J. T., Reccius, C. H., Cross, J. D., and Craighead, H. G. (2006). Conformational analysis of single DNA molecules undergoing entropically induced motion in nanochannels, *Biophys. J.*, **90**(12), 4538–4545.

196. Reccius, C. H., Mannion, J. T., Cross, J. D., and Craighead, H. G. (2005). Compression and free expansion of single DNA molecules in nanochannels, *Phys. Rev. Lett.*, **95**(26).

197. Volkmuth, W. D., and Austin, R. H. (1992). DNA electrophoresis in microlithographic arrays, *Nature*, **358**(6387), 600–602.

198. Duke, T. A. J., and Austin, R. H. (1998). Microfabricated sieve for the continuous sorting of macromolecules, *Phys. Rev. Lett.*, **80**(7), 1552.

199. Chou, C.-F., Bakajin, O., Turner, S. W. P., Duke, T. A. J., Chan, S. S., Cox, E. C., Craighead, H. G., and Austin, R. H. (1999). Sorting by diffusion: an asymmetric obstacle course for continuous molecular separation, *Proc. Natl. Acad. Sci. USA*, **96**(24), 13762–13765.

200. Huang, L. R., Tegenfeldt, J. O., Kraeft, J. J., Sturm, J. C., Austin, R. H., and Cox, E. C. (2002). A DNA prism for high-speed continuous fractionation of large DNA molecules, *Nat. Biotechnol.*, **20**(10), 1048–1051.

201. Huang, L. R., Cox, E. C., Austin, R. H., and Sturm, J. C. (2004). Continuous particle separation through deterministic lateral displacement, *Science*, **304**(5673), 987–990.

202. Bethe, H. A. (1994). Theory of diffraction by small holes, *Phys. Rev.*, **66**(7/8), 163–182.

203. Moran-Mirabal, J. M., Torres, A. J., Samiee, K. T., Baird, B. A., and Craighead, H. G. (2007). Cell investigation of nanostructures: zero-mode waveguides for plasma membrane studies with single molecule resolution, *Nanotechnology*, **18**(19).

204. Foquet, M., Samiee, K. T., Kong, X., Chaudhuri, B. P., Lundquist, P. M., Turner, S. W., Freudenthal, J., and Roitman, D. B. (2008). Improved fabrication of zero-mode waveguides for single-molecule detection, *J. Appl. Phys.*, **103**(3), 034301–9.

205. Rigneault, H., Capoulade, J., Dintinger, J., Wenger, J., Bonod, N., Popov, E., Ebbesen, T. W., and Lenne, P. F. (2005). Enhancement of single-molecule fluorescence detection in subwavelength apertures, *Phys. Rev. Lett.*, **95**(11), 117401.

206. Moran-Mirabal, J. M., and Craighead, H. G. (2008). Zero-mode waveguides: sub-wavelength nanostructures for single molecule studies at high concentrations, *Methods*, **46**(1), 11–17.

207. Lenne, P. F., Rigneault, H., Marguet, D., and Wenger, J. (2008). Fluorescence fluctuations analysis in nanoapertures: physical concepts and biological applications, *Histochem. Cell Biol.*, **130**(5), 795–805.

208. Korlach, J., Marks, P. J., Cicero, R. L., Gray, J. J., Murphy, D. L., Roitman, D. B., Pham, T. T., Otto, G. A., Foquet, M., and Turner, S. W. (2008). Selective aluminum passivation for targeted immobilization of single DNA polymerase molecules in zero-mode waveguide nanostructures, *Proc. Natl. Acad. Sci.*, **105**(4), 1176–1181.

209. Korlach, J., Bibillo, A., Wegener, J., Peluso, P. Pham, T. T., Park, I., Clark, S., Otto, G. A., and Turner, S. W. (2008). Long, processive enzymatic DNA synthesis using 100% dye-labeled terminal phosphate-linked nucleotides, *Nucleosides, Nucleotides Nucleic Acids*, **27**(9), 1072–1082.

210. Eid, J., Fehr, A., Gray, J., Luong, K., Lyle, J., Otto, G., Peluso, P., Rank, D., Baybayan, P., Bettman, B., *et al.* (2009). Real-time DNA sequencing from single polymerase molecules, *Science*, **323**(5910), 133–138.

211. Samiee, K. T., Fouquet, M., Guo, L., Cox, E. C., and Craighead, H. G. (2005). Lambda-repressor oligomerization kinetics at high concentrations using fluorescence correlation spectroscopy in zero-mode waveguides, *Biophys. J.*, **88**(3), 2145–2153.

212. Wenger, J., Gerard, D., Pierre-Francios, L., Rigneault, H., Dintinger, J., Ebbesen, T. W., Boned, A., Conchonaud, F., and Marguet, D. (2006). Dual-color fluorescence cross-correlation spectroscopy in a single nanoaperture: towards rapid multicomponent screening at high concentrations, *Opt. Express*, **14**(25), 12206–12216.

213. Samiee, K. T., Moran-Mirabal, J. M., Cheung, Y. K., and Craighead, H. G. (2006). Zero mode waveguides for single-molecule spectroscopy on lipid membranes, *Biophys. J.*, **90**(9), 3288–3299.

214. Wenger, J., Rigneault, H., Dintinger, J., Marguet, D., and Pierre-Francios, L. (2006). Single-fluorophore diffusion in a lipid membrane over a subwavelength aperture, *J. Biol. Phys.*, **32**(1), SN1–SN4.

215. Wenger, J., Conchonaud, F., Dintinger, J., Wawrezinieck, L., Ebbesen, T. W., Rigneault, H., Marguet, D., and Pierre-Francios, L. (2007). Diffusion analysis within single nanometric apertures reveals the ultrafine cell membrane organization, *Biophys. J.*, **92**(3), 913–919.

216. Dekker, C. (2007). Solid-state nanopores, *Nat. Nanotechnol.*, **2**(4), 209–215.

217. Storm, A. J., Storm, C., Chen, J., Zandbergen, H., Joanny, J.-F., and Dekker, C. (2005). Fast DNA translocation through a solid-state nanopore, *Nano Lett.*, **5**(7), 1193–1197.

218. Storm, A. J., Chen, J. H., Zandbergen, H. W., and Dekker, C. (2005). Translocation of double-strand DNA through a silicon oxide nanopore, *Phys. Rev. E*, **71**(5).

219. Chansin, G. A. T., Mulero, R., Hong, J., Kim, K. J., deMello, A. J., and Edel, J. B. (2007). Single-molecule spectroscopy using nanoporous membranes, *Nano Lett.*, **7**(9), 2901–2906.

220. Nils, G. W. (2007). Future of biomedical sciences: single molecule microscopy, *Biopolymers*, **85**(2), 103–105.

221. Whitcombe, D., Theaker, J., Guy, S. P., Brown, T., and Little, S. (1999). Detection of PCR products using self-probing amplicons and fluorescence, *Nat. Biotechnol.*, **17**(8), 804–807.

222. Braslavsky, I., Herbert, B., Kartalov, E., and Quake, S. R. (2003). Sequence information can be obtained from single DNA molecules, *Proc. Natl. Acad. Sci. USA*, **100**(7), 3960–3964.

223. Meldrum, D. (2000). Automation for genomics, part two: sequencers, microarrays, and future trends, *Genome Res.*, **10**(9), 1288–1303.

224. Sauer, M., Angerer, B., Ankenbauer, W., Földes-Papp, Z., Göbel, F., Han, K. T., Rigler, R., Schulz, A., Wolfrum, J., and Zander, C. (2001). Single molecule DNA sequencing in submicrometer channels: state of the art and future prospects, *J. Biotechnol.*, **86**(3), 181–201.

225. Harris, T. D., Buzby, P. R., Babcock, H., Beer, E., Bowers, J., Braslavsky, I., Causey, M., Colonell, J., DiMeo, J., Efcavitch, J. W., *et al.* (2008). Single-molecule DNA sequencing of a viral genome, *Science*, **320**(5872), 106–109.

226. Kartalov, E. P., and Quake, S. R. (2004). Microfluidic device reads up to four consecutive base pairs in DNA sequencing-by-synthesis, *Nucleic Acids Res.*, **32**(9), 2873–2879.

227. Ryan, D., Rahimi, M., Lund, J., Mehta, R., and Parviz, B. A. (2007). Toward nanoscale genome sequencing, *Trends Biotechnol.*, **25**, 385–389.

Chapter 8

Particle Engineering for Inhalational Drug Delivery

Basma Ibrahim,[a] **Yan Yang,**[a,b] **and Yoon Yeo**[a,c]

[a]*College of Pharmacy, Purdue University, West Lafayette, IN 47907, USA*
[b]*Research Center for Drug Metabolism, College of Life Science, Jilin University, Changchun 130021, P. R. China*
[c]*Weldon School of Biomedical Engineering, Purdue University, West Lafayette, IN 47907, USA*
yyeo@purdue.edu

Inhalable dry particle is an attractive modality for local therapy of pulmonary diseases as well as systemic drug delivery. For effective inhalational drug delivery, physical properties of the particles are critical. First, for efficient deposition in the lungs, the particles should have an ideal particle size and morphology that provide optimal aerodynamic performance. Second, unless macrophages are an intended target for drug delivery, the particles should avoid uptake by alveolar macrophages. Moreover, the ideal dry particles should maintain physical and chemical stability during storage. Various types of microparticles have been engineered to meet these requirements, leading to several dry-particle systems in the market. With the recent advances in nanomedicine, nanoparticles (NPs) have also been considered as an inhalable drug-delivery system. However, the utility of dry NPs for inhalational drug delivery is limited because of the unfavorable aerodynamic properties. Recent efforts have addressed this challenge by engineering microparticulate agglomerates of NPs. This review discusses recent advances in particle engineering for inhalational drug delivery, with a specific focus on various techniques for creating micro- and nanoparticle formulations suitable for inhalation.

Biomedical Nanosensors
Edited by Joseph Irudayaraj

ISBN 978-981-4303-03-3 (Hardcover), 978-981-4303-04-0 (eBook)
www.panstanford.com

8.1 Dry Particles for Inhalational Therapy

8.1.1 Inhalable Drug-Delivery Systems

The lung is a logical site of drug delivery for pulmonary diseases such as asthma and chronic obstructive pulmonary diseases (COPD), tuberculosis, chronic lung infections, cystic fibrosis lung disease, and lung cancer. The advantage of aerosol therapy is that it is simple to administer yet it delivers the drugs directly to the desired site of action, diminishing side effects and decreasing the need for systemic, especially parenteral, therapy. Inhalable drug-delivery systems have become a standard therapeutic option for treating a variety of respiratory diseases, especially asthma and COPD [1].

The lung is also considered a unique portal for systemic drug delivery because of the large surface area for drug absorption (~100 m^2 in adults) [2], good epithelial permeability, and a rich blood supply [3, 4]. Furthermore, drugs deposited in the lungs face relatively lower concentrations of drug-metabolizing enzymes than those delivered orally [2, 5]. This property is particularly beneficial for the delivery of protein drugs, whose administration is currently limited to parenteral routes. Accordingly, the lung has gained increasing attention as a potential site for the systemic delivery of drugs with poor oral bioavailability, as evident from recent advancement of inhalable insulin [6–8]. Although it was withdrawn from the market in 2007, a year after launch, patient satisfaction with inhaled insulin was quite high compared with injections, particularly among patients with needle anxiety [9].

Three forms of inhalable delivery systems are widely used in pulmonary drug delivery: pressurized metered dose inhalers (pMDI), nebulizers, and dry powder inhaler (DPI).

8.1.1.1 Pressurized metered dose inhaler

Drugs delivered by pMDIs are prepared as suspensions or solutions in a propellant, which often contain cosolvents or surfactants to aid in the dispersion of drugs (Fig. 8.1a) [10]. A drug aerosol is created as follows: a small volume of pressurized drug dispersion is isolated in a metering chamber and then released through a spray orifice. As the released drug dispersion begins to equilibrate with the atmospheric pressure, it is propelled from the container, forming a spray of droplets [10]. Several pMDIs have been commercialized, providing a convenient system for the management of asthma and COPD [11]. pMDI has been a gold standard inhaler devices since its introduction in the 1950s [12] and currently delivers 80% of inhalation therapies worldwide [11]. However, there have been concerns regarding the use of chlorofluorocarbon (CFC), an environmentally incompatible propellant, and its potential adverse effects on patients [11]. These concerns prompted the use of alternative propellants, such as hydrofluoroalkane (HFA) as well as propellant-free pMDIs [13–17]. A weakness of pMDIs is that optimal drug delivery requires a good inhalation technique, which many patients find difficult [11, 12]. Typical lung deposition achieved by pMDIs is 5–20% of the delivered dose [18–20] due to the high velocity, which results in extensive deposition in the oropharynx [12, 20, 21].

8.1.1.2 Nebulizer

Nebulizers generate droplets of a drug dispersion (usually an aqueous solution or suspension [22]) using energy from compressed air or piezoelectric ceramics (Fig. 8.1b)

[10]. Small liquid droplets are delivered to the patients' lungs on their inspiratory flow [10]; therefore, nebulizers are less dependent on the user's technique [23]. Nebulizers are thus suitable for the treatment of young and elderly patients and the emergency treatment of severely dyspneic patients [19]. On the other hand, nebulizers are more time consuming than pMDIs or DPIs and require hygienic maintenance of the equipment [23]. Compressed-air nebulizers are less efficient than pMDIs in production and delivery of the aerosol [19]. Typically, 1% to 5% of the dose delivered by nebulizers reaches the lower airways [19]. The portability of classic nebulizer systems is limited due to the requirement for a source of compressed air, but small, portable devices are now available [23–25].

8.1.1.3 Dry-powder inhaler

Dry powder is a collection of dry particles contained in an inhaler device (Fig. 8.1c). The powder is usually composed of a drug and carrier excipients, which are added to improve the dispersibility of the drug. An aerosol of dry powder is created as it is exposed to inspiratory airflow, which then carries the small drug particles to the lung [10]. Depending on the inspiratory flow rate, the dose deposited in the lung ranges from 5% to 40% [13, 26]. DPIs overcome several limitations of pMDIs and nebulizers. They are portable, avoid problems associated with the use of propellants, and do not require skillful coordination of actuation and inhalation [27]. For this reason, the DPI market is rapidly expanding [12, 20, 27]. Most DPIs in the market are used for asthma and COPD treatment [27], but insulin DPIs have also been developed. With advances in particle-engineering technology, opportunities in DPI-based pulmonary drug delivery will continue to grow. This chapter focuses on dry-powder formulations and technologies.

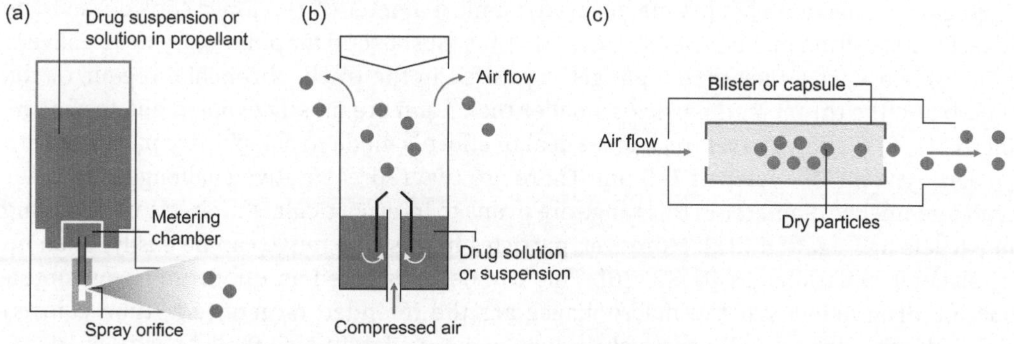

Figure 8.1 An illustration of devices for pulmonary drug delivery: (a) pMDI, (b) nebulizer, and (c) DPI. (Modified from ref. 10.)

8.1.2 Dry-Powder Inhaler Devices

An inhalable dry-powder system consists of two components — a dry-powder formulation and the inhaler device [28]. While the formulation determines physicochemical properties of the dry powder, the device governs the aerodynamic and flow properties [28]. The device is responsible for separating the drug particles from the carrier, deagglomerating particles, and dispersing the particles prior to inspiration [27].

Therefore, the performance of the dry powder is very much dependent on the inhaler device. New formulation development is often accompanied by a parallel effort to develop an optimal device for the powder formulation [18, 26]. Ideally, the inhaler devices should be inexpensive, easy to use, convenient to carry, able to contain multiple doses, stabilize the actives from environmental conditions, and deliver a consistent dose throughout the inhalation process [29]. Demonstrating the reliable delivery of a uniform dose is one of the most challenging tasks in device development [27]. The devices currently on the market are mostly "breath-actuated" passive devices, which require patients' effort and ability to inhale dry powder with a force needed to disperse the powder optimally [12]. The performance of passive devices is thus highly variable among patients [13]. Power-assisted "active" devices are recently developed to provide consistent performance. Various types of current and future inhaler devices are reviewed elsewhere [12, 13, 18, 27, 30].

8.1.3 Requirements of Inhalable Dry Particles

The deposition of dry particles in the airways is in part determined by the aerodynamic properties of the particles, which depend on the size, density, shape, and surface properties of the particles [31]. It is noteworthy that the size of an inhalable particle is expressed in the aerodynamic diameter (d_a), which is the geometric diameter of a particle with a unit mass density (1 g/cm^3) that would fly at the same velocity as the given particle [32]. d_a is related to the geometric diameter (d) of the particle according to the following relationship [32]: $d_a = (\rho)^{1/2}d$, where ρ is the particle mass density. In most cases, d_a is different from d because most particles do not necessarily possess a unit mass density and/or fly as aggregates of particles [21]. A mean aerodynamic diameter of 1–5 μm is considered to be ideal for deposition in the lower airways and thus most useful for pulmonary drug delivery [33, 34]. Particles larger than 5 μm get deposited in the tracheobronchial region, or the mouth and the throat, and particles smaller than 1 μm are mostly exhaled due to the low inertia [1, 6, 27, 35]. Therefore, a great deal of effort is made to achieve dry particles with an aerodynamic diameter of 1–5 μm. There are two representative challenges. Particles with a geometric diameter in this range are prone to interparticulate interactions resulting in particle aggregation [32]. Moreover, particles in this size range can be easily taken up by alveolar macrophages [6, 32, 36]. This can be considered an opportunity for target-specific drug delivery if the macrophages are the intended targets (e.g., tuberculosis) [37–39]; otherwise, it will negatively influence the availability of inhaled particles.

8.1.4 Dry Particle Formulations

Existing particle-engineering technologies have addressed these challenges by controlling the interparticulate forces or making inherently dispersible particles [30, 40, 41]. A representative effort is to blend small drug particles with large (30–300 μm) carrier particles (lactose or glucose) and allow the drug particles to adhere to the surface of carrier particles [28, 40]. The resulting "ordered mixture" formulation will behave like a collection of large particles during production and at inhalation. On the other hand, the binding of drug particles to the carrier particles is weak enough to allow their dissociation in the airflow [28]. Upon inhalation, the drug particles are separated from the carrier particles and reach the lower airways whereas the carrier particles end up in the mouthpiece of the inhaler or the oropharynx of the patient [10]. Diskus®/Diskhaler® formulations employ this strategy.

Alternatively, small carrier particles (2–4 μm) and drug particles (or drug particles alone) are tumbled together (cospheronized) to form loose agglomerates of a larger size (~1 mm) [28, 42]. The resulting agglomerates are free flowing but dispersed into constituent particles as they pass through a deaggregation zone in the device [20]. This method is used in Turbuhaler® formulations.

Another example is a large porous particle that has a large geometric diameter but a low mass density. While the increased particle size decreases the particle aggregation and their phagocytic uptake, the low mass density allows the large particle to possess an adequate aerodynamic diameter [32, 43]. AIR® particle technology (Alkermes, Cambridge, Massachusetts, United States) is based on this principle.

With recent advances in nanomedicine, there has been increasing interest in the potential application of NPs in pulmonary drug delivery [31, 44–46]. Nanoparticulate drug delivery systems (<200 nm) have the potential to improve the delivery of water-insoluble drugs and overcome cellular barriers to the entry of macromolecular therapeutics [31, 47]. While mainly explored for systemic applications [48–51], NPs have potential advantages for pulmonary drug delivery as well. Due to the small size, NPs are less likely to be taken up by alveolar macrophages as compared to micron-sized particles [52]. Adequately engineered NPs are also efficient in penetrating the mucosal layer in the airways [53–55], which has been recognized as a significant barrier to the pulmonary drug delivery [56]. In spite of these advantages, the use of dry NPs in pulmonary drug delivery is quite challenging as they are prone to aggregation via various interparticulate forces and are likely to be exhaled even if they manage to reach the airways [31, 57]. Hence, NPs for pulmonary drug delivery are typically formulated as suspensions and delivered via nebulizers [58]. Recently, new particle-engineering technologies are developed to enable the delivery of NPs in a dry-particle form. For example, NPs are formulated into inhalable, micron-scale agglomerates, which are readily redispersed into NPs after deposition in the lung [59–63]. This chapter will discuss recent advances in dry-particle engineering, especially in the nanometer scale, which will further improve the efficiency of drug delivery to the lungs and create new opportunities in local and systemic therapy via the lungs. Prior to the nanoparticulate dry-particle engineering, traditional microparticulate systems will be briefly reviewed with a focus on the manufacturing technologies.

8.2 Traditional Particle Manufacturing for Inhalational Drug Delivery

A number of review articles provide detailed information about existing DPIs, which have translated to marketed products or are undergoing clinical trials [13, 18, 20, 26, 27, 40]. Dry particle products ideal for inhalation should have a uniform size distribution, adequate physical and chemical stability, excellent reproducibility, low interparticulate forces, and optimal aerodynamic properties represented by a high fine-particle fraction (FPF) — the mass fraction of particles with efficient cut-off diameters between 0.5 μm and 5 μm — and emitted dose (ED) — the amount of particles leaving the inhalation device and, therefore, that can be inhaled by the patient [64, 65]. These requirements have prompted the development of various techniques for dry-particle engineering.

8.2.1 Micronization

Micronization is a traditional method of producing drug particles within the size range adequate for inhalation. The principle of this method is to apply mechanical forces on the particles, resulting in particle-particle or particle-equipment collisions [66]. During micronization, coarse particles are shattered into small pieces until desirable crystalline particles are achieved [41]. Different types of equipment, such as ball milling, jet milling, media milling, and pearl milling, are used to create such forces [65, 67–69]. A disadvantage of micronization methods is that there is limited control over the particle properties, including size, size distribution, shape, crystallinity, morphology, surface properties, and electrostatic charges [67, 68, 70–72]. High energy input for the mechanical comminution may cause an excessively charged and/or amorphous surface, resulting in particle agglomeration and broad size distribution [41, 65, 68, 73, 74]. Since the amorphous surface is thermodynamically unstable, undesirable physical and chemical changes in particles are likely to occur during the long-term storage [65, 67, 75, 76].

Some of the issues associated with the micronization process can be solved by specialized milling processes [70]. Media milling utilizes liquid as a medium, which maintains low temperatures during the milling process to prevent the heat-induced degradation of the drug [68, 70]. For example, a fluid propellant that can be cooled to -70°C is used to suspend drug particles during pearl milling [77]. Once the milling process is completed, the propellant is evaporated to yield dry micronized particles. This method is suitable for handling macromolecular drugs like proteins.

Due to the high surface energy of micronized particles, it is often necessary to employ another means to improve the flowability and dispersibility of the particles. By far the most common method is to blend the drug particles with coarse carrier particles [66, 70, 78, 79]. The carrier particles provide high-energy active sites on the surface for the drug particles to adhere to [70, 80]. The carrier particles also serve as a bulking agent for the delivery of potent drugs [81, 82]. Representative carrier particles include coarse lactose particles (30–60 μm) [65], mannitol, sucrose, sorbitol, magnesium stearate, glucose, cyclodextrin, leucine, lecithin, raffinose, trehalose, and xylitol [70, 80, 83–86]. Properties of carrier particles play an important role in the resulting products [81]. Lactose carriers with higher crystallinity show better aerodynamic performances than amorphous ones [65, 79, 87–90]. Carrier particles with a rough surface may entrap drug particles and resist separation of the drug particles from the carrier surface during inhalation [65, 91, 92]. The presence of fine carrier particles or the third additives facilitate particle deaggregation during inhalation [70, 83, 84] because they saturate the high-energy active sites on the coarse carrier surface and leave relatively less active sites for drug particles [80, 83, 93, 94].

8.2.2 Crystallization and Precipitation

In the crystallization procedure, drug particles are precipitated from the solution by the addition of a nonsolvent [66, 95, 96]. This method has a good control over the particle properties, stability, density, and surface energy [41]. On the other hand, the particle size distribution is usually broad [68] as the small particles can grow into large crystals [65, 89]. Various technologies were developed to control the particle size of crystals during crystallization and precipitation processes [97, 98]. Crystals with a well-controlled size can be formed by adding an antisolvent in the presence of an additive that retards the growth of the crystals (micronization technique) [65, 67, 99–103]. The particle size can also be controlled by the "microprecipitation" procedure [70]. The precipitation procedure employs a small chamber in which jet streams of drug solution and an antisolvent meet and create drug crystals [70]. Recently, the sono-crystallization technique has been applied to facilitate the mixing of the antisolvent and the drug solution [70, 104]. Another technique is the evaporative precipitation into aqueous solution (EPAS) process [67, 105, 106]. A solution of a drug dissolved in an organic solvent, heated above the boiling point, is sprayed into a heated aqueous solution containing stabilizing agents. As the temperature is maintained above the boiling point of the organic solvent, solvent evaporation occurs immediately, resulting in drug precipitation. The resulting aqueous drug suspension is either spray-dried or spray freeze-dried [67].

The formation of different polymorphs during crystallization is another potential problem in the crystallization and precipitation technique [68, 107]. The identification and purification of stable polymorphs are important from the regulatory perspective due to their influence on the bioavailability of a drug [68, 108]. The solution-enhanced dispersion by supercritical fluid (SEDS) technology is shown to be useful for purifying specific polymorphs [68, 107, 109].

8.2.3 Spray Drying

The spray-drying process is relatively simple and easy to scale up [110] and able to rapidly form particles with desirable flowability and aerodynamic properties in a reproducible manner [41, 67, 68, 70, 71, 78]. Therefore, this technology has become a popular method for the preparation of inhalable particles.

In the spray-drying process, a feed solution is atomized into fine droplets, which are rapidly dried to form dry particles due to the high surface area, high vapor pressure of the solvent, and warm air current [68, 70]. The feed solution can be prepared in a variety of solvents, including cosolvent systems, as a solution, suspension or a single/multiple emulsion [41, 70]. Organic solvents are often used to dissolve excipients and facilitate evaporation. The particle properties can be controlled by operating parameters of the spray-drying process and the formulation of a feed solution (solvent system and excipients) [41, 65, 66, 68, 70, 111, 112]. Although a relatively high drying temperature is needed in spray drying, most heat-sensitive components survive this procedure because exposure to the drying temperature is quite brief [66]. Moreover, the actual evaporating temperature that the drying droplets experience is lower than the operating drying temperature, due to heat loss during the evaporation [41, 66, 68, 70].

The first inhalable insulin powder (Nektar/Pfizer) was produced by the spray-drying method [7]. An aqueous buffered solution of insulin and amorphous sugar was spray-dried to produce dry particles in which insulin and the amorphous sugar (glass) are homogeneously distributed. The glass reduces molecular motions and chemical reactivity of the insulin, improving the stability of insulin during storage. The spray-drying process is also used to produce PulmoSphere® (Nektar Therapeutics) particle platforms. Here, inhalable microparticles with a particle size ranging from 1 μm to 4 μm are prepared by spray-drying an emulsion of fluorocarbon in an aqueous drug solution/suspension stabilized by a monolayer of phospholipid [13, 41, 110, 113]. The volatile fluorocarbon serves as a "blowing agent" or "inflation agent" and creates a hollow structure in particles during the spray-drying process [113]. The ratio of a blowing agent to phospholipid in the emulsion greatly influences the porosity and density of resulting particles [12, 113]. Inhalable particles of cromolyn sodium, albuterol sulfate, formoterol fumarate [113], and budesonide [114] have been produced by the PulmoSphere technology.

8.2.4 Spray Freeze-Drying

In a typical spray freeze-drying (SFD) technique, a drug solution is sprayed into a vessel filled with a freezing medium (cryogenic liquids or halocarbon refrigerants) [41, 65, 68, 70, 115–118]. Because of the low boiling point of the freezing medium and high heat-transfer rates, the liquid droplets freeze instantly upon contact with the freezing medium [65, 68, 117–121]. Once the atomization process is finished, the frozen solids is lyophilized to remove the freezing medium [65]. This process is able to create the particles with inhalable size range (<5 μm), nanometer-scale or porous properties [41, 65, 117, 118].

8.2.5 Supercritical Fluid Technology

Supercritical fluids (SCF) have properties of both gases and liquids, whose balance depends on temperature and pressure [65, 66, 68, 122]. Various precipitation methods rely on the versatility of SCF for producing inhalable particles with narrow particles size distribution, controllable morphology, and a relatively low surface charge [70, 72, 122]. In pharmaceutical applications, supercritical carbon dioxide is the most widely used because of its low supercritical temperature, moderate pressure, low cost, lack of toxicity, and nonflammable properties [65, 67, 68, 123]. SCF methods can typically be divided into four categories [68, 72, 122]:

- The rapid expansion of supercritical solutions (RESS) process, where SCF are used as solvents. The drug is dissolved in an SCF, followed by a rapid depressurization leading to a decrease in the solvation power and precipitation of the drug.
- Gas antisolvent (GAS) process: SCF are used as antisolvents. The drug is initially dissolved in an organic liquid, which is then saturated with carbon dioxide to decrease the solvation power of the solvent and precipitate the drug. Moderate pressures are used in this process to make carbon dioxide more soluble in organic solvents.
- Supercritical antisolvent (SAS) processes: Similar to GAS, the SAS process is developed to improve the mixing of SCF and the solvent in the GAS process. A relatively high pressure is employed to completely mix the SCF and the solvent.
- Precipitation from gas-saturated solution (PGSS) process: SCF are mixed with a melted solute, and the resulting mixture is expanded and depressurized via a nozzle. The vaporization of carbon dioxide during the expansion process cools the solute and causes its precipitation.

8.2.6 Large Porous/Hollow Particles

Large porous microparticles, characterized by large geometric diameters (>5 μm) and low density (ρ < 0.4 g/cm^3) [32, 43], exhibit desirable aerodynamic diameters and lung deposition profiles [26, 43, 124–127]. Because of the small surface-to-volume ratio, the large porous particles are less prone to aggregation and easily dispersed as compared with nonporous small particles [12, 43]. The large geometric diameters of these particles reduce their uptake by alveolar macrophages, resulting in longer residence time of the particles in the lungs [43, 128, 129]. The use of biodegradable polymers such as poly(lactic-co-glycolic acid) (PLGA) can further prolong the pulmonary residence of drugs, increasing the local bioavailability of the drugs [32, 43]. Several methods are developed to produce the large porous particles.

8.2.6.1 Spray drying

Spray drying is widely used to create large porous particles as it easily accommodates diverse excipients with favorable aerodynamic properties and biocompatibility such as dipalmitoylphosphatidylcholine (DPPC) and albumin. Spray drying of these excipients results in large hollow particles with a "crumpled-paper" structure [65, 110]. This unique morphology is attributed to the low aqueous solubility and low diffusion coefficient of DPPC, which make the DPPC rapidly saturated on the droplet surface and unable to catch up with the receding surface of a drying droplet [110]. Albumin is not absolutely required for forming the hollow structure [34, 60, 130, 131] but facilitates the change in particle morphology and decreases the particle density [132–135].

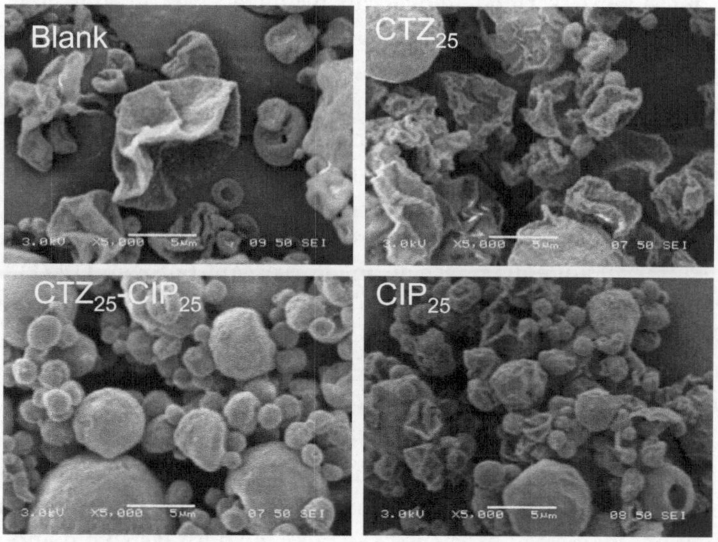

Figure 8.2 Scanning electron micrographs of spray-dried microparticles. CTZ and CIP are microparticles containing ceftazidime and/or ciprofloxacin, respectively. Subscripts indicate the percentages of ceftazidime or ceftazidime in the microparticles by weight. The remainder was DPPC, albumin, and lactose (60:20:20, respectively, by weight %). All pictures were taken at 5000 × magnification (scale bar = 5 μm). (Reprinted with permission from ref. 138.)

The DPPC-albumin system has been used to deliver a number of drugs, including low-molecular-weight drugs [128, 129, 131, 136] to insulin [137]. We have employed this particle platform for delivering ciprofloxacin and ceftazidime simultaneously [138]. Figure 8.2 shows that the presence of DPPC, albumin, and lactose in the formulation resulted in crumpled microparticles or rough spheroids. Ciprofloxacin tended to increase the FPF of the particles, which benefited the delivery of codelivered ceftazidime. Coencapsulation of the antibiotics in microparticles ensured their codeposition at desired ratios, improved the FPF of the particles as compared to microparticles with an equivalent amount of ceftazidime alone, and achieved the additive antipseudomonal activity [138].

8.2.6.2 Emulsification techniques

Standard double- and single-emulsification solvent-evaporation procedures followed by a freeze-drying step can be used to produce large porous particles [43, 110]. These particles have been prepared with a water-insoluble biodegradable polymer, PLGA. Additional procedures or excipients are employed to increase the porosity and size of the otherwise solid and small polymeric microparticles. For example, aqueous solutions of osmogens [139–142] or oil porogens [143] are added to the internal phase as pore-forming agents. With osmogens, porous structures are generated as water in the external phase moved into the internal phase due to the osmotic pressure [139–142]. With oil porogens, porous particles form as the oil phase is extracted into the aqueous phase [143]. Supercritical carbon dioxide is also used as a porogen [112, 144, 145]. Here, the supercritical carbon dioxide penetrates the amorphous polymer matrix and expands to form pores upon isothermal depressurization.

While the methods described above can produce large porous polymeric microparticles, there is a limitation in drug encapsulation. The use of osmogens and oil porogens can interfere with efficient drug encapsulation [139, 141–143] because drugs tend to diffuse out during pore-forming processes, which depend on diffusional mass exchange between discontinuous (polymer droplets) and continuous (external water) phases. Recently, our laboratory showed that the use of an effervescent salt can prevent this problem [146]. PLGA microparticles (average diameter 10–20 μm) were made by the double-emulsion method. To impart favorable aerodynamic properties, ammonium bicarbonate (ABC) was included in the internal aqueous phase as an effervescent porogen [147]. ABC induced a highly porous structure (Fig. 8.3a) in the PLGA particles as it decomposed into ammonia and carbon dioxide. This morphology was clearly contrasted with nonporous small microparticles (Fig. 8.3b). The FPF of the microparticles increased as a function of the ratio of ABC to PLGA. Two model drugs, lysozyme (a model of therapeutic macromolecules) and doxorubicin hydrochloride (an example of small-molecular-weight drugs), were encapsulated in the highly porous large PLGA microparticles. Unlike other encapsulation methods based on osmogens or extractable porogens, this method could encapsulate both drugs, with high encapsulation efficiency (~100% for both lysozyme and doxorubicin), in the porous PLGA microparticles with a desirable mass median aerodynamic diameter (MMAD) and FPF. The high encapsulation efficiency is attributable to the fact that pore formation depends more on the effervescence of decomposition products rather than diffusional mass exchange and that gas production accelerated the solidification of the polymer phase.

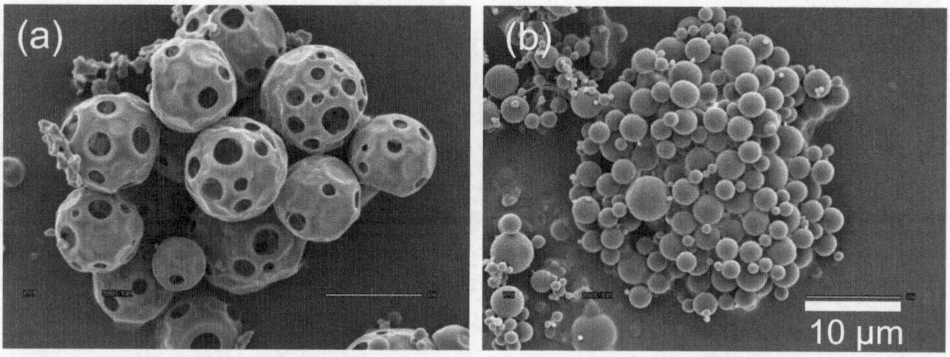

Figure 8.3 Scanning electron microscopic images of PLGA particles (a) made with 2.5%w/w ABC and (b) without an effervescent porogen. All particles were made with PLGA (50:50, acidic end group, inherent viscosity = 0.18 dL/g). (Reprinted with permission from ref. 146.)

8.3 NPs for Inhalational Drug Delivery

8.3.1 NPs for Pulmonary Drug Delivery

NPs for pharmaceuticals are defined as colloidal solid particles that can be used therapeutically as drug carriers, in which a therapeutic agent is dissolved, entrapped, or encapsulated or to which the therapeutic agent is adsorbed or attached [148]. NPs offer several advantages for pulmonary drug delivery. The small size of NPs provides a large surface area per mass for drug adsorption or modification as well as the abilities to avoid phagocytic uptake [52] and to penetrate the lung epithelium [149, 150]. The NP surface can also be used to attach target-cell-specific ligands, which will improve the cellular uptake of the NPs by the target cells [151–153].

By effectively avoiding phagocytic uptake, NPs can increase the pulmonary retention time of a drug and, thus, provide a sustained pharmacological effect [154, 155]. For example, NPs were used to encapsulate frontline antituberculosis drugs, and a single nebulization of the NPs provided sustained drug levels in the lungs up to 11 days [156]. When appropriately formulated, NPs can improve the lung deposition of a drug. A recent clinical study that examined the lung deposition of radiolabeled particles by scintigraphy reported that blend formulations of nanoparticulate salbutamol (Nano-SBS) and lactose showed better lung deposition than micronized salbutamol blends [157]. Nano-SBS particles showed a 2.3-fold increase in the total lung deposition and lower oropharyngeal deposition (25.3 ± 4.5%) compared to a micronized SBS formulation (58.4 ± 6.1%) [157]. The higher lung deposition of Nano-SBS is attributable to the reduced interaction with the lactose carrier, which increases the FPF, and the continued movement of the NPs deposited in the lungs into the target area [157].

8.3.2 Nebulizer, a Conventional Way of Delivering NPs to the Lungs

NPs are typically delivered by a nebulizer [61]. Table 8.1 shows examples of NPs delivered by nebulizers. Despite the advantages of nebulizers mentioned in Section 8.1.1.2, there are some limitations in nebulizer-based NP delivery. The stability of NPs in suspension can be negatively influenced during storage. Polymers used for NP formation can undergo hydrolytic degradation in aqueous solutions and gradually aggregate over time. Difficulties in maintaining the original size and resuspending the aggregates have also been reported [158]. The addition of stabilizers such as mannitol [159], polysorbate, or polyoxamer [160] may reduce the aggregation of NPs, but they are often needed in large quantities and may burden the lungs. Moreover, performances of existing nebulizers are variable, and their portability is limited [161–163]. A great deal of efforts has been made to improve the reproducibility and portability of nebulizers, but development of new devices is still in its infancy.

Table 8.1 Examples of NP formulations delivered by nebulizers

Active pharmaceutical ingredient (API)	Encapsulating material	Method of NP preparation	Type of nebulizer	Results and remarks	Ref.
Rifampicin, isoniazid, and pyrazinamide	PLGA	Emulsion-solvent diffusion method	Compressor nebulizer	Sustained levels of three drugs are obtained for as long as 6–8 days in the plasma and up to 11 days in the lungs.	[156]
Insulin	PLGA	Emulsion-solvent diffusion method	Sieve-type ultrasonic nebulizer	Blood glucose level was significantly reduced and hypoglycemia maintained over 48 hours in guinea pigs.	[164]
Rifampicin, isoniazid, and pyrazinamide	Alginate	Ionic gelation	Compressor nebulizer	Therapeutic levels of drugs are observed in the lungs, liver, and spleen up to 15 days after nebulization.	[165]
Rifampicin, isoniazid, and pyrazinamide	Lectin-functionalized PLGA	Emulsion-solvent diffusion method followed by the conjugation of lectin to the NP	Compressor nebulizer	Sustained plasma levels of rifampicin are obtained up to 6–7 days and of isoniazid and pyrazinamide for 13–14 days. All drugs achieve therapeutic levels in the lung, spleen, and liver for 15 days.	[166]
Elcatonin	Chitosan-modified PLGA	Emulsion-solvent diffusion method	Sieve-type ultrasonic nebulizer	There is prolonged pharmacological action for 24 hours due to the retention of NPs adhered to the bronchial mucus and lung tissues.	[167]
Itraconazole	Poloxamer 407 and Polysorbate 80	EPAS and SFL	Aeroneb® Professional micropump nebulizer	Therapeutic concentration is maintained for >24 hours by single nebulization.	[168]

Active pharmaceutical ingredient (API)	Encapsulating material	Method of NP preparation	Type of nebulizer	Results and remarks	Ref.
Itraconazole	Mannitol and lecithin	URF	Aeroneb® Professional micropump nebulizer	Bioavailability is highly enhanced by increasing the dissolution of itraconazole 27 times the crystalline solubility.	[159]
Nanocrystal suspension of budesonide		Not available	Jet nebulizer	Compared to Pulmicort Respules®, nanobudesonide has a shorter nebulization time, higher C_{max}, a similar AUC but a more rapid rate of drug delivery when tested on 16 healthy volunteers.	[169]
carboxyfluorescein (a model hydrophilic drug)	DEAPA-PVAL-g-PLGA	Solvent displacement technique	Vibrating mesh nebulizer Aeroneb®	An *ex vivo* lung model showed burst release of the dye.	[170]

Abbreviations: AUC, the area under the plasma concentration time curve; DEAPA-PVAL-g-PLGA, diethylaminopropylamine poly(vinyl alcohol)-grafted-poly(lactic-co-glycolic acid); EPAS, evaporative precipitation into aqueous solution; PLGA, poly(lactic-co-glycolic acid); SFL, spray-freezing into liquid; URF, ultra-rapid freezing

8.3.3 Current Approaches for Dry Inhalable Nanomedicine

Due to these limitations in NP delivery with nebulizers, dry-powder formulations have been considered as an alternative. However, there are several challenges in delivering NPs in dry-powder formulations. First, NPs tend to aggregate due to high surface energy, making them difficult to handle and disperse [171]. Second, even if it is possible to disperse NPs, it is hard to achieve good lung deposition of dry NPs due to the low inertia [6]. Several approaches have been developed to overcome these challenges.

8.3.3.1 NPs delivered as spray-dried microparticles

8.3.3.1.1 *NPs delivered as spray-dried microparticles*

Tsapis *et al.* have first reported NPs delivered as spray-dried microparticles (NP/SDMPs) (Fig. 8.4a), also known as "Trojan," large-porous nanoparticles (LPNPs), or porous nanoparticle-aggregate particles (PNAPs) [60]. This novel technique creates large porous micron-sized particles consisting of NPs, which can be delivered to the airways as microparticles but disintegrate into the constituent NPs upon airway deposition. Therefore, PNAPs have features of both large porous microparticles and NPs. The former provide desirable aerodynamic properties for inhalational delivery and proper particle handling, and the latter facilitate drug release and prolong the residence time of the particles in the lung by effectively escaping the phagocytic and mucociliary clearance [60].

PNAPs are prepared by the simple spray-drying process. Unlike a solution of lactose alone, which produces dense spherical particles, a suspension of polystyrene (PS) NPs in a lactose solution forms hollow spherical microparticles with a thin wall (400 nm) and a geometric diameter of 8 ± 3 μm and MMAD of 4 ± 2 μm. This unique shape and density of PNAPs can be explained by the high Peclet number of the NP suspension. The Peclet number is defined as $P_e = \kappa/8D$, where κ is the evaporation rate of the spray droplets and D is the diffusion coefficient of the solute(s) [110]. A low Peclet number favors the formation of dense spherical particles, while a high Peclet number results in particles with shells that may deform in various ways [110]. A lactose solution has a relatively small Peclet number (~10) due to the high diffusion coefficient of lactose. On the other hand, NPs have a much higher D and, thereby, a larger Peclet number (~2000) [60].

A recent *in vivo* study shows that the promise of the PNAP strategy translated to pharmacokinetic consequences. PLGA PNAPs delivered to guinea pigs by intratracheal insufflations provided the extended systemic and local exposure of rifampicin [59]. The bioavailability of inhaled rifampicin/PNAPs (F 0.72–0.73) was significantly higher than that of oral rifampicin (F 0.17) [59]. The inhaled rifampicin/PNAPs maintained much higher rifampicin concentrations in the lungs for an extended period (>8 hours) compared to the intravenous or oral groups, although the exact fate of PNAPs in the lungs (macrophage uptake versus intact particles in the lung tissues) remains to be seen.

The idea of delivering NPs in the microparticle form was also investigated by another group [62]. NPs made of either gelatin (242 ± 14 nm) or polybutylcyanoacrylate (PBCA) (173 ± 63 nm) were spray-dried with lactose. Microparticles containing the NPs (gelatin or PBCA) showed aerodynamic properties (FPF of 40% and MMAD of 3.0 μm) comparable to those of lactose alone. Unlike PNAPs [60], whose geometric particle size and shape were dependent on the presence of the incorporated NPs, the size and shape of the microparticles described in this study were not influenced by the NPs [62]. Such a difference may stem from the use of different solvent systems: in PNAPs [60], the NPs and carrier excipients were dispersed in 70% ethanol, whereas the gelatin or PBCA NPs were in water [62]. The former might have relatively higher increase in the Peclet number during drying due to the high κ, which resulted in a more dramatic deformation of the particles.

8.3.3.1.2 *Examples and variations of NP/SDMPs*

A variety of drugs were effectively loaded in the NP/SDMPs for local or systemic applications. For example, insulin was formulated as NP/SDMPs with a high loading capacity (65–80%) [123]. Insulin-loaded chitosan NPs were incorporated in a matrix of lactose-mannitol mixture by spray-drying [123]. The resulting NP/SDMPs showed a hollow spherical shape and was readily dispersed achieving an MMAD of 2–3 μm. Chitosan NPs maintained a consistent size and zeta potential after spray drying [123]. In another example, doxorubicin was encapsulated in PBCA NPs and formulated as inhalable lactose microparticles (MMAD of 3.41 ± 0.22 μm) [172]. The microparticles were prepared by the spray freeze-drying technique. Doxorubicin NPs were precoated by polysorbate 80 prior to spray freeze-drying to facilitate the disintegration of the NPs after lung deposition [172].

Chougule *et al.* used a similar strategy for the delivery of amiloride hydrochloride [173] and tacrolimus [174] for the local therapy of cystic fibrosis and lung transplantation, respectively. In both studies, the drugs were first loaded in nanoliposomes (NLs) and spray-dried with sugars and polyols to produce microparticles. The NL dispersion was stabilized using amino acids (glycine [173] or leucine [174]) prior to the spray drying. Microparticles containing amiloride hydrochloride NLs, mannitol, and glycine showed an FPF of 67.6 ± 0.6% and an MMAD of 2.3 ± 0.1 μm [173], and those with tacrolimus NLs, trehalose, and leucine had an FPF of 71.1 ± 2.5% and an MMAD of 2.2 ± 0.1 μm [174]. In the albino rat model, the microparticles of amiloride hydrochloride and tacrolimus administered intratracheally showed the maximum residence time of 12 and 24 hours in the lungs, respectively [173, 174].

Recently, NP/SDMPs were prepared by a one-step spray-drying process by the aid of a four-fluid nozzle [175]. Mannitol microparticles containing rifampicin/PLGA NPs were prepared by cospraying an organic solution of rifampicin and PLGA and an aqueous solution of mannitol through two nozzles. Cellular uptakes of NP/SDMPs and conventional microparticles were compared on both *in vitro* and *in vivo* levels using alveolar macrophages. While the *in vitro* uptake of NP/SDMPs by alveolar macrophages was lower than that of conventional microparticles, the *in vivo* drug levels in the macrophages were higher for NP/SDMPs [175]. The *in vivo* result was probably due to the mucociliary movement, which removed microparticles more rapidly than NP/SDMPs [175]. The long residence time of NP/SDMPs in the lung would be advantageous for the local delivery of antituberculosis drugs.

In another variation of the NP/SDMP formulation, effervescent excipients were included to facilitate NP release from the SDMPs [63]. Ciprofloxacin or PBCA NPs were spray-dried with lactose, leucine, PEG 6000, and polysorbate 80. Additionally, sodium carbonate and citric acid were added as effervescent excipients and ammonium hydroxide was added to keep the pH high enough to prevent a premature effervescence reaction. The resulting dry particles showed a desirable FPF of 46.47 ± 15% and an MMAD of 2.17 ± 0.42 μm [63]. The formation of gas bubbles upon exposure to humidity was observed by confocal microscopy. The release of ciprofloxacin from the effervescent particles was higher than that from the lactose carrier (56 ± 8% versus 32 ± 3%) [63]. The gas formation also enhanced the dispersion of PBCA NPs upon exposure to water [63, 176]. The safety of effervescent particles was evaluated *in vivo* [176]. PBCA NPs were formulated into effervescent microparticles by spray freeze-drying. Five mice tolerated the inhaled effervescent particles without any changes in the morbidity score (measured from physiologic parameters, such as appearance, body weight, food intake, clinical signs, and behavior) [176].

8.3.3.1.3 *Control of drug release from NP/SDMPs*

Drug release from spray-dried NP clusters was investigated using aspirin or salbutamol sulfate encapsulated in polyacrylate NPs, which were then spray-dried [177]. The drug release rate was more dependent on the morphology of the resulting microparticles (size, degree of hollowness) rather than the size of the NPs. The NP clusters with a thicker shell showed slower drug release because the constituent NPs were more resistant to dispersion (thus the total surface area was smaller). The NP size influenced the drug release by affecting the drug-loading capacity.

Hydrophobic excipients can be added to microparticle components to control the release of hydrophilic drugs from NPs [178]. Nanoparticulate terbutaline sulphate (TS), a model hydrophilic drug, was loaded into spray-dried microspheres of glyceryl behenate

60% and tripalmitin 40% (GBTP) or hydrogenated palm oil (HPO) [178]. Additionally, DPPC (0.2%w/w) was included to enhance the aerosolization and reduce the macrophage uptake of the particles [179, 180]. The microparticles had an FPF of 46.5 ± 1.8% and an MMAD of 3.93 ± 0.12 μm. Of the encapsulated TS, 32.7% was released over 180 minutes in a phosphate buffer and simulated lung fluids without an initial burst release [178]. In contrast, without encapsulation in the hydrophobic excipients, 100% of the TS was dissolved in the buffer in 15 minutes.

8.3.3.2 NP agglomerates by controlled flocculation (Fig. 8.4b)

Another approach to incorporate NPs into inhalable microparticles is to form controlled flocculates of NPs (nanoclusters) [181]. PLGA NPs coated with a cationic surfactant are used as a platform for the delivery of protein drugs. The adsorption of a protein like ovalbumin on the NP surface leads to the flocculation of the NPs, which can be lyophilized to form micron-sized NP agglomerates suitable for inhalation. The NPs instantly release the protein upon exposure to buffer. The level of NP flocculation is governed by the surface charge of the NPs, the ionic strength of the NP suspension, the shear force, the protein-to-particle ratio, and the total protein-particle concentration.

A similar strategy was used to make inhalable agglomerates of paclitaxel nanoprecipitates (alone or in combination with cisplatin) for the local therapy of lung cancers [182]. Leucine was added as a flocculating agent to trigger the agglomeration of paclitaxel nanoprecipitates. The MMAD of lyophilized agglomerates were suitable for lung delivery (1–2 μm). The dissolution rate of paclitaxel from this agglomerate formulation was much higher (60% in 8 hours) than that of micronized paclitaxel particles (18% in 8 hours) and further enhanced by the presence of water-soluble cisplatin [182].

8.3.3.3 NP as a dispersant of microparticles (Fig. 8.4c)

NPs were used to improve the aerosolization of microparticles or influence the dissolution of a drug. The surface of micronized panlukast hydrate particles was coated with hydroxypropylmethylcellulose phthalate (HPMCP) NPs [183]. The hydrophilic NPs reduced cohesive interactions of the drug particles and improved their dispersibility and detachment from the lactose carrier particles. A similar strategy was used to deposit PLGA NPs on budesonide and triamincinolone acetonide microparticles [184]. Coating of the drug particle surface with NPs resulted in more sustained dissolution of the drug than uncoated particles. This resulted in a significant improvement in the pulmonary delivery of the drug particles in a rat model [184].

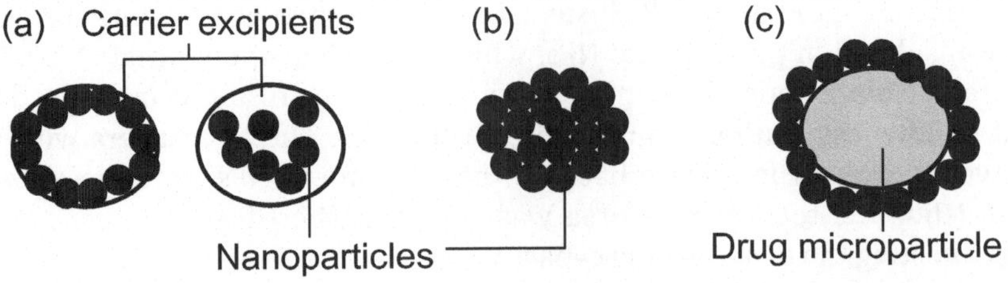

Figure 8.4 Schematic diagrams of NP assemblies for inhalation. (a) NPs delivered by spray-dried microparticles; (b) NP agglomerates by controlled flocculation; (c) NP as a dispersant of microparticles.

8.3.4 Further Consideration of Inhalable NPs

The features of NPs that make them attractive for inhalational drug delivery are also considered as a potential risk to human health. NPs can easily pass through the epithelial cells, enter the circulation, and distribute throughout the body parts. NPs with a small particle size (<100 nm) have the ability to distribute all over the respiratory tract. In particular, NPs <60 nm can deeply penetrate into the bronchotracheal region [185, 186]. Moreover, the large surface area per mass may increase the biological activity of the NPs and induce inflammatory responses [187–189]. A number of review articles discuss the nanotoxicology in more details [186, 190–195]. Here, we will focus on the toxic effects of NPs on pulmonary tissues and the factors to be considered in preparation of NPs for inhalation.

8.3.4.1 General mechanisms of NP toxicity

Among the many mechanisms proposed to explain the proinflammatory effects of NPs, the most widely accepted one is their ability to produce reactive oxygen species (ROS) [196, 197]. Although experimental proof is limited to specific materials, this has been the general consensus since the mid-1990s [186]. One of the supporting observations is that the free-radical production measured by DNA damage was significantly higher upon exposure to ultrafine TiO_2 (20 nm) as compared to larger TiO_2 particles (500 nm) [198]. Figure 8.5 describes the potential mechanism by which the NPs interact with cells to induce toxicity [186]. NPs can cause oxidative stress, increasing the cytosolic calcium concentration [186]. The oxidative stress can be caused by (i) the direct interaction of NPs with the cell surface, (ii) the activation of cellular receptors such as epidermal growth factor receptor, or (iii) NP endocytosis followed by the stimulation of the NADPH oxidase. Oxidative stress and the increased cytosolic calcium concentration can promote calcium influx to the nucleus, which stimulates the transcription of proinflammatory genes via NF-kB activation [199]. In addition, NPs can enter the cells directly via passive diffusion and interact with the mitochondria, interfering with the electron transport. NPs can also directly interact with DNA, causing genotoxicity [186].

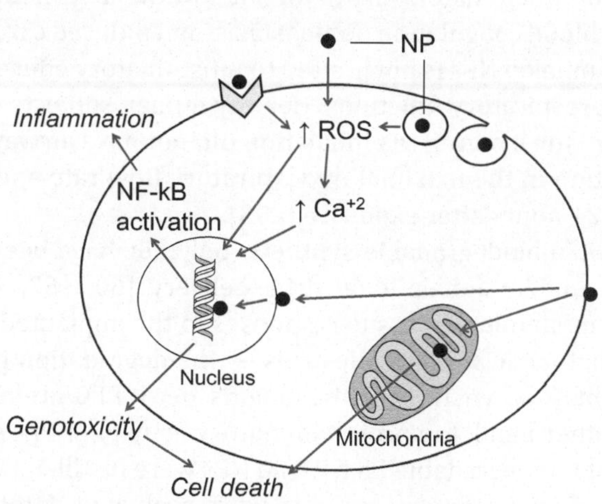

Figure 8.5 Potential interactions of NPs with a cell that can cause toxicity. (Modified from ref. 186.)

8.3.4.2 Factors affecting the toxic effects of NP on pulmonary tissues

Biodegradability, hydrophobicity, and charge density of the materials constituting the NPs are important considerations for the toxicological evaluation of the system [187]. Biodegradable PLGA NPs are less inflammatory than nondegradable PS NPs [187]. Hydrophilic materials showed relatively slow inflammation than hydrophobic ones [200]. Pulmonary inflammation was higher for positively charged PS NPs than for those with negative charges [201].

Particle size, size distribution, the total surface area, the crystal structure, the inherent pH, and the shape of the particles can also play a role in regulating inflammatory reactions [194, 202]. There is increasing evidence for the potential role of the total surface area of NPs in stimulating inflammatory reactions in the lung. According to a comparison of three different sizes of PS NPs, the particles with the largest total surface area induced the most significant increase in inflammatory reactions in the lung [188]. Consistent with this observation, another study found that the number of polymorphonucleocytes (PMNs) present in the bronchial alveolar lavage fluid (BALF) of Balb/C mice, measured 14 days after instillation, was higher with 75 nm PS NPs than with 220 nm ones [187]. Similarly, Oberdörster *et al.* showed that the intratracheal administration of ultrafine TiO_2 (~20 nm) induced significantly higher inflammatory effects as compared to larger ones (~250 nm) [189].

8.3.4.3 Toxicity of inhalable NP

Notwithstanding the increasing interest in nanotechnology and its application in drug delivery, there are only a handful of toxicological studies of inhaled therapeutic NPs [190]. Most toxicological studies involving human subjects were done with ultrafine particles of inherently toxic and/or nondegradable materials; thus, the conclusions from these studies may not be directly applicable to biodegradable NPs, which are more clinically relevant [194, 203].

An epidemiological study reported that ultrafine carbon NPs (~20 nm) in polluted air induced the release of inflammatory mediators, led to the deterioration of the diseased lungs, influenced the blood coagulation homeostasis, and induced cardiovascular diseases [204]. The toxicity of inhaled NPs is not limited to inflammatory effects. Ultrafine NPs may also negatively affect respiratory functions. Healthy human subjects exposed to ultrafine carbon NPs (50 $\mu g/m^2$ for 2 hours) via inhalation did not have airway inflammations but experienced a reduction in the maximal midexpiratory flow rate and carbon monoxide-diffusing capacity at 21 hours after exposure [203].

NPs made of PLGA, a biodegradable synthetic polymer, have been used in numerous studies for systemic and/or inhalational drug delivery [60, 153, 156, 158, 166, 167, 205–207]. Given the inflammatory tissue responses to the implanted PLGA devices [208, 209], it is possible that lactic and glycolic acids — the degradation products of PLGA — would induce inflammatory reactions in the lung tissues if PLGA remains in the lungs for a long time. On the other hand, the proinflammatory activity of PLGA NPs was shown to be milder than that of nondegradable PS NP. The NPs were instilled to the lungs of Balb/C mice, and the levels of inflammatory markers were evaluated. After 1 and 14 days, the PLGA NP group did not show any increase in PMN recruitment and lactate dehydrogenase

(LDH) release, an indicator of cytotoxicity, in the BALF, while nondegradable PS NP did [187]. Both PLGA NPs and PS NPs led to a twofold increase in the mRNA levels of macrophage inflammatory protein-2, a chemotactic factor associated with the presence of inflammatory responses, in the lavage cells, as compared to the glucose control. However, it remains to be seen how the slowly degrading PLGA NPs would influence the lungs inflicted with obstructive diseases, which often prolong the retention of NPs [210, 211]. The residence time of PLGA NPs in unobstructed lungs also remains a question. A study reported that intratracheally instilled PLGA NPs were rapidly absorbed in the blood (within 0.25 hours); therefore, the long-term tissue response was minimal [212].

In addition to PLGA, other biodegradable polymers can be used for inhalational drug delivery. Examples include poly(ether-anhydrides) [213], chitosan [123], and branched polyesters such as diethylaminopropylamine poly(vinyl alcohol)-grafted-poly(lactic-co-glycolic acid) (DEAPA-PVAL-g-PLGA) [170, 187, 214]. Unfortunately, not much is known for their *in vivo* toxicological effects. DEAPA-PVAL-g-PLGA was synthesized by grafting polyvinyl alcohol (PVAL) and diethylaminopropylamine (DEAPA) to PLGA. The resulting polymer carried cationic charges and degraded more rapidly than PLGA. The DEAPA-PVAL-g-PLGA NPs induced significantly fewer cell deaths than PS NPs. However, there was no significant difference between the PLGA NPs and DEAPA-PVAL-g-PLGA NPs in the *in vivo* LDH release after the intratracheal instillation of the NP [187].

8.4 Conclusions

Microparticles with an aerodynamic diameter adequate for inhalation are important carriers for drug delivery to the lungs for both local and systemic therapies. Inhalable dry particles have widely been used for asthma and COPD. While the majority of commercial products are based on micronized drugs blended with lactose carriers, a number of new technologies are being developed to provide particles with a more reliable and consistent performance. NPs hold great promise as new drug carriers for inhalational therapy. The interest in the nanotechnology in pulmonary drug delivery will continue to grow with advances in new strategies to facilitate the delivery of NPs. On the other hand, little toxicological data is available for the potential effects of inhaled NPs on the patients, especially those with pulmonary diseases. The development of new inhalable drug carriers should be accompanied by toxicological evaluation of their local and systemic effects.

References

1. Kleinstreuer, C., Zhang, Z., and Donohue, J. F. (2008). Targeted drug-aerosol delivery in the human respiratory system, *Annu. Rev. Biomed. Eng.*, **10**, 195–220.
2. Patton, J. S. (1996). Mechanisms of macromolecule absorption by the lungs, *Adv. Drug Delivery Rev.*, **19**(1), 3–36.
3. Schanker, L. L. S. (1978). Drug absorption from the lung, *Biochem. Pharmacol.*, **27**(4), 381–385.
4. Patton, J. S., Fishburn, C. S., Weers, J. G. (2004). The lungs as a portal of entry for systemic drug delivery, *Proc. Am. Thorac. Soc.*, **1**(4), 338–344.

5. Tronde, A., Nordén, B., Marchner, H., Wendel, A. K., Lennernäs, H., and Bengtsson, U. H. (2003). Pulmonary absorption rate and bioavailability of drugs in vivo in rats: structure-absorption relationships and physicochemical profiling of inhaled drugs, *J. Pharm. Sci.*, **92**(6), 1216–1233.
6. Patton, J. S., and Byron, P. R. (2007). Inhaling medicines: delivering drugs to the body through the lungs, *Nat. Rev. Drug Discovery*, **6**, 67–74.
7. White, S. S., Bennett, D. B., Cheu, S. S., Conley, P. W., Guzek, D. B., Gray, S. S., Howard, J., Malcolmson, R., Parker, J. M., and Roberts, P. (2005). EXUBERA: pharmaceutical development of a novel product for pulmonary delivery of insulin, *Diabetes Technol. Ther.*, **7**(6), 896–906.
8. Pearson, J. (2006). Inhalation technologies – a breath of fresh air, *Drug Del. Rep.*, Spring/Summer, 19–21.
9. Brunton, S. (2008). Insulin delivery systems: reducing barriers to insulin therapy and advancing diabetes mellitus treatment, *Am. J. Med.*, **121**(6 Suppl), S35–41.
10. Hickey, A. J. (2005). Pulmonary drug delivery: pharmaceutical chemistry and aerosol technology, in *Drug Delivery Principles and Applications* (ed. Wang, B., Siahaan, T., and Soltero, R. A.), John Wiley & Sons, Hoboken, New Jersey, pp. 341–361.
11. McDonald, K. J., and Martin, G. P. (2000). Transition to CFC-free metered dose inhalers – into the new millennium, *Int. J. Pharm.*, **201**(1), 89–107.
12. Newman, S. P. (2004). Dry powder inhalers for optimal drug delivery, *Expert Opin. Biol. Ther.*, **4**(1), 23–33.
13. Son, Y.-J., and McConville, J. T. (2008). Advancements in dry powder delivery to the lung, *Drug Dev. Ind. Pharm.*, **34**(9), 948–959.
14. Leach, C. L., Davidson, P. J., and Boudreau, R. J. (1998). Improved airway targeting with the CFC-free HFA-beclomethasone metered-dose inhaler compared with CFC-beclomethasone, *Eur. Respir. J.*, **12**(6), 1346–1353.
15. Acerbi, D., Brambilla, G., and Kottakis, I. (2007). Advances in asthma and COPD management: delivering CFC-free inhaled therapy using Modulite® technology, *Pulmonary Pharmacol. Ther.*, **20**(3), 290–303.
16. Zeidler, M. M., and Corren, J. J. (2004). Hydrofluoroalkane formulations of inhaled corticosteroids for the treatment of asthma, *Treat. Respir. Med.*, **3**(1), 35–44.
17. Hendeles, L., Colice, G. L., and Meyer, R. J. (2007). Withdrawal of albuterol inhalers containing chlorofluorocarbon propellants, *N. Engl. J. Med.*, **356**(13), 1344–1351.
18. Atkins, P. J. (2005). Dry powder inhalers: an overview, *Respir. Care*, **50**(10), 1304–1312; discussion 1312.
19. Johnson, C. C. E. (1989). Principles of nebulizer-delivered drug therapy for asthma, *Am. J. Hosp. Pharm.*, **46**(9), 1845–1855.
20. Smith, I. J., and Parry-Billings, M. M. (2003). The inhalers of the future? A review of dry powder devices on the market today, *Pulm. Pharmacol. Ther.*, **16**(2), 79–95.
21. Van Campen, L., and Venthoye, G. (2006). Inhalation: dry powder, *Encyclopedia Pharm. Technol.*, **3**(1), 2077–2091.
22. O'Riordan, T. G. (2002). Formulations and nebulizer performance, *Respir. Care*, **47**(11), 1305–1312; discussion 1312.

23. Geller, D. E. (2005). Comparing clinical features of the nebulizer, metered-dose inhaler, and dry powder inhaler, *Respir. Care*, **50**(10), 1313–1321; discussion 1321.

24. Dalby, R., Spallek, M., and Voshaar, T. (2004). A review of the development of Respimat® Soft Mist(TM) Inhaler, *Int. J. Pharm.*, **283**(1–2), 1–9.

25. Watts, A. B., McConville, J. T., and Williams, R. O. (2008). Current therapies and technological advances in aqueous aerosol drug delivery, *Drug Dev. Ind. Pharm.*, **34**(9), 913–922.

26. Newman, S. P., and Busse, W. W. (2002). Evolution of dry powder inhaler design, formulation, and performance, *Respir. Med.*, **96**(5), 293–304.

27. Islam, N., and Gladki, E. (2008). Dry powder inhalers (DPIs)–a review of device reliability and innovation. *Int. J. Pharm.*, **360**(1–2), 1–11.

28. Borgstrom, L., Bisgaard, H., O'Callaghan, C., and Pedersen, S. (2002). Dry-powder inhalers, in *Drug Delivery to the Lung* (ed. Bisgaard, H., O'Callaghan, C., and Smaldone, G. C.) Marcel Dekker, New York, pp. 421–448.

29. Leach, C. L. (2007). Inhalation aspects of therapeutic aerosols, *Toxicol. Pathol.*, **35**(1), 23–26.

30. Prime, D., Atkins, P. J., Slater, A., and Sumby, B. (1997). Review of dry powder inhalers, *Adv. Drug Delivery Rev.*, **26**(1), 51–58.

31. Yang, W., Peters, J. I., and Williams, R. O., 3rd (2008). Inhaled nanoparticles–a current review. *Int. J. Pharm.*, **356**(1–2), 239–247.

32. Edwards, D. A., Ben-Jebria, A., and Langer, R. (1998). Recent advances in pulmonary drug delivery using large, porous inhaled particles, *J. Appl. Physiol.*, **85**(2), 379–385.

33. Bosquillon, C., Lombry, C., Preat, V., and Vanbever, R. (2001). Comparison of particle sizing techniques in the case of inhalation dry powders, *J. Pharm. Sci.*, **90**(12), 2032–2041.

34. Bosquillon, C., Lombry, C., Preat, V., and Vanbever, R. Influence of formulation excipients and physical characteristics of inhalation dry powders on their aerosolization performance, *J. Controlled Release*, **70**(3), 329.

35. Sosnowski, T. R., Moskal, A., and Gradon, L. (2007). Mechanism of aerosol particle deposition in the oro-pharynx under non-steady airflow, *Ann. Occup. Hyg.*, **51**(1), 19–25.

36. Chono, S., Tanino, T., Seki, T., and Morimoto, K. (2006). Influence of particle size on drug delivery to rat alveolar macrophages following pulmonary administration of ciprofloxacin incorporated into liposomes, *J. Drug Targeting*, **14**(8), 557–566.

37. Muttil, P., Kaur, J., Kumar, K., Yadav, A. B., Sharma, R., and Misra, A. (2007). Inhalable microparticles containing large payload of anti-tuberculosis drugs, *Eur. J. Pharm. Sci.*, **32**(2), 140–150.

38. Sharma, R., Saxena, D., Dwivedi, A. K., and Misra, A. (2001). Inhalable microparticles containing drug combinations to target alveolar macrophages for treatment of pulmonary tuberculosis, *Pharm. Res.*, **18**(10), 1405–1410.

39. Suarez, S., O'Hara, P., Kazantseva, M., Newcomer, C. E., Hopfer, R., McMurray, D. N., and Hickey, A. J. (2001). Airways delivery of rifampicin microparticles for the treatment of tuberculosis, *J. Antimicrob. Chemother.*, **48**(3), 431–434.

40. Chougule, M. B., Padhi, B. K., Jinturkar, K. A., and Misra, A. Development of dry powder inhalers, *Recent Pat. Drug Deliv. Formul.*, **1**(1), 11–21.

41. Weers, J. G., Tarara, T. E., and Clark, A. R. (2007). Design of fine particles for pulmonary drug delivery, *Expert Opin. Drug Delivery*, **4**(3), 297–313.

42. Ariyananda, P. L., Agnew, J. E., and Clarke, S. W. (1996). Aerosol delivery systems for bronchial asthma, *Postgrad. Med. J.*, **72**(845), 151–156.

43. Edwards, D. A., Hanes, J., Caponetti, G., Hrkach, J., Ben-Jebria, A., Eskew, M., Mintzes, J., Deaver, D., Lotan, N., and Langer, R. (1997). Large porous particles for pulmonary drug delivery, *Science*, **276**(5320), 1868–1871.

44. Smola, M. M., Vandamme, T. T., and Sokolowski, A. A. (2008). Nanocarriers as pulmonary drug delivery systems to treat and to diagnose respiratory and non respiratory diseases, *Int. J. Nanomed.*, **3**(1), 1–19.

45. Gelperina, S., Kisich, K., Iseman, M. D., and Heifets, L. (2005). The potential advantages of nanoparticle drug delivery systems in chemotherapy of tuberculosis, *Am. J. Respir. Crit. Care Med.*, **172**(12), 1487–1490.

46. Rogueda, P. G., and Traini, D. (2007). The nanoscale in pulmonary delivery. Part 1: deposition, fate, toxicology and effects, *Expert Opin. Drug. Deliv.*, **4**(6), 595–606.

47. Durcan, N., Murphy, C., and Cryan, S. A. Inhalable siRNA: potential as a therapeutic agent in the lungs, *Mol. Pharm.*, **5**(4), 559–566.

48. Farokhzad, O. C., and Langer, R. Impact of nanotechnology on drug delivery, *ACS Nano*, **3**(1), 16–20.

49. Peer, D., Karp, J. M., Hong, S., Farokhzad, O. C., Margalit, R., and Langer, R. (2007). Nanocarriers as an emerging platform for cancer therapy, *Nat. Nano*, **2**(12), 751–760.

50. Alexis, F., Pridgen, E., Molnar, L. K., and Farokhzad, O. C. (2008). Factors affecting the clearance and biodistribution of polymeric nanoparticles, *Mol. Pharm.*, **5**(4), 505–515.

51. Pridgen, E. M., Langer, R., and Farokhzad, O. C. (2007). Biodegradable, polymeric nanoparticle delivery systems for cancer therapy, *Nanomedicine*, **2**(5), 669–680.

52. Tabata, Y., and Ikada, Y. (1998). Effect of the size and surface charge of polymer microspheres on their phagocytosis by macrophage, *Biomaterials*, **9**(4), 356.

53. Suk, J. S., Lai, S. K., Wang, Y.-Y., Ensign, L. M., Zeitlin, P. L., Boyle, M. P., and Hanes, J. (2009). The penetration of fresh undiluted sputum expectorated by cystic fibrosis patients by non-adhesive polymer nanoparticles, *Biomaterials*, **30**(13), 2591–2597.

54. Lai, S. K., O'Hanlon, D. E., Harrold, S., Man, S. T., Wang, Y.-Y., Cone, R, and Hanes, J. (2007). Rapid transport of large polymeric nanoparticles in fresh undiluted human mucus, *Proc. Natl. Acad. Sci.*, **104**(5), 1482–1487.

55. Lai, S. K., Wang, Y.-Y., and Hanes, J. (2009). Mucus-penetrating nanoparticles for drug and gene delivery to mucosal tissues, *Adv. Drug Delivery Rev.*, **61**(2), 158–171.

56. Hanes, J., and Demeester, J. (2009). Drug and gene delivery to mucosal tissues: the mucus barrier, *Adv. Drug Delivery Rev.*, **61**(2), 73–74.

57. Rogueda, P. G., and Traini, D. (2007). The nanoscale in pulmonary delivery. Part 2: formulation platforms, *Expert Opin. Drug Deliv.*, **4**(6), 607–620.

58. Rytting, E., Nguyen, J., Wang, X., and Kissel, T. (2008). Biodegradable polymeric nanocarriers for pulmonary drug delivery, *Expert Opin. Drug Delivery*, **5**(6), 629–639.

59. Sung, J. C., Padilla, D. J., Garcia-Contreras, L., Verberkmoes, J. L., Durbin, D., Peloquin, C. A., Elbert, K. J., Hickey, A. J., and Edwards, D. C. (2009). Formulation and pharmacokinetics of self-assembled Rifampicin nanoparticle systems for pulmonary delivery, *Pharm. Res.*, **26**(8), 1847–1855.

60. Tsapis, N., Bennett, D., Jackson, B., Weitz, D. A., and Edwards, D. A. (2002). Trojan particles: large porous carriers of nanoparticles for drug delivery, *PNAS*, **99**(19), 12001–12005.

61. Sung, J. C., Pulliam, B. L., and Edwards, D. A. (2007). Nanoparticles for drug delivery to the lungs, *Trends Biotechnol.*, **25**(12), 563–570.

62. Sham, J. O., Zhang, Y., Finlay, W. H., Roa, W. H., Löbenberg, R. (2004). Formulation and characterization of spray-dried powders containing nanoparticles for aerosol delivery to the lung, *Int. J. Pharm.*, **269**, 457–467.

63. Ely, L., Roa, W., Finlay, W. H., and Lobenberg, R. (2007). Effervescent dry powder for respiratory drug delivery, *Eur. J. Pharm. Biopharm.*, **65**(3), 346–353.

64. Shekunov, B. Y., Feeley, J. C., Chow, A. H. L., Tong, H. H. Y., and York, P. (2003). Aerosolisation behaviour of micronised and supercritically-processed powders, *J. Aerosol Sci.*, **34**(5), 553–568.

65. Chow, A. H. L., Tong, H. H. Y., Chattopadhyay, P., and Shekunov, B. Y. (2007). Particle engineering for pulmonary drug delivery, *Pharm. Res.*, **24**(3), 411–437.

66. Gombotz, W. R., Healy, M. S., Brown, L. R. (inventors) Enzytech (assignee) (1991). Very low temperature casting of controlled release microspheres, US Patent 5019400.

67. Rasenack, N., and Muller, B. W. (2004). Micron-size drug particles: common and novel micronization techniques, *Pharm. Dev. Technol.*, **9**(1), 1–13.

68. Shoyele, S. A., and Cawthome, S. (2006). Particle engineering techniques for inhaled biopharmaceuticals, *Adv. Drug Delivery Rev.*, **58**(9–10), 1009–1029.

69. Salekigerhardt, A., Ahlneck, C., and Zografi, G. (1994). Assessment of disorder in crystalline solids, *Int. J. Pharm.*, **101**(3), 237–247.

70. Mahavir, B., Chougule, B. K. P., Jinturkar, K. A., and Misra, A. (2007). Development of dry powder inhalers, *Recent Pat. Drug Deliv. Formul.*, **1** (1), 11–21.

71. Thi, T. H. H., Danede, F., Descamps, M., and Flament, M. P. (2008). Comparison of physical and inhalation properties of spray-dried and micronized terbutaline sulphate, *Eur. J. Pharm. Biopharm.*, **70**(1), 380–388.

72. Malcolmson, R. J., and Embleton, J. K. (1998). Dry powder formulations for pulmonary delivery, *Pharm. Sci. Technol. Today.*, **1**(9), 394–398.

73. Ward, G. H., and Schultz, R. K. (1995). Process-induced crystallinity changes in albuterol sulfate and its effect on powder physical stability, *Pharm. Res.*, **12**(5), 773–779.

74. Ticehurst, M. D., Basford, P. A., Dallman, C. I., Lukas, T. M., Marshall, P. V., Nichols, G., and Smith, D. (2000). Characterisation of the influence of micronisation on the crystallinity and physical stability of revatropate hydrobromide, *Int. J. Pharm.*, **193**(2), 247–259.

75. Kitamura, S., Miyamae, A., Koda, S., and Morimoto, Y. (1989). Effect of grinding on the solid-state stability of cefixime trihydrate, *Int. J. Pharm.*, **56**(2), 125–134.

76. Waltersson, J., and Lundgren, P. (1985). The effect of mechanical comminution on drug stability, *Acta Pharm. Suec.*, **22**(5), 291–300.

77. Irngartinger, M., Camuglia, V., Damm, M., Goede, J., and Frijlink, H. W. (2004). Pulmonary delivery of therapeutic peptides via dry powder inhalation: effects of micronisation and manufacturing, *Eur. J. Pharm. Biopharm.*, **58**(1), 7–14.

78. Chan, H. K., Clark, A., Gonda, I., Mumenthaler, M., and Hsu, C. (1997). Spray dried powders and powder blends of recombinant human deoxyribonuclease (rhDNase) for aerosol delivery, *Pharm. Res.*, **14**(4), 431–437.

79. Harjunen, P., Lehto, V. P., Martimo, K., Suihko, E., Lankinen, T., Paronen, P., and Järvinen, K. (2002). Lactose modifications enhance its drug performance in the novel multiple dose Taifun (R) DPI, *Eur. J. Pharm. Sci.*, **16**(4–5), 313–321.
80. Chan, H. K. (2005). Dry powder aerosol delivery systems: current and future research directions, *J. Aerosol Med.*, **19**(1), 21–27.
81. Daniher, D. I., and Zhu, J. (2008). Dry powder platform for pulmonary drug delivery, *Particuologly*, **6**(4), 225–238.
82. Kohlhaufl, M., Haidl, P., Voshaar, T., Haussinger, K., and Kohler, D. (2004). Dry powder inhalers – a review, *Dtsch. Med. Wochenschr.*, **129**(39), 2048–2052.
83. Lucas, P., Anderson, K., and Staniforth, J. N. (1998). Protein deposition from dry powder inhalers: fine particle multiplets as performance modifiers, *Pharm. Res.*, **15**(4), 562–569.
84. Patton, J. S., Bukar, J., and Nagarajan, S. (1999). Inhaled insulin, *Adv. Drug Delivery Rev.*, **35**(2–3), 235–247.
85. Begat, P., Morton, D. A., Shur, J., Kippax, P., Staniforth, J. N., and Price, R. The role of force control agents in high-dose dry powder inhaler formulations, *J. Pharm. Sci.*, **98**(8), 2770–2783.
86. Jones, M. D., and Price, R. (2006). The influence of fine excipient particles on the performance of carrier-based dry powder inhalation formulations, *Pharm. Res.*, **23**(8), 1665–1674.
87. Larhrib H., Martin, G. P., Prime, D., and Marriott, C. (2003). Characterisation and deposition studies of engineered lactose crystals with potential for use as a carrier for aerosolised salbutamol sulfate from dry powder inhalers, *Eur. J. Pharm. Sci.*, **19**(4), 211–221.
88. Heng, P. W. S., Chan, L. W., and Lim, L. T. (2000). Quantification of the surface morphologies of lactose carriers and their effect on the in vitro deposition of salbutamol sulphate, *Chem. Pharm. Bulletin.*, **48**(3), 393–398.
89. Vanderbist, F., Wery, B., Moyano-Pavon, I., and Moes, A. J. (1999). Optimization of a dry powder inhaler formulation of nacystelyn, a new mucoactive agent, *J. Pharm. Pharmacol.*, **51**(11), 1229–1234.
90. Zeng, X. M., Martin, G. P., Marriott, C., and Pritchard, J. (2001). Lactose as a carrier in dry powder formulations: the influence of surface characteristics on drug delivery, *J. Pharm. Sci.*, **90**(9), 1424–1434.
91. Flament, M. P., Leterme, P., and Gayot, A. (2004). The influence of carrier roughness on adhesion, content uniformity and the in vitro deposition of terbutaline sulphate from dry powder inhalers, *Int. J. Pharm.*, **275**(1–2), 201–209.
92. Zeng, X. M., Martin, G. P., Marriott, C., and Pritchard, J. (2001). The use of lactose recrystallised from carbopol gels as a carrier for aerosolised salbutamol sulphate, *Eur. J. Pharm. Biopharm.*, **51**(1), 55–62.
93. Zeng, X. M., Martin, G. P., Tee, S. K., Ghoush, A. A., and Marriott, C. (1999). Effects of particle size and adding sequence of fine lactose on the deposition of salbutamol sulphate from a dry powder formulation, *Int. J. Pharm.*, **182**(2), 133–144.
94. Zeng, X. M., Pandhal, K. H., and Martin, G. P. (2000). The influence of lactose carrier on the content homogeneity and dispersibility of beclomethasone dipropionate from dry powder aerosols, *Int. J. Pharm.*, **197**(1–2), 41–52.
95. Hu, T. T., Zhao, H., Jiang, L. C., Le, Y., Chen, J. F., and Yun, J. (2008). Engineering pharmaceutical fine particles of budesonide for dry powder inhalation (DPI), *Ind. Eng. Chem. Res.*, **47**(23), 9623–9627.

96. Murnane, D., Martin, G. P., and Marriott, C. (2009). Dry powder formulations for inhalation of fluticasone propionate and salmeterol xinafoate microcrystals, *J. Pharm. Sci.*, **98**(2), 503–515.
97. Murnane, D., Marriott, C., and Martin, G. P. (2008). Comparison of salmeterol xinafoate microparticle production by conventional and novel antisolvent crystallization, *Eur. J. Pharm. Biopharm.*, **69**(1), 94–105.
98. Murnane, D., Marriott, C., and Martin, G. P. (2008). Polymorphic control of inhalation microparticles prepared by crystallization, *Int. J. Pharm.*, **361**(1–2), 141–149.
99. Rasenack, N., Steckel, H., and Muller, B. W. (2004). Preparation of microcrystals by in situ micronization, *Powder Technol.*, **143**:291–296.
100. Rasenack, N., Hartenhauer, H., and Muller, B. W. (2003). Microcrystals for dissolution rate enhancement of poorly water-soluble drugs, *Int. J. Pharm.*, **254**(2), 137–145.
101. Rasenack, N., Steckel, H., and Muller, B. W. (2003). Micronization of anti-inflammatory drugs for pulmonary delivery by a controlled crystallization process, *J. Pharm. Sci.*, **92**(1), 35–44.
102. Rasenack, N., and Muller, B. W. (2002). Dissolution rate enhancement by in situ micronization of poorly water-soluble drugs, *Pharm. Res.*, **19**(12), 1894–1900.
103. Yang, Z. Y., Le, Y., Hu, T. T., Shen, Z. G., Chen J. F., and Yun, J. (2008). Production of ultrafine sumatriptan succinate particles for pulmonary delivery, *Pharm. Res.*, **25**(9), 2012–2018.
104. Dhumal, R. S., Biradar, S. V., Paradkar, A. R., and York, P. (2009). Particle engineering using sonocrystallization: salbutamol sulphate for pulmonary delivery, *Int. J. Pharm.*, **368**(1–2), 129–137.
105. Sarkari, M., Brown, J., Chen, X. X., Swinnea, S., Williams, R. O., and Johnston, K. P. (2002). Enhanced drug dissolution using evaporative precipitation into aqueous solution, *Int. J. Pharm.*, **243**(1–2), 17–31.
106. Chen, X. X., Young, T. J., Sarkari, M., Williams, R. O., and Johnston, K. P. (2002). Preparation of cyclosporine A nanoparticles by evaporative precipitation into aqueous solution, *Int. J. Pharm.*, **242**(1–2), 3–14.
107. Pasquali, I., Bettini, R., and Giordano, F. (2006). Solid-state chemistry and particle engineering with supercritical fluids in pharmaceutics, *Eur. J. Pharm. Sci.*, **27**(4), 299–310.
108. Davey, R. J., Blagden, N., Potts, G. D., and Docherty, R. (1997). Polymorphism in molecular crystals: stabilization of a metastable form by conformational mimicry, *J. Am. Chem. Soc.*, **119**(7), 1767–1772.
109. Edwards, A. D., Shekunov, B. Y., Kordikowski, A., Forbes, R. T., and York, P. (2001). Crystallization of pure anhydrous polymorphs of carbamazepine by solution enhanced dispersion with supercritical fluids (SEDS (TM)), *J. Pharm. Sci.*, **90**(8), 1115–1124.
110. Vehring, R. (2008). Pharmaceutical particle engineering via spray drying, *Pharm. Res.*, **25**(5), 999–1022.
111. Stahl, K., Claesson, M., Lilliehorn, P., Linden, H., and Backstrom, K. (2002). The effect of process variables on the degradation and physical properties of spray dried insulin intended for inhalation, *Int. J. Pharm.*, **233**(1–2), 227–237.
112. Kraft, K. S., and Grant, M. (2009). Preparation of macromolecule-containing dry powders for pulmonary delivery, *Methods Mol. Biol.*, **480**, 165–174.
113. Dellamary, L. A., Tarara, T. E., Smith, D. J., Woelk, C. H., Adractas, A., Costello, M. L., Gill, H., and Weers, J. G. (2000). Hollow porous particles in metered dose inhalers, *Pharm. Res.*, **17**(2), 168–174.

114. Duddu, S. P., Sisk, S. A., Walter, Y. H., Tarara, T. E., Trimble, K. R., Clark, A. R., Eldon, M. A., Elton, R. C., Pickford, M., Hirst, P. H., Newman, S. P., and Weers, J. G. Improved lung delivery from a passive dry powder inhaler using an engineered PulmoSphere (R) powder, *Pharm. Res.*, **19**(5), 689–695.

115. Rogers, T. L., Johnston, K. P., and Williams, R. O. (2001). Solution-based particle formation of pharmaceutical powders by supercritical or compressed fluid CO_2 and cryogenic spray-freezing technologies, *Drug Dev. Ind. Pharm.*, **27**(10), 1003–1015.

116. Maa, Y. F., Ameri, M., Shu, C., Payne, L. G., and Chen, D. X. (2004). Influenza vaccine powder formulation development: spray-freeze-drying and stability evaluation, *J. Pharm. Sci.*, **93**(7), 1912–1923.

117. Sweeney, L. G., Wang, Z. L., Loebenberg, R., Wong, J. P., Lange, C. F., and Finlay, W. H. (2005). Spray-freeze-dried liposomal ciprofloxacin powder for inhaled aerosol drug delivery, *Int. J. Pharm.*, **305**(1–2), 180–185.

118. Maa, Y. F., Nguyen, P. A., Sweeney, T., Shire, S. J., and Hsu, C. C. (1999). Protein inhalation powders: spray drying vs spray freeze drying, *Pharm. Res.*, **16**(2), 249–254.

119. Webb, S. D., Golledge, S. L., Cleland, J. L., Carpenter, J. F., and Randolph, T. W. (2002). Surface adsorption of recombinant human interferon-gamma in lyophilized and spray-lyophilized formulations, *J. Pharm. Sci.*, **91**(6), 1474–1487.

120. Costantino, H. R., Firouzabadian, L., Hogeland, K., Wu, C. C., Beganski, C., Carrasquillo, K. G., Córdova, M., Griebenow, K., Zale, S. E., and Tracy, M. A. (2000). Protein spray-freeze drying. Effect of atomization conditions on particle size and stability, *Pharm. Res.*, **17**(11), 1374–1383.

121. Costantino, H. R., Firouzabadian, L., Wu, C. C., Carrasquillo, K. G., Griebenow, K., Zale, S. E., and Tracy, M. A. (2002). Protein spray freeze drying. 2. Effect of formulation variables on particle size and stability, *J. Pharm. Sci.*, **91**(2), 388–395.

122. Martin, A., and Cocero, M. J. (2008). Micronization processes with supercritical fluids: fundamentals and mechanisms, *Adv. Drug Delivery Rev.*, **60**(3), 339–350.

123. Grenha, A., Seijo, B., and Remunan-Lopez, C. (2005). Microencapsulated chitosan nanoparticles for lung protein delivery, *Eur. J. Pharm. Sci.*, **25**(4–5), 427–437.

124. Courrier, H. M., Butz, N., and Vandamme, T. F. (2002). Pulmonary drug delivery systems: recent developments and prospects, *Crit. Rev. Ther. Drug Carrier Sys.*, **19**(4–5), 425–498.

125. Vanbever, R., Mintzes, J. D., Wang, J., Nice, J., Chen, D. H., Batycky, R., Langer, R., and Edwards, D. A. (1999). Formulation and physical characterization of large porous particles for inhalation, *Pharm. Res.*, **16**(11), 1735–1742.

126. Dunbar C., Scheuch G., Sommerer K., DeLong M., Verma A., and Batycky, R. (2002). In vitro and in vivo dose delivery characteristics of large porous particles for inhalation, *Int. J. Pharm.*, **245**(1–2), 179–189.

127. Ben-Jebria, A., Eskew, M. L., and Edwards, D. A. (2000). Inhalation system for pulmonary aerosol drug delivery in rodents using large porous particles, *Aerosol Sci. Technol.*, **32**(5), 421–433.

128. Wang, J., Ben-Jebria, A., and Edwards, D. A. (1999). Inhalation of estradiol for sustained systemic delivery, *J. Aerosol Med.-Deposition Clearance Eff. Lung*, **12**(1), 27–36.

129. Ben-Jebria, A., Chen, D. H., Eskew, M. L., Vanbever, R., Langer, R., and Edwards, D. A. (1999). Large porous particles for sustained protection from carbachol-induced bronchoconstriction in guinea pigs, *Pharm. Res.*, **16**(4), 555–561.

130. Bosquillon, C., Rouxhet, P. G., Ahimou, F., Simon, D., Culot, C., Preat, V., and Vanbever, R. (2004). Aerosolization properties, surface composition and physical state of spray-dried protein powders, *J. Controlled Release*, **99**(3), 357–367.

131. Tsapis, N., Bennett, D., O'Driscoll, K., Shea, K., Lipp, M. M., Fu, K., Clarke, R. W., Deaver, D., Yamins, D., Wright, J., *et al.* (2003). Direct lung delivery of para-aminosalicylic acid by aerosol particles, *Tuberculosis*, **83**(6), 379–385.

132. Chew, N. Y. K., and Chan, H. K. (2001). Use of solid corrugated particles to enhance powder aerosol performance, *Pharm. Res.*, **18**(11), 1570–1577.

133. Faldt, P., and Bergenstahl, B. (1994). The surface-composition of spray-dried protein lactose powders, *Colloids Surf., A*, **90**(2–3), 183–190.

134. Chew, N. Y. K., Tang, P., Chan, H. K., and Raper, J. A. (2005). How much particle surface corrugation is sufficient to improve aerosol performance of powders? *Pharm. Res.*, **22**(1), 148–152.

135. Conte, U., Giunchedi, P., Maggi, L., and Torre, M. L. (1994). Spray-dried albumin microspheres containing nicardipine, *Eur. J. Pharm. Biopharm.*, **40**(4), 203–208.

136. Sung, J. C., Garcia-Contreras, L., VerBerkmoes, J. L., Peloquin, C. A., Elbert, K. J., Hickey, A. J., and Edwards, D. A. (2009). Dry powder nitroimidazopyran antibiotic PA-824 aerosol for inhalation, *Antimicrob. Agents Chemother.*, **53**(4), 1338–1343.

137. Vanbever, R., Ben-Jebria, A., Mintzes, J. D., Langer, R., and Edwards, D. A. (1999). Sustained release of insulin from insoluble inhaled particles, *Drug Dev. Res.*, **48**(4), 178–185.

138. Tsifansky, M. D., Yeo, Y., Evgenov, O. V., Bellas, E., Benjamin, J., and Kohane, D. S. (2008). Microparticles for inhalational delivery of antipseudomonal antibiotics, *AAPS J.*, **10**(2), 254–260.

139. Ungaro, F., Bianca, R., Giovino, C., Miro, A., Sorrentino, R., Quaglia, F., and La Rotonda, M. I. (2009). Insulin-loaded PLGA/cyclodextrin large porous particles with improved aerosolization properties: in vivo deposition and hypoglycaemic activity after delivery to rat lungs, *J. Controlled Release*, **135**(1), 25–34.

140. Ungaro, F., De Rosa, G., Miro, A., Quaglia, F., and La Rotonda, M. I. (2006). Cyclodextrins in the production of large porous particles: development of dry powders for the sustained release of insulin to the lungs, *Eur. J. Pharm. Sci.*, **28**(5), 423–432.

141. Lee, E. S., Kwon, M. J., Na, K., and Bae, J. H. (2007). Protein release behavior from porous microparticle with lysozyme/hyaluronate ionic complex, *Colloids Surf. B Biointerfaces*, **55**(1), 125–130.

142. Kwona, M. J., Baea, J. H., Kima, J. J., Nab, K, and Lee, E. S. (2007). Long acting porous microparticle for pulmonary protein delivery, *Int. J. Pharm.*, **333**(1–2), 5–9.

143. Arnold, M. M., Gonnan, E. M., Schieber, L. J., Munson, E. J., and Berkland, C. (2007). NanoCipro encapsulation in monodisperse large porous PLGA microparticles, *J. Controlled Release*, **121**(1–2), 100–109.

144. Koushik, K., and Kompella, U. B. (2004). Preparation of large porous deslorelin-PLGA microparticles with reduced residual solvent and cellular uptake using a supercritical carbon dioxide process, *Pharm. Res.*, **21**(3), 524–535.

145. Koushik, K., Dhanda, D. S., Cheruvu, N. P. S., and Kompella, U. B. (2004). Pulmonary delivery of deslorelin: large-porous PLGA particles and HP beta CD complexes, *Pharm. Res.*, **21**(7), 1119–1126.

146. Yang, Y., Bajaj, N., Xu, P., Ohn, K., Tsifansky, M. D., and Yeo, Y. (2009). Development of highly porous large PLGA microparticles for pulmonary drug delivery, *Biomaterials*, **30**(10), 1947–1953.

147. Kim, T. K., Yoon, J. J., Lee, D. S., and Park, T. G. (2006). Gas foamed open porous biodegradable polymeric microspheres, *Biomaterials*, **27**(2), 152–159.

148. Swarbrick, J., and Boylan, J. C. (eds) (1994). *Encyclopedia of Pharmaceutical Technology*, New York, Marcel Dekker.

149. Grenha, A., Carrion-Recio, D., Teijeiro-Osorio, D., Seijo, B., and Remunan-Lopez, C. (2008). Nano- and micro-particulate carriers for pulmonary drug delivery, *Handb Part Drug Delivery*, **2**,165–192.

150. Alonso, M. J. (2004). Nanomedicines for overcoming biological barriers, *Biomed. Pharmacother.*, **58**(3), 168–172.

151. Moghimi, S. M., Hunter, A. C., and Murray, J. C. (2001). Long-circulating and target-specific nanoparticles: theory to practice, *Pharmacol. Rev.*, **53**(2), 283–318.

152. Brannon-Peppas, L., and Blanchette, J. O. (2004). Nanoparticle and targeted systems for cancer therapy, *Adv. Drug Delivery Rev.*, **56**(11), 1649–1659.

153. Li, S.-D., and Huang, L. (2006). Surface-modified LPD nanoparticles for tumor targeting, *Ann. N. Y. Acad. Sci.*, **1082**(Oligonucleotide Therapeutics), 1–8.

154. Wendorf, J., Singh, M., Chesko, J., Kazzaz, J., Soewanan, E., Ugozzoli, M., and O'Hagan, D. (2006). A practical approach to the use of nanoparticles for vaccine delivery, *J. Pharm. Sci.*, **95**(12), 2738–2750.

155. Duncan, G., Jess, T. J., Mohamed, F., Price, N. C., Kelly, S. M., and van der Walle, C. F. (2005). The influence of protein solubilization, conformation and size on the burst release from poly(lactide-co-glycolide) microspheres, *J. Controlled Release*, **110**(1), 34–48.

156. Pandey, R., Sharma, A., Zahoor, A., Sharma, S., Khuller, G. K., and Prasad, B. (2003). Poly (DL-lactide-co-glycolide) nanoparticle-based inhalable sustained drug delivery system for experimental tuberculosis, *J. Antimicrob. Chemother.*, **52**(6), 981–986.

157. Bhavna, Ahmad, F. J., Mittal, G., Jain, G. K., Malhotra, G., Khar, R. K., and Bhatnagar, A. (2009). Nano-salbutamol dry powder inhalation: a new approach for treating broncho-constrictive conditions, *Eur. J. Pharm. Biopharm.*, **71**(2), 282–291.

158. Saez, A., Guzman, M., Molpeceres, J., and Aberturas, M. R. (2000). Freeze-drying of polycaprolactone and poly(d,l-lactic-glycolic) nanoparticles induce minor particle size changes affecting the oral pharmacokinetics of loaded drugs, *Eur. J. Pharm. Biopharm.*, **50**(3), 379–387.

159. Yang, W., Tam, J., Miller, D. A., Zhou, J., McConville, J. T., Johnston, K. P., and Williams, R. O., 3rd (2008). High bioavailability from nebulized itraconazole nanoparticle dispersions with biocompatible stabilizers, *Int. J. Pharm.*, **361**(1–2), 177–188.

160. Vaughn, J. M., McConville, J. T., Burgess, D., Peters, J. I., Johnston, K. P., Talbert R. L., and Williams, R. O., 3rd (2006). Single dose and multiple dose studies of itraconazole nanoparticles, *Eur. J. Pharm. Biopharm.*, **63**(2), 95–102.

161. Loffert, D. T., Ikle, D., and Nelson, H. S. A. (1994). Comparison of commercial jet nebulizers, *Chest*, **106**(6), 1788–1792.

162. Smith, E. C., Denyer, J., and Kendrick, A. H. (1995). Comparison of twenty three nebulizer/compressor combinations for domiciliary use, *Eur. Respir. J.*, **8**(7), 1214–1221.

163. Le Brun, P. P. H., de Boer, A. H., Gjaltema, D., Hagedoorn, P., Heijerman, H. G. M., and Frijlink, H. W. (1999). Inhalation of tobramycin in cystic fibrosis part 1: the choice of a nebulizer, *Int. J. Pharm.*, **189**(2), 205–214.

164. Kawashima, Y., Yamamoto, H., Takeuchi, H., Fujioka, S., and Hino, T. (1999). Pulmonary delivery of insulin with nebulized DL-lactide/glycolide copolymer (PLGA) nanospheres to prolong hypoglycemic effect, *J. Controlled Release*, **62**(1–2), 279–287.

165. Zahoor, A., Sharma, S., and Khuller, G. K. (2005). Inhalable alginate nanoparticles as antitubercular drug carriers against experimental tuberculosis, *Int. J. Antimicrob. Agents*, **26**(4), 298–303.

166. Sharma, A., Sharma, S., and Khuller, G. K. (2004). Lectin-functionalized poly (lactide-co-glycolide) nanoparticles as oral/aerosolized antitubercular drug carriers for treatment of tuberculosis, *J. Antimicrob. Chemother.*, **54**(4), 761–766.

167. Yamamoto, H., Kuno, Y., Sugimoto, S., Takeuchi, H., and Kawashima, Y. (2005). Surface-modified PLGA nanosphere with chitosan improved pulmonary delivery of calcitonin by mucoadhesion and opening of the intercellular tight junctions, *J. Controlled Release*, **102**(2), 373–381.

168. McConville, J. T., Overhoff, K. A., Sinswat, P., Vaughn, J. M., Frei, B. L., Burgess, D. S., Talbert, R. L., Peters, J. I., Johnston, K. P., and Williams, R. O., 3rd (2006). Targeted high lung concentrations of itraconazole using nebulized dispersions in a murine model, *Pharm. Res.*, **23**(5), 901–911.

169. Kraft, W. K., Steiger, B., Beussink, D., Quiring, J. N., Fitzgerald, N., Greenberg, H. E., and Waldman, S. A. (2004). The pharmacokinetics of nebulized nanocrystal budesonide suspension in healthy volunteers, *J. Clin. Pharmacol.*, **44**(1), 67–72.

170. Beck-Broichsitter, M., Gauss, J., Packhaeuser, C. B., Lahnstein, K., Schmehl, T., Seeger W, Kissel, T., and Gessler, T. (2009). Pulmonary drug delivery with aerosolizable nanoparticles in an ex vivo lung model, *Int. J. Pharm.*, **367**(1–2), 169–178.

171. Hinds, W. C. (1998). *Aerosol Technology: Properties, Behavior, and Measurement of Airborne Particles*, John Wiley, New York.

172. Azarmi, S., Tao, X., Chen, H., Wang, Z., Finlay, W. H., Lobenberg, R., and Roa, W. H. (2006). Formulation and cytotoxicity of doxorubicin nanoparticles carried by dry powder aerosol particles, *Int. J. Pharm.*, **319**(1–2), 155–161.

173. Chougule, M. B., Padhi, B. K., and Misra, A. (2006). Nano-liposomal dry powder inhaler of amiloride hydrochloride, *J. Nanosci. Nanotechnol.*, **6**(9–10), 3001–3009.

174. Chougule, M., Padhi, B., and Misra, A. (2007). Nano-liposomal dry powder inhaler of tacrolimus: preparation, characterization, and pulmonary pharmacokinetics, *Int. J. Nanomed.*, **2**(4), 675–688.

175. Ohashi, K., Kabasawa, T., Ozeki, T., and Okada, H. (2009). One-step preparation of rifampicin/poly(lactic-co-glycolic acid) nanoparticle-containing mannitol microspheres using a four-fluid nozzle spray drier for inhalation therapy of tuberculosis, *J. Controlled Release.*, **135**(1), 19–24.

176. Azarmi, S., Lobenberg, R., Roa, W. H., Tai, S. S., and Finlay, W. H. (2008). Formulation and in vivo evaluation of effervescent inhalable carrier particles for pulmonary delivery of nanoparticles, *Drug Dev. Ind. Pharm.*, **34**(9), 943–947.

177. Hadinoto, K., Zhu, K., and Tan, R. B. (2007). Drug release study of large hollow nanoparticulate aggregates carrier particles for pulmonary delivery, *Int. J. Pharm.*, **341**(1–2), 195–206.

178. Cook, R. O., Pannu, R. K., and Kellaway, I. W. (2005). Novel sustained release microspheres for pulmonary drug delivery, *J. Controlled Release.*, **104**(1), 79–90.

179. Evora, C., Soriano, I., Rogers, R. A., Shakesheff, K. N., Hanes, J., and Langer, R. (1998). Relating the phagocytosis of microparticles by alveolar macrophages to surface chemistry: the effect of 1,2-dipalmitoylphosphatidylcholine, *J. Controlled Release*, **51**(2–3), 143–152.

180. Jones, B. G., Dickinson, P. A., Gumbleton, M., and Kellaway, I. W. (2002). Lung surfactant phospholipids inhibit the uptake of respirable microspheres by the alveolar macrophage NR8383, *J. Pharm. Pharmacol.*, **54**(8), 1065–1072.

181. Peek, L. J., Roberts, L., and Berkland, C. (2008). Poly(DL-lactide-co-glycolide) nanoparticle agglomerates as carriers in dry powder aerosol formulation of proteins, *Langmuir*, **24**(17), 9775–9783.

182. El-Gendy, N., and Berkland, C. (2009). Combination chemotherapeutic dry powder aerosols via controlled nanoparticle agglomeration, *Pharm. Res.*, **26**(7), 1752–1763.

183. Kawashima, Y., Serigano, T., Hino, T., Yamamoto, H., and Takeuchi, H. (1998). A new powder design method to improve inhalation efficiency of pranlukast hydrate dry powder aerosols by surface modification with hydroxypropylmethylcellulose phthalate nanospheres, *Pharm. Res.*, **15**(11), 1748–1752.

184. Talton, J. D., Fitz-Gerald, J. M., Singh, R. K., and Hochhaus, G. (2000). Nano-thin coating for improved lung targeting of glucocorticoid dry powder: in-vitro and in-vivo charactristics, *Respir. Drug Delivery*, **7**, 67–74.

185. Asgharian, B., and Price, O. T. (2007). Deposition of ultrafine (nano) particles in the human lung, *Inhal. Toxicol.*, **19**(13), 1045–1054.

186. Oberdorster, G., Stone, V., and Donaldson, K. (2007). Toxicology of nanoparticles: a historical perspective, *Nanotoxicology*, **1**(1), 2–25.

187. Dailey, L. A., Jekel, N., Fink, L., Gessler, T., Schmehl, T., Wittmar, M., Kissel, T., and Seeger, W. (2006). Investigation of the proinflammatory potential of biodegradable nanoparticle drug delivery systems in the lung, *Toxicol. Appl. Pharmacol.*, **215**(1), 100–108.

188. Brown, D. M., Wilson, M. R., MacNee, W., Stone, V., and Donaldson, K. (2001). Size-dependent proinflammatory effects of ultrafine polystyrene particles: a role for surface area and oxidative stress in the enhanced activity of ultrafines, *Toxicol. Appl. Pharmacol.*, **175**(3), 191–199.

189. Oberdorster, G., Finkelstein, J. N., Johnston, C., Gelein, R., Cox, C., Baggs, R., and Elder, A. C. (2000). Acute pulmonary effects of ultrafine particles in rats and mice, *Res. Rep. Health, Eff. Inst.*, (96), 5–74, disc 75–86.

190. Card, J. W., Zeldin, D. C., Bonner, J. C., and Nestmann, E. R. (2008). Pulmonary applications and toxicity of engineered nanoparticles, *Am. J. Physiol.*, **295**(3, Pt. 1), L400–L411.

191. Oberdorster, G., Oberdorster, E., and Oberdorster, J. (2005). Nanotoxicology: an emerging discipline evolving from studies of ultrafine particles, *Environ. Health Perspect.*, **113**(7), 823–839.

192. Hoet, P. H., Bruske-Hohlfeld, I., and Salata, O. V. (2004). Nanoparticles – known and unknown health risks, *J. Nanobiotechnol.*, **2**(1), 12.

193. Borm, P. J., and Kreyling, W. (2004). Toxicological hazards of inhaled nanoparticles–potential implications for drug delivery, *J. Nanosci. Nanotechnol.*, **4**(5), 521–531.

194. Oberdorster, G., Maynard, A., Donaldson, K., Castranova, V., Fitzpatrick, J., Ausman, K., Carter, J., Karn, B., Kreyling, W., Lai, D., *et al.* (2005). Principles for characterizing the potential human health effects from exposure to nanomaterials: elements of a screening strategy, *Part Fibre Toxicol.*, **2**, 8.

195. De Jong, W. H., and Borm, P. J. (2008). Drug delivery and nanoparticles: applications and hazards, *Int. J. Nanomed.*, **3**(2), 133–149.

196. Oberdorster, G., Sharp, Z., Atudorei, V., Elder, A., Gelein, R., Kreyling, W., and Cox, C. (2004). Translocation of inhaled ultrafine particles to the brain, *Inhal. Toxicol.*, **16**(6–7), 437–445.

197. Rancan, F., Rosan, S., Boehm, F., Cantrell, A., Brellreich, M., Schoenberger, H., Hirsch, A., and Moussa, F. (2002). Cytotoxicity and photocytotoxicity of a dendritic C(60) mono-adduct and a malonic acid C(60) tris-adduct on Jurkat cells, *J. Photochem. Photobiol. B*, **67**(3), 157–162.

198. Donaldson, K., Beswick, P. H., and Gilmour, P. S. (1996). Free radical activity associated with the surface of particles: a unifying factor in determining biological activity? *Toxicol. Lett.*, **88**(1–3), 293–298.

199. Dolmetsch, R. E., Xu, K., and Lewis, R. S. (1998). Calcium oscillations increase the efficiency and specificity of gene expression, *Nature*, **392**(6679), 933–936.

200. Oberdorster, G. (2001). Pulmonary effects of inhaled ultrafine particles, *Int. Arch. Occup. Environ. Health*, **74**(1), 1–8.

201. Hamoir, J., Nemmar, A., Halloy, D., Wirth, D., Vincke, G., Vanderplasschen, A., Nemery, B., and Gustin, P. (2003). Effect of polystyrene particles on lung microvascular permeability in isolated perfused rabbit lungs: role of size and surface properties, *Toxicol. Appl. Pharmacol.*, **190**(3), 278–285.

202. Warheit, D. B., Webb, T. R., Reed, K. L., Frerichs, S., and Sayes, C. M. (2007). Pulmonary toxicity study in rats with three forms of ultrafine-TiO_2 particles: differential responses related to surface properties, *Toxicology*, **230**(1), 90–104.

203. Pietropaoli, A. P., Frampton, M. W., Hyde, R. W., Morrow, P. E., Oberdorster, G., Cox, C., Speers, D. M., Frasier, L. M., Chalupa, D. C., and Huang, L. S. (2004). Pulmonary function, diffusing capacity, and inflammation in healthy and asthmatic subjects exposed to ultrafine particles, *Inhal. Toxicol.*, **16**, 59–72.

204. Seaton, A., Macnee, W., Donaldson, K., and Godden, D. (1995). Particulate air-pollution and acute health-effects, *Lancet*, **345**(8943), 176–178.

205. Eldridge, J. H., Staas, J. K., Meulbroek, J. A., Tice, T. R., and Gilley, R. M. (1991). Biodegradable and biocompatible poly(DL-lactide-co-glycolide) microspheres as an adjuvant for staphylococcal enterotoxin B toxoid which enhances the level of toxin-neutralizing antibodies, *Infect. Immun.*, **59**(9), 2978–2986.

206. O'Hara, P., and Hickey, A. J. (2000). Respirable PLGA microspheres containing rifampicin for the treatment of tuberculosis: manufacture and characterization, *Pharm. Res.*, **17**(8), 955–961.

207. Suarez, S., O'Hara, P., Kazantseva, M., Newcomer, C. E., Hopfer, R., McMurray, D. N., and Hickey, A. J. (2001). Respirable PLGA microspheres containing rifampicin for the treatment of tuberculosis: screening in an infectious disease model, *Pharm. Res.*, **18**(9), 1315–1319.

208. Putnam, D. (2008). Drug delivery: the heart of the matter, *Nat. Mater.*, **7**(11), 836–837.

209. Grayson, A. C., Voskerician, G., Lynn, A., Anderson, J. M., Cima, M. J., and Langer, R. (2004). Differential degradation rates in vivo and in vitro of biocompatible poly(lactic acid) and poly(glycolic acid) homo- and co-polymers for a polymeric drug-delivery microchip, *J. Biomater. Sci. Polym. Ed.*, **15**(10), 1281–1304.

210. Chalupa, D. C., Morrow, P. E., Oberdorster, G., Utell, M. J., and Frampton, M. W. Ultrafine particle deposition in subjects with asthma, *Environ. Health Perspect.*, **112**(8), 879–882.

211. Anderson, P. J., Wilson, J. D., and Hiller, F. C. (1990). Respiratory tract deposition of ultrafine particles in subjects with obstructive or restrictive lung disease, *Chest*, **97**(5), 1115–1120.

212. Hara, K., Tsujimoto, H., Tsukada, Y., Huang, C. C., Kawashima, Y., and Tsutsumi, M. (2008). Histological examination of PLGA nanospheres for intratracheal drug administration, *Int. J. Pharm.*, **356**(1–2), 267–273.

213. Fiegel, J., Fu, J., and Hanes, J. (2004). Poly(ether-anhydride) dry powder aerosols for sustained drug delivery in the lungs, *J. Controlled Release*, **96**(3), 411–423.

214. Dailey, L. A., Kleemann, E., Wittmar, M., Gessler, T., Schmehl, T., Roberts, C., Seeger, W., and Kissel, T. (2003). Surfactant-free, biodegradable nanoparticles for aerosol therapy based on the branched polyesters, DEAPA-PVAL-g-PLGA, *Pharm. Res.*, **20**(12), 2011–2020.

Chapter 9

Devices and Sensors for Bioelectric Monitoring and Stimulation

Benjamin Moody, Mathew K. Zachek, and Gregory S. McCarty

Department of Biomedical Engineering, North Carolina State University and University of North Carolina at Chapel Hill, Raleigh, NC, 27695, USA

bpmoody@ncsu.edu

9.1 Introduction

The application of man-made structures to effectively monitor and actuate biological systems has long been a goal of biomedical engineering. Much progress has been made over the last 50 years with man-made actuation devices, such as pacemakers and cochlear implants, now used routinely in patients. The recent advances in the nanosciences and bionanotechnology have created new hope of generating innovative and powerful devices for both clinical and research applications. This chapter highlights some of the recent advances made in applying nanoscience to create systems capable of monitoring and stimulating bioelectric signals, with a specific emphasis on application to the brain and heart.

9.1.1 History

Electrophysiology, in general, seeks to monitor the ionic movements that result from action potentials. This is accomplished most notably through the use of electrode systems that detect the minute charge displacements that occur as ions move into and out of the electrode's sphere of influence. These systems can be made of silver wires, solutions of gold nanoparticles, arrays of carbon fibers, or any number of arrangements that allow for

Biomedical Nanosensors
Edited by Joseph Irudayaraj

ISBN 978-981-4303-03-3 (Hardcover), 978-981-4303-04-0 (eBook)
www.panstanford.com

the transfer of electrons. Thus, due to the wide variety of applicable systems, electrodes have long been used to study everything, from benchtop chemical reactions to *in vivo* biological events. Even before the term "electrode" was first used, the famous anatomist Luigi Galvani used electrodes to study the effects of electricity on muscles, noting that the muscles in a frog's leg are capable of both conducting and detecting electrical currents [1].

Since those important experiments in the late eighteenth century, biologically relevant charge-based experiments have become more robust, sensitive, and precise. Not until 1949 did researchers begin using pulled glass capillaries and metal electrodes to examine single-cell-membrane potentials [2]. Even the patch clamp technique, a profound electrophysiological advance in 1976 that dramatically refined the exploration of action potentials [2], has since been improved upon many times over. Research and innovations that have come out of the computer industry's evolution have advanced electronics and enabled better measurements with such tools as lock-in amplifiers, operational amplifiers, and micromachining techniques. Additionally, electrodes themselves are just beginning to profit from the nanotechnology revolution — especially in the field of biomedical nanosensors and nanoelectrodes, where submicron-sized devices are being explored for possible advantages. Such sensors are expected to facilitate advances in machine-biological interfacing.

9.1.2 Perspective

In order to gain an understanding of the requirements for monitoring bioelectric signals and actuating cells and tissues, something of the origin and physiology of bioelectric signals must be known. The earliest bioelectric signals and still the bulk of bioelectric measurements that are collected are based on variations in electrical potential that arise from ion movements about cells or tissues. Viewed simplistically as a single whole entity, the intracellular space of mammalian cells is a heterogeneous environment comprised of parts and components enclosed within a phospholipid membrane called the plasma membrane. This intracellular environment in excitable cells, such as muscle and neuronal cells, can be polarized with respect to the extracellular environment, that is to say, there is an imbalance of potassium, calcium, and sodium ions between the intracellular and extracellular spaces. This imbalance creates a potential gradient or a voltage across the plasma membrane. Most, if not all, ionic movements and electrical propagation in the body are controlled to some extent by the level of polarization that exists between the intra- and extracellular spaces. A case in point, the fundamental physiological event that is monitored in electrophysiological measurements is an action potential, Fig. 9.1 [3]. An action potential is a transient event in an excitable cell's life where the normally polarized cell membrane becomes depolarized. The cellular depolarization can be caused by any number of mechanical, electrical, or chemical stimuli and leads to an opening and closing of specific protein channels along the cell membrane. These protein channels allow a flux of ions both into and out of the cell, thus generating and propagating the electrical signal along the cell. This ionic propagation of an action potential is primarily due to voltage-gated Na^+ and K^+ channels. During the action potential, ionic species rush toward a chemical and electrical equilibrium that can be described by the Nernst equation. It is almost exclusively within these moments that excitable cells produce their action. For example, neurons release neurotransmitters and muscle cells contract.

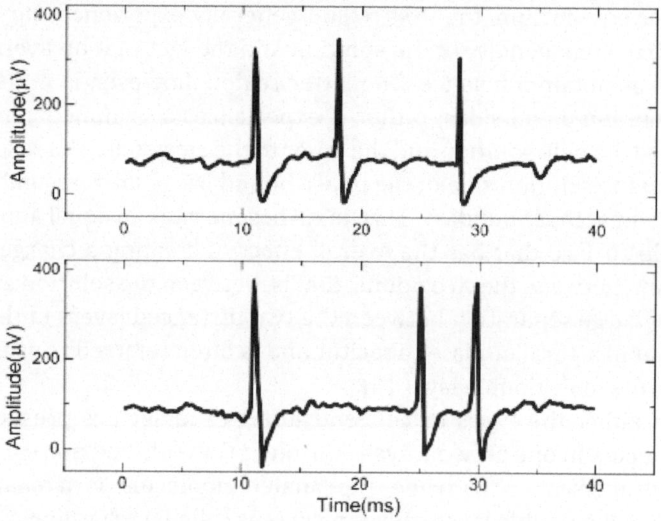

Figure 9.1 Action potentials recorded by a tungsten nanoelectrode inserted in a cat neuron. (Reprinted with permission from IOP Publishing Ltd., ref. [3].)

The occurrence of an action potential in one cell can set up a chain reaction of action potentials in other cells in close physical proximity. This occurs when the ionic fluxes attributed to the depolarization of one cell cause a change in the local chemical environment that subsequently depolarizes neighboring cells. Thus, if properly timed and connected, a single stimulus at one point in the body can cause action potentials to propagate along long lines of cells in order to relay a message to another point. Such signal propagation is a prominent method of intercellular communication and is one method used by nerve cells to transmit input information from sensory cells to the brain or to deliver command signals from the brain. Specific scenarios regarding nerve and muscle depolarization and signal propagation will be discussed in more detail in later sections.

9.2 Electrode Basics

This section will describe the basics of a single electrode and its interaction with ionic solutions. As coined by Michael Faraday in 1834, an electrode is the surface at which matter is bound as it decomposes into electric current [4]. Today, perhaps a more modern and generic description could be that an electrode is the interface of any conductor with a nonconductor. An electrode can be thought of as the surface of an electrical conductor that acts as a transducer that allows machines and electronics to interface with an external environment via electrical interaction. Building upon this general definition, it is also necessary to describe the remaining component of this electrode system, in our case the electrolyte, and the relationship between the two. In the case where an electrode exists within an electrolytic solution, the system can be approximated as two ionic regions within the solution — one that is composed of the bulk solution far away from the electrode and one that exists immediately upon the electrode surface at the phase boundary. Under equilibrium, the first region generally experiences no net electric field due to the relative homogeneity of the solution and the fact that on average surrounding electrical forces maintain a balance. The latter region, however, is not surrounded by a homogenous solution on all sides; rather, it experiences a "symmetry break" where one side is exposed to the bulk solution and the other to the electrode. The result is a net charge that develops on the solution side of the phase boundary as the force imbalance creates a net dipole gradient at that boundary. This, in turn, generates an equal and opposite charge on the electrode surface that has the overall effect of creating a charge separation or a potential gradient between the two points, that is, between the solution and the electrode. In essence, the charge separation between the two distanced layers in this system can be compared to that of a parallel-plate capacitor and is often referred to as the "double-layer capacitance" or the ionic double layer [5].

Variation in either the types or concentrations of molecules present at this double layer can be detected in one of two ways — through Faradaic and non-Faradaic processes. Non-Faradaic processes occur when the analyte molecule is a charged ion whose movement within the double layer causes a current called a "charging current" to flow in the electrode. Such a signal is typically what is collected in the case of electrophysiological measurements of cellular action potentials.

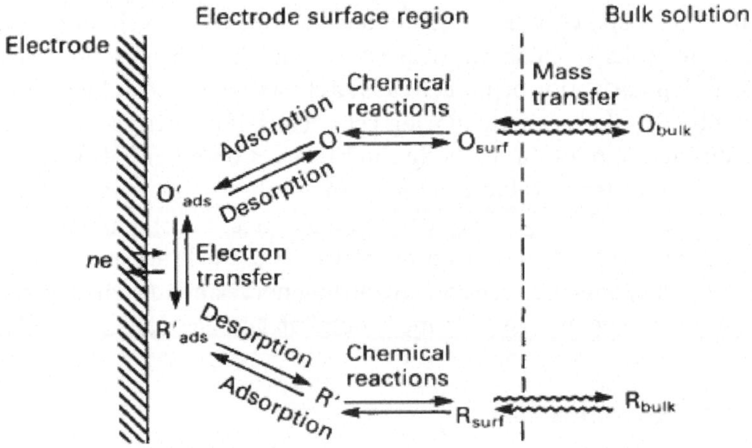

Figure 9.2 Generalized depiction of Faradaic reactions at an electrode interface. (Reprinted with permission from John Wiley & Sons, ref. [6].)

A Faradaic reaction, or process, is the term used to describe an electrochemical reaction occurring at the electrode surfaces (Fig. 9.2) [6]. These reactions occur when there is an electron transfer between the electrode and analyte that causes the analyte to undergo a chemical reduction or oxidation, in other words, a "redox" reaction at an electrode surface. These redox reactions can result in a detectable current at the electrode and are the topic of study in electrochemistry. These ideas lead to two distinct classes of signals that are commonly collected by electrodes in biological systems — electrophysiological measurements, whose origins have already been briefly discussed, and electrochemical measurements, which will be introduced now.

9.3 Electrochemical Measurements

The ability to monitor the electrochemical reactions at an electrode surface enables some insight into the numbers and types of molecules at that surface. In general, electrochemical measurements in biology are used to monitor the concentration of electroactive species, a chemical species that can undergo redox reactions, in the local environment near the electrode. In general, these measurements are accomplished by setting up a two or three electrode electrochemical cells, with the simplest case consisting of a working electrode, a reference electrode, and the electrolyte. A variety of materials can be used as a working electrode, including solid conductors (gold, platinum, tungsten, etc.), semiconductors (silicon, gallium, nitride, etc.), and liquid metals (mercury, amalgams, etc.). The electrolytic portion of the setup is usually an aqueous solution containing various ionic species (H^+, Na^+, Cl^-, etc.) [6]. In biological electrochemical measurements, this ionic aqueous solution typically surrounds or is the sample. In addition to a working electrode and the electrolyte, the electrochemical cell is completed through the use of a counter or auxiliary electrode, in the case of a three-electrode system, and reference electrodes (Fig. 9.3). The small size and, therefore, small currents at micro- and nano-scale working electrodes remove the need for a counter or auxiliary electrode; so they will not be discussed. The reference electrode, as the name implies, acts as a reference point. The reference electrode is a nonpolarizable electrode. In other words, it has a constant potential. This type of electrode can source or sink current without changing the potential.

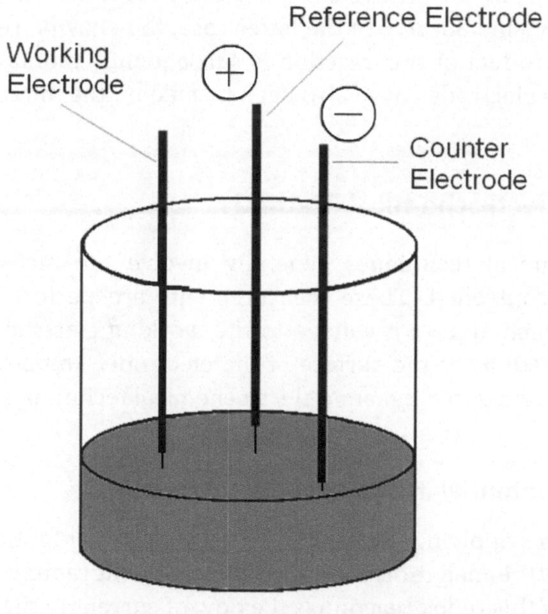

Figure 9.3 A schematic presentation of a typical three-electrode electrochemical cell.

The field of bioanalytical electrochemistry is primarily concerned with the relationship of electric current and the chemical redox reactions that occur at the electrode surface-solution interface. These reactions can be observed using different electrochemical techniques, which can be divided into two categories — static methods (where current (i) is 0) like potentiometry and dynamic methods (where i is not 0) such as amperometry, coulometry, and voltammetry.

9.3.1 Static Electrochemical Methods

Static electrochemical measurements utilized in potentiometry are accomplished by measuring the current through the use of a potentiometer, which ideally holds the current in an electrochemical cell at zero, thus not altering the electrolyte composition inside the cell while the cell potential is observed. The electrochemical potential is passively observed using this technique and is described mathematically through the Nernst equation:

$$E = E^0 - \frac{RT}{\eta F}\log Q \tag{9.1}$$

where E is the total cell potential of the cell, E^0 is the standard cell potential, R is the gas constant, T is the temperature, η is the number of electrons involved in the redox reaction, F is faraday's constant, and Q is the redox reaction quotient (the relative mixture of oxidized and reduced products in the cell).

Generally, potentiometric measurements are coupled with an ion-selective working electrode designed to observe only a specified ion such as H_3O^+, whose concentration relates to the pH of a solution. Biological sensors implementing the potentiometric technique are involved in gas sensing or involve enzyme electrodes in which enzymes are trapped within an ion-selective membrane. In the latter case, the enzyme reacts with an analyte of interest and the product of this reaction is subsequently monitored at the electrode surface. Ion-selective electrodes have also been formed on the micro- and nanoscale (for a review see [7]).

9.3.2 Dynamic Electrochemical Methods

Dynamic electrochemical techniques generally involve the measurement of current as the potential is controlled. These measurements are performed through the use of a potentiostat, which applies a voltage to the working electrode (or the reference), promoting redox reactions at the surface of the electrode. Amperometry, voltammetry, and coulometry are considered dynamic electrochemical techniques.

9.3.2.1 Constant potential measurements

Amperometry involves applying (or stepping to) a constant potential that is sufficient to oxidize or reduce only the analyte of interest, without inducing redox reactions of interfering species. The result of this redox reaction is the flow of current in the electrochemical cell, which is monitored at the electrode with respect to time. Amperometric traces of current versus time can be integrated to resolve the total charge and subsequently the amount of oxidized/reduced product via Faraday's law:

$$Q = nFN \tag{9.2}$$

and

$$Q = \int_{t=0}^{T} i(t)dt \tag{9.3}$$

Coulometric measurements are very similar to amperometric measurements; however, instead of transient observations of current flow, coulometry is primarily based on the complete oxidation or reduction of the analyte of interest. These traces can also be analyzed with the above methods to quantify a number of parameters, such as the total reaction charge and the amount of oxidized species.

Commonly used biosensors constructed using a constant potential platform are the Clark electrode (used to monitor oxygen) and enzyme modified sensors, the most common of which is used to monitor glucose. This type of electrode uses an enzyme to convert the nonelectrochemically active molecules of glucose to hydrogen peroxide, which is easily oxidizable and thus detectable. Other electrodes employ similar schemes to monitor nonelectroactive molecules, such as choline and glutamate [8].

9.3.2.2 Dynamic potential measurements

Voltammetry is a dynamic measurement in respect to both current and voltage. The voltage at the working electrode in the electrochemical cell is controlled with a potentiostat and can be varied or "ramped" in a number of ways. The advantage of voltammetry is that multianalyte systems (as routinely found in biology) can be analyzed simultaneously with some degree of selectivity. This selectivity stems from the different redox potentials associated with a particular analyte.

One form of voltammetry, called background-subtracted fast-scan cylic voltammetry (FSCV), is able to resolve the oxidation and reduction of analytes on the millisecond timescale. As the name suggests, in FSCV, the potential is ramped very fast (>100 V/s), causing a large non-Faradaic charging current. Over the course of the experiment, this charging current becomes stable and can be digitally eliminated from the signal, enabling the direct observation and identification of transient changes in the concentration of electroactive species, such as electroactive neurotransmitter release. Another advantage of using a fast scan rate is that the sensitivity of the electrode is proportional to the scan rate. Therefore, in addition to an improved time response, a faster scan rate will allow quantitative measurements of very small concentrations to be made. These advantages of background-subtracted FSCV have enabled the direct observation of biological analytes *in vivo* and *in vitro* using carbon-based ultramicroelectrodes [9, 10].

9.3.3 Other Methods of Detection

9.3.3.1 Electrophysiology

Electrophysiology, as mentioned, is used to monitor action potentials and is the most commonly collected type of bioelectric signal in research. It should be noted, as described above, that electrophysiology is technically a potentiometry measurement in that electrodes measure changes in the electrophysiological voltage induced from single or multiple action potentials within an electrochemical cell.

9.3.3.2 Conductometry

Less commonly, electrolytic solutions can also be analyzed using conductometry. This method is unlike those previously mentioned in that it is less concerned with the Faradaic and non-Faradaic processes occurring immediately at the electrode surface but is instead more a measure of ion mobility in the bulk solution. As the name implies, conductometry measures the conductivity (or the resistance) between two electrodes in solution — a known current, or voltage, is applied to a solution to obtain the resistance. The inverse of the resistance is the conductance of the solution via Ohm's law.

$$V = iR \tag{9.4}$$

This information can be used to infer the ion concentrations since both ionic mobility and concentration contribute to a solution's conductivity [11].

9.4 Biosensor Electrodes

9.4.1 Biocatalytic and Bioaffinity Sensors

Electrophysiological and electrochemical methodologies have been explored for developing bioanalytical sensors and biosensors. The technical definition of a biosensor is a sensor with a biological recognition element. *Bioanalytical sensor* is a more general term referring to any sensor designed to make biological measurements, regardless of its composition. In an extension of this, the difference between an ordinary electrochemical sensor and an electrochemical biosensor is often only that the biosensor is functionalized with some sort of biological recognition element, the recognition or activation of which produces a specific, detectable change in current, potential, or solution conductivity. In general, electrochemical biosensors can be loosely classified into one of two categories — bioaffinity sensors or biocatalytic sensors — each operating as its name implies. *Bioaffinity* sensors use sensors coupled to recognition or capture molecules designed to bind to specific biological analytes. Bioaffinity-based sensors attach biorecognition molecules directly onto the electrode in order to selectively bring certain ions or redox components to interact at the ionic double layer. Complementary DNA strands or antibodies and antigens are common examples of molecules that have been immobilized onto electrodes acting as bioaffinity sensors. *Biocatalytic* sensors, on the other hand, use sensors coupled to catalytic molecules that selectively initiate or otherwise enhance certain reactions at the double layer. Enzymes are probably the most widely used biocatalyst [11]. It should be noted that while both approaches can be used to attain quantitative sample concentrations, both also rely heavily on the researcher's ability to immobilize molecules onto the electrode surface. The immobilization of the biological recognition element has long been an issue in biosensors and due to its importance, will be briefly reviewed.

9.4.2 Immobilization of Biological Recognition Elements

Immobilization is accomplished through three main methods: adsorption (physisorption), covalent bonding (chemisorption), and mechanical encapsulation in polymers, gels, or pastes. Adsorption is a simple and relatively weak method by which molecules can be bound to an electrode surface, generally through electrostatic or polar interactions. Because adsorption results in a weak interaction between molecule and substrate, it is mostly used for experiments that are meant to be fast or those in which a reversible binding or desorption is appropriate [11]. Where a more stable surface structure is required, covalent bonding is used in place of adsorption. With covalent bonding, molecules containing "tether groups" are employed to form a stable attachment between the electrode and the molecule to be immobilized. Thiol-functionalized groups, for instance, are extensively used to form covalently bound structures on gold electrodes. Other tether groups include amine, hydroxyl, carboxyl, and phenolic groups [11].

A third immobilization scheme is encapsulation. In this scheme, the biological recognition element is typically encapsulated into a polymer coating on the electrode surface. Often, the electrode surface is coated with a conductive, adsorptive film such as a carbon paste or an ion-exchange polymer, such as Nafion. Nafion has been used to encase oxidase enzymes for the detection of glutamate release in the brains of anesthetized rats [12]. Carbon paste adsorption has the extra advantage of being compatible with macroscale adsorption — not only can it be used to bind proteins, enzymes, and antibodies to an electrode, carbon paste has also been used to adsorb small organs of animals [11]. Though there are numerous different recipes of making carbon paste, in general, the paste is made from a combination of carbon powders and a hydrophobic binder such as silicon oil, mineral oil, liquid paraffin, or bromoform [11, 13, 14].

9.5 Advances in Electrodes for Biological Measurements

9.5.1 Overview

Electrodes for the actuation and collection of biological signals have been continually improved for over a hundred years. The advent and implementation of nanotechnology has accelerated these electrode improvements and has the potential for revolutionizing how bioelectric signal are measured and mimicked.

9.5.1.1 Advantages of microelectrodes

Experiments and devices implementing macroelectrodes are becoming increasingly overshadowed by those employing electrodes in the micro- and nanoscale regime because of the host of advantages microelectrodes present. Specifically, the improved mass transport of analytes to and from microelectrodes leads to improved performance such as larger current densities and faster time responses. For example, it has been shown that as an electrode's size decreases, so does the time it takes for electrochemical reactions at the electrode surface to reach steady-state. Much of an electrode's ability to monitor electroactive analytes is limited by the diffusion of species to the electrode surface. Therefore, a faster steady-state response means increased electrode sensitivity and signal-to-noise ratio [15]. This improved performance, in turn, makes it easier to test transient variations in an analyte concentration [16]. The decreased area also reduces the magnitude of the collected current and limits the amount of background charging during voltammetric experiments such as fast-scan cyclic voltammetry [17]. Very small electrodes are characterized by low Ohmic (IR) drops and small time constants in the hundreds of nanoseconds [18].

Microelectrodes have much higher current densities and smaller collected currents (usually in the nano- to picoampere range [11] that are typically nondestructive to living cells [19]. Shorter response times versus potential or concentration changes mean not only a faster response but also give researchers the ability to rapidly scan through applied potentials. Further, because the measured currents are lower, the voltage drop (IR drop) caused by solution resistance becomes negligible [16] and voltammetric readings can be taken without the three-electrode setup by simply applying voltage between the working and reference electrodes. Microelectrodes can also be used to collect data from very small volumes, including intracellular volumes [2, 19, 20], and can thus be used to directly monitor or stimulate local *in vivo* environments [2]. Additionally, a variety of biologically compatible microelectrodes have been constructed from noble metals and carbon. By insulating all but the tip of a fine wire and then polishing the tip down to micrometer dimensions, simple disk microelectrodes can be easily fashioned. Other common shapes include spheres, hemispheres, bands, and cylinders. Different geometries have distinct differences concerning effective diffusion areas and current flow and resistance properties.

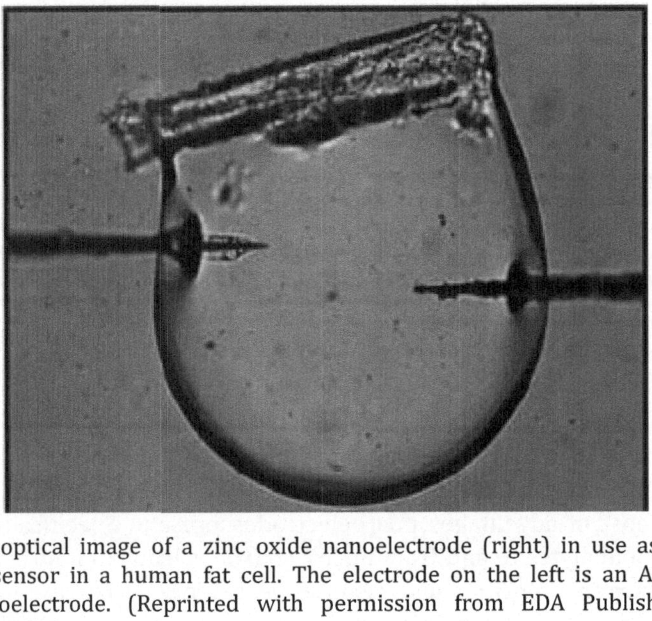

Figure 9.4 The optical image of a zinc oxide nanoelectrode (right) in use as an intracellular pH sensor in a human fat cell. The electrode on the left is an Ag/AgCl reference microelectrode. (Reprinted with permission from EDA Publishing Association, ref. [20].)

However, there are a few drawbacks to reducing the size of electrodes. Specifically, external noise can be problematic with microelectrodes because of the small currents being measured. Measurements can be optimized (less distorted) by using low-noise operational amplifiers and damping and filtering units and by performing experiments in a grounded Faraday cage [11]. Another common problem with electrodes that are used for biological measurements is the deleterious adsorption of proteins, lipids, and many kinds of cellular debris. This common problem, known as biofouling, disrupts the wanted chemistry at the electrode surface and can completely inactivate the electrode. Metal electrodes are especially prone to biofouling, and for this reason, carbon microelectrodes have gained popularity for electrochemical experiments in biological environments [19].

9.5.1.2 Advantages of nanoelectrodes

The advantages in the nanosciences are often an extension of the advantages enabled by microsciences. So, nanoelectrodes are expected to have even greater signal-to-noise ratios, higher current densities, faster response times, shorter steady-state lags, and smaller Ohmic drops than micro- or macroelectrodes. Nanoelectrodes can also be used to probe smaller experimental volumes such as synaptic gaps, alluding to the possibility of increased selectivity through specific placement. Nanoelectrodes will also cause less tissue damage than larger electrodes when being placed in sensitive areas such as the brain.

Though a more thorough explanation of the individual fabrication processes of specific nanoelectrodes will be discussed later, as an overview, electrode structures can either be "grown" on suitable substrates or they can be "carved" from such substrates. Each technique has advantages and disadvantages with respect to how much the process costs, how much control it gives over critical dimensions, how much time it takes, its degree of precision and reproducibility, and the yield and quality of the produced devices. To date, a wide variety of nanolithographic techniques have been developed, optimized, and utilized mostly with the computer industry in mind. Nonetheless, the same techniques hold merit for creating biological nanosensors. For a more in-depth analysis of nanolithography, several references are suggested [21, 22].

9.5.2 Advances in Gold Electrodes

9.5.2.1 Overview

As research in nanoelectrodes has only recently begun, many of the experiments have been attempted utilizing relatively inert noble metals. Gold has many characteristics that qualify it for creating excellent biomedical electrodes: gold is not considered cytotoxic, it integrates well into *in vivo* systems [23–25], and it can easily be functionalized with antibodies [26, 27] and other proteins [28]. Its well-defined surface chemistry can be used to control everything, from the "bottom-up" growth of gold-based nanostructures to the attachment of various useful enzymes [29, 30], pharmaceuticals, and miscellaneous biomolecules [31, 32]. Gold is a noble metal, which means it does not oxidize as quickly as other metals. It is cheaper than platinum, is a relatively good conductor, and has been extensively used with standard microfabrication techniques. Compared to carbon fiber electrodes, gold electrodes have faster electron kinetics, which gives it the capability of detecting electrolytic species in a smaller potential window [33]. However, gold has also been shown to foster greater adsorption and chemical fouling than carbon fiber electrodes and may be less sensitive when using certain voltammetric techniques [33]. Additionally, the varieties of architectures gold can achieve make it useful in producing a multitude of electrode systems, including nanoelectrodes.

9.5.2.2 Gold nanoparticles and clusters

One of the simpler yet most modern incarnations of the gold nanoelectrode is essentially a cluster of gold nanoparticles. It has been shown that gold nanoparticles can be converted to electrodes in mostly one of two ways. One, the nanoparticles in a solution are complexed with short alkanethiolate monolayers. This prevents aggregation by stabilizing the core nanoparticle, and also lends the nanoparticle a molecular capacitance that gives rise to the double-layer capacitance characteristic of planar electrodes [34–36]. These so-called monolayer-protected clusters (MPCs) are typically based around nanoparticles with gold cores of between 1.1 to 5 nm in diameter. Diameters below this average tend to undergo redox themselves instead of experiencing conductance in a more passive manner [35]. Particles within this size range, however, exhibit significant electron-hopping conductivity [37] and participate in single-electron-tunneling events known collectively as the Coulomb staircase [36]. It is the Coulomb staircase phenomenon that purportedly allows electrons to flow from outside the protective monolayer to the gold core within and thus gives MPCs their molecular capacitance. Unfortunately, though not unexpectedly, previous detection studies with MPCs in dopamine solutions show that the alkanethiolate monolayer decreases electron-transfer kinetics of the overall system as compared to standard gold electrodes, possibly by serving as a blocking layer between the electrolyte and the electrode [36].

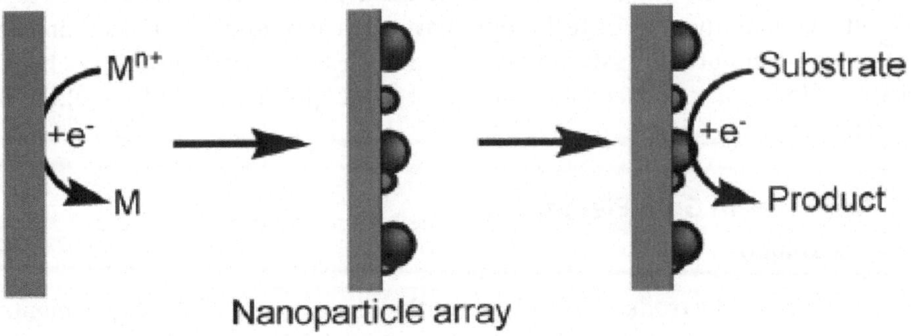

Figure 9.5 A schematic diagram of metal nanoparticles being electrodeposited onto an electrode surface to form a nanoparticle array. (Reprinted with permission from Elsevier, ref. [17].)

Another method to use gold nanoparticles is to apply them as a mediating electrolytic reaction surface between a standard electrode and the solution interface. In this scenario, gold nanoparticles of varying sizes (1.4–15 nm) are self-assembled onto an electrode surface, often through amino- or thiol-related chemistries, in order to enhance the macroelectrode's chemical or electrical relationship with the solution [17, 29, 38, 39]. A schematic demonstrating this process is shown in Fig. 9.5. For instance, Tian *et al.* grew traditional citrate-capped gold nanoparticles on a boron-doped diamond (BDD) film.

BDD films are known to have "excellent" electrochemical properties, and in this case, the film was activated with amino groups to overcome BDD's otherwise inert surface. Researchers found that adding nanoparticles was much like roughening the electrode in that the nanoparticles increased the electrode's surface area by about 22% (with a six-hour assembly time). As a result, the background current density of the nanoparticle-BDD electrode was lower than that of a traditional gold macroelectrode. Additionally the percent coverage of nanoparticles could be controlled by varying the amount of time nanoparticles were allowed to assemble onto the BDD film, which, in turn, adjusted the signal-to-noise ratio [39].

Xiao *et al.* are using gold nanoparticles to mediate electron transfer between the redox enzyme apo-glucose oxidase and a gold macroelectrode [29]. In conjunction with the enzyme's cofactor, flavin adenine dinucleotide, the system boasts an electron transfer turnover rate over seven times greater than that without the gold nanoparticles. Xiao *et al.* believe the increase comes from two effects — (1) the gold nanoparticles help enzymes align themselves into a position that properly exposes the enzymatic active sites and (2) the nanoparticles act as electron relays between the macroelectrode and the enzymatic redox site [29].

9.5.2.3 Nanowires and nanotubes

Besides nanoparticles, gold can also be constructed into nanotubes or nanowires — three-dimensional cylinders with diameters on the nanometer scale leading to extremely high aspect ratios. This has been achieved primarily through polycarbonate (PC) template-assisted techniques. In some cases, nanowires have been created though electrochemical etching, sol-gel encapsulation, or polymer-assisted dispersion with impressive results [40, 41]. In the case of PC template assistance, a dissolvable and porous PC membrane is attached to the surface of a macroelectrode and then immersed in an electroplating solution. Gold is then electrodeposited onto the template and throughout the template's nanopores. After deposition, the template is removed through dissolution with chloroform, leaving only the gold nanowires that filled the pores. Thus, electrodes produced by this technique have diameters approximately the size of the template's nanopores, which can range from 100 nm to 12 μm [42].

As a proof of concept, researchers used the PC template method to create and then test biosensors for glucose. The sensors were manufactured by covering gold electrodes with a spread array of nanowires and then immobilizing glucose oxidase onto the resulting surface [42, 43]. One report states that the nanowire addition increased the electrode surface area by five times and allowed for fast response times, high current densities, and a good affinity for glucose [42]. The nanoelectrodes were reported to be linear, stable, and reproducible though not necessarily more sensitive than glucose sensors produced without nanometer dimensions [43], most likely due to the previously discussed sensitivity considerations concerning gold.

Similarly, an array of nanotubes on a gold macroelectrode was also tested with an amperometric detection of hydrogen peroxide (H_2O_2) [44]. Figure 9.6 shows an SEM of a gold nanotube array created with PC. In the presence of hydroquinone, H_2O_2 is involved in an electrochemical reaction that results in a net consumption of two electrons from the electrode. With hydroquinone as a mediator, the nanotube-modified electrode showed significant enhancement in sensitivity for this reaction [44]. Furthermore, with the immobilization of horseradish peroxidase acting as the catalyst, the modified electrode exhibited an improved limit of detection of H_2O_2 and an increased selectivity over a wide potential range [44].

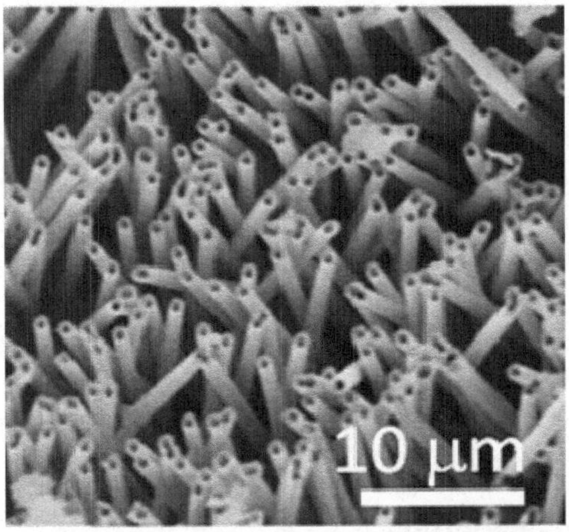

Figure 9.6 A scanning electron microscope (SEM) image of gold nanotubes created through the PC membrane template method. (Reprinted with permission from Elsevier, ref. [44].)

One final template technique should also be mentioned. Nanopore-based electrodes involve a similar idea of depositing noble metals into nanopore templates. However, in this case, the entire process generally involves a more complex sequence of photolithographic microfabrication followed by drilling with a focused electron beam to form the nanopores and a vacuum deposition of the electrode metal. In 2006, Krapf *et al.* demonstrated how this overall technique can be used to form gold electrodes with radii down to 2 nm on a gold substrate insulated with silicon nitride [45]. By comparison, electrodes created with PC membranes generally have diameters on the order of tens to hundreds of nanometers [42, 43].

9.5.2.4 Nanogap electrodes

In addition to enhancing the properties of larger electrodes, nanowires and nanotubes can be further fashioned into structures that have become known as nanogaps. As the term implies, nanogaps can be considered any structure comprising two smaller structures that at some point are separated by a nanometer-sized space. Though new and more efficient methods of nanogap fabrication are being investigated [46, 47], traditional nanogaps have been created through electromirgration [48], electron beam lithography [49], and the focused electron beam milling of small metal features like nanotubes [50].

Recent research has begun to utilize nanogaps for monitoring electronic and optical variations that occur with the addition or removal of molecules at the nanogap. For example, nanogaps have been used to sculpt a 2 nm gap in a 150 nm wide gold line, which was then used to trap and study a gold nanoparticle [50]. This process is shown schematically in Fig. 9.7. In another recent study, electrodes patterned through nanoimprint lithography with spacing of less than 10 nm demonstrated the ability to monitor individual nucleic acid strands flowing through a nanogap sensor. The structure in this study has about two orders of magnitude better throughput than a "nanopore," a commonly used research technique for single-oligonucleotide monitoring, and therefore highlights one of the capabilities of solid state nanogaps — single-molecule resolution. The report also demonstrates the need for a sub-10 nm structure to make these types of measurements [51]; a 9 nm nanogap was shown to have single-molecule resolution, while gaps above 13 nm were not sensitive to single molecules.

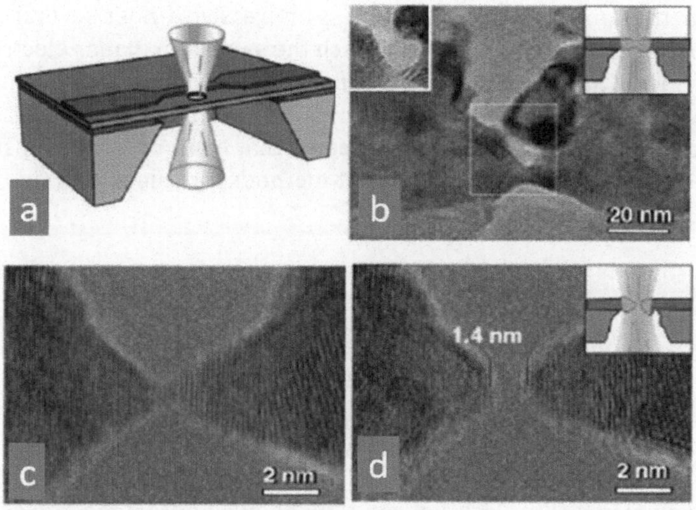

Figure 9.7 A schematic (a) and TEM (b–d) representation of the manufacturing of nanogaps in 150 mm platinum lines. (Adapted with permission from the American Chemical Society, ref [50].) See also Color Insert.

Nanogaps have also been utilized as surface-enhanced Raman spectroscopy (SERS) substrates. Both electromigration [48, 52] and molecular lithography [53] have been used to fabricate nanogaps for utilization as SERS substrates. The modification of the nanogap by adsorbing molecules in close proximity to the nanogap could then be achieved using SERS. Nanogaps fabricated with molecular lithography enabled the addition of ssDNA to be observed [53]. Coupling nanogaps as high-enhancement-factor substrates with SERS is expected to create a very sensitive and molecule-specific monitoring technique.

The idea is that such nanogaps could eventually be used to investigate small biologically relevant molecules like neurotransmitters or assorted pharmacological compounds. The electrochemical aspects of the technique are initially more like conductimetry, though voltammetric nanogap systems are not difficult to imagine. Unfortunately, nearly all nanogap fabrication techniques suffer from reproducibility problems [50] that have as yet kept nanogap electrodes from maturing into sophisticated biomedical devices.

9.5.2.5 Nanowells

One final gold nanoelectrode that is likely yet to see its full potential is the gold nanowell. Again, as the name suggests, gold nanowells are small recessions of nanometer dimensions formed into gold substrates. Since the late 1980s, researchers have been using template-based techniques to produce nanowell-type structures. The process is very much an antithesis of the template-assisted model used for creating nanowires. Here, instead of using a nanopore template, researchers deposit gold over what has been called an "anopore" template, or a template that has nanometer diameter protrusions rather than nanometer diameter holes. Initially, these templates were created by depositing silver into nanopore templates, dissolving the nanopore templates, depositing gold, and then dissolving the silver. In 2002, Evans *et al.* published a description of a variation of this technique used to manufacture 600 nm deep nanowells that were 200 nm in diameter. Electrochemically, the results from the nanowell surface were less impressive. Under several conditions, the nanowell nanoelectrode array performed much the same as a planar electrode though, in general, the nanowell had a higher background current using cyclic voltammetry and the residual presence of the nanopore membrane may have manifested itself in an oxygen-stripping peak on the voltamagram. It is expected that nanowells add substantial surface area to the electrode though several different methods for calculating this addition were inconclusive [54].

9.5.3 Silver

Of course there is no reason to suggest that only gold may be used to construct nanoelectrodes. Silver, for instance, is a better conductor than gold, has electrocatalytic abilities, and has already been used to produce nanodisks [55] and nanowire [56, 57] arrays for electrochemical detection purposes. As with gold nanowires, silver can be electrodeposited into porous templates (often aluminum oxide) to produce arrays of varying sized nanowires (Fig. 9.8) that can then be fixed onto a macroelectrode. Though silver nitrate was initially used as the electroplating solution, it has been shown that such a deposition forms inferior metal structures when compared to those created from solutions in which the silver has been complexed with ligands like thiosulfate, cyanide, or hydroxylamine. Depending on the electrodeposition parameters, namely the potential, silver nanowire arrays can be made into different sizes, densities, aspect ratios, and independent orientations [56].

Figure 9.8 SEM images of silver nanowires formed with an aluminum membrane template. (Reprinted with permission from IOP Publishing Ltd., ref. [56].)

Silver nanostructures can also be grown on conductive substrates or simply deposited in colloidal solutions to varying degrees of success [55, 58]. By far, the most popular method of including silver is the latter. Though several reports discuss the electrochemical improvements that silver nanoparticles can have on macroelectrodes, relatively few have been published regarding such schemes applied to biosensors. A welcome exception is Wang *et al.*'s work using silver nanoparticle-modified electrodes to detect H_2O_2. Wang's design is similar to the gold H_2O_2 sensor discussed above, except that Wang immobilizes silver nanoparticles onto the macroelectrode instead of gold nanotubes. Additionally, the silver nanoparticles are functionalized with a layer of DNA before adding peroxidase. Results show that DNA provides a more biocompatible environment for the peroxidase enzyme than would the silver alone, thus increasing peroxidase immobilization and the sensor's overall sensitivity to H_2O_2 [59] A similarly interesting work was published in 2004, describing a sol-gel/silver/peroxidase coating on a macroelectrode [60]. The resulting detection limits and chemical affinities in both papers are promising.

9.5.4 Miscellaneous Metal Electrodes

Much of the proof of concept research accomplished in the past has utilized gold or silver as the conducting portion of the electrode. This is due to the beneficial and well-understood surface properties of gold and silver as well as the abundance of tools available for processing these metals. Other materials, however, also have beneficial properties that make them interesting candidates for further study. Platinum, for example, is often used as either the macroelectrode to which nanoparticle enhancements are made or as the nanoparticle that is used to make such enhancements. Platinum can also be etched into thin wires, as discussed in a 2006 paper where researchers used the change in platinum's apparent capacitance during protein adsorption to detect hemoglobin molecules. The electrode in this example was a platinum wire that had been electrochemically etched to a diameter of under 400 nm [61]. Similar research has been performed using tungsten. Tungsten was electrochemically etched and then insulated with electrophoretic paint to create nanoelectrodes for electrophysiology measurements, Fig. 9.9 [3].

Intracellular pH sensors have also seen advancements by the nanostructuring of tungsten [62] and zinc oxides [20]. These sensors were formed by electrochemical etching and by a chemical growth mechanism, respectively, and were both viable in pH ranges from at least 4.0 to 12.0. Both formed nanotube probes, and both were shown to avoid cell apoptosis. The chemically grown zinc oxide electrode boasted diameters of down to 60 nm, while the tungsten electrode diameters ranged from 500 nm to 800 nm. Besides these few examples, however, metallic nanoelectrodes are largely constructed from gold or silver.

9.5.5 Carbon Electrodes

Similar to the situation with nongold metallic nanoelectrodes, it is not uncommon for carbon nanostructures to be applied as a modification of a microelectrode. There has been a flurry of work using carbon nanotubes (CNTs) as functioning nanoelectrodes ever since their initial discovery in 1991; however, to date, most successful applications have used CNTs only to modify or amplify an existing microelectrode as opposed to using the nanotubes as the sole electrode. Glassy carbon electrodes, on the other hand, have been widely used but rarely miniaturized.

The advantages of carbon electrodes over metallic electrodes are that carbon is cheaper than the noble metals and more resistant to biofouling. Slower electron transfer kinetics at the surface, perhaps the greatest disadvantage to using carbon electrodes, is often overshadowed by carbon's ability to remain viable in overly adsorptive solutions that would quickly invalidate metal electrodes. Carbon is also biocompatible, and though not generally as easy to integrate into traditional nanolithography, carbon can be formed into single-walled nanotubes (SWNTs) and multiwalled nanotubes as well as nanofiber arrays.

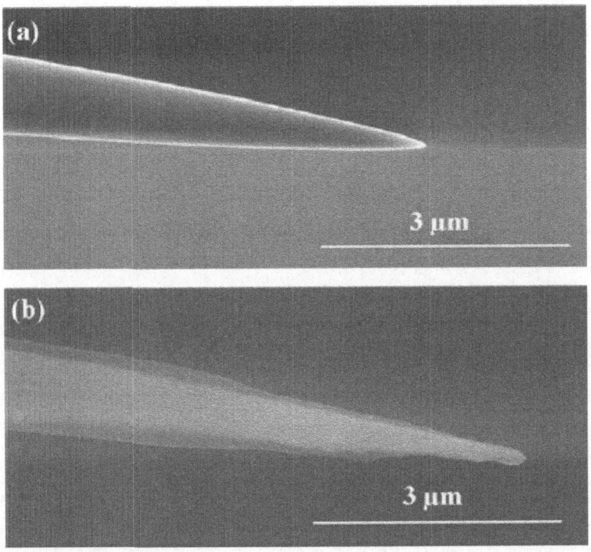

Figure 9.9 SEM images of tungsten wires formed by electrochemical etching (a, b) and coated with a layer of cathodic electrophoretic paint (b). (Reprinted with permission from IOP Publishing Ltd., ref. [3].)

9.5.5.1 CNT electrodes

Multiwalled CNTs were first discovered in 1991 and SWNTs two years later. Conceptually, CNTs are composed of graphite sheets "rolled" into a cylinder where SWNTs represent a single graphite sheet and multiwalled nanotubes are formed from multiple graphite sheets layered in concentric cylinders. Each has been extensively studied to determine how such unique structures translate into unique properties — chemical, mechanical, electrical, and so forth. Interestingly, theory predicts that CNTs will act as either metals or semiconductors depending on which sheet direction the graphite sheets are rolled [63]. Furthermore, for the semiconductor type, the band gap or energy gap can be tailored by manipulating the tube diameter given that the energy gap decreases with increasing fiber diameter [47]. Nonetheless, it has been reported that the nanotube achieves its excellent electrolytic properties not because of a particularly unique surface but more so because of the precarious placement of "dangling" oxygenated species at the end of the tubes [64]. Because there are already several good reviews on the history of CNTs along with staggering amounts of miscellaneous literature describing characterization, this section will limit discussion only to CNT's use in electrochemistry and specifically as biosensors. Suffice it to say that nanotube and carbon fiber–modified electrodes have achieved equal sensitivity and selectivity as gold electrodes under the appropriate conditions [33].

For example, in 1996, researchers were able to create an electrode by mixing CNTs and bromoform into a paste and then packing the paste into a small glass tube [14]. The study used pulse voltammetry to detect dopamine [14]. Since those early experiments, paste electrodes consisting of multiwalled nanotubes and mineral oil have been used to detect ascorbic acid, dopac and uric acid, dopamine, and H_2O_2 [13]. Nanotubes provide a significant enhancement in reversibility with dopamine reactions and a 5,000% increase in sensitivity as compared to paste electrodes made of powder carbon [13].

Multiwalled CNTs can also be converted into electrodes through a method engineered by Nugent *et al.* [65]. In 2001, Nugent created multiwalled nanotube "bundles" by exposing a graphitized carbon electrode to an electric arc discharge. The micron-sized bundles were then removed from the carbon electrode and fixed onto a copper wire with silver paint. An SEM image of the nanotube bundle electrode is shown in Fig. 9.10. Again, the nanotube bundle exhibited faster electron transfer kinetics than a traditional carbon paste electrode [65]. Three years later, Lin *et al.* created a more orderly array of aligned CNTs on a silicon substrate. After immobilizing glucose oxidase onto the nanotube ends, Lin's device was used to detect glucose down to a limit of 80 μM [66].

Figure 9.10 SEM images of the tip of a whole nanotube bundle electrode (a) and of a smaller area on the electrode showing the individual multiwalled nanotubes (b). (Reprinted with permission from the American Chemical Society, ref. [65].)

Multiwalled nanocomplexes can be used on their own or as redox mediators on a macroelectrode's surface. Specifically, carbon macroelectrodes (glassy carbon, graphite, and activated carbon) are difficult to use in biological settings because of the high potentials that are required to encourage redox at their surfaces. CNTs overcome this complication and expand the use of carbon electrodes by assisting charge transfer between the electrode and the analyte, thus allowing the use of lower potentials [67]. Carbon electrodes modified in this manner have been used to detect nicotinamide adenine dinucleotide [68], peroxidases, proteins, epinephrine, and hydrogen peroxide [64]. In 2003, a research group used such a system to simultaneously determine both dopamine and serotonin while experiencing little interference from large concentrations of ascorbic acid [69]. Ascorbic acid is often considered an interferent during *in vivo* electroanalytical studies because it is an electroactive molecule that is easily oxidized and is found in relatively large concentrations *in vivo* [70].

SWNT-modified electrodes have also shown catalytic activity with biologically relevant compounds like ascorbic acid, dopamine, epinephrine, norepinephrine [71], and assorted proteins [72]. Additionally, in 2008, Yan *et al.* reported that depositing 2–3 nm platinum nanoparticles onto glucose oxidase and CNT-modified carbon electrodes exhibited good sensitivity to glucose, even better than what was achieved with nanoparticles or nanotubes alone [73].

9.5.6 Electrode Arrays

9.5.6.1 Metal electrode arrays

The development of electrode arrays was the logical next step in electrode evolution. If a key disadvantage to decreasing the amount of an electrode's surface area is decreasing the amount of space available for mass transfer and electrolytic reactions, the logical solution is to maintain the small electrode dimensions so as to take advantage of the aforementioned micro/nano- properties but to somehow add more reaction area. The result of this reasoning was the assembly of multiple independent-channel microelectrodes into patterned arrays. First realized in 1972, researchers used photolithography to pattern thirty 7 μm wide gold lines insulated by photoresist onto a glass substrate [74]. Throughout the 1970s, scientists continued to improve on the initial design and began experimenting with multielectrode arrays (MEAs) for the purposes of detecting and stimulating action potentials of neurons. The 1980s and 1990s expanded on this work to include MEA applications on brain slices, explanted animal retinas, and *in vivo* subthreshold action potentials [75]. Now, with the twenty-first century's rise of curiosity and practical possibility in nanotechnology, experiments have begun exploring arrays of nanoelectrodes for potentials' roles in similar biological applications.

Nanoelectrodes can theoretically be made from any conductive material that can be patterned on the nanoscale. In practice, this limits the options mostly to metals and carbon. Nanoelectrodes can be made using metal nanowires and nanoparticles or using CNTs and nanowires. Individual electrodes are wired in parallel, adsorbed or bound in a slurry to a macroelectrode, or simply left suspended in an electrolytic solution. Regardless, nanoelectrodes can be spaced in arrays such that each electrode maintains an independent diffusion layer and the entire array can be used to analyze a distribution of points in a solution or on a surface. A contrasting technique that spaces electrodes closer to one another can be used to overlap each electrode's diffusion layer, thus creating a single semiplanar diffusion layer that simulates that of a macroelectrode yet maintains a better signal-to-noise ratio and uses less material or potentially a combination of advantageous materials [17]. The remainder of this chapter will highlight a variety of specific examples of nanoelectrodes and nanoelectrode arrays before detailing practical and proposed biological applications for such devices.

9.5.6.2 Carbon nanofiber arrays

Much like nanotube arrays, nanofiber arrays are predominantly used as redox mediators that enhance a parent electrode's electrochemical activity. Nanofiber arrays differ from nanotube arrays in that where nanotubes are made chemically and then deposited onto a surface. Nanofibers are typically either etched from a larger carbon fiber or grown directly onto a conductive surface. Both paths have been used in biosensing. Researchers in China were able to flame etch a single carbon fiber down to several hundred nanometers and then use the nanoelectrode (coupled with glucose oxidase) to detect glucose in real-time monitoring of rabbit blood. The nanoelectrode achieved a detection limit of 4 μM, a vast sensitivity improvement over conventional gold electrodes [76].

Finally, research presented at the 2005 10th Annual Conference of the International Functional Electrical Stimulation Society shows the next evolution in carbon nanofiber technology — dual-sensing and stimulating carbon nanofiber arrays. The research showed that carbon nanofibers could be deposited through plasma-enhanced chemical vapor deposition (PECVD) onto microcircuits to form an electrode. From there, depending on how much insulation (SiO_2) was added, these electrodes could be used to determine whether the array was to be used for detecting electric currents or for stimulating them. Researchers also showed that coating the arrays with 24 nm of polypyrrole caused the electrode's capacitance to increase by 100 times and displayed negligible Ohmic resistance as compared to solution. When grown into nanodisk arrays, the CNF electrodes were able to detect dopamine at 60 nM. Researchers propose that such electrodes could be used to investigate neurotransmitter releases and to improve deep-brain stimulation (DBS) [77]. A schematic of the fabrication process is shown in Fig. 9.11.

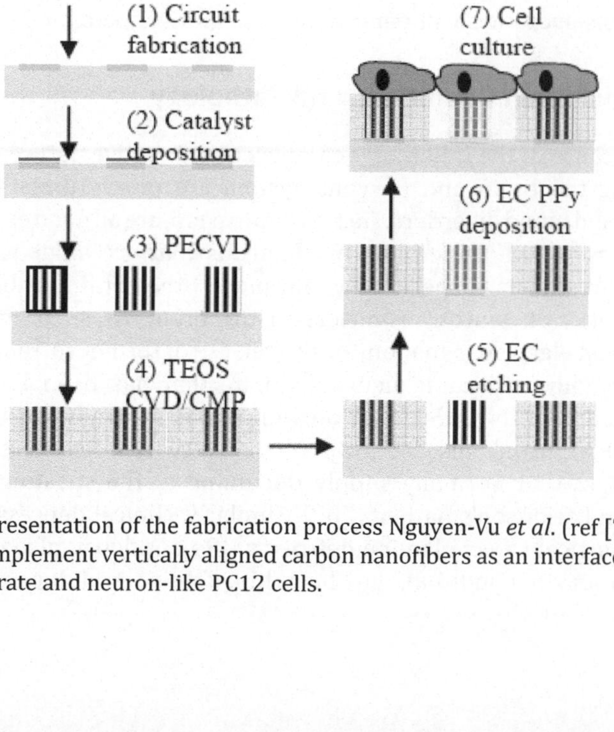

Figure 9.11 A representation of the fabrication process Nguyen-Vu *et al.* (ref [77]) used to create and implement vertically aligned carbon nanofibers as an interface between a silicon substrate and neuron-like PC12 cells.

9.6 Applications in the Brain

Apart from detecting relevant neurobiological molecules like dopamine, ascorbic acid, and seretonin, nanobioelectronics have begun to take the next step — direct *in vitro* and *in vivo* observation of the brain and individual neurons. For years, researchers have been utilizing electrodes to measure action potentials from electrogenic cells like neurons but high *in vivo* noise and low longevity have been limiting factors [78]. Since the 1980s, electroencephalography (EEG) and magnetoencephalography (MEG) have utilized electrodes placed on the outside surface of the skull to successfully record general brain activity. However, direct brain stimulation and specific electrochemical monitoring must be done using electrodes within the brain and, therefore, requires invasive implantation surgery [79]. Furthermore, the precise positioning of electrodes within the brain is difficult and creates questions as to what exact volume is being monitored or stimulated [80, 81]. The unique benefits of nanoelectrodes address many issues currently at hand in neuroscience research.

9.6.1 Neurophysiology

Neurochemical signals are transmitted and relayed throughout the brain and nervous system by either nerve transmission or nerve conduction. The former occurs when a chemical neurotransmitter is excreted from one nerve cell across a junction to another nerve cell. Whatever the function of the target cell, the neurotransmitter invariably causes a change in its ion permeability and can cause an action potential, the release of another neurotransmitter, and/or electrochemical conduction down the target cell's axons [82]. Through this mechanism, muscles are commanded to move, blood vessels to dilate, gastric juices to begin production, and all other manners of normal physiological function. A disruption in this mechanism, therefore, can have any number of negative consequences that manifest themselves physically, emotionally, cognitively, etc.

9.6.2 Abnormal Brain Conditions and Pathology

Unfortunately, because of the inherently complex and sensitive nature of the brain, many diseases involving the brain and nervous system are poorly understood. Parkinson's disease, epilepsy, and mood disorders, such as depression, are all good examples of common neurological diseases with underlying mechanisms that are inadequately understood. However, these three disease specifically and many other brain problems can often be traced to imbalances of neurotransmitters in the brain. As such, many neuroscience studies have utilized electrodes to monitor the release of the neurotransmitter dopamine. Dopamine is a readily oxidizable neuromodulate that has been linked to numerous diseases. It was found in the 1950s that dopamine was severally lacking in the striatum of patients with Parkinson's disease [83]. Further research discovered that neurons in the substantia nigra that normally supply dopamine to the striatum mysteriously die off prematurely in Parkinson's patients [82]. Similarly, clinical depression has long been linked to a neurological catecholamine deficiency [84]. Epilepsy, also often considered a "dopamine underactivity condition," has been linked to several dopaminergic receptors, including an anticonvulsant receptor in the forebrain and a proconvulsant receptor in the midbrain [85].

9.6.3 Neuromodulation and Interfacing

As with electrodes in solution, an electrode placed adjacent to electrogenic cells like neurons can be used to detect electrical activity along and within the cells. Though the solution/electrode interface and the cell/electrode junction are not exactly alike, the same general sequence occurs — electric activity in the cell or along the cell membrane causes a detectable polarization at the electrode surface. Conversely, a voltage applied to the electrode is detected by voltage-sensitive proteins suspended in the cell membrane. Such proteins control ion channels and, therefore, control ion flux and cellular action potentials [78]. Cheung *et al.* describe the cell/electrode interface as equivalent circuits containing a "space charge layer, charge transfer resistance, Warburg impedances, spreading resistance, shunt resistance, and other parasitic capacitances" [86]. This ability both to detect electrical activity and to stimulate chemical activity from cells is the linchpin of several emerging clinical and research techniques, namely, neuromodulation, brain/computer interfacing, and neuroprosthetics.

9.6.3.1 Deep-brain stimulation

DBS (Fig. 9.12) falls more broadly under the category of neuromodulation, the field in which devices implanted in the nervous system tune neuronal activity through electrical and/or chemical means [87]. This tuning has shown therapeutic effects with ailments such as epilepsy, chronic pain, Tourette syndrome, bladder and sphincter dysfunction, Parkinson's disease, and mood disorders such as depression and obsessive-compulsive disorder [87–89]. The standard procedure for DBS is to insert a 1 mm diameter electrode into the brain through a small hole in the skull and then pass a current through the electrode in order to stimulate surrounding neurons. Though there are several theories as to why this works, paradoxical findings and conflicting results have hampered a complete understanding of the underlying mechanisms [90].

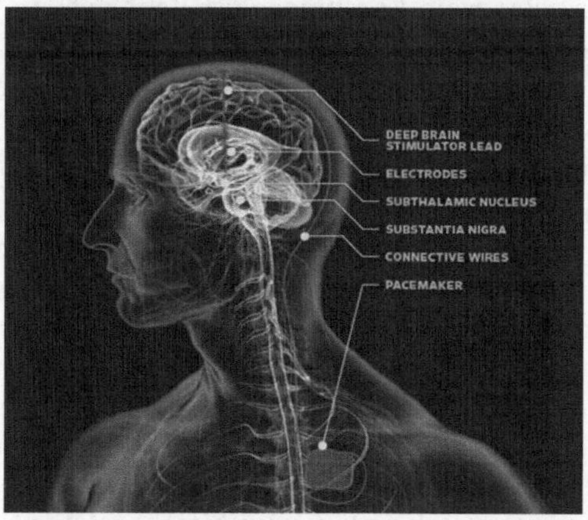

Figure 9.12 A representation of the standard DBS system components and placement.

For Parkinson's treatment, a vague description is that DBS somehow compensates for the decreased inhibitory activity that the globus pallidus externus (GPe) normally exhibits on the subthalmic nucleus and by extension also on the globus pallidus internus (GPi) [90]. To treat Parkinson's disease, the electrode is slowly inserted into the subthalamic nucleus while the patient is still awake. Placement is optimized to a position at which the patient feels the most relief [88]; however, the size of the electrode alone prevents a precise stimulation of only the appropriate neurons. Since most neurons are about 20 μm in diameter, a stimulatory electrode larger than this size will stimulate multiple neurons and nearby axons. The large size also limits the number of electrodes that can practically be placed in the brain and increases the risk of brain hemorrhaging at the site of insertion [88]. The body's foreign-body response compounds complications as inflammatory cells adhere to and ultimately encapsulate the electrode in proteins and fibrous scar tissue. This scar tissue both disrupts the electrode's intended electrolytic functions and inhibits neurite growth in the surrounding area [79].

Additionally, DBS electrodes stimulate with relatively large currents that are supplied from battery packs implanted within the patient's chest or abdomen. The battery implantation is also invasive as are the follow-up surgeries that are required every three to five years to replace the battery. Infections at the battery site as well as corrosion and rupture of the battery itself enhance the risk of morbidity [88].

Fortunately, advances in nanoelectrodes hold promise for eliminating or at least vastly reducing several of these risk factors. For example, a group in Berkeley, California, used a reactive ion etch to create 300 nm tall polysilicon nanopillars with diameters ranging from 20 nm to 100 nm [86]. The small size is predicted to reduce tissue damage, thus decreasing implant site morbidity and provoking less of an immune response. Because nanoelectrodes and nanoelectrode arrays can be easily fabricated at or below the size of a neuron, more precise stimulation is possible though optimal positioning may remain a problem.

At the forefront of this multidisciplinary research are CNTs and carbon nanofibers. In 2006, the Nanotechnology Group at NASA Ames Research Center showed that vertically aligned multiwalled CNT arrays coated with an electrically conductive polymer, polyprrole (PPy), exhibit capacitances of over a 1,000-fold increase compared to an analogous platinum microelectrode. Combined with decreased impedances characteristic of nanoarrays, such an electrode would require a lower current to achieve the same effect in neuromodulation, subsequently reducing the need for large traditional battery packs. The group's CNT array–supported neuronal growth were able to detect dopamine at 64 nM and exhibited short response times when used in conjunction with differential pulse voltammetry [88]. Research progressed further as the group created a second type of array similar to the first except now insulated with silicon oxide. The tips of these electrodes are mechanically polished, leaving only the carbon-fiber ends exposed. Again, the Ames group grew PC12 cells, neron-like cells derived from an immortalized adrenal cell line, on the insulated electrodes and was able to simultaneously measure nanomolar concentrations of dopamine and glutamate with a millisecond temporal resolution [80, 89]. The next step in DBS evolution will be to combine the stimulatory electrodes and the sensory electrodes into a closed circuit feedback loop capable of internally controlling electrical activity and neurotransmitter levels within a localized area of the brain. Such an advance would be a dramatic breakthrough in long-term DBS strategies [87, 89].

Another breakthrough could come from how the electrodes are delivered. A research team at the NYU School of Medicine is experimenting with neurovascular implantation in which a small (600 nm diameter) platinum electrode is inserted into the bloodstream and gets carried to the brain intravascularly. The group's results show that both electrical stimulation and monitoring can occur across capillary vessels in the brain. Future plans are to replace the platinum electrodes with electrically conductive polymer nanofibers to increase biocompatibility and enhance electrical properties as previously mentioned [79].

9.6.3.2 Computer/brain interfacing

Of course, apart from manipulating interneuron communication, electrodes can theoretically also be used to transmit information from the brain externally and to collect and integrate externally information into the brain. A burgeoning science referred to as brain/computer interfacing (BCI) is focused on the former while neuroprosthetics deal with the latter. For example, a collaborative BCI project centered around the Harvard School of Medicine published a 2006 paper in *Nature* describing the use of implanted microelectrodes to decode brain activity from a tetraplegic subject and then transmit those signals to a computer. With practice and focus, the subject was able to move the computer cursor with his thoughts alone. The final intent of this type of interface is the creation of prosthetic limbs that respond to the user's neural commands [91].

9.6.3.3 Neuroprosthetics

By contrast, neuroprosthetic electrodes serve the brain as an aid to damaged or dysfunctional sensory neurons. In other words, the difference between neuroprosthetics and BCI is that BCIs are intended to interpret physical intent from brain activity alone whereas neuroprosthetics serve as sensory relays designed to collect, transmit, or interpret external information to the brain. Thus, neuroprosthetics use similar technology as DBS and BCI but to different ends. Cochlear implants are perhaps one of the most recognizable forms of neuroprosthetics. For these devices, implanted electrodes are used to relay externally detected sound waves by directly stimulating auditory nerves in the cochlea. Analogous devices are being prepared for retinal implants, those involving the visual cortex, and optical nerve stimulators [92]. However, because of the novelty of many of these applications, nanoscience has largely failed to make inroads into BCIs or neuroprosthetics. Fortunately though, the appropriate technology exists and it is only a matter of time before the fields merge. Cochlear implants, for one, would likely see vast improvement if thousands of nanoelectrodes were used to stimulate cochlear nerves as opposed to the mere dozens used in current devices. Similarly, BCIs and optical implants would surely benefit from the increased sensitivity and enhanced spatial and temporal resolutions CNT arrays provide [93].

9.7 Applications to the Heart and Muscles

Building upon earlier research by Walter Gaskell and Augustus D. Waller, the first electrocardiograph was published by Willem Einthoven in 1902. Einthoven used a string galvanometer to follow the electrical depolarization and subsequent contraction of cardiac muscles in the atria and into ventricles. Over 50 years later, the first fully internal human pacemakers were implanted in an attempt to regulate this rhythmic depolarization. Since its inception six years before, the pacemaker had shrunk from a "breadbox"-sized machine to something more along the lines of a thick, heavy pocket watch. Pacemakers have gone from using a single lead implanted in the right ventricle to multiple leads able to send stimuli to both the atria and the ventricles. The associated pulse generators now weigh only grams and use negative feedback control loops to accurately and intermittently control the heart's operation [94].

9.7.1 Cardiophysiology

Muscles contract when electrical and/or chemical stimuli cause an increase in intracellular calcium concentrations that eventually leads to a chemical cross-linking reaction between muscle fibers. In a healthy heart, however, no external stimulus is necessary as a small bundle of cardiac cells called the sinus node spontaneously and rhythmically depolarizes. From this initial depolarization, the electrical wave spreads throughout the atria and eventually terminates in the ventricles, causing heart muscles to contract as it passes.

9.7.2 Electrochemistry and Stimulation

Very few experiments have been reported examining the uses of nanoelectrodes with cardiac tissue or cardiac myocytes. Perhaps muscular stimulation in general requires a much coarser stimulatory system and, therefore, existing macro- and microtechnology suffice for most cardiac medical needs. On the other hand, coordinating the system of muscular contraction and feedback controls is overly complex and at present confounds attempts at intricate manipulation. Regardless, with a little imagination, one can visualize a myriad of potential uses that could develop with nanosensors and nanoelectrodes. Analogous to the advantages of nanoelectrodes in DBS, precise control of the heart's pacemaker node could optimize the way pacemakers and defibrillators operate resulting in smaller, more efficient batteries and better overall results for the patients. Nanoelectrodes coupled to understimulated muscles could be combined with the BCI applications discussed above to overcome miscellaneous neuromuscular defects. Finally, nanoelectrode sensor/stimulator combinations could be implemented as negative feedback control loops to improve atrophied muscle reflex behaviors. The list is endless. As a generalized proof of concept, recent myocyte studies show that both individual muscle cells and confluent sheets of cells can be electrically stimulated with a variety of different electrode systems. In these studies, researchers have used conductive polymer films [95], metal electrodes [32, 96], light-addressable indium tin oxide leads [97], and carbon fiber sensors [98]. Cardiac myocytes specifically have been both monitored [96–100] and stimulated [32, 95] individually. Examples of electrochemical cells for stimulating cardiac myocytes are shown in Figs. 9.13 and 9.14.

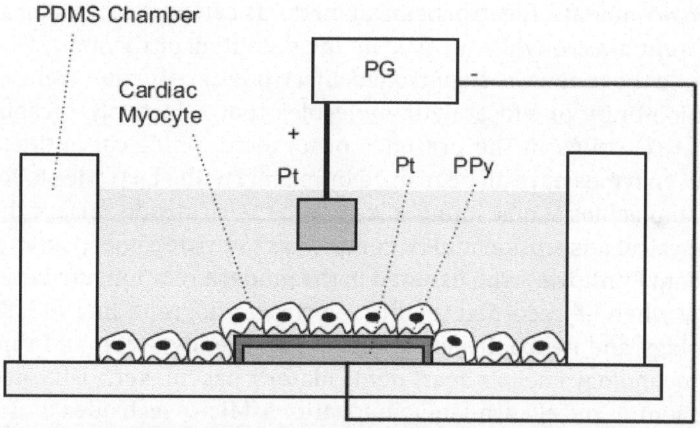

Figure 9.13 A schematic representation of an electrochemical cell designed to stimulate cardiac myocytes with a current pulse generator. (Reprinted with permission from Elsevier, ref. [95].)

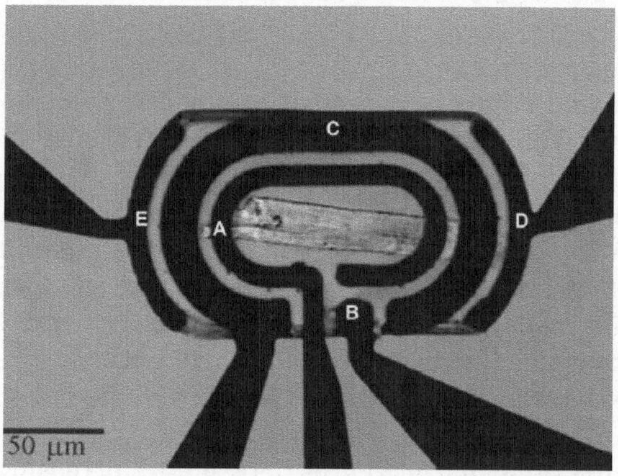

Figure 9.14 An image of a cardiomocyte growing within a shallow microchamber and surrounded by various microelectrodes (A–E). (Reprinted with permission from RSC Publishing Ltd., ref. [100].)

9.8 Conclusion

As with nanotechnology in general, the true benefits of nanoelectrodes and small-scale precision biosensors are predominantly still to come. Progress thus far highlights both the historic interplay between electrode sensors and physiological measurements and further underlines the enormous potential these applications have in clinical medicine. Galvani's early pioneering of experimental electrophysiological techniques combined with Michael Faraday's contributions in electrode theory and behavior now form the basis of modern electrochemistry in physiology. Typical modern experiments involve a standard three-electrode electrochemical cell assembled in a variety of configurations and used to observe the electron transfer of Faradaic reactions in biological environments. Electrochemical methods can be static, using a potentiometer to hold the current at zero while measuring the potential, or dynamic, where the current is observed while the potential is controlled. Electrodes are further utilized as biosensors by attaching bioaffinity or biocatalytic molecules that selectively recognize or catalyze electrochemical reactions in the presence of targeted biological entities. DNA strands, antibodies, and enzymes have all been used in concert with electrodes to form biosensors. These types of molecules can be immobilized onto the electrode surface through methods as simple as physical adsorption and as complex as multistep encapsulations.

It was William Einthoven who ushered in the modern relationship between electrodes and physiology when he recorded the first electrocardiograph in 1902. Since then, both cardiac physiology and neurophysiology have seen the invention and rapid evolution of bioelectronic technology such as heart defibrillators, pacemakers, DBS, neuroprosthetics, and an assortment of purely academic innovations. Microelectrodes and nanoelectrodes represent the latest advances for these innovations. Their increased signal-to-noise ratios, decreased background charging, and enhanced mass transport profiles improve the response time and the ability to detect transient analytes. They also require lower currents to operate, making them easier and less harmful to utilize in living organisms. Both micro- and nanoelectrodes wired in parallel arrays can be used as sensitive detectors in small-scale environments like cells and bacteria or, using an appropriately large volume and density, they can be used to create a single semiplanar diffusion layer similar to that of a macroelectrode yet with increased sensitivity and a response that is characteristic of smaller electrodes.

Though most clinical applications currently utilize macro- or microelectrodes, there is a large potential for nanotechnology in biomedical areas. Implanting nanoscale systems in general is proposed to decrease physical damage at the implant site and improve interfacing between the electronics and the brain or heart. In particular, existing neuromodulation techniques such as DBS would benefit greatly by the use of smaller, more sensitive and precise electrodes. Likewise, it is predicted that applying nanotechnology to emerging fields like neuroprosthetics could revolutionize cochlear and retinal implants and potentially lead to brain-controlled prosthetic limbs. Increasingly in-depth research in cardiac physiology will also find substantial value with using nanosensors and nanoelectrodes as mechanisms of precise stimulation and sensitive detection.

While this chapter has attempted to highlight the pure science and research achieved to date in the area of nanoscale-biosensing technology, it has largely excluded predictions as to how such advances will affect the scientific community and society on whole. The potential influence of metal and carbon-based electrodes is not limited to the few fields discussed here. Bioelectric monitoring will have applications in not only clinical disease detection, pathology, and treatment but also in areas of advanced physiology, chemistry, psychology, psychiatry, biology, pharmacology, and even physics research. Ultimately biosensors provide the potential for precision understanding of the underlying mechanisms behind how life itself is fundamentally experienced, sustained, and improved.

References

1. Galvani, L. (1954). In *Commentary on the Effects of Electricity of Muscular Motion*, (ed. Fulton, J. F., and Stanton, M. E.), Burndy Library.

2. McKnight, T. E., Melechko, A. V., Griffin, G. D., Guillorn, M. A., Merkulov, V. I., Doktycz, M. J., Ericson, M. N., and Simpson, M. L. (2007). Cellular interfacing with arrays of vertically aligned carbon nanofibers and nanofiber-templated materials, in *Nanotechnology in Biology and Medicine: Methods, Devices, and Applications* (ed. Vo-Dinh, T.), CRC Press, New York, pp. 22–1.
3. Qiao, Y., J. Chen, Guo, X., Cantrell, D., Ruoff, R., and Troy, J. (2005). Fabrication of nanoelectrodes for neurophysiology: cathodic electrophoretic paint insulation and focused ion beam milling, *Nanotechnology*, **16**, 1598.
4. Faraday, M. (1834). Experimental researches in electricity. Seventh series, *Philos. Trans. R. Soc. London*, **124**, 77.
5. Rubenstein, I. (1995). Fundamentals of physical electrochemistry, in *Physical Electrochemistry: Principles, Methods, and Applications* (ed. Rubenstein, I.), Marcel Dekker, New York, p. 1.
6. Bard, A. J., and Falkner, L. R. (2001). Introduction and overview of electrode processes, in *Electrochemical Methods: Fundamentals and Applications*, John Wiley & Sons, New York, p. 1.
7. Bakkera, E., and Pretsch, E. (2008). Nanoscale potentiometry, *TrAC, Trends Anal. Chem.*, **27**, 612.
8. Burmeister, J. J., Pomerleau, F., Huettl, P., Gash, C. R., Werner, C. E., Bruno, J. P., and Gerhardt, G. A. (2008). Ceramic-based multisite microelectrode arrays for simultaneous measures of choline and acetylcholine in CNS, *Biosens. Bioelectron.*, **23**, 1382.
9. Zachek, M. K., Takmakov, P., Moody, B., Wightman, R. M., and McCarty, G. S. (2009). Simultaneous decoupled detection of dopamine and oxygen using pyrolyzed carbon microarrays and fast-scan cyclic voltammetry, *Anal. Chem.*, **81**(15), 6258–6265.
10. Wightman, M. R. (2006). Probing cellular chemistry in biological systems with microelectrodes, *Science*, **311**, 1570.
11. Grundler, P. (2007). *Chemical Sensors: An Introduction for Scientists and Engineers*, Springer, New York, p. 273.
12. Burmeister, J. J., and Gerhardt, G. A. (2001). Self-referencing ceramic-based multisite microelectrodes for the detection and elimination of interferences from the measurement of l-glutamate and other analytes, *Anal. Chem.*, **73**(5), 1037.
13. Rubianes, M. D., and Rivas, G. A. (2003). Carbon nanotubes paste electrode, *Electrochem. Commun.*, **5**, 689.
14. Britto, P. J., Santhanam, K. S. V., and Ajayan, P. M. (1996). Carbon nanotube electrode for oxidation of dopamie, *Bioeletrochem. Bioenerg.*, **41**, 121.
15. Berduque, A., Lanyon, Y. H., Beni, V., Herzog, G., Watson, Y. E., Rodgers, K., Stam, F., Alderman, J., and Arrigan, D. W. M. (2007). Voltammetric characterisation of silicon-based microelectrode arrays and their application to mercury-free stripping voltammetry of copper ions, *Talanta*, **71**, 1022.
16. Stulik, K., Amatore, C., Holub, K., Marecek, V., and Kutner, W. (2002). Microelectrodes. Definitions, characterization, and applications, *Pure Appl. Chem.*, **72**(8), 1483.
17. Compton, R. G., Wildgoose, G. G., Rees, N. V., Streeter, I., and Baron, R. (2008). Design, fabrication, characterisation and application of nanoelectrode arrays, *Chem. Phys. Lett.*, **459**, 1.
18. Bard, A. J., and Falkner, L. R. (2001). Basic potential step methods, in *Electrochemical Methods: Fundamentals and Applications*, John Wiley & Sons, New York, p. 156.
19. Hietpas, P. B., Gilman, S. D., Lee, A. R., Wood, M. R., Winograd, N., and Ewing, A. G. Developement of voltammetric methods, capillary electrophoresis and TOF SIMS imaging for constituent

analysis of single cells, in *Nanofabrication and Biosystems: Integrating Materials Science, Engineering, and Biology* (ed. Hoch, H. C., Jelinski, L. W., and Craighead, H. G.), Cambridge University Press, New York, p. 139.

20. Al-Hilli, S., Al-Mofarji, R., Klason, P., Gutman, N., Sa'ar. A., Ost, A., Stralfors, P., and Willander, M. (2007). *Zinc oxide nanorods as an intracellular pH sensor*, ENS'07, 38.
21. Moody, B., and McCarty, G. S. (2008). Nanolithographic techniques in tissue engineering, in *Nanotechnology and Tissue Engineering: The Scaffold* (ed. Laurencin, K. T., and Nair, L. S.), CRC Press, p. 261.
22. Lanyon, Y. H., and Arrigan, D. W. M. (2008). Top-down approaches to the fabrication of nanopatterned electrodes, in *Nanostructured Materials in Electrochemistry* (ed. Eftekhari, A.), Wiley-VCH Verlag GmbH & Co. KGaA, Germany, p. 187.
23. Kneipp, J., Kneipp, H., Rice, W. L., and Kneipp, K. (2005). Optical probes for biological applications based on surface-enhanced Raman scattering from indocyanine green on gold nanoparticles, *Anal. Chem.*, **77**(8), 2381–2385.
24. Kneipp, K., Haka, A. S., Kneipp, H., Badizadegan, K., Yoshizawa, N., Boone, C., Shafer-Peltier, K. E., Motz, J. T., Dasari, R. R., and Feld, M. S. (2002). Surface-enhanced Raman Spectroscopy in single living cells using gold nanoparticles, *Appl. Spectrosc.*, **56**(2), 150–154.
25. West, J. L., and Halas, N. J. (2003). *Engineered nanomaterials for biophotonics applications: improving sensing, imaging, and therapeutics*, **5**, 285.
26. Kramarcy, N. R., and Sealock, R. (1991). Commercial preparations of colloidal gold-antibody complexes frequently contain free active antibody, *J. Histochem. Cytochem.*, **39**(1), 37.
27. Hainfeld, J. F., and Furuya, F. R. (1992). A 1.4-nm gold cluster covalently attached to antibodies improves immunolabeling, *J. Histochem. Cytochem.*, **40**(2), 177.
28. Shah, D. S., Thomas, M. B., Phillips, S., Cisneros, D. A., Le Brun, A. P., Holt, S. A., and Lakey, J. H. (2007). Self-assembling layers created by membrane proteins on gold, *Biochem. Soc. Trans.*, **35**, 522–526.
29. Xiao, Y., Patolsky, F., Katz, E., Hainfeld, J. F., and Willner, I. (2008). Plugging into enzymes: nanowiring of redox enzymes by a gold nanoparticle, *Science*, **299**, 1877.
30. Crumbliss, A. L., Perine, S. C., Stonehuerner, J., Tubergen, K. R., Zhao, J., Henkenst, R. W., and O'Daly, J. P. (1992). Colloidal gold as a biocompatible immobilization matrix for the fabrication of enzyme electrodes by electrodeposition, *Biotechnol. Bioeng.*, **40**, 483.
31. Farrell, N. (1989). Transition metal complexes as drugs and chemotherapeutic agents, in *Catalysis by Metal Complexes* (ed. James, B. R., and Ugo, R.), Kluwer, Dordrecht.
32. Chen, Y., Tsai, C., Huang, P., Chang, M., Cheng, P., Chou, C., Chen, D., Wang, C., Shiau, A., and Wu, C. (2007). Methotrexate conjugated to gold nanoparticles inhibits tumor growth in a syngeneic lung tumor modelmethotrexate conjugated to gold nanoparticles inhibits tumor growth in a syngeneic lung tumor model, *Mol. Pharmaceutics.*, **4**(5), 713.
33. Zachek, M. K., Hermans, A., Wightman, M. R., and McCarty, G. S. (2008). Electrochemical dopamine detection: comparing gold and carbon fiber microelectrodes using background subtracted fast scan cyclic voltammetry, *J. Electroanalytic. Chem.*, **614**, 113.
34. Green, S. J., Stokes, J. J., Hostetler, M. J., Pietron, J., and Murray, R. W. (1997). Three-dimensional monolayers: nanometer-sized electrodes of alkanethiolate-stabilized gold cluster molecules, *J. Phys. Chem. B*, **101**(14), 2663.

35. Chen, S., Ingram, R. S., Hostetler, M. J., Pietron, J. J., Murray, R. W., Schaaff, T. G., Khoury, J. T., Alvarez, M. M., and Whetten, R. L. (1998). Gold nanoelectrodes of varied size: transition to molecule-like charging, *Science*, **280**, 2098.

36. Zhang, J. Z., Wang, Z., Liu, J., Chen, S., and Liu, G. (2004). Electrochemical properties of nanoparticle assemblies, in *Self-Assembled Nanostructures*, Springer, New York, p. 271.

37. Hostetler, M. J., Green, S. J., Stokes, J. J., and Murray, R. W. (1996). Monolayers in three dimensions: synthesis and electrochemistry of omega-functionalized alkanethiolate-stabilized gold cluster compounds, *J. Am. Chem. Soc.*, **118**(17), 4212.

38. Wang, X., Zhong, H., Lv, Y., and Chen, H. (2003). Fabricatin of nanoelectrode ensembles of porous gold nanoshells and direct electrochemistry of horseradish peroixdase immobilized on the electrode, *Chem. Lett.*, **32**(11), 1054.

39. Tian, R., and Zhi, J. (2007). Fabrication and electrochemical properties of boron-doped diamond film-gold nanoparticle array hybrid electrode, *Electrochem. Commun.*, **9**, 1120.

40. Wang, X. L., and Jin, W. R. (2003). A simple method for fabrication of gold nanoelectrodes, *Chinese Chem. Lett.*, **14**(7), 728.

41. Slevin, C. J., Gray, N. J., Macpherson, J. V., Webb, M. A., and Unwin, P. R. (1999). Fabrication and characterisation of nanometre-sized platinum electrodes for voltammetric analysis and imaging, *Electrochem. Commun.*, **1**, 282.

42. Liu, Y., Zhu, Y., Zeng, Y., and Xu, F. (2009). An effective amperometric biosensor based on gold nanoelectrode arrays, *Nanoscale Res. Lett.*, **4**, 210.

43. Vastarella, W., Seta, L. D., Masci, A., Maly, J., De Leo, M., Moretto, L. M., and Pilloton, R. (2007). Biosensors based on gold nanoelectrode ensembles and screen printed electrodes, *Intern. J. Environ. Anal. Chem.*, **87**(10–11), 701.

44. Delvaux, M., Walcarius, A., and Demoustier-Champagne, S. (2004). Electrocatalytic H_2O_2 amperometric detection using gold nanotube electrode ensembles, *Anal. Chim. Acta*, **525**, 221.

45. Krapf, D., Wu, M., Smeets, R. M. M., Zandbergen, H. W., Dekker, C., and Lemay, S. G. (2006). Fabrication and characterization of nanopore-based electrodes with radii down to 2 nm, *Nano Lett.* **6**, 105.

46. McCarty, G. S. (2004). Molecular lithography for wafer-scale fabrication of molecular junctions, *Nano Lett.*, **4**, 1391.

47. Saito, R., Fujita, M., Dresselhaus, G., and Dresselhaus, M. S. (1992). Electronic structure of chiral graphene tubules, *Appl. Phys. Lett.*, **60**(18), 2201.

48. Ward, D. R., Grady, N. K., Levin, C. S., Halas, N. J., Wu, Y. P., Nordlander, P., and Natelson, D. (2007). Electromigrated nanoscale gaps for surface-enhanced Raman spectroscopy, *Nano Lett.*, **7**(5), 1396–1400.

49. Subramanian, A., Dong, L., and Nelson, B. J. (2005). *Assembly of arrays of individual lateral nanotube emitters on nanoelectrodes*, Proceedings of 2005 5th IEEE Conference on Nanotechnology, Nagoya, Japan.

50. Zandbergen, H. W., van Duuren, R. J. H. A., Alkemade, P. F. A., Lientschnig, G., Vasquez, O., Dekker, C., and Tichelaar, F. D. (2005). Sculpting nanoelectrodes with a transmission electron beam for electrical and geometrical characterization of nanoparticles, *Nano Lett.*, **5**(3), 549.

51. Liang, X., and Chou, S. Y. (2008). Nanogap detector inside nanofluidic channel for fast real-time label-free DNA analysis, *Nano Lett.*, **8**(5), 1472–1476.

52. Ward, D. R., Halas, N. J., Ciszek, J. W., Tour, J. M., Wu, Y., Nordiander, P., and Natelson, D. (2008). Simultaneous measurements of electronic conduction and Raman response in molecular junctions, *Nano Lett.*, **8**(3), 919–924.
53. Moody, B., and McCarty, G. S. (2009). Solid state nanogaps for differential measurements of molecular properties, *Appl. Phys. Lett.*, **94**, 122140–122141.
54. Evans, U., Colavita, P. E., Doescher, M. S., Schiza, M., and Myrick, M. L. (2002). Construction and characterization of a nanowell electrode array, *Nano Lett.*, **2**(6), 641.
55. Maillard, M., Huang, P., and Brus, L. (2003). Silver nanodisk growth by surface plasmon enhanced photoreduction of adsorbed [Ag^+], *Nano Lett.*, **3**(11), 1611.
56. Riveros, G., Green, S., Cortes, A., Gomez, H., Marotti, R. E., and Dalchiele, E. A. (2006). Silver nanowire arrays electrochemically grown into nanoporous anodic alumina templates, *Nanotechnology*, **17**, 561.
57. Kong, C., Hu, Z., Zhao, H., Yang, Y., Shang, X., Ren, L., and Wang, Y. (2007). Preparation of Ag nanowire array electrode by transplantation and its electrochemical activities, *Int. J. Electrochem. Sci.*, **2**, 133.
58. Xia, G., Hu, X. Y., Wang, C. Y., Jin, G. D., and Guo, R. (2002). Preparation of nanoelectrode ensembles by assembly of nanosilver colloid on gold surface, *Chinese Chem. Lett.*, **13**(2), 159.
59. Wang, F., Yuan, R., Chai, Y., and Tang, D. (2007). Probing traces of hydrogen peroxide by use of a biosensor based on mediator-free DNA and horseradish peroxidase immobilized on silver nanoparticles, *Anal. Bioanal. Chem.*, **387**, 709.
60. Xu, J., Zhang, Y., Lia, G., and Zhu, J. (2004). An electrochemical biosensor constructed by nanosized silver particles doped sol–gel film, *Mater. Sci. Eng. C*, **24**(6–8), 833.
61. Wang, C., Chen, Y., Hu, X., and Guo, R. (2006). Effect of Hemoglobin Adsorption on Apparent Capacitance Density of Platinum Disk Nanoelectrode, *Int. J. Electrochem. Sci.*, **1**, 139.
62. Yamamoto, K., Shi, G., Zhou, T., Xu, F., Zhu, M., Liu, M., Kato, T., Jin, J., and Jin, L. (2003). Solid-state pH ultramicrosensor based on a tungstic oxide film fabricated on a tungsten nanoelectrode and its application to the study of endothelial cell, *Anal. Chim. Acta*, **480**, 109.
63. Baughman, R. H., Zakhidov, A. A., and de Heer, W. A. (2002). Carbon nanotubes — the route toward applications, *Science*, **297**, 787.
64. Gooding, J. J. (2005). Nanostructuring electrodes with carbon nanotubes: A review on electrochemistry and applications for sensing, *Electromchim. Acta*, **50**, 3049.
65. Nugent, J. M., Santhanam, K. S. V., Rubio, A., and Ajayan, P. M. (2001). Fast electron transfer kinetics on multiwalled carbon nanotube micobundle electrodes, *Nano Lett.*, **1**, 87.
66. Lin, Y., Lu, F., Tu, Y., and Ren, Z. (2004). Glucose biosensors based on carbon naotube nanolectrode ensembles, *Nano Lett.*, **4**(2), 191.
67. Wring, S. A., and Hart, J. P. (1992). Chemically modified, carbon-based electrodes and their application as electrochemical sensors for the analysis of biologically important compounds. A review, *Analyst*, **117**, 1215.
68. Musameh, M., Wang, J., Merkoci, A., and Lin, Y. (2002). Low-potential stable NADH detection at carbon-nanotube-modified glassy carbon electrodes, *Electrochem. Commun.*, **4**, 743.
69. Wu, K., Fei, J., and Hu, S. (2003). Simultaneous determination of dopamine and serotonin on a glassy carbon electrode coated with a film of carbon nanotubes, *Anal. Biochem.*, **318**, 100.
70. Gonon, F. G., Fombarlet, C. M., Buda, M. J., and Pujol, J. F. Electrochemical treatment of pyrolytic carbon fiber electrodes, *Anal. Chem.*, **53**, 1386.

71. Wang, J., Li, M., Shi, Z., Li, N., and Gu, Z. (2002). Electrocatalytic oxidation of norepinephrine at a glassy carbon electrode modified with single wall carbon nanotubes, *Electroanalysis*, **14**, 225.

72. Wang, J., Li, M., Li, N., Shi, Z., and Gu, Z. (2002). Direct electrochemistry of cytochrome c at a glassy carbon electrode modified with single-wall carbon nanotubes, *Anal. Chem.*, **74**(9), 1993.

73. Yan, J., Zhou, H., Yu, P., Su, L., and Mao, L. (2008). Rational functionalization of carbon nanotubes leading to electrochemical devices with striking applications, *Adv. Mater.*, **20**, 2899.

74. Thomas, C. A. J., Springer, P. A., Loeb, G. E., Berwald-Netter, Y., and Okun, L. M. (1972). A miniature microelectrode array to monitor the bioelectric activity of cultured cells, *Exp. Cell Res.*, **74**, 61.

75. Pine, J. (2006). A history of MEA development, in *Advances in Network Electrophysiology: Using Multi-Electrode Arrays* (ed. Taketani, M., and Baudry, M.), Springer, New York, p. 3.

76. Fei, J., Wu, K., Wang, F., and Hu, S. (2005). Glucose nanosensors based on redox polymer/glucose oxidase modified carbon fiber nanoelectrodes, *Talanta*, **65**, 918.

77. Nguyen-Vu, T. B., Chen, H., Cassell, A., Koehne, J., Purewal, H., Meyyappan, M., Andrews, R., and Li, J. (2005). *Carbon nanofiber nanoelectrode array for closed-loop electrical stimulation*, 10th Annual Conference of the International FES Society.

78. Cheran, L., Benvenuto, P., and Thompson, M. (2008). Coupling of neurons with biosensor devices for detection of the properties of neuronal populations, *Chem. Soc. Rev.*, **37**, 1229.

79. Llinas, R. R., Walton, K. D., Nakao, M., Hunter, I., and Anquetil, P. A. (2005). euro-vascular central nervous recording/stimulating system: Using nanotechnology probes, *J. Nanoparticle Res.*, **7**, 111.

80. Li, J., and Andrews, R. J. (2007). Trimodal nanoelectrode array for precise deep brain stimulation: prospects of a new technology based on carbon nanofiber arrays, *Acta Neurochir. Suppl.*, **97**, 537.

81. Leary, S. P., Liu, C. Y., and Apuzzo, M. L. J. (2006). Toward the emergence of nanoneurosurgery: part III – nanomedicine: targeted nanotherapy, nanosurgery, and progress toward the realization of nanosurgery, *Neurosurgery*, **58**(6), 1009.

82. Grubbs, R. D., and Sundberg, D. K. (2003). Chemical messenger systems, in *Neuroscience in Medicine* (ed. Conn, P. M.), Humana Press, New Jersey, p. 457.

83. Cooper, G., and Rodnitzky, R. L. (2003). Parkinson's disease, in *Neuroscience in Medicine* (ed. Conn, P. M.), Humana Press, New Jersey, p. 485.

84. Weiss, J. M., Glazer, H. I., and Pohorecky, L. A. (1976). Coping behavior and neurochemical changes, in *Animal Models in Human Psychobiology* (ed. Serban, G., and Kling, A.), Plenum Press, New York, p. 141.

85. Starr, M. S. (1996). The role of dopamine in epilepsy. Animal models in human psychobiology, *Synapse*, **22**, 159.

86. Cheung, K. C., Choi, Y., Kubow, T. M., and Lee, L. P. (2002). Nanostructured electrodes for improved neural recording, *Mat. Res. Soc. Symp. Proc.*, **729**, U4.11.1.

87. Sakas, D. E., Panourias, I. G., and Simpson, B. A. (2007). An introduction to neural networks surgery, a field of neuromodulation which is based on advances in neural networks science and digitised brain imaging, *Acta Neurochir. Suppl.*, **97**(2), 3.

88. Andrews, R. J. (2007). Neuroprotection at the nanolevel – part II: nanodevices for neuromodulation – deep brain stimulation and spinal cord injury, *Ann. N. Y. Acad. Sci.*, **1122**, 185.

89. Andrews, R. J., Li, J., Cassell, A., Koehne, J., Meyyappan, M., Nguyen-Vu, B., Huang, N., and Chen, L. (2006). In *The NASA Nanoelectrode Array for Deep Brain Stimulation: Monitoring Neurotransmitters and Electrical Activity Plus Precise Stimulation* (ed. Kanno, T.), Springer: Japan, 2006; pp 212.

90. Vitek, J. L. (2008). Deep brain stimulation: How does it work? *Cleveland Clin. J. Med.*, **75**, S59.

91. Hochberg, L. R., Serruya, M. D., Friehs, G. M., Mukand, J. A., Saleh, M., Caplan, A. H., Branner, A., Chen, D., Penn, R. D., and Donoghue, J. P. (2006). Neuronal ensemble control of prosthetic devices by a human with tetraplegia, *Nature*, **442**, 164.

92. Leuthardt, E. C., Schalk, G., Moran, D., and Ojemann, J. G. (2006). The emerging world of motor neuroprosthetics: a neurosurgical perspective, *Neurosurgery*, **58**, 1.

93. Andrews, R. J. (2007). Neuroprotection at the nanolevel – part I: introduction to nanoneurosurgery, *Ann. N. Y. Acad. Sci.*, **1122**, 169.

94. Jeffrey, K. (2001). *Machines in Our Hearts: The Cardiac Pacemaker, The Implantable Defibrillator, and American Health Care*, The Johns Hopkins University Press, Baltimore.

95. Nishizawa, M., Nozaki, H., Kaji, H., Kitazume, T., Kobayashi, N., Ishibashi, T., and Abe, T. (2007). Electrodeposition of anchored polypyrrole film on microelectrodes and stimulation of cultured cardiac myocytes, *Biomaterials*, **28**, 1480.

96. Offenhausser, A., Ingebrandt, S., and Mayer, D. (2005). Interfacing biology with electronic devices, *Solid State Phenom.*, **108–109**, 789.

97. Bucher, V., Schubert, M., Kern, D., and Nisch, W. (2001). Light-addressed sub-um electrodes for extracellular recording and stimulation of excitable cells, *Microelectron. Eng.*, **57–58**, 705.

98. Malinski, T., Grunfeld, S., Taha, Z., and Tomboulian, P. (1994). Monitoring metal concentrations in tissues and single cells using ultramicrosensors, *Environ. Health Perspect.*, **102**, 147.

99. Shimaoka, T., Lee, T., Morishima, M., Wang, Y., Miyamoto, S., and Ono, K. (2008). Chronic magnesium deficiency causes ion channel remodelings in cardica myocytes, *J. Mol. Cell Cardiol.*, **44**, 435.

100. Cheng, W., Klauke, N., Sedgwick, H., Smith, G. L., and Cooper, M. (2006). Metabolic monitoring of the electrically stimulated single heart cell within a microfluidic platform, *Lab Chip*, **6**, 1424.

Chapter 10

Microelectromechanical Systems for *in vivo* Therapeutics

Masaru P. Rao

Department of Mechanical Engineering, Materials Science and Engineering Program, and Department of Bioengineering, University of California, Riverside, Bourns Hall A313, University of California, Riverside, CA 92521, USA

mprao@engr.ucr.edu

The continuing need for the enhancement of efficacy, safety, and/or functionality for *in vivo* therapeutics provides immense opportunity for microelectromechanical systems (MEMS). Motivated by this opportunity, most efforts to date have focused on applications in drug delivery and neuro/muscular prosthetic interfaces. However, the potential for the exploitation of advantageous characteristics, such as miniaturization, massive parallelization, and high-density multifunctional integration, is now driving the exploration of a far broader application space as well. This chapter will begin with a brief overview of the advantageous characteristics of MEMS for medical applications within the context of *in vivo* therapeutics. The materials and fabrication processes that make up the MEMS "tool kit" will then be introduced, followed by a discussion of the recent progress in selected therapeutic application areas. This will then be followed by a discussion of applications enabled by the recently developed titanium micromachining techniques that provide a means for addressing many of the limitations of conventional silicon-based MEMS. The chapter will then conclude with a brief commentary on the future outlook for these and other applications of MEMS for *in vivo* therapeutics.

Biomedical Nanosensors
Edited by Joseph Irudayaraj

ISBN 978-981-4303-03-3 (Hardcover), 978-981-4303-04-0 (eBook)
www.panstanford.com

10.1 Introduction

The advent of manufacturing techniques that enable the fabrication of microscale devices and systems with highly integrated electrical and mechanical functionalities (among others) has provided a unique opportunity for the development of sophisticated miniaturized transducers and other microdevices that now find use in commercial products within a myriad of areas, including automotive, aerospace, telecommunications, and consumer electronics applications [1]. However, the broader potential of these so-called MEMS has yet to be fully realized, thus driving significant efforts focused on the further exploration and advancement of the limits of this technology. One area in particular where there is immense opportunity is in the application of MEMS to the life sciences, that is BioMEMS. This is clearly evidenced by the current commercial success and future growth potential of the BioMEMS market, which exceeded $0.5 billion in revenues in 2005 and is forecasted to reach $2.3 billion by 2012 [2].

As detailed in numerous BioMEMS reviews published over the past decade [3–11], areas of current impact and emerging potential within the biomedical realm include diagnostics, therapeutics, and/or surgical devices and systems. *In vitro* diagnostics and monitoring have seen the most development thus far, due in part to a reduced regulatory burden relative to *in vivo* applications, as well as lessened demands upon biocompatibility. However, significant opportunity also exists for the application of MEMS in medical therapeutics; hence the focus of this chapter on such applications. For example, therapeutic MEMS may provide a means for improving vaccine delivery, which has been identified as one of the grand challenges in global health [12]. They could also provide a means for facilitating the development of various enabling strategies for personalized medicine, such as devices for targeted drug delivery, constructs for tissue engineering, and instrumentation for drug discovery, all of which have been identified as critical components of the National Academy of Engineering's Grand Challenge for Engineering of Better Medicines [13].

In this chapter, the discussion will be focused on MEMS devices intended for use in therapeutic applications on, or within, the body. The intent is to provide a sampling of topics within this area rather than an exhaustive review. Devices and/or advances enabled by bottom-up or self-assembly fabrication techniques more commonly associated with the field of nanotechnology will not be discussed. However, it is important to acknowledge the often-enabling role MEMS serve at the interface between the nano- and macroscale realms [14, 15].

10.2 Advantageous Features of MEMS

The most advantageous features of MEMS, with regard to device performance and functionality, can be generally ascribed to three fundamental hallmarks: (1) miniaturization, (2) massive parallelization, and (3) high-density multifunctional integration. All are largely enabled by the fabrication processes used for MEMS manufacture, which will be further elaborated upon in Section 10.3. A generalized discussion of the opportunities afforded by these hallmarks, within the context of *in vivo* therapeutics, is presented herein, while specific examples associated with selected therapeutic MEMS device applications will be presented in Sections 10.4 and 10.5.

Miniaturization in MEMS is primarily enabled by reliance upon photolithographic patterning techniques, developed and refined by the integrated circuit (IC) industry, which provide an opportunity for the routine fabrication of devices with micro- to nanometer-scale features in a reproducible manner. This allows the exploitation of the favorable scaling of various phenomena, such as increased heat and mass transfer, higher surface-area-to-volume ratios, and greater electric field intensities and gradients. Potential benefits arising from such scaling include higher performance (e.g., faster operation, higher sensitivity, and greater selectivity), greater efficiency (e.g., decreased sample size and chemical usage and reduced voltage and power), and novel functionality (e.g., dielectrophoretic separation, mixing, or pumping based upon electric field gradients). Miniaturization also allows for a reduced device and structural feature size, which can minimize invasiveness (e.g., minimized pain and trauma) and provide the potential for interaction with the body at physiologically relevant spatial scales (e.g., modulation of cellular and molecular interactions).

Photolithography is also a key enabler of massive parallelization, which provides an opportunity for the simultaneous fabrication of large arrays of micromechanical structures with identical (or unique) features. This allows for increased throughput via parallelization of identical device operations (e.g., high flow rates at low pressures via multiplicity of flow channels) or the parallelization of distinct device operations (e.g., the simultaneous screening of unknown samples against large libraries of known recognition elements). Massive parallelization also provides an opportunity for enhanced reliability and safety through redundancy (e.g., redundant flow paths or recognition elements).

Multifunctionality arises from the versatility of many MEMS-processing techniques, which allow the realization of microscale device structures that are capable of interacting with their surroundings across multiple domains (e.g., electrical, mechanical, and fluidic). This enables the integration of previously discrete macroscale system elements (e.g., sensors, actuators, control electronics, and signal conditioning) within close proximity of one another at the microscale level. As such, an opportunity is presented for the fabrication of miniaturized systems that can modulate their activity in response to physiological stimuli (i.e., closed-loop control) and can do so with a superior temporal resolution relative to conventional macroscale systems. Moreover, close proximity enables the reduction of parasitic losses, thus increasing the signal quality and decreasing the power consumption.

An additional hallmark of MEMS, with respect to cost effectiveness, and thus market viability, is the batch fabrication paradigm borrowed from the IC industry. This allows the fabrication of large numbers of devices on a single wafer simultaneously, which provides the potential for scalability to low-cost/high-volume manufacturing. When coupled with the superior performance and/or unique functionalities provided by MEMS, this has allowed rapid penetration in many market segments and even commoditization for selected applications, such as accelerometers for airbag deployment and microphones for cellular phone handsets. However, in some therapeutic MEMS applications, the unit volume demand may not be adequate to fully leverage this opportunity. In such cases, market viability will, therefore, be largely determined by whether MEMS can help fulfill a critical unmet need or provide sufficient improvement in performance and/or functionality relative to prevailing technologies to justify any added cost.

10.3 The MEMS "Tool Kit": Common MEMS Materials and Fabrication Techniques

As stated earlier, many of the advantageous features afforded by MEMS are fundamentally enabled by the fabrication processes used for their manufacture. Many of these fabrication processes are taken directly from the IC industry while others have emerged in response to unique demands imposed by specific types of MEMS device applications. Similarly, while the MEMS process infrastructure is predominantly dedicated to silicon, owing to its origins in the IC industry, numerous alternative micromechanical materials systems have emerged to address the limitations of silicon. This diversity of processes and materials provides significant flexibility with regard to the types of devices and designs that may be realized relative to conventional IC fabrication. It also may provide a competitive advantage with regard to commercialization in many instances, due to an increased opportunity for intellectual property differentiation and generation of barriers to market entry based upon novel materials and processes and the unique devices and designs they enable. However, this diversity also hampers standardization, thus placing greater burden upon designers to understand the capabilities and limitations of a far broader set of processes and materials than is typically required for ICs. Moreover, the lack of standardization continues to constrain the growth of the MEMS market as a whole, since new MEMS device development efforts often require development and thorough optimization of new custom fabrication processes as well.

In this section, common MEMS materials and fabrication techniques are discussed within the context of their capabilities and limitations for structural design, device performance, and other important considerations. For the sake of brevity, only general descriptions are provided, and the reader is encouraged to consult any one of a number of excellent introductory texts for further details [16–18]. It is also important to note that there are many other techniques, not discussed here, which may be used for the fabrication of microscale medical devices. Examples include laser micromachining [19], microelectrodischarge machining [20], and ultraprecision machining [21]. Many of these techniques offer feature size resolutions approaching those of MEMS fabrication processes. Moreover, some also offer unique capabilities that cannot be matched by MEMS techniques (e.g., the ability to work with nonplanar substrates and easily create arbitrarily complex three-dimensional structural geometries). However, most offer only limited opportunity for multifunctional integration. Since this is one of the fundamental hallmarks of MEMS, such techniques are excluded from further discussion.

10.3.1 MEMS Fabrication Routes: Surface vs. Bulk Micromachining

Despite the broad diversity of MEMS fabrication processes that have been developed thus far, most can be conveniently classified into one of two distinct categories based upon whether the process is primarily additive or subtractive in nature. The additive approach, typically referred to as surface micromachining [22], initially proceeds in a manner similar to conventional IC-manufacturing approaches. Beginning with a planar substrate (i.e., a wafer), the device is fabricated by sequential steps consisting of the deposition and selective removal of various thin-film materials. Typical processes used in this approach include thin-film growth/deposition, photolithographic patterning, and thin-film etching, each of which is discussed in further detail in Sections 10.3.3–10.3.5. As shown in Fig. 10.1, however, the final stages of this approach depart from traditional IC-manufacturing schemes by creating suspended mechanical elements through the selective removal of the underlying sacrificial layers. Surface micromachining offers a number of advantages, such as multiple structural material levels, the potential for several electrically isolated conductor levels, and reduced constraint of cross-section geometry. Moreover, surface micromachining forms the basis of some of the only commercially available complete MEMS fabrication processes with standardized design rules developed to date, for example, the MEMSCAP Multi-User MEMS Processes (MUMPs®) [23] and the Sandia National Laboratories Sandia Ultra-planar, Multi-level MEMS Technology 5 (SUMMiT V™) [24]. Finally, there are numerous examples of successful surface micromachined MEMS devices in the marketplace today, including the Analog Devices ADXL 330 accelerometer, currently used in the controllers for the Nintendo® Wii™ video game console, and the Texas Instruments DLP® spatial light modulator, which is used widely in rear-projection televisions and digital projectors.

However, surface micromachining is not without its limitations. For example, it requires thorough optimization of the thin-film-deposition parameters to ensure uniformity, quality, and reproducibility of the properties of each film layer within the device, as well as the balancing of their intrinsic and thermal expansion mismatch-induced stresses to minimize undesirable device deformation. Moreover, the high costs, long process times, and/or capability limitations associated with most thin-film-deposition techniques often make it prohibitively expensive to deposit layers with thickness greater than a few micrometers. The total number of layers that can be deposited is also often limited to five structural layers or less by process and cost considerations. As a result, creating surface-micromachined devices with structural heights greater than 10–20 μm is typically not feasible.

One consequence of this limitation is that these types of devices are often highly compliant in the out-of-plane direction. When coupled with narrow vertical gap distances, also resulting from layer thickness limitations (i.e., sacrificial layer thickness), this can increase the potential for adhesion of suspended structures to the underlying substrate. Known as stiction, this adhesion can occur during device processing when the stiffness of the suspended member is insufficient to overcome the surface tension of the aqueous solutions typically used as release etchants and final rinses. It can also occur during service due to moisture condensation or electrostatic charge accumulation in the dielectric layers. Stiction is undesirable since it can constrain the motion of suspended structures, thus preventing proper device function. It can be largely mitigated by any number of techniques (e.g., use of CO_2 critical-point drying after device release and/or surface passivation via self-assembled monolayers prior to packaging), but each adds complexity and cost.

Limited structural heights also limit the capacitance in electrostatic transducers that rely on vertical device surfaces as opposing capacitor plates (e.g., the interdigitated electrode transducer pictured in Fig. 10.1), thus reducing sensitivity or actuation force. This can be addressed by increasing the number of capacitor plates in the device (e.g., increasing the number of stator and rotor tines in the device pictured in Fig. 10.1). However, doing so frequently results in an increased overall device size, a reduced device count per wafer, and a higher unit cost since structural, process, and performance constraints typically limit the spatial density of such structures (e.g., stator and rotor tine widths, lengths, and separation distances). Finally, in many cases, the vertical displacement range of suspended elements is also constrained by a structural height limitation and/or a narrow vertical gap distance, which limits the performance and functionality (e.g., the tilt angle in micromirror devices).

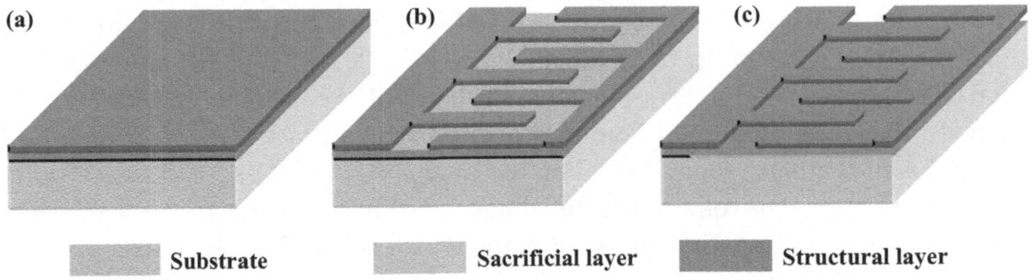

Figure 10.1 A schematic representation of an abbreviated fabrication process for a surface-micromachined electrostatic-interdigitated electrode transducer. (a) Sacrificial and structural layer thin-film depositions. (b) The selective removal of the structural layer via photolithographic patterning and anisotropic etching. (c) The selective removal of the sacrificial layer via isotropic etching to create suspended mechanical elements. The structural element on the left (i.e., stator) remains rigidly fixed to the substrate via the residual sacrificial layer, while the structure on the right (i.e., rotor) is mobile and connected to the substrate via mechanical elements not shown in schematic (e.g., flexural beam elements). Note: The vertical scale is exaggerated for clarity. Details regarding each of the aforementioned process techniques are provided in Sections 10.3.3–10.3.5. See also Color Insert.

As shown in Fig. 10.2, an alternate, subtractive approach for fabricating MEMS devices relies on the machining of the substrate itself; that is, devices are created by machining into the substrate rather than building up from its surface. This is achieved using many of the same techniques used for surface micromachining (e.g., thin-film growth/deposition, photolithography, and thin-film etching); however, it also requires the use of processes specifically developed for etching deeply into the substrate (e.g., anisotropic wet etching and deep reactive-ion etching [DRIE]), which will be discussed in Section 10.3.5. Referred to as bulk micromachining [25], this approach enables the fabrication of devices with a far greater structural height, thus providing the opportunity for increased transducer sensitivity, force, and vertical displacement range, as well as reduced susceptibility to stiction. Moreover, the common use of single-crystal-silicon (SCS) substrates for such devices also provides a structural material with high quality, reproducible properties, and minimal residual stress [26]. Bulk-micromachined pressure sensors were the first MEMS to reach the commercial market (in 1972), followed soon after by inkjet printheads (in 1977), each of which now represent a substantial portion of the overall MEMS market (18% and 40%, by revenue in 2004, respectively) [27].

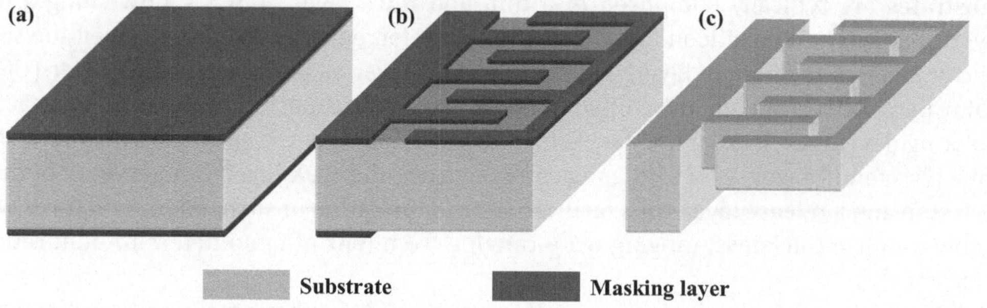

Figure 10.2 A schematic representation of an abbreviated fabrication process for a bulk-micromachined electrostatic-interdigitated electrode transducer. (a) Front- and backside masking layer thin-film depositions. (b) The selective removal of front and backside masking layers via photolithographic patterning and anisotropic etching. (c) Front- and backside anisotropic substrate etching, followed by masking layer removal by etching, to create suspended mechanical elements. The stator (left) remains rigidly fixed to the substrate, while the rotor (right) is mobile and connected to the substrate via mechanical elements not shown in the schematic (e.g., flexural beam elements). Details regarding each of the aforementioned process techniques are provided in Sections 10.3.3–10.3.5. See also Color Insert.

Despite its relative maturity, bulk micromachining suffers from a number of limitations. For example, structural material selection has been largely limited to conventional semiconductor substrates, such as SCS. As will be discussed shortly, although silicon is a highly desirable micromechanical material for many applications [26], its intrinsic brittleness may be a limiting factor for many medical applications. Cross-section geometry complexity in bulk-micromachined devices is also often limited to three-dimensional prismatic projections of the two-dimensional mask patterns, which can limit performance, functionality, and cost effectiveness in some applications. The concurrent development of wafer-bonding techniques, which will be discussed in greater detail in Section 10.3.6, has provided added capability in this regard. However, process and cost considerations typically limit the number of wafers that can be bonded, thus limiting the structural complexity that can be achieved.

Although the limitations of surface and bulk micromachining limit their individual application spaces, the opportunity for utilizing elements of both approaches in a single fabrication process allows the exploitation of the most advantageous features of each. However, process and cost considerations often constrain the extent to which this may be realized, thus providing impetus for continuing efforts to develop new MEMS fabrication techniques and process schemes that afford greater capability.

10.3.2 Typical Substrate Materials

Owing to their origins in IC manufacturing, most MEMS fabrication processes begin with planar SCS wafer substrates. These substrates are available in a wide variety of sizes, thicknesses, crystallographic orientations, and doping types and levels. More recently, silicon-on-insulator (SOI) wafers have also become widely available. These unique substrates are typically composed of a thin and thick layer of SCS sandwiching a thin layer of SiO_2. The thin silicon layer is commonly referred to as the device layer, the thick silicon layer as the handle layer, and the thin SiO_2 layer as the buried oxide layer (BOX). SOI wafers were originally developed to reduce parasitic capacitance in ICs by electrically isolating devices defined in the device layer from the bulk of the substrate. However, they have also found favor as MEMS substrates because the BOX layer can serve as both an etch stop and a release layer, thus facilitating the fabrication of suspended structures with highly uniform thickness, ranging from ranging from tens of nanometers to hundreds of micrometers.

SCS and SOI wafers offer numerous benefits as MEMS substrates, such as widespread availability and direct compatibility with IC fabrication infrastructure. Moreover, when used as a structural material in bulk-micromachined devices, SCS offers high strength and well-defined, reproducible mechanical properties [26]. However, its intrinsic brittleness, arising from low fracture toughness (K_{IC} = 0.5 MPa·m$^{1/2}$), can detrimentally affect mechanical reliability. This is a particularly key consideration with regard to safety of therapeutic MEMS, in which device structures are required to interact directly with the physiological environment (i.e., device structures are not protected by packaging). Silicon's predisposition for catastrophic fracture provides the potential for generation of atomically sharp fragments within the body and complete loss of device functionality upon overloading. Moreover, although SCS and SOI wafers are largely commoditized, their cost may still be prohibitive for many applications where disposability is required. This is particularly so for SOI wafers, which are considerably more expensive than SCS wafers due to the added complexity of their manufacture.

Glass substrates are also widely used in MEMS, particularly in microfluidic devices, for bioanalytical, biomedical, and pharmaceutical applications. This is due in large part to their optical transparency as well as the widespread acceptance and understanding of the behavior of silica-based surfaces when in contact with biological, biomedical, and pharmacological solutions. Glass substrates are available in many of the same-size formats as Si wafers, which provides compatibility with Si process equipment, assuming contamination control constraints are not a concern (e.g., sodium contamination restrictions for complementary metal-oxide-semiconductor [CMOS] process equipment). Similar wafer-size formats also facilitate heterogeneous integration with Si devices via wafer bonding, as will be discussed in Section 10.3.6. Glass wafers are also available in a wide variety of compositions that provide enhanced properties for specific applications. For example, fused silica wafers (i.e., high-purity amorphous SiO_2) offer extended transparency into the UV spectrum and lower autofluorescence for devices that rely on fluorescence-based interrogation methods while borosilicate glass wafers offer a thermal expansion coefficient matching with silicon to facilitate heterogeneous integration for microfluidics and packaging applications. However, similarly to silicon, glass substrates and devices made from them are limited by their brittleness and cost. Moreover, they are limited in the complexity of the device geometries that can be realized, due to reduced process capability relative to silicon. Emerging photostructurable glass-ceramics may provide new capability in this regard, however [28, 29].

Polymeric substrates have begun to see increasing use in MEMS applications, particularly for microfluidics. This is primarily due to their low cost and fabrication ease relative to silicon and glass, which enables the realization of inexpensive disposable devices. Disposability is desirable in many bioanalytical and biomedical applications since it reduces potential for cross contamination, clogging, and fouling. Elastomeric polymers, such as polydimethylsiloxane (PDMS), also offer capability for elastic strains of 500% or more. This provides a means for the realization of devices with extreme flexibility and better compliance matching to physiological tissues. It also provides opportunities for unique functionalities that would be difficult to achieve using more conventional micromechanical materials, such as pneumatically driven valving and pumping via highly deformable membranes. As will be discussed in Section 10.3.7.1, fabrication ease arises from the fact that most polymer MEMS can be fabricated via molding or casting rather than conventional micromachining approaches. Despite these advantages, polymers are often limited by their large thermal expansion coefficients and low moduli, strength, hardness, and melting/decomposition temperatures.

Metallic materials offer a wide variety of mechanical and physical properties that would be highly advantageous relative to conventional micromechanical materials for therapeutic MEMS applications. For example, their high fracture toughness relative to silicon and glass provides an opportunity for improved mechanical reliability and safety by ensuring graceful failure via plastic deformation rather than catastrophic fracture (i.e., metal devices can be designed to bend before they break). Moreover, their significantly higher moduli relative to polymers affords the capability for the design of implantable devices with sufficient stiffness in small cross-section areas to ensure desired performance and efficacy. However, despite these advantages, metallic materials have seen only limited use in therapeutic MEMS applications thus far. This is largely due to a lack of scalable and cost-effective fabrication processes that provide capability for creating complex multifunctional devices with physiologically relevant features sizes in biocompatible metals. However, as will be discussed in Section 10.5, emerging fabrication techniques that allow fabrication of titanium-based MEMS are beginning to address this limitation.

10.3.3 Thin-Film Deposition

Most MEMS fabrication processes typically begin with the deposition or growth of a thin-film material on a substrate. This deposited material may serve one or more of the following functions:

- Structural or sacrificial layer for surface-micromachined devices, for example, polysilicon and phosphosilicate glass, respectively
- Dielectric layer for electrical insulation, for example, SiO_2 or Si_3N_4
- Conductive layer for electrically active structures, for example, Al, Cu, Au, or doped polysilicon
- Masking layer for the protection of underlying materials during subsequent etching steps, for example, Si, SiO_2, Si_3N_4, Ni, or Cr Or
- Active layer for mechanical transducers, for example, shape memory alloys (such as NiTi) or piezoelectrics (such as lead zirconate titanate, or PZT) and aluminum nitride, or AlN)

The selection of materials best suited for the intended application is driven by a number of factors, such as device performance (e.g., stiffness, electrical conductivity/

resistivity, or piezoelectric coefficient) and fabrication process requirements (e.g., thermal, chemical, and contamination constraints). Materials selection is also often driven by the type of deposition techniques that may be used, since many important film characteristics are dictated by the capabilities and limitations of these techniques. Common thin-film-deposition techniques used in MEMS fabrication fall into three general classes: oxidation, chemical vapor deposition, and physical vapor deposition.

Oxidation processes are most often used to grow thin layers of SiO_2 on SCS substrates using tube furnaces that heat the substrates to temperatures ranging from 900°C to 1,200°C in an oxidizing ambient (i.e., O_2 or H_2O vapor). Oxidation yields high film quality (e.g., uniformity, stoichiometry, conformality, and adhesion), excellent control, and reproducibility of film thickness and allows batch processing of large number of wafers, which reduces the process cost per wafer. However, the relatively slow growth kinetics of SiO_2, coupled with costs associated with extended high-temperature processes, often makes it prohibitively expensive to grow films with thickness greater than a few micrometers. Moreover, the high process temperatures required for oxidation exceed the thermal budget of most other materials used in typical MEMS processes, thus limiting the compatibility to wafers that have seen limited prior processing.

Chemical vapor deposition (CVD) processes are most commonly used in MEMS fabrication for the deposition of silicon (amorphous or polycrystalline), SiO_2, or Si_3N_4 films via the chemical reaction of gaseous reactants. The two most common CVD processes are low-pressure CVD (LPCVD) and plasma-enhanced CVD (PECVD). LPCVD is most commonly used for the deposition of structural and sacrificial layers in surface-micromachining processes. It is performed using tube furnaces that heat the wafers to 500°C to 900°C under flowing source gas environments (e.g., pure or N_2-diluted SiH_4 for Si, SiH_4 and O_2 for SiO_2, or SiH_4 and NH_3 for Si_3N_4). The use of reduced pressures (0.1–1 Torr) minimizes undesirable gas-phase reactions, thus ensuring good film uniformity. Like oxidation, LPCVD yields high film quality, allows batch processing, and is limited to thin films, although there is less restriction due to lower process temperatures. Lower temperatures also expand the process compatibility relative to oxidation but still exceed the thermal budget of some important MEMS materials (e.g., many metals used for electrical structures).

PECVD provides a means for addressing these limitations through the use of radio frequency (RF)-generated plasma to reduce the thermal energy required for film formation. Process temperatures as low as room temperature are possible, which provides compatibility with most conventional MEMS materials as well as alternative materials, such as compound semiconductors, metals, and polymers. While PECVD film quality is often inferior to LPCVD, it is generally sufficient for many of the thin-film applications described earlier. PECVD also provides a means for more cost-effective deposition of thicker films than LPCVD, which has benefited applications such as integrated optical waveguides.

Physical vapor deposition (PVD) is commonly used in MEMS fabrication for the deposition of metal films for electrical structures, although many dielectrics and other materials are also possible, e.g., AlN- or PZT-active layers for transducers. The two most widely used PVD processes are evaporation and sputtering. In evaporation, the source material to be deposited is heated (typically by electron beam) under ultrahigh vacuum, which results in the formation of vapor that is then deposited on the substrate of interest.

In sputtering, the source material is bombarded by plasma-generated high-energy inert ions (typically Ar) under ultrahigh vacuum, which results in the removal of neutral atoms and clusters that then deposit on the substrate of interest. Evaporation typically yields higher-purity films, but sputtering provides the capability for deposition of a wider variety of materials with better conformality, uniformity, adhesion, and control of film properties.

10.3.4 Photolithography

Once a thin film has been deposited, portions are typically removed to begin defining the desired device structures (e.g., interdigitated electrode structures shown in Fig. 10.1b and 10.2b). This is commonly achieved using photolithographic patterning to protected selected areas from subsequent etch processes (which will be discussed in Section 10.3.5). The photolithographic process begins with the deposition of a thin layer of photosensitive polymer (i.e., photoresist) on the substrate, usually via spin-coating. The photoresist layer is then selectively exposed using UV light passed through a photomask that contains the desired exposure pattern. This induces a chemical reaction in the exposed areas of the photoresist, thus modifying their solubility during subsequent development. So-called positive photoresists become more soluble upon UV exposure, resulting in the direct reproduction of the original photomask pattern. Conversely, "negative" photoresists become less soluble, producing inverse images of the photomask pattern. Positive photoresists are more widely used since they often provide higher fidelity. However, negative photoresists often find use in so-called lift-off processes that allow the selective deposition of thin films, eliminating the need for subsequent pattern definition by etching.

10.3.5 Etching

Once the deposited thin films have been photolithographically patterned, the pattern is transferred into the underlying film using etching (e.g., Fig. 10.1b and 10.2b). These etched thin films can then serve as a structural (e.g., Fig. 10.1b) or sacrificial layer in surface-micromachining processes, or as a structural or masking (e.g., Fig. 10.2b) layer for substrate etching in bulk-micromachining processes. Etching is also used to release structural members in surface-micromachined devices by the selective removal of the underlying sacrificial layer (e.g., Fig. 10.1c). This diversity of applications has motivated the development of numerous etching processes, which can be broadly classified into two categories, wet etching and dry etching, which will be discussed shortly. The choice of whether to use wet or dry etching and which specific process to use within each of these categories, is typically determined based on requirements such as etch rate, selectivity, anisotropy, process compatibility, and cost. Selectivity describes the ratio of preferential etching of the desired material relative to the masking material. High selectivity is typically desired in most applications since it reduces the thickness of the masking material needed, which reduces the process time and cost associated with mask material deposition and etching. Anisotropy describes the profile of the etch front within the material being etched. As illustrated in Fig. 10.3, profiles may be isotropic or anisotropic, depending on the etching mechanism. Anisotropic vertical etches (e.g., Fig. 10.3c) are generally preferred for applications in which the faithful reproduction of mask features is paramount, while isotropic etches provide the capability for releasing structural members by lateral undercutting (e.g., Fig. 10.1c).

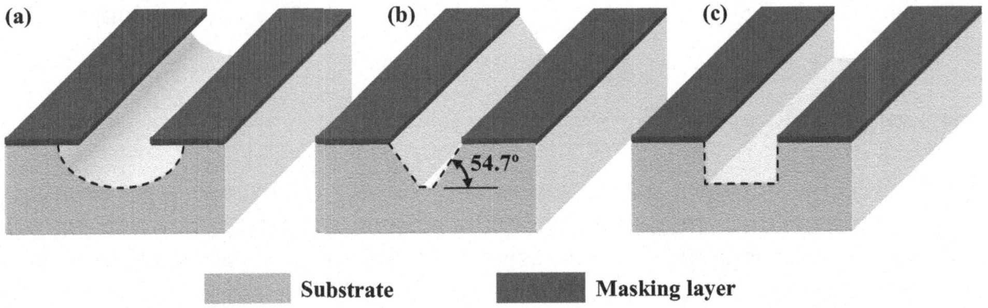

Figure 10.3 Schematic representations of common etch profiles. Note: Dotted lines in each schematic demarcate the etch boundary in the substrate cross-section plane. (a) Isotropic etching produced by wet etchants or dry etching with strong chemical component. Note the significant degree of masking material undercutting. (b) Anisotropic etching produced by orientation-dependent silicon wet etchants on (100) wafers. Note that the angular sidewall facets correspond to slow etching (111) crystalline faces. (c) Anisotropic etching produced by dry etching with strong physical component. See also Color Insert.

Wet etches rely on a chemical attack to facilitate material removal using acidic or basic solutions. Most wet etchants are isotropic (i.e., etch rates are similar in all directions), yielding the mask undercutting and rounded etch profile illustrated in Fig. 10.3a. Typical wet isotropic etchants for silicon include combinations of hydrofluoric acid (HF) and ammonium fluoride (NH_4F). There are also notable examples of anisotropic wet etchants that find wide use in MEMS fabrication, including potassium hydroxide (KOH), tetramethylammonium hydroxide (TMAOH), and ethylenediaminepyrocatechol (EDP) for orientation-dependent etching of SCS. Through strongly preferential etching of the (100) and (110) planes of Si, these etchants yield the opportunity for significant etch anisotropy in SCS wafers, as illustrated in the angularly facetted etch profile for (100) Si wafers shown in Fig. 10.3b. They also provide excellent selectivity to heavily doped Si (p++ B-doping), which can be spatially localized on or within the vicinity of the SCS substrate surface via ion implantation or thermal diffusion. The coupling of orientation- and doping-dependent wet etching was, in fact, a key enabler of the first MEMS devices because it allowed the definition of thin Si membranes for pressure sensors with superior control of thickness and uniformity. Wet etchants typically offer high selectivity due to their chemically based etching mechanism and can be implemented with low capital cost (i.e., minimal hardware requirements). However, undercutting in isotropic etching can degrade pattern transfer fidelity, and geometric constraints associated with the tapering sidewall profiles in anisotropic Si wet-etching limit the device count per wafer, which can result in higher unit cost. Williams *et al.* have tabulated an extensive list of wet etchants and their relevant performance metrics (e.g., etch rate and selectivity) for many materials used in MEMS fabrication [30].

Dry etches rely on either chemical and/or physical material removal mechanisms using vapor phase or plasma etchants. Typical vapor phase etchants used in MEMS fabrication include xenon difluoride (XeF_2) and hydrofluoric acid (HF) for etching of SCS and SiO_2, respectively. Both are chemically based etches and are, therefore, highly selective and isotropic in nature (Fig. 10.3a). They are commonly used for the release of structural members and can be implemented with minimal hardware and low capital cost. However, dedicated vapor phase etching systems that more precisely regulate gas flow and distribution typically afford superior process control, uniformity, and/or safety.

Plasma-based dry etching can be chemical or physical in nature and is often composed of varying degrees of both material removal mechanisms. In general, plasma etching relies on RF excitation-induced ionization of process gases to generate highly reactive species (typically neutral radicals) that react with the material to be etched (i.e., substrate or thin film), thus forming volatile by-products that can be pumped away. Typical process gases used for etching of MEMS materials include those based on fluorinated (e.g., CF_4 and/or CHF_3 for Si, SiO_2, and Si_3N_4), chlorinated (e.g., Cl_2 for Si), or brominated (e.g., HBr for Si) chemistries. While numerous plasma-etching techniques have been developed, those most commonly used in MEMS fabrication for etching of thin films and bulk substrates are based on reactive-ion etching (RIE). In RIE, ionic species within the plasma are directed toward the substrate in a direction normal to the substrate plane, typically by self-bias induced by the plasma sheath (in capacitively coupled systems) or by externally applied bias (in inductively coupled systems). The physical bombardment of ions promotes or enhances the chemical reaction between the reactive species and the substrate material, enhancing the etch rate. This enhancement provides an opportunity for preferential etching on surfaces normal to the ion flux (i.e., surfaces parallel to the substrate plane), thus yielding highly anisotropic etch profiles with vertical sidewalls (Fig. 10.3c). RIE-based plasma etching is used extensively in MEMS fabrication because it provides an opportunity for highly anisotropic etching of a wide variety of materials with minimal dependence on crystalline orientation (unlike anisotropic wet etching). However, capital costs are significantly higher relative to wet and vapor etching due to the sophisticated hardware requirements for plasma process systems. Moreover, limited selectivity (typically $<5:1$ for most common MEMS etch processes) largely precludes anisotropic plasma-based dry etching to depths exceeding ~ 10 μm.

Etch depths beyond the limitations of typical RIE systems are desired in many MEMS applications. For example, as mentioned earlier, the capacitance (and, thus, the performance) of electrostatic transducers that rely on vertical surfaces as capacitor plates (e.g., Fig. 10.1 and 10.2) is directly proportional to the surface area of the plates. Furthermore, capacitance is also inversely proportional to the gap distance between the plates. As such, the need for highly anisotropic etching of deep and narrow trenches with vertical sidewalls is immediately apparent. DRIE techniques provide a means for the realization of such high-aspect-ratio (HAR) structures (i.e., structures with vertical dimensions that far exceed their lateral dimensions). The most widely used DRIE technique is the so-called the Bosch process for the deep etching of silicon, which was developed in the early 1990s by Robert Bosch GmbH [31]. It relies on the cyclic repetition of a short sulfur hexafluoride (SF_6) isotropic-etching step followed by a short fluorocarbon deposition step based on octofluorocyclobutane (C_4F_8) polymerization to provide HAR-etching capability with excellent selectivity (i.e., typically >75:1 for photoresist or SiO_2 masks). Highly directional bombardment by SF_x ions during the etch step clears the fluoropolymer passivation on the trench floors while leaving the sidewall passivation intact, thus ensuring that only the trench floor is etched by the fluorine radicals during each successive etch/deposition cycle. Si-DRIE is now used extensively in MEMS fabrication for device structures requiring significant etch depths with vertical sidewalls despite a high capital cost relative to wet etching and vapor-phase etching. Moreover, unlike anisotropic wet etching, the capability for deep etching with vertical sidewalls minimizes constraint upon the areal density of device structures. Finally, DRIE can also be used for the singulation of MEMS devices either during or after the completion of the fabrication process, thus reducing the potential for damage of delicate released device structures.

10.3.6 Wafer Bonding

The stacking and bonding of wafer-based substrates atop one another, typically referred to as wafer bonding, enables the fabrication of complex micromechanical device structures that would be difficult or impossible to realize with surface- or bulk-micromachining techniques alone. Wafer bonding first found use in MEMS as a means for creating hermetically sealed cavities with controlled environments that served as reference pressure chambers for pressure sensors. It has since evolved into a fundamental enabler for a broad array of MEMS applications, including SOI wafer manufacturing, wafer-level packaging, and the heterogeneous integration of silicon with other substrates (e.g., glass, compound semiconductors, polymers, or metals). Extensive reviews of the fundamentals and applications of wafer bonding to MEMS (and many other fields) can be found in the texts by Gösele and colleagues [32, 33].

A wide variety of wafer-bonding techniques have been developed for use in MEMS to meet the often-unique requirements and restrictions of each device application and fabrication process. These techniques can be broadly categorized on the basis of whether an intermediate layer is required for bonding. Direct-bonding techniques do not require an intermediate layer and are often preferred in applications where the hermeticity of the bond interface is paramount. This is achieved through the initiation of covalent bonding across the mating interface. Examples of direct bonding of most relevance to MEMS fabrication include fusion and anodic bonding.

In fusion bonding, substrates to be bonded are brought into intimate contact and then heated to complete the bonding process. The bonding of SCS wafers to one another is often performed in this manner, as is the manufacture of certain types of SOI wafers. However, the high temperatures required to ensure the completion of the covalent-bonding process (typically >800°C for Si-to-Si bonding) exceed the thermal budget of many MEMS materials, thus limiting the types of processing that can be performed on the wafers prior to bonding. Moreover, high temperatures also largely preclude fusion bonding of substrates with dissimilar coefficients of thermal expansion (CTEs) since this produces large residual stresses that often result in device or substrate degradation, delamination, or cracking. Finally, owing to the high moduli of semiconductor MEMS materials and their limited plasticity, minimization of surface roughness and maximization of surface planarity is required to ensure bonding uniformity, strength, and yield. Consequently, a significant demand is placed upon the quality of the wafers and their mating surfaces (e.g., wafer bow <60 μm and surface roughness <1 nm RMS) [33]. A stringent demand is also placed upon the cleanliness of the mating surfaces since surface particulates may prevent bonding over areas one to two orders of magnitude larger than the particle diameter itself.

A number of techniques for reducing the temperatures required for fusion bonding have been developed, one of the most promising of which is plasma activation. This technique relies on the use of low-energy plasmas (often based on O_2, N_2, or NH_3) to remove surface contaminants. More importantly, however, plasma exposure increases the kinetics of the interfacial reactions underlying the covalent bond formation between the mating substrates although the exact mechanism by which this occurs has yet to be definitively elucidated. A wide variety of plasma-activation techniques have been reported for use with an even wider range of materials and applications, with many demonstrating a capability for producing full-strength bonding using annealing temperatures at or below 200°C. Using a proprietary variation on this theme, Ziptronix has commercialized wafer-bonding processes that are reportedly capable of yielding covalent bonding of MEMS substrates (and others) at room temperatures [34].

Anodic bonding provides another means of reducing the temperatures required for direct bonding, although it is most commonly limited to bonding of sodium-rich glass substrates to silicon substrates. Under typical process conditions, the wafer stack is brought into contact and heated to 300–400°C and a large DC voltage (~1 kV) is applied across the stack to promote bonding. While the exact nature of the bonding mechanism is not yet completely understood, it is generally presumed that the applied field drives sodium ions away from the bonding interface. It is suggested that this then creates a depletion layer, which drives strong electrostatic attraction between the glass and silicon wafers and facilitates the formation of covalent bonding across the interface. In addition to reduced temperature, anodic bonding also offers relaxed wafer/surface quality requirements relative to fusion bonding due to the strong electrostatic forces and softening of the glass at bonding temperatures. Anodic bonding was among the first wafer-bonding techniques used for MEMS, serving as a fundamental enabler for the fabrication of pressure sensors. It is now also used widely for device-packaging applications since it offers compatibility with substrates containing released structures and/or metallizations.

The relatively high temperatures required for most types of direct bonding, coupled with their often-stringent wafer/surface quality requirements and/or constrained material selection capability, have driven the development of numerous alternative bonding techniques that rely on the use of intermediate layers to circumvent these limitations. Common examples used in MEMS include techniques based on low-melting point glass frits, thermocompression bonding of gold-coated substrates, eutectic bonding of Au-coated and Si surfaces, and adhesive bonding using various polymers (e.g., benzocyclobutene, SU-8, or polyimide). While these techniques typically offer reduced process temperatures and an increased process latitude relative to direct bonding, their implementation can be constrained by challenges associated with compatibility and integration of the intermediate layer materials. Moreover, many of these techniques are incapable of hermetic sealing, require the application of large forces for bonding, and/or produce bond strengths well below those of direct bonding, further limiting utility for certain applications.

10.3.7 Alternative Microfabrication Techniques

As demonstrated in the previous sections, significant process capability now exists for the fabrication of micromechanical structures with complex designs and geometries. This has provided an opportunity for the realization of sophisticated MEMS devices used in a wide variety of applications. However, the majority of this capability is currently directed toward the fabrication of silicon-based devices. As discussed earlier, silicon's intrinsic brittleness may constrain utility for many therapeutic MEMS applications due to concerns regarding device reliability and safety. Moreover, the cost of silicon-based devices may exceed the threshold required for disposability in many applications. Consequently, these issues, and others, have motivated the exploration and development of alternative microfabrication techniques that enable the use of a broader set of materials for microdevice applications. The most common of these include micromolding- or casting-based processes for polymers and electroplating-based processes for metals.

10.3.7.1 Polymeric micromolding/casting

Unlike the inorganic materials commonly used in microfabrication, polymers are composed of large numbers of molecular chains with strong atomic bonding along the chain backbones but often weak secondary and/or sparse strong bonding between the individual chains. This unique microstructural basis affords the opportunity for the formation of polymeric micromechanical structures through low-cost molding and casting techniques at temperatures well within the thermal budgets of most MEMS materials. For example, thermoplastic polymers, such as polymethylmethacrylate (PMMA), are able to flow plastically at low applied stress above their glass transition temperatures (T_g), and then harden in their newly formed shape once cooled below their T_g. This behavior arises from the reversible "melting" of the weak interchain bonds, which enables rapid and low-cost forming via techniques such as microinjection molding and hot embossing. Thermoplastic polymers find increasing use in microfluidic devices for DNA separation and detection, lab-on-a-chip, and biomolecular assays and are now beginning to also find favor as low-cost alternatives to conventional metals and ceramics for MEMS device packaging [35].

Thermoset polymers, in contrast, are not amenable to such techniques since they possess varying degrees of strong interchain bonding that is retained irrespective of temperature, thus preventing significant plastic deformation. These materials are most often formed via casting and curing of precursors. One of the most common thermosets used in MEMS is PDMS. It is typically formed by the mixing of liquid precursors (i.e., a linear polymer resin and a crosslinking agent), casting onto (or into) a micromold, curing via heating (or UV exposure in photo-definable PDMS), and removal via peeling from the mold. Such processes can be repeated many times using the same mold, increasing throughput and lowering the cost. Moreover, complex multilayer devices are easily realized through the bonding of multiple PDMS layers, which can be either reversible or irreversible in nature. PDMS-based devices find wide use in MEMS, particularly in microfluidics for chemical, biological, biomedical, and pharmaceutical applications due to their low cost, fabrication ease, and/or the potential for simplified valving and pumping. A particularly compelling example in this regard is the ongoing work of Quake *et al.*, who have proposed the opportunity for the realization of large-scale integration in microfluidic devices reminiscent of early ICs, wherein many common fluidic processes (e.g., separations, mixing, reactions, and assays) may be miniaturized, parallelized, multiplexed, and/or automated at an unprecedented scale [36].

10.3.7.2 Electroplating-based approaches

One of the most commercially successful metallic MEMS devices developed to date, the Texas Instruments DLP® device, is fabricated using a surface-micromachining approach based on PVD-based deposition of aluminum and aluminum/titanium alloy [37]. However, fabrication techniques based on electroplating are also used widely for metallic MEMS since they offer opportunity for far greater layer thicknesses. These techniques rely on the electrochemical reduction of metallic cations from a metal salt solution to form metals on conductive substrate surfaces immersed in the solution. A number of electroplating-based microfabrication techniques have been developed thus far, with the distinction between them arising primarily in the nature of the molding in which the plated metal is deposited.

The oldest of these techniques, developed in Germany in the early 1980s, is the LIGA process, which is an acronym for lithographie (lithography), galvanoformung (electroforming), and abformtechnik (molding). In this process, high-aspect-ratio molds are created via synchrotron radiation-based lithographic patterning of thick PMMA resists. Metals plated into these molds may remain attached to the substrate after the removal of the mold material, or they may be removed from the substrate to form free-standing microparts. The molded parts may serve as the basis of a device or as a mold or mold master for the subsequent micromolding of other materials, such as thermoplastic polymers. Although many application areas have been explored (e.g., accelerometers, microfluidics, and microoptical components), the high cost and limited availability of synchrotron exposure sources have remained significant barriers to widespread adoption. However, so-called pseudo-LIGA processes based on emerging thick photoresists (e.g., SU-8) may soon address this limitation. These unique photoresists, which can be patterned using conventional photolithographic equipment, are beginning to provide capability approaching that of LIGA at a fraction of the cost, thus suggesting potential for greater uptake in the future.

An additional limitation of many LIGA-based processes, with regard to device design freedom, is their constraint to fabrication of prismatic structures that are often simple three-dimensional projections of the two-dimensional mask patterns — that is, the device cross-section geometry is constant in the thickness direction. However, this limitation has been addressed by the EFAB® technology, which was first developed at the University of Southern California in the mid 1990s and has since been commercialized by Microfabrica® Inc [38]. The EFAB® technology is essentially a surface-micromachining process in which electroplated metals are used as both structural and sacrificial layers. Briefly, a sacrificial metal (e.g., Cu) is selectively deposited into areas defined by a photomask, followed by the blanket deposition of a structural material layer (e.g., Ni), and planarization of the resulting stack. The process is then repeated to build the device layer by layer in a manner reminiscent of many rapid prototyping schemes. Once the device structure is completed, a selective etch process is performed to remove the sacrificial metal, thus yielding the final device structure. The retention of the sacrificial metal throughout the build process provides mechanical support for the subsequently deposited layers, thus allowing largely arbitrary variation of cross-section geometry through the thickness. It also allows net shape fabrication of intricate interconnected assemblies, thus minimizing the need for often-laborious and costly assembly operations. The opportunity for the realization of complex metallic miniaturized medical devices using this technology is driving the company's current efforts in applications such as interventional cardiology, minimally invasive surgery, and drug delivery (among others).

Although electroplating-based fabrication techniques begin to provide greater material selection capability in medical MEMS applications, their ultimate utility may be nonetheless constrained by limitations associated with the biocompatibility and/or cost of metals that are amenable to plating. For example, many widely used metals are cytotoxic (e.g., Ni and Cu) while many others are precious metals (e.g., Au, Ag, Pd, Pt, and Rh), which reduces cost effectiveness. These issues may be addressed through the use of inexpensive base metals for the bulk of the device structure (e.g., Ni) in conjunction with thin precious-metal coatings (e.g., Au). However, utility may be limited nonetheless, particularly in chronically implanted devices, by the adverse consequences associated with coating failure. Such failures would expose the underlying base metal to the physiological environment, which could result in cytotoxic ion release and/or corrosion-induced mechanical degradation, thus compromising device safety and/or reliability. Alternative materials, such as the proprietary Ni-Co alloy developed by Microfabrica® (Valloy™ 120), show greater promise in this regard, as evidenced by their favorable performance in short-term toxicity and irritation studies [38]. However, these materials have yet to be tested for safety and efficacy in chronic implantation, and potential may exist for the release of cytotoxic Ni ions with prolonged exposure to physiological conditions. Consequently, the need still exists for the development of microfabrication techniques that provide access to well-proven biocompatible metals, allow the fabrication of high-aspect-ratio structures with micrometer- to nanometer-scale resolution in such metals, and are scalable to low-cost/high-volume manufacturing. As will be discussed in Section 10.5, recently developed techniques for the deep etching of titanium may provide a means for addressing this need.

10.3.8 Packaging and Biocompatibility Considerations

In many MEMS applications, raw devices must be protected from the surrounding environment to prevent contamination and damage and minimize the affect of external conditions on device functionality and performance. This is typically achieved through the use of a package, which is a ceramic, metallic, or polymeric enclosure that surrounds and isolates the device from the external environment. However, access to the outside world over multiple interaction domains (e.g., electrical, mechanical, and fluidic) is often required to provide connectivity to off-chip components (e.g., for power, control, data, and fluidic transport). Access to on-chip transduction elements is also often required (e.g., sensors or actuators). As such, when coupled with the diversity of MEMS materials, devices, and functionalities, these often-conflicting demands significantly increase the complexity and cost of MEMS packages relative to conventional packages for ICs. These same drivers also frequently necessitate the development of unique packaging solutions for each new MEMS device and/or application, thus adding further challenge. Elaboration of these issues, and many others relevant to MEMS packaging, is provided in the texts by Tummala [39], Gilleo [40], and Hsu [41].

Packaging of MEMS for *in vivo* therapeutic applications creates even greater challenge. In addition to the needs discussed above, packages for use on or within the body must also meet stringent requirements that are frequently well beyond the scope and/or magnitude of other types of MEMS devices. For example, *in vivo* packaging must ensure safety and biocompatibility by isolating device materials that are toxic or produce otherwise adverse tissue responses. Moreover, the package material and design must not induce a physiological response that adversely affects the patient or the device performance. Packages must also ensure reliability, stability, and longevity of device function, as well as hermiticity in many cases. This can be especially challenging for chronic implantation in dynamic and often-aggressive *in vivo* environments, since the consequences associated with device failure may be severe and device retrieval or replacement is often infeasible. Implantable device applications also frequently require the minimization of package size to facilitate placement near physiological tissues/organs of interest and minimize trauma and tissue/organ displacement. Finally, packages, and the devices contained therein, must also be capable of withstanding the harsh conditions associated with conventional sterilization techniques (e.g., autoclave, ethylene oxide, or gamma irradiation). The review by Velten *et al.* provides further discussion of the issues involved in packaging of MEMS for biological and biomedical applications, as well as examples of relevant technologies and solutions developed thus far [42].

While the preceding discussion is relevant to many therapeutic MEMS applications, in many others, enclosure-based packaging routes are not possible. For example, as will be discussed in Section 10.5.3, chronically implanted neuroprostheses require direct interaction with neurological tissues as well as minimized size to reduce the potential for functional degradation via reactive tissue response. These requirements preclude the use of enclosure-based packaging, thus placing significant restriction on the materials that may be utilized. Materials utilized in such environments must maintain biocompatibility over the lifetime of the device and/or patient, i.e., they must not adversely affect, or be affected by, the physiological environment within which they operate.

In the United States (and many other countries), the biocompatibility of medical devices is evaluated using the ISO-10993 family of standards, which seek to ensure that neither device materials nor any products that might be released from them (e.g., extractables or degradation products) produce adverse local or systemic effects, carcinogenic effects, or adverse reproductive or developmental effects [43]. The ISO-10993 family is composed of 20 distinct standards, each with a focus on a specific subset of requirements for establishing biocompatibility. Per Food and Drug Administration (FDA) regulation [43], all medical devices must undergo testing for cytotoxicity, irritation, and sensitization, regardless of materials or intended application (i.e., ISO 10993-5 [Tests for Cytotoxicity] and ISO 10993-10 [Tests for Irritation and Sensitization]). Other testing requirements are dictated by the device type, with test selection guidance elaborated in ref. [43].

Using selected elements of ISO-10993 (i.e., 10993-5 and 10993-6 [Tests for Local Effects after Implantation]), Kotzar *et al.* have demonstrated minimal cytotoxicity and irritancy for a number of common MEMS materials, including SCS, LPCVD-based P-doped polysilicon, thermally grown SiO_2, LPCVD-based Si_3N_4, CVD-based 3C-SiC, PVD-based Ti, and spin-coated SU-8 [44]. However, while these results serve as a valuable first step for screening purposes, it is important to recognize that these results are insufficient in themselves for the full validation of biocompatibility, at least from a regulatory perspective. For example, testing did not include ISO 10993-10, which is a requisite for all device types. Testing was also largely performed on as-deposited materials whereas most MEMS devices will be subjected to subsequent fabrication processes (e.g., photolithographic patterning and etching), which may alter testing outcomes, as might changes in the deposition conditions themselves (e.g., different deposition rates, temperatures, and source gases). Moreover, results from testing of constituent materials independently may not be representative of outcomes in MEMS devices that will be composed of various combinations of these materials (and possibly others). Finally, the intended device use will dictate which specific types of additional testing will be required. For example, devices that will be subject to contact with circulating blood, such as the endovascular pressure sensors discussed in Section 10.4.3, will require additional assessment of hemocompatibility (i.e., ISO 10993-4 [Selection of Tests for Interactions with Blood]) whereas other devices may not. As such, the need for evaluation of devices in their final form is clearly apparent. ISO 10993-12 (Sample Preparation and Reference Materials) codifies this need by recommending testing of devices in their final product form and condition (e.g., same device design, functionality, materials, and processing) whenever possible. Further discussion of biocompatibility and reliability issues associated with implantable microsystems can be found in the review by Pan [45]. Overviews of the ISO 10993 standards can be found in the informative series of articles authored by Wallin *et al.* [46] (the first article in the series; links to others are provided therein).

10.4 Selected Examples of Application Areas within *in vivo* Therapeutics

In this section, selected examples of MEMS application areas within *in vivo* therapeutics are presented. These areas include drug delivery, microsurgery, and endovascular intervention. It should be noted that the intent in each is to provide a sampling of the breadth of application areas in which the use of MEMS is reported. As such, synopses of recent review papers covering these topical areas are provided and the reader is encouraged to refer to these sources for further details and original citations. Moreover, emphasis is placed on the discussion of devices that are nearing commercialization, to demonstrate the burgeoning maturity of these application areas.

Examples of other areas of significance that are not discussed include cardiac rhythm management, neuroprostheses, particle engineering, and tissue engineering. Cardiac rhythm management (e.g., chronically implanted pacemakers and defibrillators) is not discussed due to its long-standing commercial maturity. However, devices used in such applications now routinely benefit from the integration of accelerometers for the rate-adaptive regulation of electrical stimulus based on physiologic activity level, as well as wireless telemetry for remote patient monitoring [47]. Neuroprostheses, which provide means for direct interfacing with neural tissues, are not discussed since they are the subject of a separate chapter. Similarly, methods for engineering of particles for drug delivery are also not discussed since they too are detailed elsewhere in this book. Tissue-engineering applications are not discussed due to space constraint. However, recent reviews by Park *et al.* [48], Puleo *et al.* [49], and Ainslie and Desai [50] highlight the broad potential for MEMS and microfabrication in the treatment and/or replacement of damaged or diseased tissues. Finally, the reader is encouraged to refer to the text by Desai and Bhatia [51], which contains broad treatment of topics of relevance to MEMS therapeutics, including cell-based therapeutics, tissue engineering, MEMS for neurosurgery, and drug delivery.

10.4.1 Drug Delivery

Numerous routes exist for the introduction of drugs into the body (e.g., gastrointestinal, parenteral, and dermal), and an even greater number of delivery methods are currently used to exploit these routes (e.g., orally administered tablets, intramuscular injections, and transdermal patches, respectively). However, considerable need still exists in many areas for the improvement of efficacy, safety, and patient compliance (among others). MEMS provide significant promise in this regard due to the opportunity for miniaturization, massive parallelization, and high-density multifunctional integration. One of the most compelling advantages within this context is the potential for considerable improvement in the spatiotemporal control of delivery. Improved spatial control enables precise targeting of therapies to diseased tissues, thus providing the opportunity for (a) the minimization of adverse side effects associated with systemic exposure, (b) the enhancement of efficacy through the utilization of higher local dosages than can be safely tolerated using conventional systemic delivery, and/or (c) the delivery of therapies that are not amenable to conventional methods due to limitations associated with systemic toxicity, degradation, or insufficient efficacy. Improved temporal control, particularly when coupled with closed-loop control afforded by the tight integration of sensors and ICs, provides an opportunity for the improved modulation of delivery kinetics, which may allow (a) timing of delivery in response to physiological stimuli to maximize therapeutic efficacy, minimize side effects, and increase drug utilization efficiency and/or (b) administration of complex delivery profiles consisting of multiple bolus and/or continuous infusion segments to closely match physiological needs while minimizing the potential for excess or insufficient dosage. Finally, other potential advantages provided by MEMS include reduced invasiveness arising from miniaturization and enhanced reliability and safety resulting from redundancy provided by massive parallelization.

The sheer diversity of needs within drug delivery, coupled with the advantageous features afforded by MEMS, has motivated intensive efforts in both academia and industry focused on exploring and exploiting the potential embodied in this technology. Microneedle-based devices, which will be discussed shortly, represent the most extensively developed application area to date. Other compelling areas that have attracted interest, but will not be discussed due to space constraints, include (a) implantable microreservoirs for the programmable and sustained delivery of therapeutic polypeptides, (b) ingestible or injectable bioadhesive microdevices for the targeted delivery of peptide- or protein-based drugs, (c) neuroprosthetic devices with integrated drug delivery capabilities to enhance long-term functional reliability, and (d) microdevices for the encapsulation and/or immunoisolation of xenografted or genetically engineered cells for cell-based therapies. Further discussion of these applications, and many others, can be found in recent MEMS-related drug delivery reviews by Nuxoll and Siegel [52], Staples *et al.* [53], Razzacki *et al.* [54], Reed and Lye [55], and LaVan, McGuire, and Langer [56].

Microneedle devices are typically composed of arrays of penetrating structures with individual lengths and cross-section dimensions on the order of tens to hundreds of micrometers, depending upon the intended application. These minute dimensions provide the basis for one of the key advantages conferred by such devices, namely minimally invasive injection. Devices demonstrated to date include those fabricated using conventional MEMS materials and processing techniques (e.g., anisotropic and isotropic etching of Si or SiO_2), as well as alternative materials and processes (e.g., micromolding of polymers and wet etching or electroplating of metals), with the latter intended to address potential safety and reliability issues associated with devices based on intrinsically brittle materials such as Si (see Section 10.5.2 for further discussion). Devices with passive and active delivery capability have been developed, with the former primarily consisting of solid needles with drugs coated on their surface (or within their bulk, in the case of dissolvable polymeric needles), and the latter consisting of needles with internal lumens for active fluidic delivery or sampling). Active devices with integrated drug reservoirs, micropumps, filters, sensors, and other functionalities have also been demonstrated. A number of recent reviews [57–63] further elaborate on the extraordinary diversity of microneedle designs that have been developed thus far, as well as the immense potential they offer for enhancing chronic disease management (e.g., diabetes and osteoporosis) and acute delivery applications (e.g., immunization).

Transdermal drug delivery constitutes the most extensively developed application area for microneedles, due in large part to the opportunity for painless injection, which arises from their diminutive nature (i.e., microneedles typically do not penetrate deeply enough to stimulate pain receptors). As will be discussed in Section 10.5.2, this provides the potential for increasing patient compliance, particularly in chronic delivery applications (e.g., insulin injection for diabetes management). The skin also represents an ideal target for vaccination due to its accessibility and high concentration of antigen-presenting cells (i.e., Langerhan's cells). Microneedles have been shown to elicit superior immune response relative to conventional intramuscular injection, often with far lower dosages (i.e., dose sparing). Moreover, unlike other alternative delivery technologies (e.g., jet injection, sonophoresis, and gene-gun), microneedles typically require minimal external hardware for application. This reduces the cost, increases transportability, and minimizes the need for maintenance or external power. Microneedles also offer enhanced ease of use, thus providing the potential for self-administration. Finally, use of dry-coated polymers as passive-release reservoirs eliminates the need for refrigerated storage, further reducing the logistical burden and cost. Collectively, these advantageous features suggest a strong promise for microneedles to serve as a fundamental enabler for low-cost mass immunizations, particularly in the developing world.

Although microneedles have yet to reach the market (for drug delivery applications), there are numerous devices in preclinical development, as well as a smaller number that are currently undergoing clinical trials. The Macroflux® transdermal microprojection drug delivery device developed by Zosano Pharma™ represents one of the most mature efforts to date. As shown in Fig. 10.4, the device relies on passive drug release from polymer coatings situated at the needle tips. The device is fabricated from thin titanium sheets using photolithographic patterning and wet etching to define the needle shanks within the substrate plane, followed by the mechanical deformation of the shanks out of plane. Recent results from Phase 2 clinical trials of the Macroflux®-based delivery of parathyroid hormone (PTH 1-34) for the treatment of severe osteoporosis have demonstrated efficacy equal to or better than conventional FORTEO® subcutaneous injection (Eli Lilly, hPTH 1-34 peptide), thus laying the foundation for forthcoming Phase 3 studies focused on the demonstration of noninferiority [64]. The company is also evaluating opportunities for the delivery of numerous other drugs, including Phase 1 trials with insulin, desmopressin (for treatment of bed-wetting), and infleunza vaccine, as well as preclinical studies with human growth hormone, Fentanyl (for the treatment of chronic pain), and tetanus vaccine [64]. In each case, the opportunity for painless injection, coupled with ease of use, convenience, and the elimination of the need for refrigeration may create strong market potential, provided sufficient safety and efficacy are demonstrated.

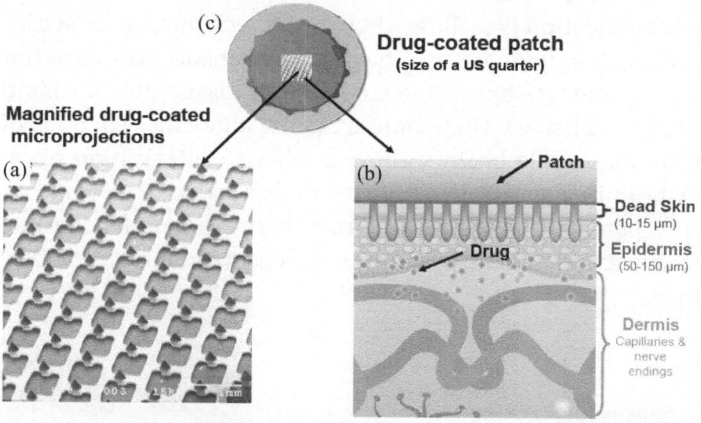

Figure 10.4 Zosano Pharma™ Macroflux® transdermal microprojection drug delivery device. (a) A scanning electron micrograph of drug-coated microprojection array; (b) a schematic representation of a microprojection array packaged with an adhesive patch; (c) a schematic representation of the transdermal drug delivery mechanism via passive diffusion from the coated microprojections. (Reprinted with permission from ref. 64.)

10.4.2 Microsurgery

The advent of minimally invasive surgical techniques using endoscopic, catheter, and guidewire-based procedures has enabled significant reduction of trauma, postoperative pain, recovery time, and cost relative to open surgery. As such, these techniques have revolutionized the diagnosis and treatment of numerous conditions ranging from gastrointestinal disorders to cardiovascular disease. Common to all such techniques is the reliance upon miniaturized surgical instruments that are mounted on small-diameter (i.e., few mm or less) tubes, wires, or rods that require only minor incisions for insertion into the body. Using various real-time imaging modalities (e.g., CCD-based video or x-ray-based fluoroscopy), these instruments are guided by the surgeon to the site of interest through the use of external manipulators. The procedure is then performed, often through the manipulation of additional external controllers that are mechanically coupled to devices situated at the tip of the instrument. Although these procedures are used widely, there is room for improvement. The opportunity provided by MEMS for the creation of miniaturized multifunctional devices with feature sizes on the order of physiologically relevant structures has driven significant research and development in this area.

In the review by Rebello [65], a number of areas of opportunity are indentified. For example, haptic feedback is currently lacking in most minimally invasive procedures, thus eliminating opportunity for the surgeon to identify different tissue types by feel (e.g., for discrimination between fatty tissue and muscle) or assess their condition by manual palpation (e.g., for determining the location of plaque-occluded sections in coronary arteries). When coupled with the reduced capability for visual inspection relative to open surgery, these represent significant limitations. Rebello discusses examples of surgical instruments with integrated microfabricated components that seek to address these issues, one of which is the "Data Knife" concept proposed by Verimetra (Fig. 10.5). In this concept, tissue identification is facilitated by both mechanical (i.e., strain and pressure) and electrical sensors (i.e., impedance). The pressure sensors also allow the measurement of the mechanical characteristics of the surrounding tissue, thus facilitating the spatial identification of desired tissues. Ultrasonic actuators allow the reduction of cutting forces and, therefore, trauma, while heaters allow localized cauterization to reduce bleeding. Although the Data Knife concept is no longer under active development [66], it clearly demonstrates the potential of MEMS for providing real-time feedback to the surgeon, extending instrument functionality, and enhancing procedural outcomes in minimally invasive surgeries.

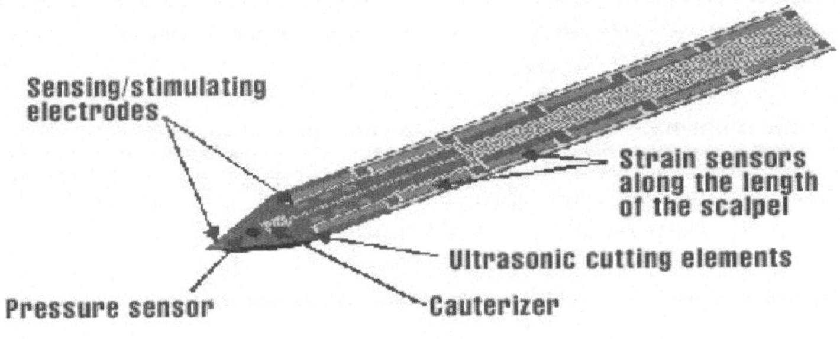

Figure 10.5 A schematic view of Verimetra "Data Knife" concept, which integrates microfabricated sensing and actuation capabilities within a conventional surgical scapel. (Reprinted with permission from ref. 67.)

10.4.3 Endovascular Intervention

Endovascular intervention is an important subset of microsurgery, which seeks to treat or repair vascular disease and damage in a minimally invasive manner from within the vessel itself. It is often used for the restoration of blood flow in vessels narrowed by the chronic accumulation of cholesterol-laden plaque (i.e., atherosclerosis) as well as repair of localized bulging of the vascular wall (i.e., an aneurysm) typically caused by disease or weakening. Common therapeutic microdevices used within this context include stents (i.e., expandable tubular metallic meshes) and stent grafts (i.e., stents with integrated fabric lumens), both of which are intended for chronic implantation. These devices are typically passive in nature, serving as mechanical scaffolding to prevent vascular recoil after dilation via balloon angioplasty (stents) or as a means for the exclusion of an aneurysm sac from the bloodstream and systemic pressure (stent grafts). However, one notable exception is the drug-eluting stent, which adds drug delivery functionality to increase therapeutic efficacy through the prevention of vessel renarrowing after implantation (see Section 10.5.4 for further details). Opportunity also exists for improving endovascular procedural outcomes and enhancing the safety and reliability of long-term monitoring through the incorporation of various sensing modalities directly within such devices or within their immediate vicinity. MEMS technology provides promise in this regard due to the opportunity for miniaturization and high-density multifunctional integration.

In the review by Potkay [68], a number of endovascular implementations of MEMS-based pressure-sensing platforms are discussed with intended applications in the monitoring of heart failure, stented arteries, or aneurysms. Most are based upon capacitive pressure sensing, and more recent devices include remote powering via induction to eliminate need for batteries, as well as passive telemetry for communication with external readout modules. The most successful of these to date is the CardioMEMS™ EndoSure® pressure sensor, pictured in Fig. 10.6, which has recently received FDA clearance for the monitoring of intrasac pressure during the endovascular repair of abdominal and thoracic aortic aneurysms [69]. The device is implanted during the stent graft deployment procedure, and is placed within the aneurysm sac, outside of the stent graft lumen wall. The device is comprised of a pair of electroplated metal inductor windings defined on internal, opposing flexible faces of a hermetically sealed reference cavity fabricated by wet etching and fusion bonding of fused silica substrates [70]. Sensory transduction is achieved by shifting of the device's resonant frequency with variation in the distance between the opposing inductor windings, which is caused by variation in the intrasac pressure acting on the device. The measurement of intrasac pressure allows the detection of stent graft leakage, thus providing means for the physiologic assessment of repair procedure success via the confirmation of aneurysm sac exclusion.

Results from a recent large-scale clinical trial of the EndoSure® device have demonstrated comparable efficacy relative to prevailing intraoperative angiography assessment techniques for the confirmation of sac exclusion in abdominal aortic aneurysms [72]. Preliminary observations also suggest the potential for reducing the need for repeated injections of contrast dye to assess stent graft deployment success, which is often required when using prevailing angiographic assessment techniques. This is a potentially significant advantage with regard to safety, since repeated contrast injections may impair kidney function. Longer-term follow-up studies of these patients will also seek to demonstrate device efficacy for the postoperative monitoring of stent graft leak tightness. If successful, these studies may demonstrate the potential for reducing the need for prevailing CT-scanning techniques for postoperative monitoring. This may be particularly advantageous since repeated CT scanning carries increased lifetime risk of contrast-induced kidney impairment, as well as greater radiation exposure, increased cost, and reduced patient compliance. The potential for the application of this compelling device technology in separate but closely related areas is also being explored, as evidenced by recently initiated clinical trials that seek to assess the efficacy for long-term blood pressure monitoring within the pulmonary artery [73]. If successful, these studies will demonstrate potential for noninvasive monitoring that can be performed by the patient to detect early warning signs of impending heart failure.

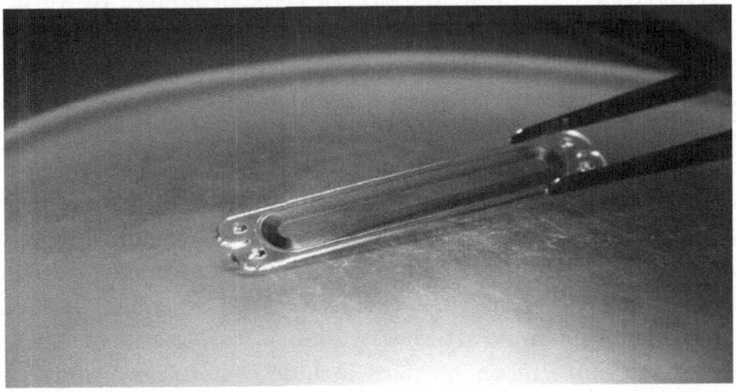

Figure 10.6 A CardioMEMS™ EndoSure® wireless pressure sensor for monitoring leakage during the endovascular repair of abdominal and thoracic aortic aneurysms. (Reprinted with permission from ref. 71.)

10.5 *In vivo* Therapeutic MEMS Enabled by Titanium Micromachining

As evidenced in the preceding section, the extraordinary opportunity afforded by MEMS has driven numerous efforts to extend its application towards *in vivo* therapeutics, some of which have already resulted in devices with significant clinical and commercial potential. However, continuing reliance upon micromechanical materials adopted from the semiconductor industry may ultimately limit the scope of what can be achieved. Many such materials suffer from poor mechanical reliability due to low fracture toughness, which results in extreme sensitivity to stress concentration and predisposition to catastrophic failure by fracture. Although this can be mitigated in some cases through robust mechanical design and packaging, this invariably increases cost and complexity. Moreover, in many emerging applications, these avenues are not available due to design constraint and/or performance restriction, thus illustrating the need for the development of viable alternatives.

In this section, recently developed titanium micromachining techniques that address this need will be introduced, followed by a discussion of several applications currently under development that demonstrate the unique opportunities they afford. In each of these applications, titanium micromachining provides a means for leveraging a host of advantageous properties that yield enhanced safety, reliability, and/or performance. As such, titanium micromachining shows considerable promise for extending the utility of MEMS for *in vivo* therapeutics.

10.5.1 Titanium Micromachining

Metals are increasingly being considered for MEMS applications due in large part to their high fracture toughness, which provides the capability for the dissipation of loading energy via plastic deformation. This often provides significant load-bearing capability, well past the yield strength of the material (i.e., the stress level associated with the onset of plasticity). Brittle materials, such as silicon, do not posses such capability; thus they fail catastrophically by fracture when overloaded. One metal of particular interest is titanium, due to its excellent biocompatibility, corrosion resistance, and fatigue properties. However, microfabrication technology that enables the realization of complex, high-aspect-ratio structures in titanium was, until recently, unavailable. A short time ago, we reported the development of techniques that enable, for the first time, the deep etching of bulk titanium substrates [74, 75]. As demonstrated in Fig. 10.7, these new techniques provide the opportunity for the fabrication of titanium-based MEMS devices with a degree of design sophistication that would be difficult if not impossible to achieve with the prevailing metal-micromachining methods. Moreover, these techniques fully leverage the existing semiconductor process infrastructure, thus ensuring scalability to low-cost/high-volume manufacturing as well as opportunity for high-density multifunctional integration.

Figure 10.7 A scanning electron micrograph of a bulk titanium–interdigitated electrode structure. The etch depth is 20 μm, and the minimum feature size is 1 μm. (Reprinted with permission from ref. 75.)

To date, two distinct titanium-micromachining techniques have been developed, the metal anisotropic reactive-ion etching with oxidation (MARIO) process [74] and the titanium ICP deep etch (TIDE) process [75]. The MARIO process relies on a cyclic etch-passivation scheme to mitigate the isotropic etch character produced by the capacitively coupled RIE used for this process. In contrast, etch anisotropy in the TIDE process is achieved through the exploitation of the high degree of ion directionality and the process control afforded by the advanced inductively coupled plasma etcher used for this process. Both processes rely on Cl- or Ar-based chemistry for etching, and oxygen is used in the passivation step of the MARIO process to form etch-resistant titanium dioxide (TiO_2) on all exposed titanium surfaces, as illustrated in Fig. 10.8. Both processes provide selectivity of ~40:1 with SiO_2 or TiO_2 masking and capability for aspect ratios well in excess of 20:1 (i.e., etched feature depth:width). However, the limited capabilities of the capacitively coupled etch system used in the MARIO process limit etch rates to ~0.5 μm/min, whereas etch rates in excess of 2–3 μm/min are possible with the TIDE process.

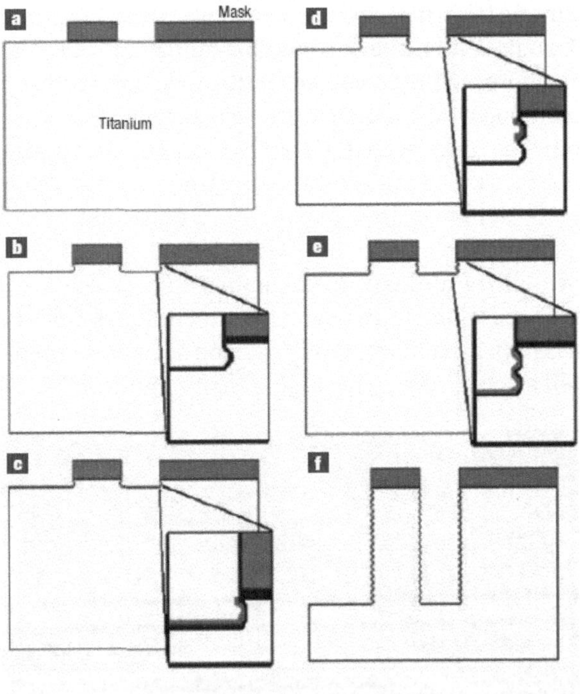

Figure 10.8 A schematic representation of the cyclic etch-passivation scheme of the MARIO process. (a) Deposition, photolithographic patterning, and etching of the masking layer. (b) The first Cl/Ar-based etch step; inset illustrates the isotropic undercutting of the mask. (c) The first oxygen-based passivation step; inset illustrates the oxidation of the exposed titanium. (d) The second etch step; inset illustrates the preferential etching of trench floors and the preservation of sidewall oxide passivation. (e) The second passivation step; inset illustrates the oxidation of newly exposed titanium surfaces. (f) The anisotropic high-aspect-ratio deep etch produced by multiple repetitions of steps (b–e). (Reprinted with permission from ref. 74.)

Although titanium micromachining shows a strong potential for the fabrication of mechanically robust analogues of current silicon MEMS devices, we believe that even greater promise lies in areas that more fully leverage the unique attributes of micro/nanoscale titanium structures. One area of particular interest is biomedical microdevices. The well-proven biocompatibility of titanium ensures stability in chronic implantation applications, while its toughness ensures mechanical reliability and safety. Its toughness also enables realization of thin, flexible devices that allow implantation in confined spaces and conformality in nonplanar applications. Titanium micromachining provides a means for the exploitation of these advantageous characteristics by enabling the fabrication of complex devices with physiologically relevant feature sizes. Moreover, the extension of this process technology to the nanoscale realm allows for surface structuring at length scales sufficient for the mediation of cellular and molecular interactions. This, therefore, provides the opportunity for the fabrication of novel biomedical microdevices with performance and functionality not attainable using other micromachining techniques.

10.5.2 Robust Microneedles for Insulin Delivery

Although the importance of insulin therapy in the management of diabetes is well established, roughly one-third of insulin-dependent patients nevertheless fail to adhere to prescribed regimens [76–79]. Recent studies have also shown that nearly one-third of insulin-naïve patients are unwilling to start insulin therapy [80, 81] and more than half express anxiety over starting [82]. The fear of injection pain features prominently among the reasons given by those not yet taking insulin [80, 83] and similar fears have been indicated by those already taking insulin as well [84, 85]. Poor insulin adherence results in poor glycemic control [76–78], which has been associated with increased hospitalizations for diabetes-related complications and greater healthcare costs [76, 86, 87]. Diabetes currently affects nearly 6% of the US population, with a total estimated cost of $174 billion in 2007 [88]. Moreover, its prevalence is predicted to double by 2050 [89]. As such, this collectively indicates a strong need for the development of technological solutions that improve adherence by mitigating fears associated with injection.

Microneedle devices have shown promise in this regard by providing potential for painless injection [90–94] while maintaining efficacy (in animal models) comparable to subcutaneous insulin injection [95, 96]. However, clinical translation may be limited by the common reliance upon silicon for such devices. The brittleness of silicon creates a predisposition to fracture, resulting in a catastrophic, but potentially imperceptible, loss of device functionality. This represents a significant safety hazard, particularly in devices intended for use in continuous infusion systems, since undetected disruption of insulin delivery for prolonged periods could lead to poor glycemic control, disease progression, and/or life-threatening complications (e.g., diabetic ketoacidosis). This, therefore, provides the impetus to continue the search for alternative micromechanical materials that enhance the reliability and safety of microneedle devices.

We have recently demonstrated titanium-based microneedles that begin to address these limitations [97]. As shown in Fig. 10.9, these needles are composed of bonded pairs of titanium shanks with internalized fluidic networks (i.e., lumens). Results from mechanical testing demonstrate that these devices are able to withstand axial loads that exceed the reported ~0.1 N force required for insertion into skin analogues for needles with similar geometry [98]. Moreover, when overloaded, needle response is characterized by graceful failure (i.e., bending) as opposed to the catastrophic failure associated with silicon (i.e., fracture), as demonstrated in Fig. 10.10. However, these tests also illustrate the need for further refinement of the bonding procedures to reduce the potential for delamination since this will degrade delivery capability in the event of severe overloading.

Collectively, this data demonstrates that plasticity prevents fragmentation and reduces the potential for the interruption of delivery during overloading, thus ensuring greater safety and reliability relative to prevailing silicon-based microneedle devices. We are currently pursuing further evaluation and development of titanium-based microneedles for use in diabetes management as well as other compelling applications, including vaccination and intraocular drug delivery.

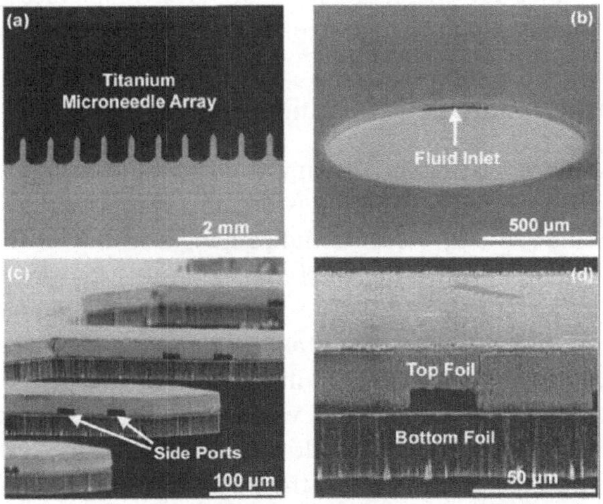

Figure 10.9 Scanning electron micrographs of titanium microneedle devices with active fluidic delivery capability. All needle shanks are 500 μm long, 100 wide, and 50 μm thick. Rectangular cross-section lumens are embedded within the shanks with 10 μm depth and variable width. (Reprinted with permission from ref. 97.)

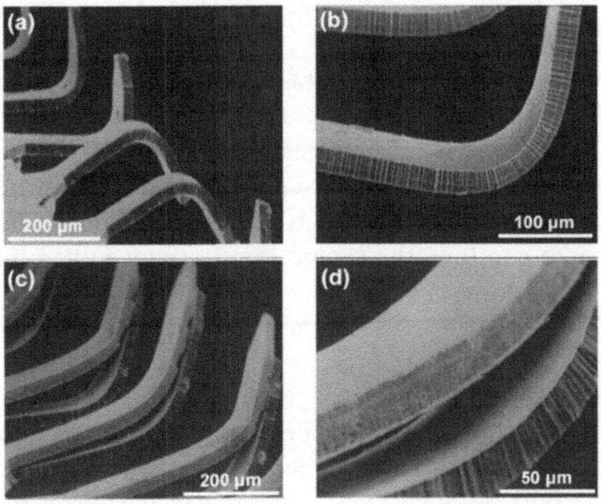

Figure 10.10 Scanning electron micrographs of titanium-based solid shank needle analogues (a and b) and bonded microneedle devices (c and d) following uniaxial compression testing. Plastic buckling is observed in all cases, as is delamination in the bonded microneedle devices. (Reprinted with permission from ref. 97.)

10.5.3 Robust Neuroprostheses

Microscale neuroprostheses, which allow the direct recording and/or stimulation of neurons, have provided unprecedented opportunity for enhancing the understanding of fundamental neurophysiological processes underlying many aspects of human function and behavior [99, 100]. They have also demonstrated immense promise for restoring neurological functions lost to disease, stroke, or injury [101]. However, current utility and future promise are limited by the common use of silicon for such devices. The brittleness of silicon creates a predisposition to catastrophic fracture, resulting in the complete loss of device functionality, as well as potential for the generation of atomically sharp debris within the brain. While some studies have shown that silicon-based neuroprostheses possess sufficient strength for insertion into cortical tissues [102, 103], in practice, breakage is commonplace. This propensity for fracture constrains the device geometries that may be realized, thus largely limiting accessibility to the cerebral cortex. Additionally, current devices are fairly difficult to insert [104], which limits their broader utilization in the research and medical community. Finally, their eventual translation to the clinic will require devices that are not only sufficiently robust to withstand rough handling by neurosurgeons [105] but also immune to safety hazards associated with device fragmentation.

Neuroprostheses based on polymeric materials, such as polyimide, benzocyclobutene, polydimethylsiloxane, parylene, and SU-8, have been proposed as a means for improving the mechanical reliability relative to Si [106–108]. However, in most polymeric devices, a trade-off must be made between flexibility and performance. In some, dimensions are kept similar to those of conventional silicon-based devices because the reactive tissue response often scales with device size [109, 110]. However, this may sacrifice stiffness and, therefore, reduce insertion reliability [111–113]. A number of methods for circumventing this trade-off have been reported, most based on the use of stiffening strategies such as dissolvable rigid polymer coatings [112], internal reinforcements [114], and fixed silicon or nickel backbone layers [111, 113, 115]. None have been demonstrated in long-term implantation thus far, and potential concerns over the effect of polymer degradation in the neurological environment, or nickel toxicity resulting from coating/encapsulation failure, may limit the ultimate utility for chronic implantation. These limitations provide impetus to continue the search for alternative micromechanical materials that enhance performance, functionality, and safety of microfabricated neuroprostheses.

We have recently demonstrated titanium-based neuroprostheses, pictured in Fig. 10.11, that may address these limitations [116, 117]. *In vitro* mechanical testing of these devices has shown that they fail gracefully via plastic deformation, rather than the catastrophic failure mode typical of silicon-based devices. Moreover, these tests have shown that titanium-based devices retain a significant portion of their recording functionality, even after the onset of plasticity, whereas the overloading of silicon-based devices results in the complete loss of functionality. *In vivo* testing in rodent models has demonstrated capability for accurate insertion and acute recording performance comparable to commercially available silicon devices. Finally, the titanium-based neuroprostheses have enabled, for the first time, simultaneous recording from both the auditory cortex and the thalamus using a single device. Due to its unique position within the neural processing hierarchy, the cortical-thalamic loop is of great interest as systems neuroscientists and neuroengineers study and attempt to reverse-engineer the brain. The ability to directly and simultaneously interrogate both the cortex and the thalamus with a single device significantly reduces surgical and experimental complexity, thereby enabling these studies. Collectively, this data demonstrates the potential for greater mechanical reliability and safety relative to prevailing devices, as well as the opportunity for compelling new measurement functionality. We are continuing the evaluation of these devices to determine their performance in chronic implantation and working toward the development of techniques to improve their long-term recording reliability.

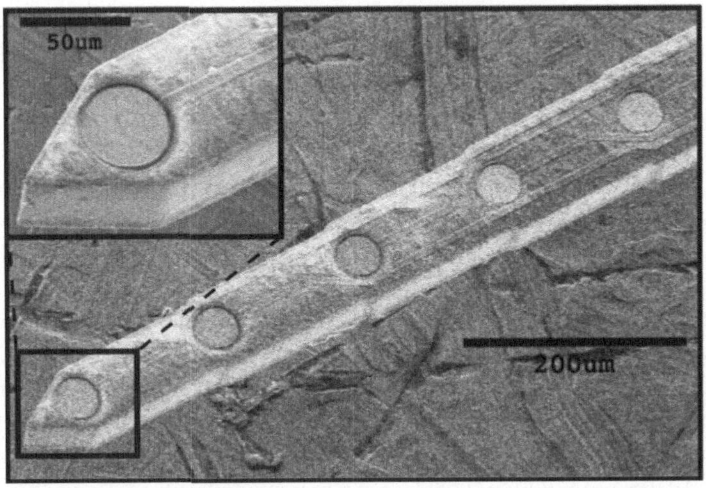

Figure 10.11 A scanning electron micrograph of a titanium-based neuroprosthetic device. (Reprinted with permission from ref 116.)

10.5.4 Rationally Designed Nanopatterned Stents

Within the past decade, drug-eluting stents have revolutionized the treatment of coronary heart disease, the leading cause of death worldwide. They have done so by providing a minimally invasive means of restoring blood flow in arteries narrowed by plaque accumulation while also minimizing restenosis (postimplantation renarrowing caused by injury-induced hyperproliferation of vascular smooth muscle cells). This has been achieved through coating-based delivery of drugs that act to suppress hyperproliferation. There is growing concern, however, over the long-term safety of such devices, due to the increased potential for late-stage thrombosis (potentially life-threatening clotting that occurs many months to years after stent implantation). Although the mechanism has yet to be fully understood, delayed healing is likely to play a significant role [118], that is, the drug-induced suppression of cellular proliferation prevents the reestablishment of the endothelium, the native inner lining of the vessel.

We have recently demonstrated preliminary evidence supporting the feasibility of an innovative new therapeutic strategy based on the facilitation of healing rather than inhibition. Prohealing strategies are not new; however, our strategy represents a departure from convention because it relies on physical rather than pharmacological or biological means for facilitating the reestablishment of the endothelium. Specifically, using our novel titanium micromachining techniques, we are exploring the potential for enhancing endothelial cell proliferation and functionality via rational design of stent surface topography at the nanoscale.

We have created bulk titanium substrate surfaces patterned with arrays of parallel grooves ranging in width and spacing from 0.75 to 100 μm. The rationale for this pattern geometry is based upon the hypothesis that such features may reduce contact inhibition, thus accelerating endothelial cell proliferation and the reestablishment of the endothelium. It is also hypothesized that such features may enhance endothelial functionality by promoting elongated cellular morphology similar to that of the native endothelium. Our recent experimental observations support both hypotheses [119] and also demonstrate the potential for the selective enhancement of desired cellular species based solely on pattern feature size [120]. As shown in Fig. 10.12, under competitive *in vitro* culture conditions, decreasing feature sizes promote the strong enhancement of endothelial proliferation over smooth muscle cells. In fact, after a three-day culture, the number of live endothelial cells on the smallest features (0.75 μm) is over 30 times greater. This stands in stark contrast to the behavior observed on representative stent surfaces (i.e., wrought Ti control samples) where endothelial proliferation is strongly inhibited relative to smooth muscle cells. Since competitive proliferation assays begin to model the complex physiological environment following stent implantation, this data clearly demonstrates the potential of nanopatterning for facilitating healing and, thus, improving outcomes in vascular stenting. Moreover, the strong suppression of smooth muscle cells on nanopatterned surfaces also suggests the potential for the mitigation of restenosis without the need for drug elution, although further study is required to validate this conjecture.

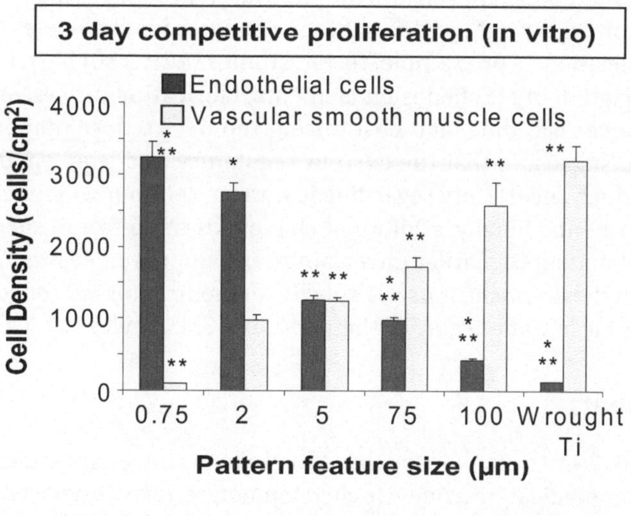

Figure 10.12 Three-day competitive cell proliferation (*in vitro*) on patterned titanium substrates as a function of the feature size. Feature sizes correspond to groove width and spacing. (Reprinted with permission from ref. 120.) See also Color Insert.

The often-beneficial effects of randomly structured nanoscale surface topography on endothelial function have been reported by others [121–126], as have the effects of microscale-patterned surface topographies on endothelial cells [127, 128]. However, we are the first to demonstrate the beneficial effects of nanoscale-patterned surfaces on endothelialization in materials of immediate relevance to vascular stents. We are also the first to demonstrate strong enhancement of competitive endothelial proliferation over vascular smooth muscle cells on nanopatterned surfaces. Finally, we are the first to demonstrate such behavior in patterned metallic surfaces created using fabrication techniques that are amenable to low-cost/high-volume manufacturing of vascular stents. This is a particularly key distinction since it demonstrates a clear path toward implementation in actual devices. We are continuing the evaluation of the potential of rationally designed nanopatterning using increasingly sophisticated *in vitro* cellular assays, in preparation for the eventual *in vivo* testing in clinically relevant animal models.

10.6 Conclusions and Outlook

As demonstrated in this chapter, the expansion of the IC-based microelectronics paradigm to a MEMS-based micro*systems* paradigm has provided extraordinary opportunity for improving the efficacy, safety, and functionality for *in vivo* therapeutics. This opportunity is fundamentally enabled by the diversity of fabrication techniques developed for MEMS, which provide opportunity for miniaturization, massive parallelization, and multifunctional integration, as well as the potential for scalability to low-cost/high-volume manufacturing. Realization of this opportunity has already begun, as evidenced by commercially available integrated microsystems for cardiac rhythm management, and the pace of new product introductions will likely quicken in the near future, with numerous therapeutic MEMS devices nearing market, such as microneedles for painless transdermal drug delivery and chronically implanted pressure sensors for endovascular monitoring.

However, despite this bright outlook, there are numerous challenges that must be faced by emerging applications. For example, recent studies [129, 130] have identified the need for the standardization of MEMS design elements, fabrication processes, and/or external interfaces to reduce risk, time, and cost during the device development process. These studies have also suggested that the current regulatory and legal environments impede market entry, as does uncertainty regarding insurance reimbursement conditions for new devices and treatments. Finally, additional challenges arise from user skepticism (from both doctors and patients), market uncertainty, and unproven economic benefits of new devices. Although these conclusions are specific to medical MEMS sensors and telemetric systems, they are largely applicable to therapeutic MEMS devices as well.

Acknowledgments

The author would like to acknowledge the efforts of and convey gratitude to those involved in the titanium biomedical microdevice development efforts discussed herein, including current and past collaborators (M. F. Aimi, E. R. Parker, A. S. Zuruzi, D. P. Bothman, N. C. MacDonald, K. L. Turner, C. D. Mienhart, T. J. Webster, and K. J. Otto) and students (P. T. McCarthy, R. Madangopal, H. P. Aguilar, J. Liu, and D. Khang).

References

1. Judy, J. W. (2001). Microelectromechanical systems (MEMS), fabrication, design and applications, *Smart Mater. Struct.*, **10**(6), 1115–1134.
2. (2008) *BioMEMS 2008 report.*
3. James, T., Mannoor, M. S., and Ivanov, D. V. (2008). BioMEMS – advancing the frontiers of medicine, *Sensors*, **8**(9), 6077–6107.
4. Receveur, R. A. M., Lindemans, F. W., and de Rooij, N. F. (2007). Microsystem technologies for implantable applications, *J. Micromech. Microeng.*, **17**(5), R50–R80.
5. Ziaie, B., Baldi, A., Lei, M., Gu, Y., and Siegel, R. A. (2004). Hard and soft micromachining for BioMEMS: review of techniques and examples of applications in microfluidics and drug delivery, *Adv. Drug Delivery Rev.*, **56**(2), 145–172.
6. Grayson, A. C. R., Shawgo, R. S., Johnson, A. M., Flynn, N. T., Yawen, L. I., Cima, M. J., and Langer, R. (2004). A BioMEMS review: MEMS technology for physiologically integrated devices, *Proc. IEEE*, **92**(1), 6–21.
7. Bashir, R. (2004). BioMEMS: state-of-the-art in detection, opportunities and prospects, *Adv. Drug Delivery Rev.*, **56**(11), 1565–1586.
8. Voldman, J., Gray, M. L., and Schmidt, M. A. (1999). Microfabrication in biology and medicine, *Annu. Rev. Biomed. Eng.*, **1**, 401–425.
9. Polla, D. L., Erdman, A. G., Robbins, W. P., Markus, D. T., Diaz-Diaz, J., Rizq, R., Nam, Y., Brickner, H. T., Wang, A., and Krulevitch, P. (2000). Microdevices in medicine, *Annu. Rev. Biomed. Eng.*, **2**, 551–576.
10. Haga, Y., and Esashi, M. (2004). Biomedical microsystems for minimally invasive diagnosis and treatment, *Proc. IEEE*, **92**(1), 98–114.
11. Panescu, D. (2006). MEMS in medicine and biology, *IEEE Eng. Med. Biol.*, **25**(5), 19–28.
12. Varmus, H., Klausner, R., Zerhouni, E., Acharya, T., Daar, A. S., and Singer, P. A. (2003). Grand challenges in global health, *Science*, **302**(5644), 398–399.
13. (2009) *Grand challenges for engineering.*
14. Staufer, U., Akiyama, T., Beuret, C., Gautsch, S., Noell, W., Schurmann, G., Stebler, C., and de Rooji, N. F. (2000). Micro-electromechanical systems for nanoscience, *J. Nanoparticle Res.*, **2**(4), 413–418.
15. Boser, B. E. (5–9 November 2006). *From Micro to Nano: MEMS as an Interface to the Nano World*, IEEE/ACM International Conference on Computer Aided Design – Digest of Technical Papers (IEEE Cat No. 06CH37762), 2 pp.|CD-ROM.
16. Maluf, N., and Williams, K. (2004). *Introduction to Microelectromechanical Systems Engineering*, Artech House Publishers, Boston, Massachusetts.
17. Madou, M. J. (2002). *Fundamentals of Microfabrication: the Science of Miniaturization*, CRC Press, New York.
18. Senturia, S. D. (2005). *Microsystem Design*, Springer, New York.
19. Kathuria, Y. P. (2005). Laser microprocessing of metallic stent for medical therapy, *J. Mater. Proc. Technol.*, **170**(3), 545–550.

20. Takahata, K., and Gianchandani, Y. B. (2002). Batch mode micro-electro-discharge machining, *J. Microelectromech. Syst.*, **11**(2), 102–110.
21. Sharon, A., Bilsing, A., Lewis, G., and Zhang, X. (19–23 January 2003) *Manufacturing of 3D Microstructures Using Novel UPSAMS Process (Ultra Precision Manufacturing of Self-Assembled Micro Systems)*, paper presented in IEEE Sixteenth Annual International Conference on Micro Electro Mechanical Systems: MEMS-03, Kyoto, Japan.
22. Bustillo, J. M., Howe, R. T., and Muller, R. S. (1998). Surface micromachining for microelectromechanical systems, *Proc. IEEE*, **86**(8), 1552–1574.
23. Calvet, P., Lapierre, H., and Charvet, J. (2000). Diversity of the Ordovician volcanism in the Pierrefitte area (hautes-pyrenees) – calc-alkaline rhyolites and alkali basalts, *Comptes Rendus des Séances de l'Academie des Sciences. Série II, Mécanique, Physique, Chimie, Sciences de l'univers, Sciences de la Terre*, **89**, 6–9.
24. Butler, R. (2000). Shear wave properties of marine sediments derived from cepstral analysis of background noise, *Geophys. Res. Lett.*, **15**(8), 836.
25. Kovacs, G. T. A., Maluf, N. I., and Petersen, K. E. (1998). Bulk micromachining of silicon, *Proc. IEEE*, **86**(8), 1536–1551.
26. Petersen, K. E. (1982). Silicon as a Mechanical Material, *Proc. IEEE*, **70**(5), 420–457.
27. (2006) *MEMS – a roadmap to technologies and applications*, Electronics.ca Research Network.
28. Karam, R. M., and Casler, R. J. (2003). *A New 3D, Direct-Write, Sub-Micron Microfabrication Process that Achieves True Optical, Mechatronic and Packaging Integration on Glass-Ceramic Substrates*, Invenios, Inc.
29. Stillman, J., Judy, J., and Helvajian, H. (2008). Aspect ratios, sizes, and etch rates in photostructurable glass-ceramic, *Proc. SPIE-Int. Soc. Opt. Eng.*, **6882**, 68820J-1–68820J-68820J-11.
30. Williams, K. R., Gupta, K., and Wasilik, M. (2003). Etch rates for micromachining processing – part II, *J. Microelectromech. Syst.*, **12**(6), 761–778.
31. Lärmer, F., and Schilp, P. (1994). *Method of anisotropically etching silicon*, German.
32. Alexe, M., and Gosele, U. M. (eds) (2004). *Wafer Bonding: Applications and Technology*, Springer.
33. Tong, Q.-Y., and Gosele, U. M. (1999). *Semiconductor Wafer Bonding: Science and Technology*, John Wiley.
34. Budd, D. A. (1988). Aragonite-to-calcite transformation during fresh-water diagenesis of carbonates – insights from pore-water chemistry, *Geol. Soc. Am. Bull.*, **100**(8), 1260.
35. Gilleo, K. (2005). *MEMS packaging update*, Advanced Packaging.
36. Thorsen, T., Maerkl, S. J., and Quake, S. R. (2002). Microfluidic large-scale integration, *Science*, **298**(5593), 580–584.
37. James, D. (2006). *Chipworks inside TI's DLP chip.*
38. Cohen, A. L. (2008). *EFAB Technology: unlocking the potential of miniaturized medical devices.*
39. Tummala, R. R. (2004). *Fundamentals of Microsystems Packaging*, McGraw-Hill.

40. Gilleo, K. (2005). *MEMS/MOEMS Packaging: Concepts, Designs, Materials, and Processes*, McGraw-Hill.

41. Hsu, T.-R. (ed) (2004). *MEMS Packaging*, INSPEC, The Institution of Electrical Engineers.

42. Velten, T., Ruf, H. H., Barrow, D., Aspragathos, N., Lazarou, P., Jung, E., Malek, C. K., Richter, M., and Kruckow, J. (2005). Wackerle packaging of bio-MEMS: strategies, technologies, and applications, *IEEE Trans. Adv. Packaging*, **28**(4), 533–546.

43. *Use of International Standard ISO-10993, "Biological Evaluation of Medical Devices Part 1: Evaluation and Testing"*, U. S. F.a.D. Administration, Editor. 1995.

44. Kotzar, G., Freas, M., Abel, P., Fleischman, A., Roy, S., Zorman, C., Moran, J. M., and Melzak, J. (2002). Evaluation of MEMS materials of construction for implantable medical devices, *Biomaterials*, **23**(13), 2737–2750.

45. Pan, J. Y. (2004). Reliability considerations for the BioMEMS designer, *Proc. IEEE*, **92**(1), 174–184.

46. Wallin, R. F., and Upman, P. J. (1998). *A practical guide to ISO 10993: part 1 — introduction to the standards*, Medical Device & Diagnostic Industry Magazine.

47. Hidefjall, P. (2006). Cardiac rhythm management: the shape of things to come, *Curr. Opin. Cardiol.*, **21**(1), 27–33.

48. Park, H., Cannizzaro, C., Vunjak-Novakovic, G., Langer, R., Vacanti, C. A., and Farokhzad, O. C. (2007). Nanofabrication and microfabrication of functional materials for tissue engineering, *Tissue Eng.*, **13**(8), 1867–1877.

49. Puleo, C. M., Yeh, H. C., and Wang, T. H. (2007). Applications of MEMS technologies in tissue engineering, *Tissue Eng.*, **13**(12), 2839–2854.

50. Ainslie, K. M., and Desai, T. A. (2008). Microfabricated implants for applications in therapeutic delivery, tissue engineering, and biosensing, *Lab Chip*, **8**(11), 1864–1878.

51. Desai, T., and Bhatia, S. (2006). *Biomems and Biomedical Nanotechnology: Volume III: Therapeutic Micro/Nanotechnology* (ed. M. Ferrari), Springer.

52. Nuxoll, E. E., and Siegel, R. A. (2009). BioMEMS devices for drug delivery improved therapy by design, *IEEE Eng. Med. Biol.*, **28**(1), 31–39.

53. Staples, M., Daniel, K., Cima, M. J., and Langer, R. (2006). Application of micro- and nano-electromechanical devices to drug delivery, *Pharm. Res.*, **23**(5), 847–863.

54. Razzacki, S. Z., Thwar, P. K., Yang, M., Ugaz, V. M., and Burns, M. A. (2004). Integrated microsystems for controlled drug delivery, *Adv. Drug Delivery Rev.*, **56**(2), 185–198.

55. Reed, M. L., and Lye, W. K. (2004). Microsystems for drug and gene delivery, *Proc. IEEE*, **92**(1), 56–75.

56. LaVan, D. A., McGuire, T., and Langer, R. (2003). Small-scale systems for in vivo drug delivery, *Nat. Biotechnol.*, **21**(10), 1184–1191.

57. Banga, A. K. (2009). Microporation applications for enhancing drug delivery, *Expert Opin. Drug Delivery*, **6**(4), 343–354.

58. Prausnitz, M. R., and Langer, R. (2008). Transdermal drug delivery, *Nat. Biotechnol.*, **26**(11), 1261–1268.

59. Arora, A., Prausnitz, M. R., and Mitragotri, S. (2008). Micro-scale devices for transdermal drug delivery, *Int. J. Pharm.*, **364**(2), 227–236.
60. Vandervoort, J., and Ludwig, A. (2008). Microneedles for transdermal drug delivery: a minireview, *Front. Biosci.*, **13**, 1711–1715.
61. Mikszta, J. A., and Laurent, P. E. (2008). Cutaneous delivery of prophylactic and therapeutic vaccines: historical perspective and future outlook, *Expert Rev. Vaccines*, **7**(9), 1329–1339.
62. Sivamani, R. K., Liepmann, D., and Maibach, H. I. (2007). Microneedles and transdermal applications, *Expert Opin. Drug Delivery*, **4**(1), 19–25.
63. Prausnitz, M. R. (2004). Microneedles for transdermal drug delivery, *Adv. Drug Delivery Rev.*, **56**(5), 581–587.
64. Daddona, P. (22 January 2009) *Zosano Drug-Coated Microprojection Technology: Rapid and Effective Transdermal Delivery of PTH (1–34) for Osteoporosis*, presentation in the 13th Annual Drug Delivery Partnerships, Las Vegas, Nevada.
65. Rebello, K. J. (2004). Applications of MEMS in surgery, *Proc. IEEE*, **92**(1), 43–55.
66. Rebello, K. J. 2009.
67. (2003) *Patent promises to place smart scalpels in surgeon's hands*, Small Times.
68. Potkay, J. A. (2008). Long term, implantable blood pressure monitoring systems, *Biomed. Microdevices*, **10**(3), 379–392.
69. (2007) *CardioMEMS, Inc. announces FDA clearance of the EndoSure™ Wireless AAA pressure measurement system for measuring intrasac pressure during thoracic aortic aneurysm (TAA) repair.*
70. Allen, M. G. (2005). *Micromachined Endovascularly-Implantable Wireless Aneurysm Pressure Sensors: From Concept to Clinic*, paper presented in The 13th International Conference on Solid-State Sensors, Actuators, and Microsystems: Transducers '05, Seoul, Korea.
71. Becker, T. J. (2006). *Heart healthy: CardioMEMS moves closer to commercializing innovative sensors for heart patients.*
72. Ohki, T., Ouriel, K., Silveira, P. G., Katzen, B., White, R., Criado, F., and Diethrich, E. (2007). Initial results of wireless pressure sensing for endovascular aneurysm repair: the APEX trial – acute pressure measurement to confirm aneurysm sac EXclusion, *J. Vasc. Surg.*, **45**(2), 236–242.
73. (2007) *CardioMEMS announces start of CHAMPION pivotal study of wireless pressure sensing for heart failure monitoring.*
74. Aimi, M. F., Rao, M. P., MacDonald, N. C., Zuruzi, A. S., and Bothman, D. P. (2004). High-aspect-ratio bulk micromachining of titanium, *Nat. Mater.*, **3**(2), 103–105.
75. Parker, E. R., Thibeault, B. J., Aimi, M. F., Rao, M. P., and MacDonald, N. C. (2005). Inductively coupled plasma etching of bulk titanium for MEMS applications, *J. Electrochem. Soc.*, **152**(10), C675–C683.
76. Morris, A. D., Boyle, D. I., McMahon, A. D., Greene, S. A., MacDonald, T. M., and Newton, R. W. (1997). Adherence to insulin treatment, glycaemic control, and ketoacidosis in insulin-dependent diabetes mellitus, *Lancet*, **350**(9090), 1505–1510.
77. Donnelly, L. A., Morris, A. D., and Evans, J. M. (2007). Adherence to insulin and its association with glycaemic control in patients with type 2 diabetes, *QJM*, **100**(6), 345–350.

78. Cramer, J. A., and Pugh, M. J. (2005). The influence of insulin use on glycemic control – how well do adults follow prescriptions for insulin? *Diabetes Care*, **28**(1), 78–83.

79. Rajagopalan, R., Joyce, A., Ollendorf, D., and Murray, F. (2003). Medication compliance in type 2 diabetes subjects: retrospective data analysis, *Value Health*, **6**(3), 328–328.

80. Polonsky, W. H., Fisher, L., Guzman, S., Villa-Caballero, L., and Edelman, S. V. (2005). Psychological insulin resistance in patients with type 2 diabetes – the scope of the problem, *Diabetes Care*, **28**(10), 2543–2545.

81. Turner, R. C., *et al.* (1998). Intensive blood-glucose control with sulphonylureas or insulin compared with conventional treatment and risk of complications in patients with type 2 diabetes (UKPDS 33), *Lancet*, **352**(9131), 837–853.

82. Korytkowski, M. (2002). When oral agents fail: practical barriers to starting insulin, *Int. J. Obesity*, **26**, S18–S24.

83. Skovlund, S. E., van der Ven, N. C., Pouwer, F., and Snoek, F. J. (2003). Appraisal of insulin treatment among Type 2 diabetes patients with and without previous experience of insulin therapy, *Diabetologia*, **46**, A89-A90.

84. Zambanini, A., Newson, R. B., Maisey, M., and Feher, M. D. (1999). Injection related anxiety in insulin-treated diabetes, *Diabetes Res. Clin. Pract.*, **46**(3), 239–246.

85. Mollema, E. D., Snoek, F. J., Heine, R. J., van der Ploeg, H. M. (2001). Phobia of self-injecting and self-testing in insulin-treated diabetes patients: opportunities for screening, *Diabetic Med.*, **18**(8), 671–674.

86. Gilmer, T. P., O'Connor, P. J., Manning, W. G., and Rush, W. A. (2007). The cost to health plans of poor glycemic control, *Diabetes Care*, **20**(12), 1847–1853.

87. Menzin, J., Langley-Hawthorne, C., Friedman, M., Boulanger, L., and Cavanaugh, R. (2001). Potential short-term economic benefits of improved glycemic control — a managed care perspective, *Diabetes Care*, **24**(1), 51–55.

88. Petersen, M., and Association, American Diabetes (2008). Economic costs of diabetes in the US in 2007, *Diabetes Care*, **31**(3), 596–615.

89. Narayan, K. M. V., Boyle, J. P., Geiss, L. S., Saaddine, J. B., and Thompson, T. J. (2006). Impact of recent increase in incidence on future diabetes burden – US, 2005–2050, *Diabetes Care*, **29**(9), 2114–2116.

90. Van Damme, P., Oosterhuis-Kafeja, F., Van der Wielen, M., Almagor, Y., Sharon, O., and Levin, Y. (2009). Safety and efficacy of a novel microneedle device for dose sparing intradermal influenza vaccination in healthy adults, *Vaccine*, **27**(3), 454–459.

91. Haq, M. I., Smith, E., John, D. N., Kalavala, M., Edwards, C., Anstey, A., Morrissey, A., and Birchall, J. C. (2009). Clinical administration of microneedles: skin puncture, pain and sensation, *Biomed. Microdevices*, **11**(1), 35–47.

92. Bal, S. M., Caussin, J., Pavel, S., and Bouwstra, J. A. (2008). *In vivo* assessment of safety of microneedle arrays in human skin, *Eur. J. Pharm. Sci.*, **35**(3), 193–202.

93. Gill, H. S., Denson, D. D., Burris, B. A., and Prausnitz, M. R. (2008). Effect of microneedle design on pain in human volunteers, *Clin. J. Pain*, **24**(7), 585–594.

94. Sivamani, R. K., Stoeber, B., Wu, G. C., Zhai, H., Liepmann, D., and Maibach, H. (2005). Clinical microneedle injection of methyl nicotinate: stratum corneum penetration, *Skin Res. Technol.*, **11**(2), 152–156.

95. Nordquist, L., Roxhed, N., Griss, P., and Stemme, G. (2007). Novel microneedle patches for active insulin delivery are efficient in maintaining glycaemic control: an initial comparison with subcutaneous administration, *Pharm. Res.*, **24**(7), 1381–1388.

96. Gardeniers, H. J. G. E., Luttge, R., Berenschot, E. J. W., de Boer, M. J., Yeshurun, S. Y., Hefetz, M., van't Oever, R., van den Berg, A. (2003). Silicon micromachined hollow microneedles for transdermal liquid transport, *J. Microelectromech. Syst.*, **12**(6), 855–862.

97. Parker, E. R., Rao, M. P., Turner, K. L., Meinhart, C. D., and MacDonald, N. C. (2007). Bulk micromachined titanium microneedles, *J. Microelectromech. Syst.*, **16**(2), 289–295.

98. Chandrasekaran, S., and Frazier, A. B. (2003). Characterization of surface micromachined metallic microneedles, *J. Microelectromech. Syst.*, **12**(3), 289–295.

99. Bak, M. J., Girvin, J. P., Hambrecht, F. T., Kufta, C. V., Loeb, G. E., and Schmidt, E. M. (1990). Visual sensations produced by intracortical microstimulation of the human occipital cortex, *Med. Biol. Eng. Comput.*, **28**(3), 257–9.

100. Gelbard-Sagiv, H., Mukamel, R., Harel, M., Malach, R., and Fried, I. (2008). Internally generated reactivation of single neurons in human hippocampus during free recall, *Science*, **322**(5898), 96–101.

101. Normann, R. A. (2007). Technology insight: future neuroprosthetic therapies for disorders of the nervous system, *Nat. Clin. Pract. Neurol.*, **3**(8), 444–452.

102. Moon, T., Ghovanloo, M., and Kipke, D. R. (17–21 September 2003) *Buckling Strength of Coated and Uncoated Silicon Microelectrodes*, paper presented in the Engineering in Medicine and Biology Society, 2003. Proceedings of the 25th Annual International Conference of the IEEE.

103. Najafi, K., and Hetke, J. F. (1990). Strength characterization of silicon microprobes in neurophysiological tissues, *IEEE Trans. Biomed. Eng.*, **37**(5), 474–81.

104. Rousche, P. J., and Normann, R. A. (1992). A method for pneumatically inserting an array of penetrating electrodes into cortical tissue, *Annals Biomed. Eng.*, **20**(4), 413–22.

105. Kipke, D. R., Shain, W., Buzsaki, G., Fetz, E., Henderson, J. M., Hetke, J. F., and Schalk, G. (2008). Advanced neurotechnologies for chronic neural interfaces: new horizons and clinical opportunities, *J. Neurosci.*, **28**(46), 11830–8.

106. Lee, K., He, J., and Wang, L. (2004). Biocompatible benzocyclobutene (BCB)-based neural implants with micro-fluidic channel, *Biosens. Bioelectron.*, **20**(2), 404–7.

107. Schuettler, M., Stiess, S., King, B. V., and Suaning, G. J. (2005). Fabrication of implantable microelectrode arrays by laser cutting of silicone rubber and platinum foil, *J. Neural. Eng.*, **2**(1), S121–8.

108. Stieglitz, T., Beutel, H., Schuettler, M., and Meyer, J.-U. (2000). Micromachined, polyimide-based devices for flexible neural interfaces, *Biomed. Microdevices*, **2**(4), 283–294.

109. Seymour, J. P., and Kipke, D. R. (2007). Neural probe design for reduced tissue encapsulation in CNS, *Biomaterials*, **28**(25), 3594–607.

110. Turner, A. M., Dowell, N., Turner, S. W., Kam, L., Isaacson, M., Turner, J. N., Craighead, H. G., and Shain, W. (2000). Attachment of astroglial cells to microfabricated pillar arrays of different geometries, *J. Biomed. Mater. Res.*, **51**(3), 430–41.

111. Lee, K. K., He, J., Singh, A., Massia, S., Ehteshami, G., Kim, B., and Raupp, G. (2004). Polyimide-based intracortical neural implant with improved structural stiffness, *J. Micromech. Microeng.*, **14**(1), 32–37.

112. Suzuki, T., Mabuchi, K., and Takeuchi, S. (20–22 March 2003) *A 3D Flexible Parylene Probe Array for Multichannel Neural Recording*, paper presented in Neural Engineering, 2003. Conference Proceedings. First International IEEE EMBS Conference.

113. Takeuchi, S., Suzuki, T., Mabuchi, K., and Fujita, H. (2004). 3D flexible multichannel neural probe array, *J. Micromech. Microeng.*, (1), 104.

114. Takeuchi, S., Ziegler, D., Yoshida, Y., Mabuchi, K., and Suzuki, T. (2005). Parylene flexible neural probes integrated with microfluidic channels, *Lab Chip*, **5**(5), 519–23.

115. Motta, P. S., and Judy, J. W. (2005). Multielectrode microprobes for deep-brain stimulation fabricated with a customizable 3-D electroplating process, *IEEE Trans. Biomed. Eng.*, **52**(5), 923–933.

116. McCarthy, P. T., Madangopal, R., Otto, K. J., and Rao, M. P. (2009). Titanium-based multi-channel, micro-electrode array for recording neural signals, *Conf. Proc. IEEE Eng. Med. Biol. Soc.*, 2062–2065.

117. McCarthy, P. T. (2009). Titanium-based multi-channel, micro-electrode arrays for brain machine interface devices, *Mech. Eng.*, Purdue University, West Lafayette.

118. Joner, M., Finn, A. V., Farb, A., Mont, E. K., Kolodgie, F. D., Ladich, E., Kutys, R., Skorija, K., Gold, H. K., and Virmani, R. (2006). Pathology of drug-eluting stents in humans – delayed healing and late thrombotic risk, *J. Am. Coll. Cardiol.*, **48**(1), 193–202.

119. Lu, J., Rao, M. P., MacDonald, N. C., Khang, D., and Webster, T. J. (2008). Improved endothelial cell adhesion and proliferation on patterned titanium surfaces with rationally designed, micrometer to nanometer features, *Acta Biomater.*, **4**(1), 192–201.

120. Rao, M. P., Lu, J., Aguilar, H. P., Khang, D., MacDonald, N. C., and Webster, T. J. (1–5 June 2008) *Rationally-Designed Surface Nanopatterning: a Potentially New Means of Enhancing Safety and Efficacy of Vascular Stents*, paper presented at the Solid-State Sensor, Actuator, and Microsystems Workshop conducted in Hilton Head Island, South Carolina.

121. Miller, D. C., Thapa, A., Haberstroh, K. M., and Webster, T. J. (2004). Endothelial and vascular smooth muscle cell function on poly(lactic-co-glycolic acid) with nano-structured surface features, *Biomaterials*, **25**(1), 53–61.

122. Choudhary, S., Berhe, M., Haberstroh, K. M., and Webster, T. J. (2006). Increased endothelial and vascular smooth muscle cell adhesion on nanostructured titanium and CoCrMo, *Int. J. Nanomed.*, **1**(1), 41–49.

123. Dalby, M. J., Riehle, M. O., Johnstone, H., Affrossman, S., and Curtis, A. S. (2002). In vitro reaction of endothelial cells to polymer demixed nanotopography, *Biomaterials*, **23**(14), 2945–2954.

124. Buttiglieri, S., Pasqui, D., Migliori, M., Johnstone, H., Affrossman, S., Sereni, L., Wratten, M. L., Barbucci, R., Tetta, C., and Camussi, G. (2003). Endothelization and adherence of leucocytes to nanostructured surfaces, *Biomaterials*, **24**(16), 2731–2738.

125. Chung, T. W., Liu, D. Z., Wang, S. Y., and Wang, S. S. (2003). Enhancement of the growth of human endothelial cells by surface roughness at nanometer scale, *Biomaterials*, **24**(25), 4655–4661.

126. Dibra, A., Kastrati, A., Mehilli, J., Pache, J., von Oepen, R., Dirschinger, J., and Schömig, A. (2005). Influence of stent surface topography on the outcomes of patients undergoing coronary stenting: a randomized double-blind controlled trial, *Catheter. Cardiovasc. Inter.*, **65**(3), 374–380.

127. Bettinger, C. J., Orrick, B., Misra, A., Langer, R., and Borenstein, J. T. (2006). Micro fabrication of poly (glycerol-sebacate) for contact guidance applications, *Biomaterials*, **27**(12), 2558–2565.

128. Fuss, C., Sprague, E. A., Bailey, S. R., and Palmaz, J. C. (2001). Surface micro grooves (MG) improve endothelialization rate in vitro and in vivo, *J. Am. Coll. Cardiol.*, **37**(2), 70a–70a.

129. Neuder, K., and Dehm, J. (2004). Technical standards for micro sensors in surgery and minimally invasive therapy, *Minim. Invasive Ther. Allied Technol.*, **13**(2), 110–113.

130. Kalanovic, D., and Schurr, M. O. (2004). Innovation requirements for telemetric sensor systems in medicine: results of a survey in Germany, *Invasive Ther. Allied Technol.*, **13**(2), 68–77.

Chapter 11

Implantable Electrochemical Biosensors: A Perspective

Sridhar Govindarajan and Bella B. Manshian

Centre for NanoHealth, Institute of Life Sciences, School of Medicine, Swansea University, Singleton Park, Swansea, SA2 8PP, UK

sri.govindarajan@gmail.com

Implantable electrochemical sensors offer great potential for improving the quality of healthcare for patients with chronic diseases (such as diabetes) and in hospital critical care. This chapter will deal with biosensors designed for continuous, real-time monitoring of analytes in the human body. Some of the recent developments and progress in the area of implantable electrochemical micro/nano biosensors have been highlighted. The focus of this work is on aspects requiring deliberation during the design of implantable biosensors, such as size, power and performance factors and additional considerations such as sterilization and *in vivo* calibration. Particularly, issues such as selectivity, biofouling, power supply, sterility, and *in vivo* calibration appear to preclude the use of biosensors as a means for routine clinical diagnosis and bedside monitoring. Despite advancements in the fields of micro/nano technology, biomaterials, and bioelectrochemistry and considerable effort to overcome these obstacles, continuous, real-time, reliable, and long-term monitoring of analytes *in vivo* still continues to be a challenge.

Biomedical Nanosensors

Edited by Joseph Irudayaraj

ISBN 978-981-4303-03-3 (Hardcover), 978-981-4303-04-0 (eBook)

www.panstanford.com

11.1 Introduction

To understand the dynamics of the human body, a system that can continuously monitor the changes in the concentrations of key biomarkers is required. This can be achieved by placing a biosensor in the body so that analytes can be measured in real time. This chapter will focus on current research trends and issues related to biosensors designed for the *in vivo* monitoring and real-time detection of clinically significant biomolecules in healthcare. Short- and long-term implantable biosensors are valuable for a variety of clinical applications, such as critical care diagnostics, surgery, and continued routine observation of analytes (such as glucose) in the body. Acquiring real-time data relating to the metabolic and pathophysiological status of a patient is of utmost importance to aid clinicians in diagnosis and take necessary action that could improve the short-term as well as the long-term outcome of patients [1]. For instance, for patients suffering from traumatic brain injury, monitoring glutamate, glucose, and the lactate-pyruvate ratio (L/Py) in the vicinity of the injury provides a superior indication about the neurochemistry and any impending clinically significant occurrences that might contribute to increasing the primary injury, such as subsequent neuronal damage or increased intracranial pressure (ICP).

In vivo biosensing is not without its complications. Electrode-tissue/blood interactions can result in the failure of a sensor within a day after implantation. Impediments associated with subcutaneous or intravascular biosensors involve tissue healing around the implanted device, platelet and cell adhesion caused by blood coagulation, and fibrous encapsulation, which are natural biological responses initiated by the host immune system. Interestingly, the level of coagulation around a biosensor has been employed as a positive technique to diagnose thrombus formation [2]. However, the aforementioned issues, along with stability, sterility, *in vivo* calibration, and toxicity, among others, have deterred advancement in this area, setting up hurdles for *in vivo* biosensors to be adopted in routine clinical practice and be commercially successful. Therefore, despite the substantial growth of the biosensor industry, continuous *in vivo* monitoring of target analytes in the human body is still in its infancy.

Micro and nano biosensors have received attention in recent years due to their potential benefits in providing rapid detection, performance, and high throughput manufacturing, resulting in low production costs when compared to laboratory-based techniques. Recent attempts have been focused on the miniaturization of biosensors assisted by advancements in technology as an obvious benefit for implantation. Instruments that provide high-resolution images, such as atomic force microscopy (AFM), scanning tunneling microscopy (STM), scanning electron microscopy (SEM), scanning electrochemical microscopy (SCEM), and transmission electron microscopy (TEM), are being used to extract valuable information that facilitates innovations in micro/nano fabrication and comprehension of the functioning of the biosensors in the nanoscale [3].

11.1.1 *In vivo* Microdialysis

Microdialysis sampling and analysis is the currently prevalent established technique for monitoring molecules *in vivo*, particularly in the brain [4]. In 1995, CMA Microdialysis

(Stockholm, Sweden) manufactured a sterile microdialysis catheter along with a simple microdialysis pump and a bedside biochemical analyzer for use in clinical settings in intensive care units (ICU) [4]. Nevertheless, this technique is not without its limitations. Some important disadvantages of the microdialysis technique are (a) low time resolution (more than 20 minutes), (b) provision of average analyte concentration over the sampling period, (c) a highly expensive setup, (d) need of sensitive analysis, (e) requirement of skilled personnel for operation, and (f) most importantly, a stationary subject.

11.2 *In vivo* Monitoring — Challenges and Progress

11.2.1 Electrochemical Biosensors

Biosensors offer a potential solution and a superior alternative to the microdialysis sampling method mentioned earlier. Biosensors can be defined as devices that incorporate a molecular recognition element in close proximity with a transducer in order to detect biologically significant molecules and metabolites. The molecular recognition element can differ depending on the application and the target substrate to be measured. Please refer to chapter 1 and the literature on biochemistry [5–7] for further information on proteins and biological recognition elements and their mechanisms of action. Regardless of the type of molecular recognition element used, electrochemical sensors produce an electrical signal that is proportional to the concentration of the target species.

Due to the simplicity of producing an electrochemical biosensor and its ease of operation in the continuous mode, it has been widely considered as the most suitable type of transduction for producing *in vivo* biosensors [8–10]. Electrochemical biosensors can be classified into three types based on the mechanisms that govern their operation; they are: potentiometric, conductometric, and voltammetric. From the vast quantity of research papers published in the last 15 years, it appears that voltammetric electrochemical biosensors are preferred for the *in vivo* monitoring of biomolecules. Voltammetric techniques involve the application of an electrical potential step or gradient to an electrode (working electrode, such as gold, platinum, glassy carbon, or graphite) relative to a reference electrode (e.g., saturated calomel electrode or Ag/AgCl) across an electrode-solution interface to oxidize or reduce electroactive species present in the solution. In the case of chronoamperometry, a constant potential step is applied to the sensing electrode for the entire duration of the experiment and the resulting current is detected. The rate of the redox reaction determines the current obtained, which depends on a number of factors, including the form of the applied potential, the size of the electrodes, and their geometry. Any change in current is directly attributed to the reaction occurring at the electrode surface due to the oxidation or reduction of the electroactive molecules. For more information on the different voltammetric techniques and the shapes of the applied potentials, readers are referred to the reviews by O'Neill *et al.* [11].

$$\text{Substrate} + \text{Enzyme (coenzyme}_{\text{ox}}) \rightarrow \text{Products} + \text{Enzyme (coenzyme}_{\text{red}}) \qquad (11.1)$$
$$\text{Enzyme (coenzyme}_{\text{red}}) + O_2 \rightarrow \text{Enzyme (coenzyme}_{\text{ox}}) + H_2O_2 \qquad (11.2)$$

A majority of the biomarkers in the body are not electroactive (the commonest example being glucose). Enzymes incorporated in biosensors can allow the binding of the electroinactive species of interest and catalyze their oxidation (consuming oxygen) to produce a reaction product and hydrogen peroxide (H_2O_2), as shown in Eq. (11.1) and (11.2). Thus, based on the method of detection of electroinactive biomolecules, enzyme-based electrochemical biosensors have been categorized into first-generation, second-generation, and third-generation biosensors [6, 11]. Primitive first-generation biosensors based on H_2O_2 or O_2 detection were easily influenced by other electroactive species surrounding the electrode. For further reading on first-, second-, and third-generation biosensors, please refer to publications by O'Neill *et al.* and Habermüller *et al.* [10, 12]. Although second-generation biosensors benefit from lowered operating potentials and eliminate the need for oxygen in the surrounding media, they are toxic and prone to leaching. In spite of their drawbacks, first-generation biosensors are the most widely used as they are robust and can be easily adapted for *in vivo* use when compared to the second-generation biosensors. Recent research has suggested that in third-generation biosensors, direct electron transfer between certain enzymes and conducting electroactive polymers, nanoparticles, and carbon nanotubes (CNTs) can occur, without the need for detecting the oxygen consumption or H_2O_2 [1, 13], thereby producing direct biosensors requiring lowered biasing potentials.

11.2.2 Design Aspects for *in vivo* Monitoring

For a biosensor to be considered clinically useful for continuous *in vivo* monitoring, it must satisfy certain design and functional requirements. These requirements emerge as a direct and indirect consequence of the challenges presented by the complex *in vivo* environment to which they will be subjected, sometimes resulting in their failure. Wisniewski and Reichert provided a schematic representation detailing the various reasons for biosensor malfunction post-implantation, as shown in Fig. 11.1 [14, 15].

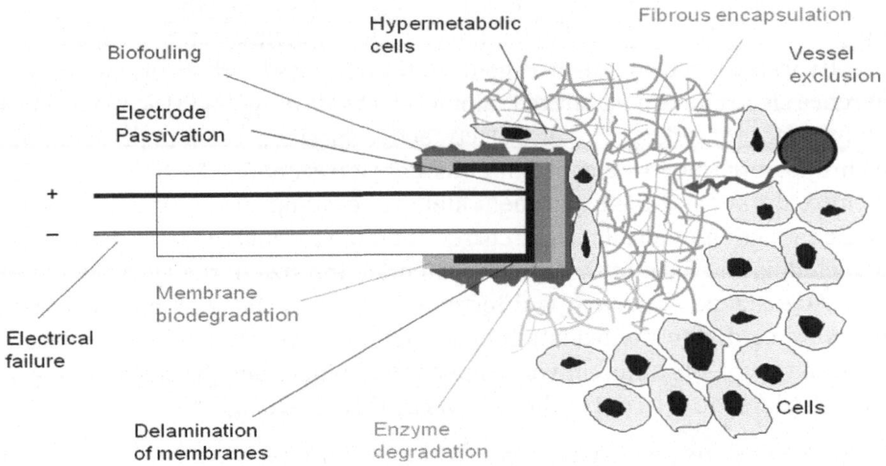

Figure 11.1 Insertion of biosensors within the subcutaneous tissue can result in sensor failure due to a multitude of reasons, as presented by Wisniewski and Reichert [15]. (Adapted and reproduced from [15] with author's permission.)

To design biosensors for implantable applications, factors that can cause sensor failure need to be addressed systematically and exhaustively to obtain accurate measurements. To detect molecules in the body on a real-time basis, considerations such as sensitivity, response times, selectivity for the target species, resistance against host response, toxicity, and long-term stability need to be deliberated. In addition, the implantation procedure and the implanted device should cause minimum discomfort to the patient during and after implantation [16]. The following aspects of biosensor design, evaluation, and fabrication will be discussed in the forthcoming sections:

- Physical aspects — (i) the size and shape of the biosensor, (ii) telemetric monitoring, and (iii) powering implantable biosensors
- Performance factors — (i) the sensitivity and response time, (ii) selectivity, and (iii) biocompatibility
- Additional considerations — (i) sterilization, (ii) *in vivo* calibration, and (iii) toxicity

11.2.2.1 Physical aspects

11.2.2.1.1 *Biosensor size and shape*

Biosensor size and shape are key factors to be considered during the design of *in vivo* biosensors to minimize patient discomfort upon implantation. Microneedles and wire-type biosensors with an outer diameter smaller than 450 μm have been stated to be ideal for glucose monitoring *in vivo* to improve acceptability [17, 18]. There is a widespread perception that smaller sensors are more sensitive, which has encouraged the scaling of biosensors to micro- and nanometer dimensions [19]. This, in turn, reduces invasiveness, tissue damage, and the development of thrombi around the sensor *in vivo.* Miniaturization of implantable biosensors can be grouped under two categories — (i) scaling down of the sensing elements and transducer and (ii) scaling down of the additional electronics necessary for power and data transmission and packaging. Miniaturized needle-type biosensors made of ultra-thin 50 μm [19] platinum alloy wires have been developed with immobilized enzymes for *in vivo* monitoring, where the applied potential and data recording were provided by an external device. Power sources and data transmission circuitry used in conjunction with implantable sensors are briefly discussed in the following sections.

Due to their nanoscale dimensions, nanoparticles and nanotubes are commonly immobilized or contained in the vicinity of the transducing electrode to improve electrical communication between the enzyme and the transducer and obtain increased sensitivity. Currently, for medical-imaging applications, nanoparticles are directly injected *in vivo* as sensing elements [20] to reach and detect intracellular molecules by penetrating through the cell membrane [21], resulting in a higher contrast. Nevertheless, the use of nanoparticles *in vivo,* either as injected sensing elements or immobilized on a transducer, is accompanied by *in vivo* toxicity concerns [20, 22], which warrant further investigation.

11.2.2.1.2 *Telemetric monitoring*

Telemetry is a system designed for transmitting information or data to a remote receiving station. Implantable biosensors can be broadly grouped into two subcategories based on the method of data transmission — (a) wired implantable biosensors and (b) implantable biosensors incorporating a wireless telemetric system. Wired implantable biosensors could be considered in situations, such as surgery, intensive care, or conditions, requiring the patient to be hospitalized, wherein the implanted sensor could be connected to a bedside measurement system for relatively short-term monitoring. In this case, the signals are monitored on a continuous basis without any time lag. In cases where the biosensor needs to be implanted in a patient continuing his or her daily routine, especially in cases such as diabetes monitoring, combining a long-term implantable biosensor with a wireless telemetry system transmitting data to a portable and wearable measuring and storing device would be preferred. The negative effect of introducing the wireless transmission system would be the increase in the time lag between the instance at which the concentration is sensed by the biosensor and the time at which the device translates the measured signal into a concentration value. If the measuring device is linked to a drug-infusion system, which would then release a small amount of drug into the body depending on the measured level of the analyte, then the time lag (of about 20 minutes in some reported cases [23]) becomes critical. Therefore, fast wireless transmitters and associated signal-processing circuitries are essential to minimize this time lag.

Wireless telemetric systems for implantable devices can be divided into (a) surface-mounted telemetry connected to an implanted sensor and (b) implantable wireless telemetry. In the surface-mounted system, the subcutaneously or intravenously implanted sensor is connected to a surface-mounted device integrating a potentiostat and a transmitter that remotely transmits the data to a portable receiver. In implantable wireless telemetry, the transmitter is also implanted within the body, connecting wirelessly to a portable receiver. Surface-mounted telemetry for short-term and implanted telemetry for long-term implantable applications have been preferred [24]. Benini *et al.* have reviewed this area describing four generations of wireless sensor networks (obtrusive, parasitic, symbiotic, and bio-inspired), including their hardware architectures and their fabrication technologies [25].

Regardless of whether the telemetric system is surface mounted or implanted, ideally, it consists of the following parts in contact with the sensor: (a) a miniature potentiostat or signal acquisition system, (b) modulation/transmitter circuit and antenna (such as radio frequency [RF] antenna [26, 27], infrared (IR) [28], Bluetooth [24], or wireless data cards [29, 30]), (c) power supply, and (d) device packaging [26]. The implanted system, which consists of a transmitter, can be powered by various means: (a) a battery [30], (b) inductive coupling [31], or (c) electromagnetic coupling [32]. Inductive or electromagnetic coupling removes the need for an implantable battery to power the sensor and the associated telemetric circuitry [32].

Rocchitta *et al.* recently developed a low-cost telemetric system based on FM transmission [27]. Using a 200 m range antenna and a receiver connected to a computer through a universal serial bus (USB) port, they demonstrated the functioning of the telemetric system *in vitro* with three different sensors (nitric oxide [NO], dopamine, and glucose). The system exhibited high stability and excellent linear response in the current ranges obtained during physiological monitoring [27]. Crespi *et al.* developed a small and lightweight one-way IR transmission system for *in vivo* sensors based on differential pulse voltammetry (DPV) and direct current amperometry (DCA) [28]. They described a novel wireless *in vivo* neurochemical monitoring of dopamine and serotonin in freely moving rodents [28]. The telemetry system was not prone to interference from electromagnetic waves, which are quite common in systems employing RF communication [28]. Another group developed a slightly more bulky and complex 16-channel surface telemetric system for use with freely moving subjects using a wireless Ethernet card using the IEEE 802.11b protocol [29]. A computer connected to a local wireless network acted as the destination for the transmitted signals, where the data was displayed and stored. The device weighing 235 g was capable of transmitting data continuously for 45 minutes, being powered by rechargeable lithium ion batteries [29]. Using a similar wireless protocol (IEEE 802.15.4-2003), Valdastri *et al.* demonstrated the telemetric monitoring of blood glucose using an optical system, in conjunction with an ultralow power bidirectional telemetric system with a lithium ion battery offering a lifetime of six months [30].

Inductive coupling has also been used to power and wirelessly transmit the data from the sensor to a receiver unit. Jacquemod *et al.*, in a to-be-published manuscript, employed a self-oscillating inductor-capacitor (L-C) tank circuit for battery-less monitoring using passive sensors [31]. Here, the capacitive sensor itself acts as the variable capacitance of the L-C circuit, thereby altering the resonance frequency based on the concentration of the detected analyte [31]. By adding a simple resistance-capacitance (R-C) circuit, they realized antiresonance cancellation, thereby increasing the distance between the sensor and the receiver to 5 cm [31]. Therefore, inductive coupling removes the need for a battery and transmitter module on the implanted system, thereby permitting the design of very small implantable sensors. However, the receiver end of the circuit that provides the necessary biasing and coupling cannot be too far away from the implanted device, typically a few centimeters, thereby restricting the mobility of the subjects. Consequently, researchers are trying to extend the function of implantable batteries used in conjunction with ultralow power electronics by designing circuits that can completely cut off the power when the sensors are not in use, especially in long-term monitoring [33]. Although completely implantable systems with built-in RF transmission circuits encapsulated in silastic or epoxy resin providing a durable and biocompatible covering for extended function have been developed [26], it is important to produce fast communication as some have reported delays of up to 20 minutes between blood glucose measurements and sensor values transmitted to a feedback-controlled insulin pumping unit [23].

In conclusion, implanted wireless telemetric monitoring could provide greater freedom to the patient and allow constant monitoring and connection to a drug delivery system administering through a feedback system. For short-term monitoring using surface-mounted telemetry, smaller and lighter units are necessary to reduce patient discomfort.

11.2.2.1.3 *Powering implantable micro/nano biosensors*

Wireless implantable sensing devices require careful consideration of power, stability, and reliability. Complete or part integration of miniaturized electronics and power supply are critical for making *in vivo* sensors, incorporating data acquisition and wireless telemetry. A major hurdle currently limiting the advancement of this technology is battery size and lifetime. Low-power systems that would preferably be self-powered using the energy extracted from the environment, micro fuel cells such as CNT fuel cells, or micro biofuel cells or powered and interrogated by handheld radio transceivers are essential [34].

Minute power sources such as biofuel cells or batteries based on implantable 7 μm carbon fibers were developed for powering autonomous sensor-transmitter systems [35]. For this purpose, tiny, membraneless glucose-oxygen CNT biofuel cells coated with glucose oxidase and laccase or bilirubin oxidase were wired with osmium-based redox hydrogels [36]. Wang and colleagues designed CNT-based biofuel cells that were membraneless in addition to being free from mediators [37]. They used electrochemical redox reactions, where gold-coated pSi/SWCNT electrodes were employed for electron transfer [37]. The CNT-based devices were direct electron transfer enzymatic biofuel cells with bioelectrodes comprised of SWCNTs deposited on porous silicon (pSi) substrates [37]. Reshetilov *et al.* have reported using bacterial cells (for improving electron transport) for the production of mediator-free biofuel cells [38]. With the growth of research in biofuel production, it has become clear that obtaining a mediatorless biofuel cell is desirable in order to overcome toxic issues and problems relating to overvoltage associated with mediated fuel cells [39].

A comprehensive list of power sources based on scavenging being proposed and developed for biosensor applications has been reviewed by Romero *et al.* [40]. A direct current (DC) piezoelectric nanogenerator was designed by Wang and colleagues [41]. The unique piezoelectric and semiconducting abilities of zinc oxide (ZnO) allowed the fabrication of vertically aligned nanowire arrays, which relied on a zigzag top electrode to generate electricity, as shown in Fig. 11.2 [41, 42].

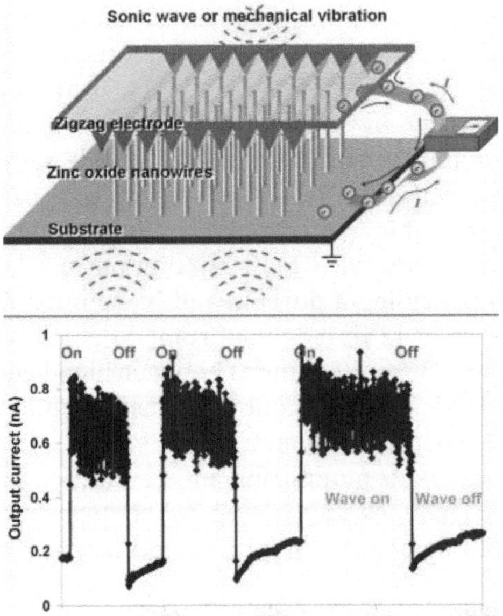

Figure 11.2 A schematic diagram showing the ZnO nanowire current generator. The lower plot shows the current obtained from a nanogenerator when the ultrasonic wave was switched on and off [41, 42]. (Courtesy of Zhong Lin Wang, Georgia Tech, United States.)

In this nanogenerator, ultrasonic waves or external mechanical agitation was employed to force the nanowires to bend, creating a piezoelectric potential along the wires, thus generating an electric current [41]. However, continuous scrubbing and sliding between the top electrode and the ZnO nanowires resulted in the wearing of the movable parts, causing increased contact resistance and instability. In an attempt to improve the stability of the nanogenerators, vertically or laterally aligned ZnO nanowire arrays were embedded into a polymer substrate instead of metallic electrodes, making these devices more robust, through which the researchers achieved a high-power-output alternating current (AC) [43]. Subsequently, the nanogenerators can be packed in a flexible enclosure for incorporation with implantable sensors [43]. These devices based on nanowires and nanobelts seem to be promising for developing and powering implantable sensor capsules for short- and long-term applications.

Many limitations still restrict energy scavenging as a practical method for generating an efficient power source for micro/nano devices. Some of these limitations include energy generation and energy density, ability to store and manage the generated energy, DC rectification, manufacturing, and finally durability of these devices [44]. Hence, in addition to the use of coin batteries (such as Energiser, Renata, and Duracell), researchers are working toward the development of smaller power sources using improved methods to obtain reliable and long-lasting batteries for future wireless implantable sensors.

11.2.2.2 Performance factors

11.2.2.2.1 *Sensitivity and response time*

Sensitivity can be defined as the minimum input of the analyte that will create a detectable output signal [45]. Mathematically, the sensitivity of a biosensor is calculated as the slope of the output characteristic curve in the linear region. Sensitivity is a key aspect of biosensor design requiring great care as decreased sensitivity has been reported post-implantation primarily due to host immune response, cell adhesion, biofouling, and the formation of a fibrotic capsule around the biosensor [14, 46, 47]. Therefore, the sensitivity and the lower detection limit of the biosensor should be sufficient to detect the physiological and pathological concentrations of the analytes of interest, particularly when measuring concentrations of a few micromolar or nanomolar such as glutamate in the brain or certain hormones in the blood. Mathematically, response time is the time taken for a biosensor output to change from its previous value to a value that lies within a tolerance band of the new steady-state value upon change in the analyte concentration. This tolerance band can be from 5% to 30% of its steady-state value at a particular concentration. Adding layers such as permselective coatings and biocompatible membranes on the transducer can restrict mass transport to the electrode, increasing the response time of the sensor, which could be over five minutes. However, a response time of 85 ms has been reported for an implantable glucose sensor of micrometer dimension [18]. Therefore, the lower the response time, the closer it is to real-time monitoring, which is the motivation for implantable sensors.

The sensitivity of a first-generation biosensor in its basic form can be relatively low unless additional coatings and elements are incorporated to increase the efficiency of electron or mass transport to the transducer. In this regard, second-generation biosensors incorporating electron transfer mediators in the sensor matrix improve the sensitivity by efficient transport of electrons to and from the electrode to the redox center of the enzyme. Since many of these mediators are toxic, O'Neill's group introduced a polycationic polymer to improve the sensitivity and decrease the oxygen dependence of glutamate biosensors for *in vivo* brain monitoring [48]. The group found that the polycationic polymer decreased the electrostatic repulsive effects between analyte molecules and the enzyme, resulting in increased analyte availability at the biosensor surface [48]. A side effect was the greater affinity of the sensor for negatively charged ascorbic acid, thereby augmenting the interference signal and decreasing its selectivity [49].

Inclusion of nanoparticles into biosensor design can improve their sensitivity as substances in nanoscale have been shown to exhibit unusual physiochemical phenomena such as enhanced reactivity, catalytic activity, and faster electron/ion transport [19]. Fluorine-doped carbon nanotubes (F-CNTs) incorporated in a biosensor was developed by Ly *et al.* [50] for monitoring glucose from human urine. This biosensor exhibited a higher sensitivity and a very low detection limit of 3.3 nM. Deng *et al.* demonstrated that layer-by-layer assembly of boron-doped carbon nanotubes (B-CNTs) and glucose oxidase on a glassy carbon electrode provides a third-generation biosensor with a good sensitivity of 111.57 μA mM^{-1} cm^{-2} to glucose [51]. In another work, CNTs mixed with chitosan and glucose oxidase was applied to glassy carbon electrode, producing a mediator-free and membrane-free glucose biosensor with very fast electron transfer, high sensitivity of 0.52 μA mM^{-1}, and stability [52]. Tian *et al.* electrodeposited gold nanoparticles (GNP) on 10 μm carbon fiber microelectrodes (CFME) to improve the electron transfer from superoxide dismutase (SOD) to the electrode producing a third-generation biosensor for superoxide [13]. With this procedure, they found an increased sensitivity to superoxide of 30 ± 2 and 25 ± 2 nA $cm^{-2}nM^{-1}min^{-1}$ at +250 and −150 mV with respect to the reference electrode, respectively, with a response time of less than five seconds [13]. In another work, the performance of glucose biosensors was found to be enhanced by the addition of platinum nanoparticles in a polycationic polymer to glucose oxidase immobilized on a chitosan matrix [53]. In addition, biosensors with enzymes immobilized on self-assembled Prussian blue coated on platinum foil electrodes and cross-linked with glutaraldehyde have been developed, providing a high sensitivity of 80 mA M^{-1} cm^{-2} to glucose [54].

Encapsulating the biorecognition element in a matrix or covalent immobilization onto the electrode surface provides a higher degree of organization, structure, and compactness to the sensor; this enhances the adhesion of the sensing layer to the transducer, resulting in improved sensitivity and long-term stability of the biosensors [55]. A novel glucose biosensor was designed by the electrochemical entrapment of glucose oxidase into porous poly(acrylonitrile-*co*-acrylic acid) on platinum electrodes, which was synthesized via the radical polymerization of acrylonitrile and acrylic acid, providing a high sensitivity of 6.82 mA $M^{-1}cm^{-2}$ for glucose with good temporal stability [56]. More recently, research into electroconductive hydrogels has shown that they can offer a receptive setting for the physical or covalent immobilization and stabilization of peptides and enzymes [1]. Along with providing biocompatibility to the sensors, the hydrogels assist mediated electron transfer between the active sites of oxidoreductase enzymes and metalloproteins with the conductive electroactive polymer and the electrode [1], eliminating the need for high biasing potentials or toxic electron transfer mediators (second-generation biosensors). Therefore, great progress has been made in enhancing the sensitivity and reducing the response time of biosensors geared toward *in vivo* monitoring.

11.2.2.2.2 *Selectivity*

The selectivity of a biosensor is crucial when monitoring an analyte from biological samples and *in vivo*. In this context, selectivity is the ability of the biosensor to discriminate between different substances [7] and sense the analyte when present among other molecules. Biorecognition elements such as enzymes, antibodies, or single-stranded DNA provide a high level of selectivity to the biosensor as they only bind to substrates with corresponding binding sites. However, with enzyme-based electrochemical biosensors, the application of a voltage to the electrodes introduces interference from surrounding electroactive species in the biological sample or *in vivo*. For instance, while monitoring molecules such as glutamate or glucose in neuroscience, electroactive species such as ascorbic acid, dopamine, or superoxide present in the extracellular space could get oxidised or reduced at the electrode surface, subsequently lowering the ability of the biosensor to detect the analyte of interest. Therefore, second- and third-generation biosensors, with their decreased electrode potential requirements, provide improved selectivity.

Along with the selectivity provided by the biorecognition element, additional permselective barriers, which block undesired electroactive species from reaching the electrode surface, are necessary to obtain outputs accurately representing the concentration of the analyte. Electropolymerized insulating or conducting polymer films such as poly(*o*-phenylenediamine) (PPD), poly(pyrrole), naphthalene derivates, and poly(aniline) or cast membranes from a liquid phase such as cellulose derivatives, nafion, polyurethanes, poly(vinyl chloride), and others have been used to coat the surface of the sensor [57–65]. An example of a biosensor incorporating an electropolymerized PPD developed by O'Neill's group [58] can be seen in Fig. 11.3. They also showed that incorporating macromolecules and lipids such as phosphatidylethanolamine (PEA) within the sensor structure improves the selectivity of the biosensors. In many cases, besides providing a selective membrane, the polymer films are also used to entrap the enzyme on the electrode. It is important to note that the permselective layer should not hinder the diffusion and binding of the analyte to the sensing element or deactivate the binding sites [58].

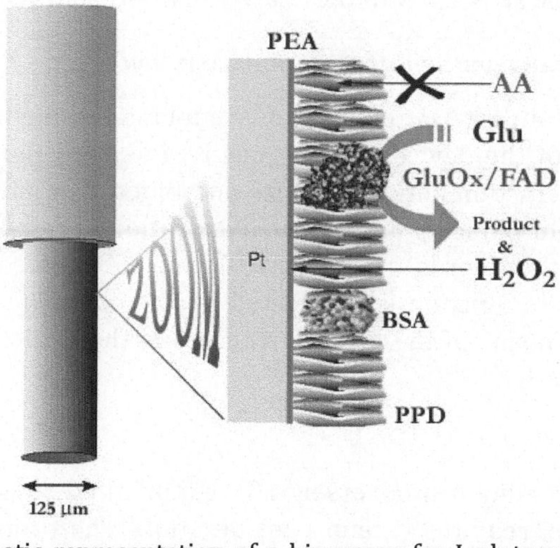

Figure 11.3 A schematic representation of a biosensor for L-glutamate developed by O'Neill's group incorporating an electropolymerized PPD and a lipid membrane (PEA) acting as interference-rejecting barriers on a platinum electrode. *Abbreviations:* AA, ascorbic acid; BSA, bovine serum albumin; FAD, flavine adenine dineucleotide; Glu, L-glutamate; GluOx, glutamate oxidase [58]. (Courtesy of Robert D. O'Neill, UCD, Dublin, Ireland.)

Recent studies have shown that the efficiency of electropolymerized films in eliminating interfering species depends on the electrode dimensions [66]. McMahon *et al.* showed that electrosynthesis of PPD on cylindrical electrodes gave an ordered structure, reducing the permeability to ascorbic acid, when compared with the synthesis on disc electrodes of the same diameter [66]. Subsequently, it was found that by decreasing the electropolymerization potential, more compact and efficient PPD films could be produced, resulting in a biosensor with higher selectivity [67].

Furthermore, in addition to decreasing the interference from electroactive species through the polymerization or membrane coatings, differential readings between a blank electrode (which contains all the elements of the biosensor except the biological sensing element) and the biosensor monitoring the analyte have been considered [63, 68]. This method theoretically improves the selectivity of the biosensor by completely eliminating the electroactive interfering species and the effect of physical parameters such as temperature and oxygen dependence, but it does have its drawbacks and complications. Despite being used since the 1980s, this technique is not always accurate due to the following reasons:

(a) Even a small variation in electrode sizes can change the observed current considerably, thereby requiring electrodes of exactly the same dimensions.
(b) The structure of the polymer film coated on the surface of the electrode may vary with the inclusion of the biological sensing element.
(c) The interaction of the electroactive species may vary between the two electrodes (which might depend on the additional electrostatic effects introduced by the biorecognition element).

Despite reports of biosensors having been fabricated with selectivities in excess of 80% in the laboratory in recent years [49, 58, 63], it is a tremendous challenge to achieve this while maintaining the sensitivity to the analyte of interest *in vivo*.

11.2.2.2.3 *Biocompatibility and sensor performance* in vivo

The biocompatibility of an arterial or a venous implantable medical device has been defined as "the ability of the device to carry out its intended function within flowing blood, with minimal interaction between device and blood that adversely affects device performance, and without inducing uncontrolled activation of cellular or plasma protein cascades" [69, 70]. In other words, an intravascular or subcutaneous implanted device is biocompatible when its function is unaffected by the components in the blood and intrinsic defense mechanisms of the host. Furthermore, the device must not introduce any health risks in the patient.

Host response and biofouling

Host response and biofouling can adversely affect implanted sensor performance and longevity and, therefore, requires careful consideration. The properties of the device, such as composition, morphology, porosity, roughness, shape, size, sterility, and surface chemistry, will affect the foreign body response [71]. In addition, it depends on the location and placement of the device in the body, whether the device has been placed intravascularly or subcutaneously. If the device is placed intravascularly, the activation of coagulation cascades occurs through protein and platelet adhesion due to continuous contact with blood, resulting in the release of additional chemicals and fibrinogen, accelerating the clotting process. Apart from potentially causing a risk to the patient in the form of an embolism through which the implanted sensor is encapsulated and transported to a different part of the body where it could clog the blood vessels, diagnostically, the sensor might not reflect the concentration of analytes in the blood due to cells covering it. In addition, the implantation of the sensor too close to the blood vessel wall has been suggested as a cause of inaccurate measurements [72].

If the sensors are implanted subcutaneously, such as needle-type glucose sensors, then the biosensor is identified as a foreign object and the immune system is activated to destroy it. The host response is initially inflammatory in nature, releasing fluids and proteins followed by the attachment of leukocytes on the surface of the sensor in order to engulf and eliminate the foreign body [47, 72]. Over the longer term, macrophages, monocytes, and lymphocytes are transported to the implant site, resulting in the formation of a foreign body capsule embedded in the tissue by a fibrotic layer that isolates the biosensor from its environment (see Fig. 11.4) [72].

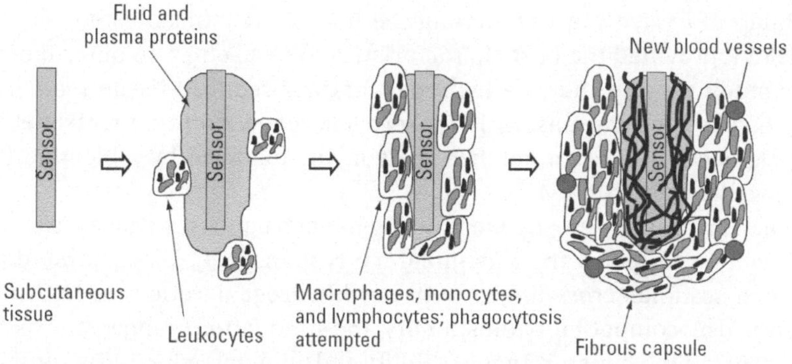

Figure 11.4 Frost and Meyerhoff provided a sequence of processes that occur following the implantation of chemical sensors, with the formation of a fibrous capsule with prolonged *in vivo* use [72]. (Reprinted with permission from ref. [72].)

The physicochemical characteristics of the biosensor play an important role in the timing, action, and intensity of these processes [73, 74]. It is reported that after about three days, the initial foreign body response either gives way to a chronic inflammatory response or initiates the development of the healing process [73]. The inflammatory phase can continue for the entire length of the implantation. Protein and cell adhesion and the fibrous capsule covering the device to a thickness of 10–100 μm substantially restrict molecular transport to and from the electrode [19]. Thus, the concentration of the analyte detected is far lower than the true concentration in blood or tissue. Nonspecific adsorption of proteins, foreign body response, and the formation of the fibrotic capsule lead to biofouling on the surface of the sensor, affecting its performance, stability, and lifetime after implantation [14, 46].

Improving biocompatibility and reducing biofouling to enhance biosensor performance

Wilson and Amman, in their review on *in vivo* biosensors, mention that it is common to obtain a 50% decrease in the sensitivity of a biosensor shortly following implantation

[9]. Therefore, the first objective would be to minimize the protein adsorption and host response on the surface. The design of a biocompatible coating depends on the intended time period of the implantation of the device. It has been suggested that the sensor longevity expected for implantable sensors in the intensive care setting could be short term, for up to a week, whereas subcutaneous sensors should retain stability and performance for weeks or months [72]. For over 25 years, membranes made of polyethylene, polyvinylchloride, polytetrafluoroethylene (PTFE), silicone elastomers, and polyurethanes [69] have been suggested to improve the biocompatibility of implanted devices. Surfactants, hydrogels, and heparin treatments have also been mentioned. Factors such as hydrophilicity, surface charge, and topography contribute to the biocompatible characteristics of the device [9]. Polyethylene glycol (PEG) and polyvinyl alcohol (PVA) hydrogel coatings have been demonstrated to be promising in reducing protein adhesion, improving the biocompatibility of implanted sensors [75].

Hydrogels based on copolymers of hydroxyethyl methacrylate (HEMA) and 2,3-dihydroxypropyl methacrylate (DHPMA), improved using *N*-vinyl-2-pyrrolidinone (VP) and ethyleneglycol dimethacrylate (EGDMA) [76], have been used to enhance the biocompatibility of *in vivo* sensors. Among the different coatings tested, Yu *et al.* found that 30 mol% VP provided the best characteristics to be used as an outer biocompatible coating. The study showed that the hydrogel coatings reduced tissue reactions around implantable needle-type biosensors [77]. Xerogels made from methyltrimethoxysilane (MTMOS) have been used to immobilize the enzyme followed by a blended PU coating [78] to improve biocompatibility.

In addition, encouraging angiogenesis using nanostructured surfaces from the sensor has been shown to suppress the inflammatory response [9]. Using three-dimensional porous collagen scaffolds cross-linked with nordihydroguaiaretic acid (NDGA), Ju *et al.* showed improved biocompatibility, biostability, and stimulation of angiogenesis around the implanted sensors [79]. They found good stability and minimal reduction in sensor function, without any degradation of the physical structure, following subcutaneous implantation in rats after four weeks [79]. Microspheres prepared using natural or synthetic polymers such as polylactic-glycolic acid (PLGA) have been proposed (either around the sensor or incorporated in the hydrogels) to deliver drugs (such as dexamethasone) and proteins to improve the biocompatibility of implanted sensors [14, 79]. Natural biomaterials such as alginate, chitosan, collagen, dextran, and hyaluronan have been used as sensor scaffolds to reduce the host response.

The administration of low-dose aspirin to patients has been shown to impede thrombus formation at implanted glucose sensors for up to two years [23]. However, systemic presence or infusion of an anticoagulant such as heparin has been revealed to be inefficient in preventing platelet adhesion and activation on implanted devices, but instead increases the risk of bleeding [72]. Recently, systemic heparin has been translated to heparin-coated catheter materials, such as polyurethanes (PU) or polyvinylchloride (PVC) for short-term monitoring [69, 80]. Immobilized active heparin on the device surface acts as a local anticoagulant, inhibiting thrombin formation, but does not affect the metabolically active cells adhered to the surface [72]. In this regard, NO has been demonstrated to be potent in blocking platelet and cell adhesion on the electrode surface, which has resulted in the design of novel materials such as NO-releasing polymers and silicone rubbers [81, 82]. Further research conducted in this area resulted in the design of lipophilic *N*-diazeniumdiolate (NO donor) incorporated PU, PVC [72, 83], and xerogels [78]. Figure 11.5(a) shows the design of an implantable sensor with an NO-releasing material, and Fig. 11.5(b) shows the difference between implanted sensors with and without the NO-releasing coating, demonstrating its importance in reducing biofouling [83]. Zhou and Meyerhoff succeeded in combining the antithrombogenic properties of heparin and antiplatelet adhesion and activation abilities of NO-releasing polymers by immobilizing both at the device/blood (device/tissue) interface to closely mimic the function of an endothelial cell [80]. Recently, copper micro/nanoparticles have also been incorporated in the outer polymeric coatings of the sensors to catalytically convert *S*-nitrosothiols in blood to NO [84], thereby fabricating NO-generating implantable sensors. For further information on NO-releasing/generating coatings, please refer to the review by Wu and Meyerhoff [85].

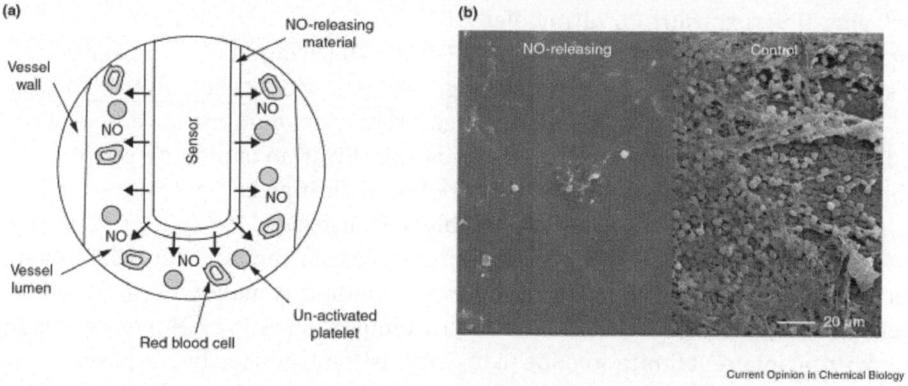

Figure 11.5 The figure shows some preliminary results of an NO-releasing intravascular sensor. (a) A prototype of an NO-releasing intravascular sensor that prevents the formation of thrombus on the surface and (b) scanning electron micrograph images of the surfaces of intravascular oxygen sensors in carotid arteries of porcine animals after 20 hours with and without NO-releasing material and without any systemic anticoagulants [83]. (Reprinted with permission from ref. [83].)

The biocompatibililty criteria for glucose sensors has been recently reviewed by Koschwanez and Reichert [86]. Sensors designed for long-term implantation are reported to be functional for up to 14 months [23]. Although many different methods have been investigated, a single method might be insufficient to obtain biomimetic interfaces that could completely eliminate the effect of cell adhesion and the formation of fibrous capsules. As explained earlier, research groups are utilizing a combination of methods to improve biocompatibility and reduce biofouling.

11.2.2.3 Additional considerations

11.2.2.3.1 *Sterilization*

Current guidelines require that "all devices that are intended to penetrate the human body and come into direct contact with the patient's first immune defense system and tissues have to be sterilized in order to prevent infections caused by pathogenic microorganisms" [87]. Bacterial contamination is suggested as the primary cause for implant-related infections after surgery [87], and it is crucial to sterilize any device intended for implantable use. An efficient sterilization method is one that has a low cost and can effectively sterilize the device without causing damage to the biosensor and leaving behind residual organic or toxic sterilization by-products that can harm the host. Determining the method to be utilized for the sterilization of biosensors should take into account two main considerations. Firstly, the device must satisfy the Pharmacopoeias' sterility assurance level (SAL) of 10^{-6} [86], which is a six-log reduction in microorganisms following sterilization. Secondly, the function of the biosensor should remain intact after sterilization [86, 88]. Therefore, considering the damaging effects of sterilization on the physical, mechanical, and biological properties of biosensors, scientists have attempted various methods to meet the sterility standards. Currently, sterilization choices include exposure of the biosensor to ethylene oxide (EtO) gas [89], immersion in alcohol solutions, autoclaving [89], exposure to ultraviolet (UV) irradiation, and electron beam (e-beam) and gamma irradiation (γ-irradiation) [89]. For the sterilization of medical devices, moist heat sterilization, also known as autoclaving, has been widely used. Autoclaving involves exposing the device to 121–132°C under high pressure conditions for a period of 20–45 min. This method is a simple and fast method of sterilization of medical devices, although the biorecognition element can be denatured at such high temperatures [90], thereby excluding it for the sterilization of implantable biosensors.

EtO gas sterilization is a very common sterilization method generally used in the medical device industry. The preference for this method is mainly because of its simple and effective handling and sterilization at low temperatures [91]. However, this method comes with an inherent limitation due to toxicity risks introduced by the gaseous residues, shown to induce carcinogenesis [88, 92]. Furthermore, a study by Voskerician has even disputed its efficiency for obtaining the SAL value [93]. In addition to toxic effects due to EtO sterilization, inconsistent results have been published [94], with some reporting total loss of sensor function [86, 92].

On the other hand, γ-irradiation sterilization involves exposure of the sensor or device to gamma rays (γ-rays) from a cobalt-60 or ceasium-137 source. Sterilization is complete when the radiation dosage reaches a given threshold. More recently, e-beam has been employed as a high-speed sterilization method [95]. E-beam sterilization has been reported to be slightly less penetrative and less destructive to biosensors [88, 96]. Besides being cold sterilization methods, γ-rays and e-beams can sterilize the product in its packaged state due to their penetrating abilities; albeit, as a consequence, they could potentially destroy a larger percentage of the biorecognition layer due to higher degrees of infiltration into the sensor matrix. Recently, high-intensity narrow spectrum (HINS) blue light has been shown to be more effective than UV sterilization, especially for methicillin-resistant *Staphylococcus aureus* (MRSA) [97].

Several groups have used a combination of two methods to optimize sterilization without affecting the functional aspects of the biosensor. Atanasov's group autoclaved the sensor body separately and then, after refilling the bioreactor, submerged the device in a sterile glutaraldehyde solution (2.5% v/v) for approximately 1 hour [98]. Finally, the device was washed in a sterile saline, resulting in a biosensor that retained performance. In a similar study, von Woedke and colleagues tested three combinations of sterilization methods for a glucose sensor [88]. In the first test, a 2% or 3% alkalinized glutaraldehyde solution combined with γ-irradiation was used [88]. This resulted in the alteration of the polymer material and *in vitro* functionality of the device after sterilization. A second test involved soaking the device in 0.6% H_2O_2 solution over four days followed by a lower level of γ-irradiation, which resulted in a slight loss of sensitivity. The third test included 300s irradiation with specially designed universal homogeneous UV rays combined with immersion in H_2O_2 with surfactants and urea (0.15% effective H_2O_2 concentration) for three days. This method was also safe in protecting the sensor function following sterilization. Even though these combination methodologies were better at preserving the functionality of the sensor compared to any individual technique, neither method was absolutely efficient in reducing *Bacillus subtilis* spore numbers to the sterility assurance levels [88].

Effect of radiation sterilization on polymers

Previous studies have shown that sterilization can affect the surface properties of polymers [3, 99]. Due to the ionization and excitation caused by high-energy radiation, physical changes in the materials could be observed, including embrittlement, discoloration, odor generation, stiffening, and softening, especially with certain polymers used as biocompatible membranes [95, 99]. The dose rate of γ-irradiation is low, consequently requiring anywhere between 7–10 hours for sterilization to occur [95, 100]. This permits the diffusion of oxygen into the samples, allowing it to react with the free radicals produced as a consequence of radiation to form peroxides or hydroperoxides, resulting in the degradation of the polymers and the sensing layer [95, 100]. However, e-beam sterilization causes less material damage due to a higher dose rate, with lowered exposure times on the order of minutes [88, 95, 99]. A higher dose rate decreases the effect of air or atmosphere on the material due to the withdrawal of the ionizing energy before oxygen diffusion can take place. Hence, e-beam sterilization could provide a twofold advantage over γ-irradiation sterilization. A review on the effect of sterilization on plastics and elastomers has been provided by Massey [99].

Therefore, in addition to affecting the biorecognition element immobilized on the transducer, sterilization methods can also alter the surface chemistry and the structure of the permselective coatings and/or modify the biorepulsive properties of the external biocompatible films. Researchers have only verified the change in sensitivity of the biosensor for the target molecule prior to and after sterilization but have not yet examined its effects on the selectivity. Despite the availability of various sterilization methods, it is important to determine the process that is least destructive to the biosensor and the biocompatible membrane to meet both sterility and functionality standards. Consequently, the most effective method to produce sterile biosensors for medical and biological applications should not only focus on the sterilization of the final product but also incorporate sterile practices and measures at every manufacturing step. In addition, multiple sterilization steps using different techniques to a lesser extent, such as the experiment by von Woedtke, could reduce the harmful effects of each of them, thereby improving the performance of the biosensor after sterilization.

11.2.2.3.2 In vivo *calibration*

Calibration of a sensor *in vivo* presents another hurdle for the development of implantable biosensors as an accurate means of determining analyte concentration *in vivo*. Techniques have now been established to calculate and correlate the measured current to the corresponding concentration value by using one-point and two-point calibration methodologies [101–103]. A one-point calibration has been used for biosensors with very high selectivity, which produces no background current at zero substrate concentration. In this method, a single determination of current for a particular concentration of the substrate is used to determine the *in vivo* sensitivity of the biosensor, which is then used to convert further current readings ($I(t)$) into concentration values ($C(t)$). The sensitivity (S) can be calculated using the equation, $S = I(t)/C(t)$. Consequently, we can determine an unknown analyte concentration by rearranging the above equation as $C(t) = I(t)/S$.

A two-point calibration method is commonly used for biosensors where the background current is significant at zero substrate concentration. In this method, along with the determination of sensitivity (S), the value of the background current (i_0) at zero substrate concentration is also determined. When a separate blank electrode without the sensing element is used to measure the background current (i_0), then the equation to obtain the sensitivity becomes $S = (I_2 - i_0)/C_2$, where I_2 is the current value corresponding to a known concentration of the substrate (C_2), determined by blood sampling. Two-point calibration is more widely used due to its higher accuracy in predicting the concentration of the substrate *in vivo* as obtaining a zero background current for zero substrate concentration at electrodes is very uncommon. In addition, regression analysis has been used to avoid the use of *in vitro* calibration and to simplify the calibration of the sensor *in vivo* [46, 86].

The effect of fibrotic encapsulation and the inflammatory responses occurring during the first day of implantation (as elaborated *a priori*) could be compensated for in part by performing a one-point recalibration of the sensor *in vivo* a day after implantation [72]. In this case, the physiology of the local environment around the sensor has to be assumed to be unchanged over time after this initial period while other groups have suggested recalibrations every week if necessary, particularly for long-term implanted sensors [23].

11.2.2.3.3 *Toxicity of implantable biosensors*

Implantable biosensors could cause toxic damage to the host tissue following implantation. This could be due to various reasons: the composition of the sensor, biocompatibility, or sterilization. An implantable biosensor should not be composed of any elements that could potentially be toxic even upon accidental contact with tissue or blood within the body. Extra care needs to be taken to verify the effect of the sensor on tissues when incorporating electron transfer mediators or nanoparticles within the sensor structure, due to their leakage and toxicity [22, 46]. Besides the toxic electron transfer mediators such as ferrocene, CNTs have also been demonstrated to induce an inflammatory response and lead to the formation of granulomas *in vivo* in addition to being compared to asbestos for their genotoxic effects [104]. Although new methods of covalent immobilization and entrapment of mediators and nanoparticles have been reported, the carcinogenic property of sensors containing such materials is still unclear [46, 104].

Toxicity evaluation forms an integral part of biocompatibility testing of materials [86]. Biocompatibility and methods of reducing inflammation and tissue reaction have been dealt with in the previous sections. It remains to be seen though if the biocompatible coatings retain their functionality after sterilization as membranes and polymers could undergo degradation due to sterilization [95, 99]. However, Zhang and colleagues reported the absence of cytotoxicity from *in vitro* cell exposure to EtO-sterilized sensors provided the device was sterilized and degassed subsequently for a determined period of time to reduce the gaseous residues that caused carcinogenesis [94]. Since this study, which was published in 1991, there appears to be a lack of research looking at the toxicity of *in vivo* biosensors as an integral system. Therefore, further focused research using *in vitro* cell/tissue culture studies that have been reported to be more sensitive than animal models [86] are essential to determine the toxicity of implantable biosensors in their final biocompatible, post-sterilized state.

11.3 Conclusions

Implantable biosensors providing continuous monitoring of a single or a range of clinically important ions and biomolecules would improve healthcare diagnostics, taking patient care to new dimensions. This is particularly true for patients in intensive care, with head injuries, or those suffering from chronic conditions such as diabetes or kidney failure. Recently, significant progress has been made in this area, especially with glucose sensors for diabetes monitoring, with some designs incorporating a feedback system to an insulin pump [16, 105]. Electrochemical biosensors comprising electrodes made from micro and nano materials offer potential in designs of small-volume systems, improved signal-to-noise ratio, low power consumption, and ease of implantation. Miniaturization and small-scale power sources have allowed the scaling down of the implantable sensor modules, which has improved patient acceptance. In addition, incorporation of the wireless systems with the biosensor, powered by nanogenerators, could improve patient comfort and independence. Attempts have been made in improving the sensitivity of biosensors by incorporating nanoparticles and polymers. Enhancement in the selectivity of implantable biosensors by means of polymer coatings has facilitated monitoring of analytes in submicromolar concentrations *in vivo* without interference from other molecules. As electrochemical biosensors are prone to variations in oxygen concentration around the implanted sensor (particularly in the case of traumatic brain injury and ischemia), Li *et al.* have developed a novel method to attenuate this effect by introducing an oxygen sensor alongside their *in vivo* glucose biosensor, enabling programmable automatic correction due to fluctuations in oxygen and temperature, thereby increasing the accuracy of the measured output [53].

Yet, biocompatibility, sterility, and toxicity are areas that still need improvement and further investigation. Advancement in biocompatible membrane technology is necessary to avoid degradation of the enzymes and fouling of the electrodes and to maintain stable responses after implantation. The reduction of biofouling at the electrodes determines their performance, stability, and longevity following implantation. In this regard, heparin-immobilized and NO-releasing polymer coatings have shown promising results. Long-term implantation of devices warrants more elaborate and extensive testing, especially those linked to the degradation of the sensor matrix with time and *in vivo* calibration.

Besides, sensitization, irritation, and tissue damage caused by the implant also need careful consideration and assessment [16, 82, 106]. The selection of a sterilization method still appears to be an arbitrary process, without any definitive protocol for sterilizing biosensors, resulting in an experimental approach to determine the least destructive technique for a particular sensor design. The reproducibility of biosensors and the transfer to mass production are additional factors that need to be considered for commercial applications.

In view of the increasing interest from the research community and the vast amount of research currently being conducted in this field driven by micro/nano technologies, it could be envisaged that before long, we may see implantable biosensors providing a feedback loop to a drug delivery system becoming a routine clinical practice.

Acknowledgments

We would like to thank the editor Professor Joseph Irudayaraj for giving us an opportunity to contribute to this volume and the reviewers for helping us improve the quality of the manuscript. We would like to express our gratitude to Professor Calum McNeil (Diagnostic and Therapeutic Technologies, Newcastle University, UK, where some of the work referred to in the chapter was conducted previously). We would like to thank Professor Robert D. O'Neill (Department of Chemistry, University College Dublin, Ireland), Professor Zhong Lin Wang (School of Materials Science and Engineering, Georgia Institute of Technology, US), and Professor William Monty Reichert and Dr. Natalie Wisniewski (alumna) (Center for Biomolecular and Tissue Engineering, Duke University, US) for giving us the permission to use their previously published schematics. The authors would also like to thank Dr. Chris Phillips (WCPC, Swansea University, UK) for proofreading the manuscript.

References

1. Guiseppi-Elie, A., *Electroconductive hydrogels: Synthesis, characterization and biomedical applications.* Biomaterials, 2010. **31**(10): p. 2701–2716.
2. Pflegerl, K., *et al.*, *Quantification of plasma-derived blood coagulation factor VIII by real-time biosensor measurements.* Journal of Chromatography B: Biomedical Sciences and Applications, 2001. **752**: p. 335–347.
3. Tiefenauer, L. and R. Ros, *Biointerface analysis on a molecular level: New tools for biosensor research.* Colloids and Surfaces B: Biointerfaces, 2002. **23**(2–3): p. 95–114.
4. Nordström, C.H. and U. Ungerstedt, *Microdialysis: principles and techniques*, in *Anaesthesia, Pain, Intensive Care and Emergency A.P.I.C.E.* 2006. p. 61–77.
5. Lehninger, A.L., D.L. Nelson, and M.M. Cox, *Lehninger Principles of Biochemistry.* 4th ed. 2005, New York: W. H. Freeman.
6. Palmisano, F., P.G. Zambonin, and D. Centonze, *Amperometric biosensors based on electrosynthesised polymeric films.* Fresenius' Journal of Analytical Chemistry, 2000. **366**(6): p. 586–601.
7. Eggins, B.R., *Chemical Sensors and Biosensors.* Analytical Techniques in the Sciences, ed. D.J. Ando. 2002, London: John Wiley & Sons, Ltd.

8. Grieshaber, D., *et al.*, *Electrochemical Biosensors – Sensor Principles and Architectures*. Sensors, 2008. **8**(3): p. 1400–1458.

9. Wilson, G.S. and M. Amman, *In Vivo Biosensors*. Federation of European Biochemical Societies Journal, 2007. **274**: p. 5452–5461.

10. O'Neill, R.D. and J.P. Lowry, *Neuroanalytical Chemistry In Vivo Using Electrochemical Sensors*, in *Encyclopedia of Sensors*, C.A. Grimes, E.C. Dickey, and M.V. Pishko, Editors. 2006, American Scientific Publishers. p. 1–23.

11. O'Neill, R.D., J.P. Lowry, and M. Mas, *Monitoring Brain Chemistry In Vivo: Voltammetric Techniques, Sensors and Behavorial Applications*. Critical Reviews in Neurobiology, 1998. **12**(1&2): p. 69–127.

12. Habermüller, K., M. Mosbach, and W. Schuhmann, *Electron-transfer mechanisms in amperometric biosensors*. Fresenius' Journal of Analytical Chemistry, 2000. **366**(6): p. 560–568.

13. Tian, Y., *et al.*, *A carbon fiber microelectrode-based third-generation biosensor for superoxide anion*. Biosensors and Bioelectronics, 2005. **21**(4): p. 557–564.

14. Wisniewski, N., F. Moussy, and W.M. Reichert, *Characterization of implantable biosensor membrane biofouling*. Fresenius' Journal of Analytical Chemistry, 2000. **366**(6–7): p. 611–621.

15. Wisniewski, N. and M. Reichert, *Methods for reducing biosensor membrane biofouling*. Colloids and Surfaces B: Biointerfaces, 2000. **18**(3–4): p. 197–219.

16. Wilson, G.S. and R. Gifford, *Biosensors for real-time in vivo measurements*. Biosensors and Bioelectronics, 2005. **20**(12): p. 2388–2403.

17. Wang, J., *In vivo glucose monitoring: Towards 'Sense and Act' feedback-loop individualized medical systems*. Talanta, 2008. **75**(3): p. 636–641.

18. Wilson, G.S. and Y. Hu, *Enzyme-Based Biosensors for in Vivo Measurements*. Chemical Reviews, 2000. **100**(7): p. 2693–2704.

19. Vaddiraju, S., *et al.*, *Emerging synergy between nanotechnology and implantable biosensors: A review*. Biosensors and Bioelectronics, 2010. **25**(7): p. 1553–1565.

20. Jain, K.K., *Applications of Nanobiotechnology in Clinical Diagnostics*. Clinical Chemistry, 2007. **53**(11): p. 2002–2009.

21. Koo Lee, Y.-E., R. Smith, and R. Kopelman, *Nanoparticle PEBBLE Sensors in Live Cells and In Vivo*. Annual Review of Analytical Chemistry, 2009. **2**(1): p. 57–76.

22. Boisselier, E. and D. Astruc, *Gold nanoparticles in nanomedicine: preparations, imaging, diagnostics, therapies and toxicity*. Chemical Society Reviews, 2009. **38**(6): p. 1759–82.

23. Renard, E., *et al.*, *Artificial β-cell: clinical experience toward an implantable closed-loop insulin delivery system*. Diabetes Metabolism, 2006. **32**: p. 497–502.

24. Olivier, N.B., *et al.*, *A comparative study of surface telemetry and implantable telemetry*. Journal of Pharmacological and Toxicological Methods, 2007. **56**(2): p. e57–e57.

25. Benini, L., E. Farella, and C. Guiducci, *Wireless sensor networks: Enabling technology for ambient intelligence*. Microelectronics Journal, 2006. **37**(12): p. 1639–1649.

26. Atanasov, P., *et al.*, *Short-term canine implantation of a glucose monitoring-telemetry device*. Medical Engineering & Physics, 1996. **18**(8): p. 632–640.

27. Rocchitta, G., *et al.*, *Development of a distributed, fully automated, bidirectional telemetry system for amperometric microsensor and biosensor applications*. Sensors and Actuators B: Chemical, 2007. **126**(2): p. 700–709.

28. Crespi, F., *Wireless In Vivo Voltammetric Measurements of Neurotransmitters in Freely Behaving Rats*. Biosensors and Bioelectronics, 2010. In Press, Accepted Manuscript.

29. Obeid, I., M.A.L. Nicolelis, and P.D. Wolf, *A multichannel telemetry system for single unit neural recordings*. Journal of Neuroscience Methods, 2004. **133**(1–2): p. 33–38.

30. Valdastri, P., *et al.*, *Wireless implantable electronic platform for blood glucose level monitoring*. Procedia Chemistry, 2009. **1**(1): p. 1255–1258.

31. Jacquemod, G., *et al.*, *Novel architecture and algorithm for remote interrogation of battery-free sensors*. Sensors and Actuators A: Physical, 2010. In Press, Corrected Proof.

32. Baù, M., *et al.*, *Contactless electromagnetic excitation of resonant sensors made of conductive miniaturized structures*. Sensors and Actuators A: Physical, 2008. **148**(1): p. 44–50.

33. Morais, R., *et al.*, *An activation circuit for battery-powered biomedical implantable systems*. Sensors and Actuators A: Physical, 2009. **156**(1): p. 229–236.

34. Barton, S.C., J. Gallaway, and P. Atanassov, *Enzymatic biofuel cells for implantable and microscale devices*. Chemical Reviews, 2004. **104**(10): p. 4867–86.

35. Heller, A., *Potentially implantable miniature batteries*. Analytical and Bioanalytical Chemistry, 2006. **385**(3): p. 469–473.

36. Chen, T., *et al.*, *A miniature biofuel cell*. Journal of the American Chemical Society, 2001. **123**(35): p. 8630–8631.

37. Wang, S.C., *et al.*, *Membrane-less and mediator-free enzymatic biofuel cell using carbon nanotube/porous silicon electrodes*. Electrochemistry Communications, 2009. **11**(1): p. 34-37.

38. Reshetilov, A., *Microbial, Enzymatic, and Immune Biosensors for Ecological Monitoring and Control of Biotechnological Processes*. Applied Biochemistry and Microbiology, 2005. **41**: p. 442–449.

39. Wu, J., *et al.*, *Electrochemical performances of C/Fe nanocomposite and its use for mediator-free glucose biosensor preparation*. Talanta, 2005. **68**(1): p. 12–18.

40. Romero, E., R.O. Warrington, and M.R. Neuman, *Energy scavenging sources for biomedical sensors*. Physiological Measurement, 2009. **30**: p. R35–R62.

41. Wang, X., *et al.*, *Direct-Current Nanogenerator Driven by Ultrasonic Waves*. Science, 2007. **316**(5821): p. 102–105.

42. Wang, Z.L. and J. Song, *Piezoelectric Nanogenerators Based on Zinc Oxide Nanowire Arrays*. Science, 2006. **312**(5771): p. 242–246.

43. Xu, S., *et al.*, *Self-powered nanowire devices*. Nature Nanotechnology, 2010. **5**(5): p. 366–373.

44. Rome, L.C., *et al.*, *Generating Electricity While Walking with Loads*. Science, 2005. **309**(5741): p. 1725–1728.

45. Carr, J.J. and J.M. Brown, *Introduction to Biomedical Equipment Technology*. 4th ed. 2000, New Jersey, USA: Prentice Hall. 743.

46. Gerritsen, M., J.A. Jansen, and J.A. Lutterman, *Performance of subcutaneously implanted glucose sensors for continuous monitoring*. The Netherlands Journal of Medicine, 1999. **54**(4): p. 167–179.

47. Gerritsen, M., *et al.*, *Influence of inflammatory cells and serum on the performance of implantable glucose sensors*. Journal of Biomedical Materials Research, 2001. **54**(1): p. 69–75.

48. McMahon, C.P., *et al.*, *Oxygen tolerance of an implantable polymer/enzyme composite glutamate biosensor displaying polycation-enhanced substrate sensitivity.* Biosensors and Bioelectronics, 2007. **22**(7): p. 1466–1473.

49. Govindarajan, S., *Development of an Implantable Biosensor Suitable for Continuous Monitoring of Glutamate in the Brain*, in *Clinical and Laboratory Sciences.* 2008, Newcastle University: Newcastle upon Tyne, UK. p. 229.

50. Ly, S.Y. and J.H. Lee, *Human-Urine Diabetes Assay and In Vivo Rat Bladder Assay Using a Fluorine-Doped Carbon Nanotube Catheter Sensor.* Annals of Biomedical Engineering, 2009.

51. Deng, C., *et al.*, *Direct electrochemistry of glucose oxidase and biosensing for glucose based on boron-doped carbon nanotubes modified electrode.* Biosensors and Bioelectronics, 2008. **23**(8): p. 1272–1277.

52. Liu, Y., *et al.*, *The direct electron transfer of glucose oxidase and glucose biosensor based on carbon nanotubes/chitosan matrix.* Biosensors and Bioelectronics, 2005. **21**(6): p. 984–988.

53. Li, C., *et al.*, *Toward real-time continuous brain glucose and oxygen monitoring with a smart catheter.* Biosensors and Bioelectronics, 2009. **25**(1): p. 173–178.

54. Liu, Y., *et al.*, *Amperometric glucose biosensor with high sensitivity based on self-assembled Prussian Blue modified electrode.* Electrochimica Acta, 2009. **54**(28): p. 7490–7494.

55. Tiwari, A., *et al.*, *An amperometric urea biosensor based on covalently immobilized urease on an electrode made of hyperbranched polyester functionalized gold nanoparticles.* Talanta, 2009. **78**(4–5): p. 1401–7.

56. Shan, D., *et al.*, *A porous poly(acrylonitrile-co-acrylic acid) film-based glucose biosensor constructed by electrochemical entrapment.* Analytical Biochemistry, 2006. **356**(2): p. 215–221.

57. Malitesta, C., *et al.*, *Glucose fast-response amperometric sensor based on glucose oxidase immobilized in an electropolymerized poly(o-phenylenediamine) film.* Analytical Chemistry, 1990. **62**(24): p. 2735–2740.

58. Ryan, M.R., J.P. Lowry, and R.D. O'Neill, *Biosensor for neurotransmitter L-glutamic acid designed for efficient use of L-glutamate oxidase and effective rejection of interference.* Analyst, 1997. **122**(11): p. 1419–24.

59. Centonze, D., *et al.*, *Interference-free glucose sensor based on glucose-oxidase immobilized in an overoxidized non-conducting polypyrrole film.* Fresenius' Journal of Analytical Chemistry, 1992. **342**(9): p. 729–733.

60. Mandolfo, S., *et al.*, *In vitro and in vivo biocompatibility of substituted cellulose and synthetic membranes.* International Journal of Artificial Organs, 1997. **20**(11): p. 603–9.

61. Heras, J.Y., A.F.F. Giacobone, and F. Battaglini, *Ascorbate amperometric determination using conducting copolymers from aniline and N-(3-propane sulfonic acid)aniline.* Talanta, 2007. **71**(4): p. 1684–1689.

62. Murphy, L.J., *Reduction of Interference Response at a Hydrogen Peroxide Detecting Electrode Using Electropolymerized Films of Substituted Naphthalenes.* Analytical Chemistry, 1998. **70**(14): p. 2928–2935.

63. Gerhardt, G.A. and J.J. Burmeister, *Ceramic-based multisite microelectrode arrays for in vivo electrochemical recordings of glutamate and other neurochemicals.* Trends in Analytical Chemistry, 2003. **22**(9): p. 498–502.

64. Wasikiewicz, J.M., *et al.*, *Polymeric barrier membranes for device packaging, diffusive control and biocompatibility*. Applied Surface Science, 2008. **255**(2): p. 340–343.

65. Yu, B., *et al.*, *A long-term flexible minimally-invasive implantable glucose biosensor based on an epoxy-enhanced polyurethane membrane*. Biosensors and Bioelectronics, 2006. **21**(12): p. 2275–2282.

66. McMahon, C.P., *et al.*, *The selectivity of electrosynthesised polymer membranes depends on the electrode dimensions: implications for biosensor applications*. Chemical Communication, 2004(18): p. 2128–2130.

67. Govindarajan, S. and C.J. McNeil, *Decreasing the Polymerization Potential Improves the Selectivity of PPD-Coated Disc Biosensors for Glutamate*. Analytical Letters, 2009. **42**(5): p. 739–745.

68. Clark, L.C., *The enzyme electrode*, in *Biosensors: Fundamentals and Applications*, A.P.F. Turner, I. Karube, and G.S. Wilson, Editors. 1987, Oxford University Press: New York. p. 3–12.

69. Williams, D.F., *On the mechanisms of biocompatibility*. Biomaterials, 2008. **29**(20): p. 2941–2953.

70. Williams, D.F., *Revisiting the definition of biocompatibility*. Medical Device Technology, 2003. **14**(8): p. 10–13.

71. Voskerician, G. and J. Anderson, *Sensor Biocompatibility and Biofouling in Real-Time Monitoring*. Wiley Encyclopedia of Biomedical Engineering. 2006: John Wiley & Sons, Inc.

72. Frost, M. and M.E. Meyerhoff, *In Vivo Chemical Sensors: Tackling Biocompatibility*. Analytical Chemistry, 2006. **78**(21): p. 7370–7377.

73. Williams, D.F., *A Model for Biocompatibility and Its Evaluation*. Journal of Biomedical Engineering, 1989. **11**(3): p. 185–191.

74. Khan, A.S. and A.C. Michael, *Invasive consequences of using micro-electrodes and microdialysis probes in the brain*. TrAC Trends in Analytical Chemistry, 2003. **22**(8): p. 503–508.

75. Bhardwaj, U., *et al.*, *PLGA/PVA hydrogel composites for long-term inflammation control following s.c. implantation*. International Journal of Pharmaceutics, 2010. **384**(1–2): p. 78–86.

76. Yu, B., *et al.*, *Use of hydrogel coating to improve the performance of implanted glucose sensors*. Biosensors and Bioelectronics, 2008. **23**(8): p. 1278–1284.

77. Wang, C., *et al.*, *Synthesis and Performance of Novel Hydrogels Coatings for Implantable Glucose Sensors*. Biomacromolecules, 2008. **9**(2): p. 561–567.

78. Oh, B.K., *et al.*, *Miniaturized glucose biosensor modified with a nitric oxide-releasing xerogel microarray*. Biosensors and Bioelectronics, 2005. **21**(5): p. 749–757.

79. Ju, Y.M., *et al.*, *A novel porous collagen scaffold around an implantable biosensor for improving biocompatibility. I. In vitro/in vivo stability of the scaffold and in vitro sensitivity of the glucose sensor with scaffold*. Journal of Biomedical Materials Research Part A, 2008. **87A**(1): p. 136–146.

80. Zhou, Z. and M.E. Meyerhoff, *Preparation and characterization of polymeric coatings with combined nitric oxide release and immobilized active heparin*. Biomaterials, 2005. **26**(33): p. 6506–6517.

81. Zhang, H., *et al.*, *Nitric oxide releasing silicone rubbers with improved blood compatibility: preparation, characterization, and in vivo evaluation*. Biomaterials, 2002. **23**(6): p. 1485–1494.

82. Gifford, R., *et al.*, *Mediation of in vivo glucose sensor inflammatory response via nitric oxide release*. Journal of Biomedical Materials Research A, 2005. **75**(4): p. 755–66.

83. Frost, M.C. and M.E. Meyerhoff, *Implantable chemical sensors for real-time clinical monitoring: progress and challenges*. Current Opinion in Chemical Biology, 2002. **6**(5): p. 633–641.

84. Wu, Y., *et al.*, *Improving blood compatibility of intravascular oxygen sensors via catalytic decomposition of S-nitrosothiols to generate nitric oxide in situ*. Sensors and Actuators B: Chemical, 2007. **121**(1): p. 36–46.

85. Wu, Y. and M.E. Meyerhoff, *Nitric oxide-releasing/generating polymers for the development of implantable chemical sensors with enhanced biocompatibility*. Talanta, 2008. **75**(3): p. 642–650.

86. Koschwanez, H.E. and W.M. Reichert, *In vitro, in vivo and post explantation testing of glucose-detecting biosensors: Current methods and recommendations*. Biomaterials, 2007. **28**(25): p. 3687–3703.

87. Brétagnol, F., *et al.*, *The effect of sterilization processes on the bioadhesive properties and surface chemistry of a plasma-polymerized polyethylene glycol film: XPS characterization and L929 cell proliferation tests*. Acta Biomaterialia, 2008. **4**(6): p. 1745–1751.

88. von Woedtke, T., *et al.*, *Sterilization of enzyme glucose sensors: problems and concepts*. Biosensors and Bioelectronics, 2002. **17**(5): p. 373–382.

89. Lee-Yong, L., K. Eugene, and K. Otilia, *γ-Irradiation of chitosan*. Journal of Biomedical Materials Research, 1998. **43**(3): p. 282–290.

90. Wachiratianchai, S., A. Bhumiratana, and S. Udomsopagit, *Isolation, purification, and characterization of L-glutamate oxidase from Streptomyces sp. 18G*. Microbial Biotechnology, 2004. **7**(3): p. 277–284.

91. United States Pharmacopeial Convention, *(1211) Sterilization and sterility assurance of compendial articles*. Pharmacopeial Forum, 2004. **30**(5): p. 1729–1739.

92. Abel, P.U. and T. von Woedtke, *Biosensors for in vivo glucose measurement: can we cross the experimental stage*. Biosensors and Bioelectronics, 2002. **17**(11–12): p. 1059-1070.

93. Voskerician, G., C.-C. Liu, and J.M. Anderson, *Electrochemical characterization and in vivo biocompatibility of a thick-film printed sensor for continuous in vivo monitoring*. IEEE Sensors Journal, 2005. **5**(6): p. 1147–1158.

94. Zhang, Y., *et al.*, *Application of cell culture toxicity tests to the development of implantable biosensors*. Biosensors and Bioelectronics, 1991. **6**(8): p. 653–661.

95. Dawes, K., L. Glover, and D. Vroom, *The Effects of Electron Beam and γ-Irradiation on Polymeric Materials*, in *Physical Properties of Polymers Handbook*, J.E. Mark, Editor. 2007, Springer Science + Business Media, LLC: New York. p. 867–887.

96. Ahmed, S., *et al.*, *Comparative assessment of chemical and γ-irradiation procedures for implantable glucose enzyme electrodes*. Biosensors and Bioelectronics 2000. **15**(3–4): p. 159–165.

97. Maclean, M., *et al.*, *High-intensity narrow-spectrum light inactivation and wavelength sensitivity of Staphylococcus aureus*. FEMS Microbiology Letters, 2008. **285**(2): p. 227–232.

98. Atanasov, P. and E. Wilkins, *In vivo rechargeable glucose biosensors*. Sensors and Actuators B: Chemical, 1996. **36**(1–3): p. 435–447.

99. Massey, L.K., *The effects of sterilization methods on plastics and elastomers*. 2nd ed. PDL handbook series. 2005, Norwich, NY: William Andrew Pub. 414.

100. Hemmerich, K.J., *Radiation Sterilization: Polymer Materials Selection for Radiation-Sterilized Products*, in *Medical Device & Diagnostic Industry Magazine*. 2000, Canon Communications LLC: Los Angeles, USA.

101. Csoeregi, E., *et al.*, *Design, Characterization, and One-Point in vivo Calibration of a Subcutaneously Implanted Glucose Electrode*. Analytical Chemistry, 1994. **66**(19): p. 3131–3138.

102. Jeong, R.-A., *et al.*, *In vivo calibration of the subcutaneous amperometric glucose sensors using a non-enzyme electrode*. Biosensors and Bioelectronics, 2003. **19**(4): p. 313–319.

103. Aussedat, B., *et al.*, *A user-friendly method for calibrating a subcutaneous glucose sensor-based hypoglycaemic alarm*. Biosensors and Bioelectronics, 1997. **12**(11): p. 1061–1071.

104. Buzea, C., I.I. Pacheco, and K. Robbie, *Nanomaterials and nanoparticles: Sources and toxicity*. Biointerphases, 2007. **2**(4): p. MR17–MR71.

105. Ward, W.K., *et al.*, *A new amperometric glucose microsensor: in vitro and short-term in vivo evaluation*. Biosensors and Bioelectronics, 2002. **17**(3): p. 181–189.

106. Gifford, R., *et al.*, *Protein interactions with subcutaneously implanted biosensors*. Biomaterials, 2006. **27**(12): p. 2587–2598.

Index

Color Insert

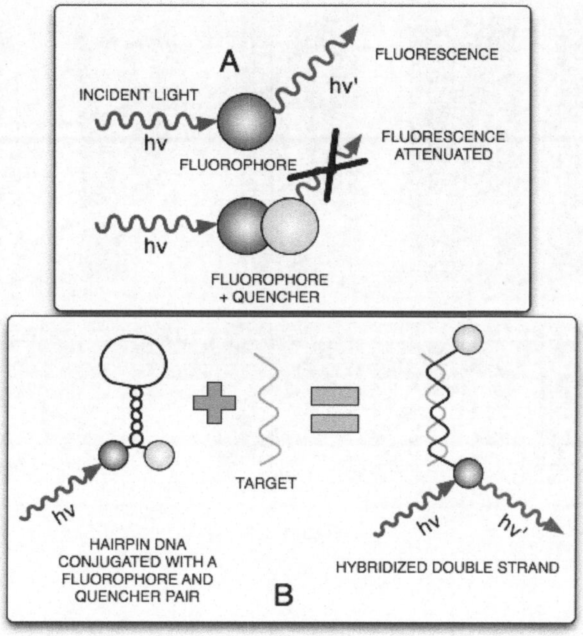

Figure 1.4

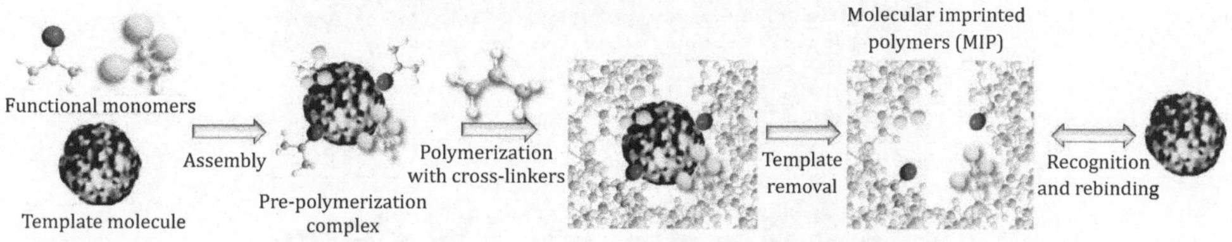

Figure 2.1

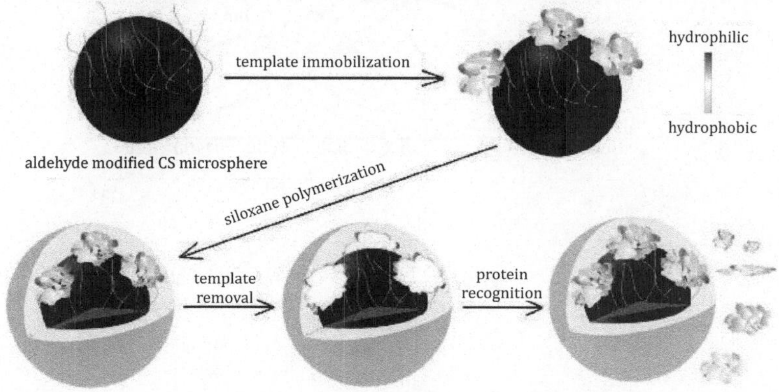

Figure 2.7

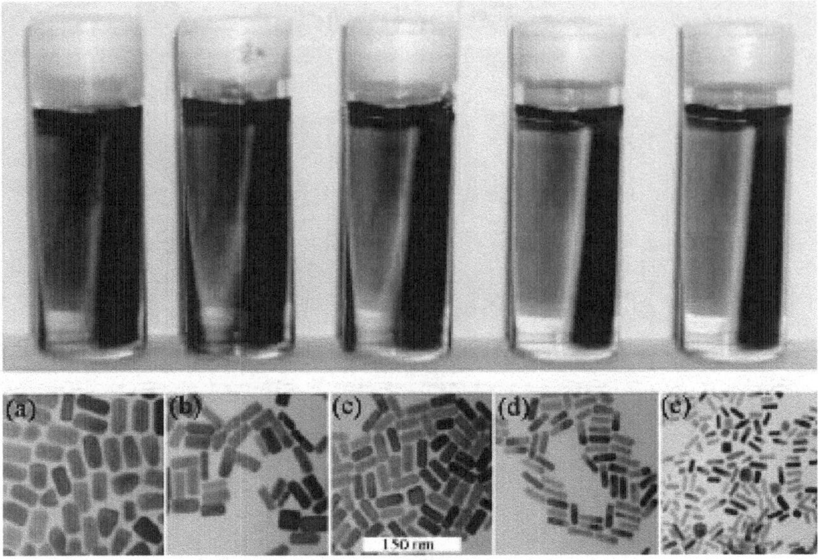

Figure 3.4

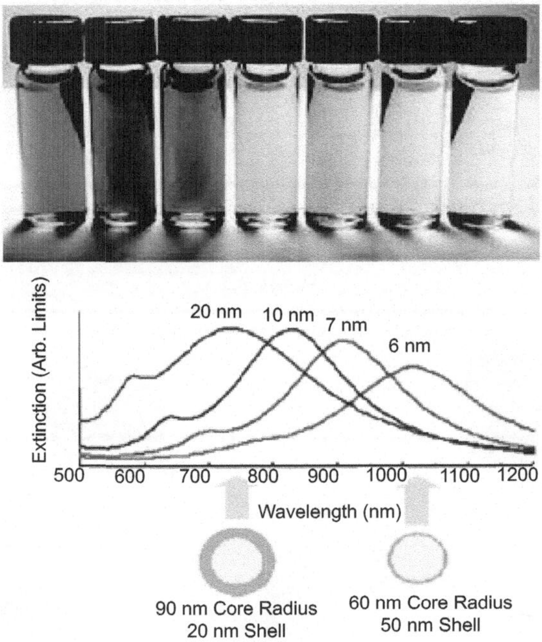

Figure 3.6

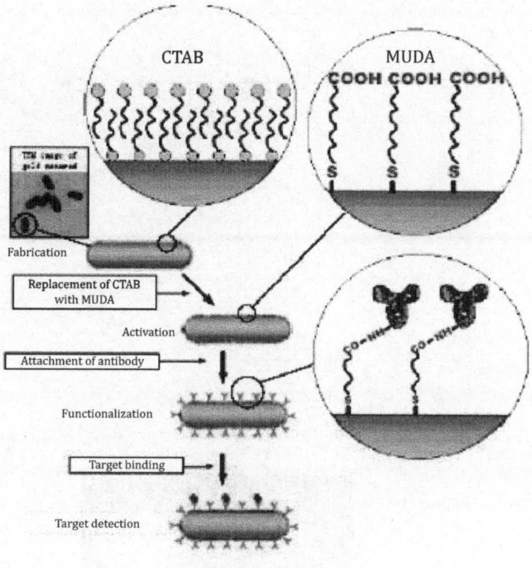

Figure 3.9

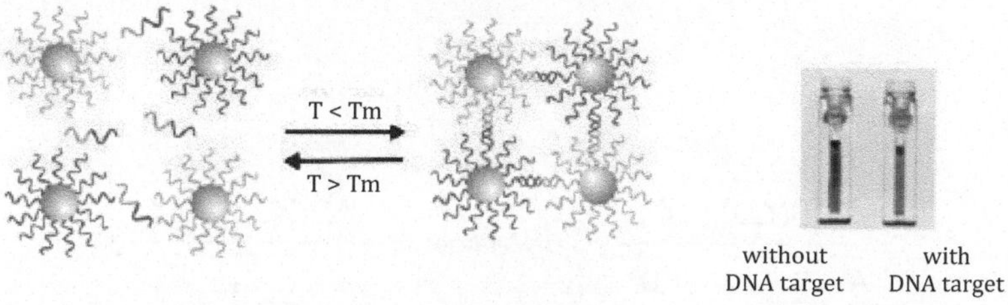

Figure 3.10

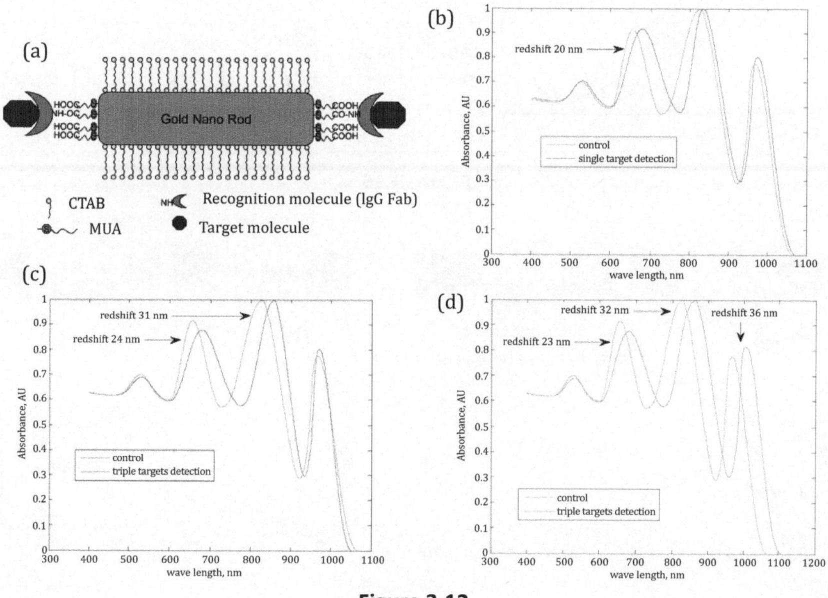

Figure 3.12

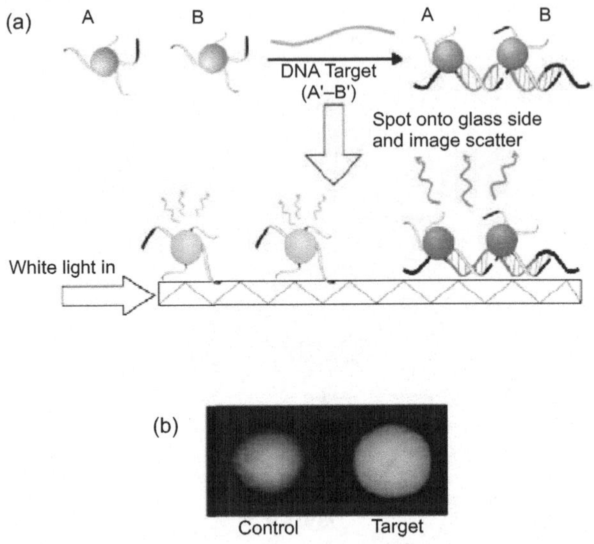

Figure 3.14

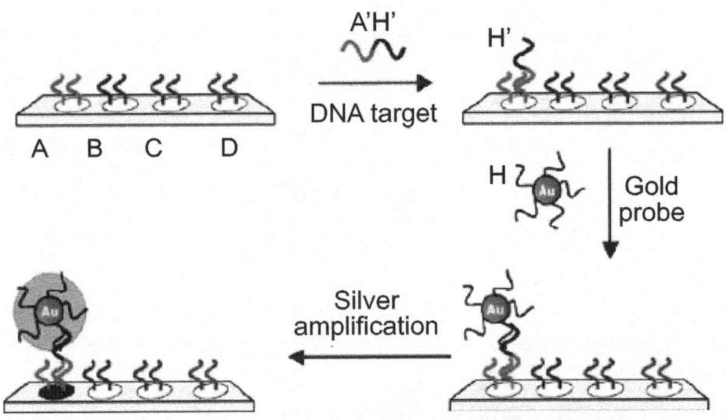

Figure 3.15

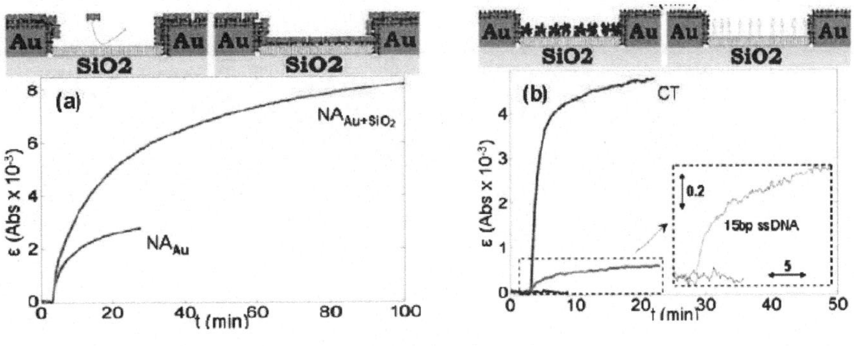

Figure 3.17

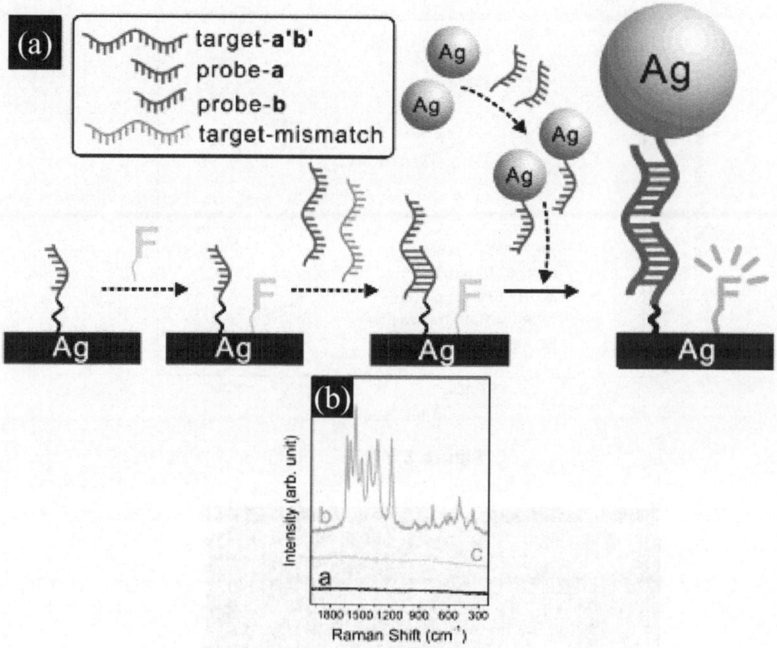

Figure 4.5

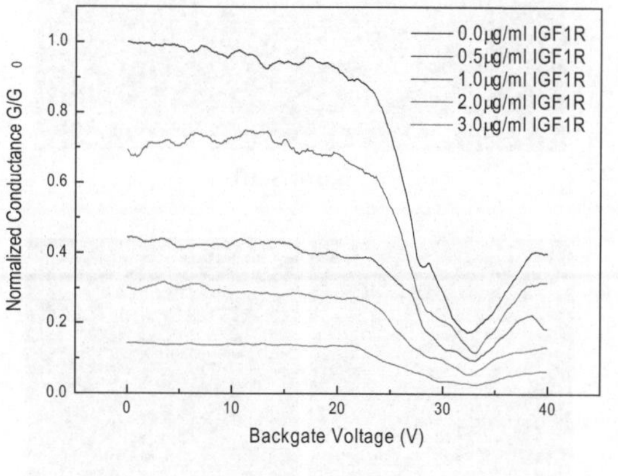

Figure 5.6

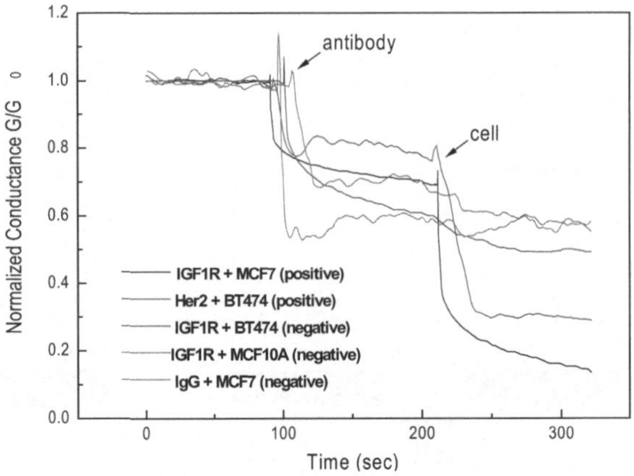

Figure 5.7

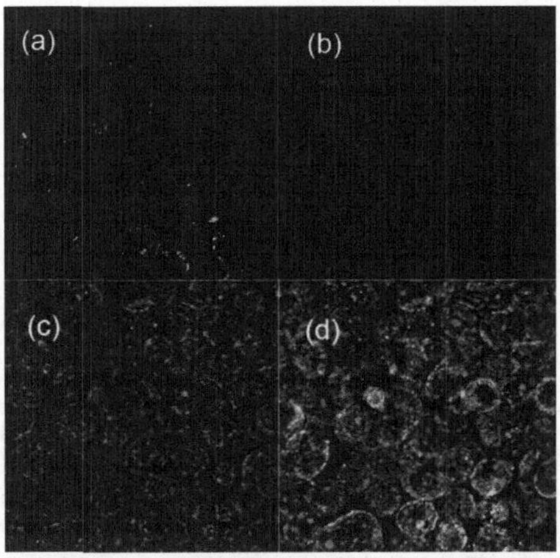

Figure 5.10

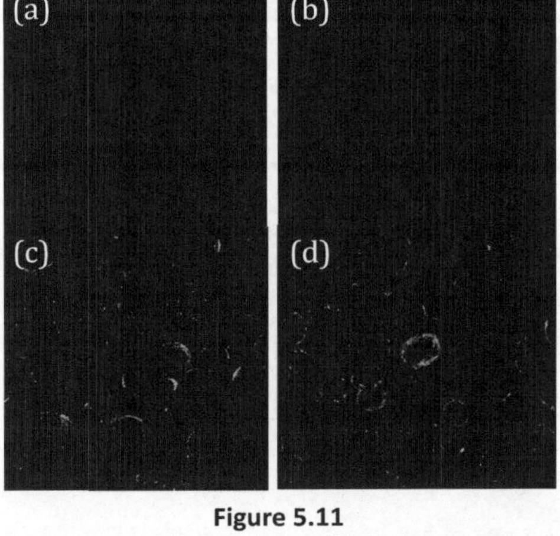

Figure 5.11

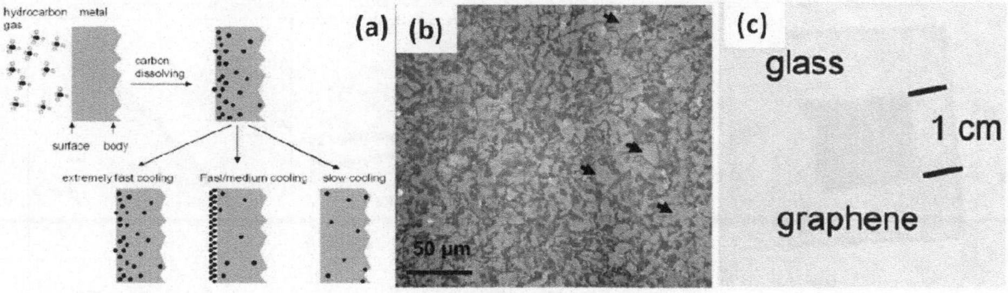

Figure 6.3

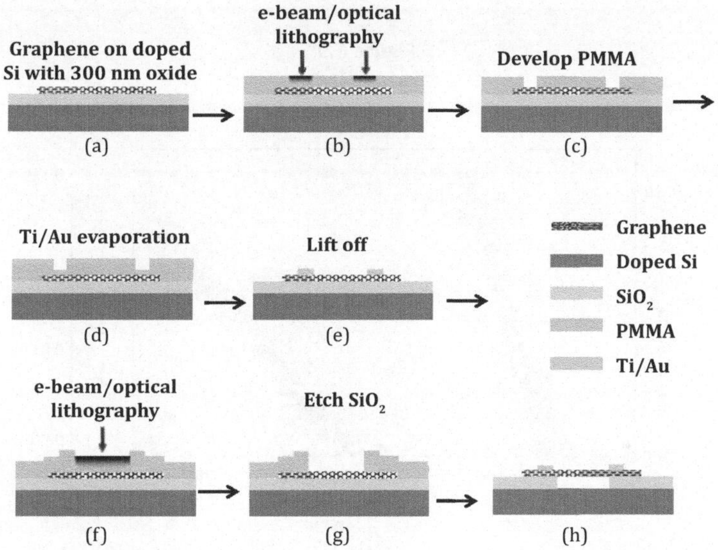

Figure 6.4

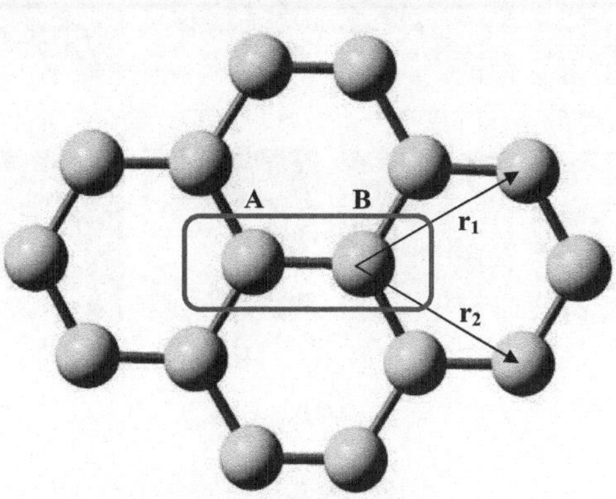

Figure 6.5

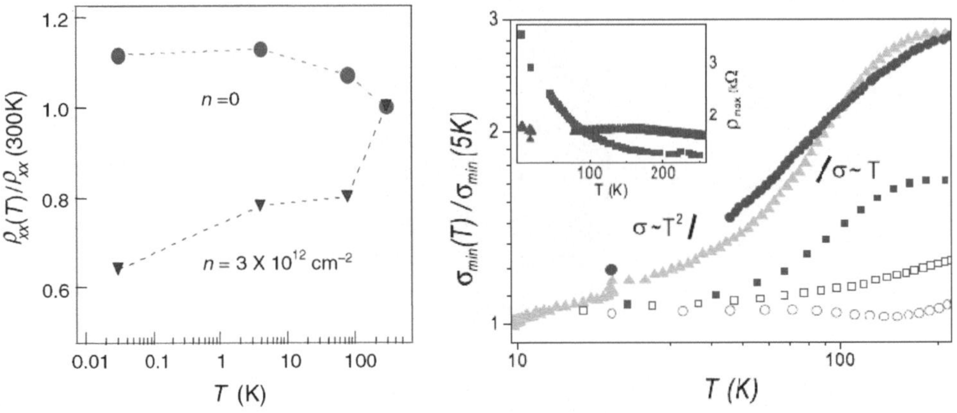

Figure 6.9

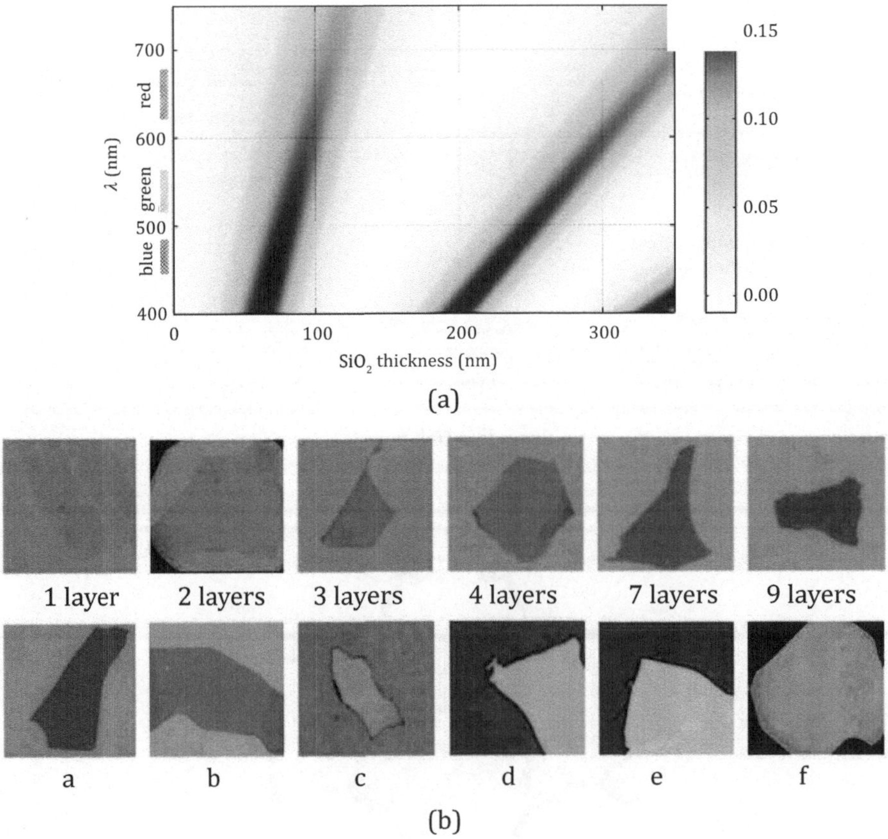

Figure 6.12

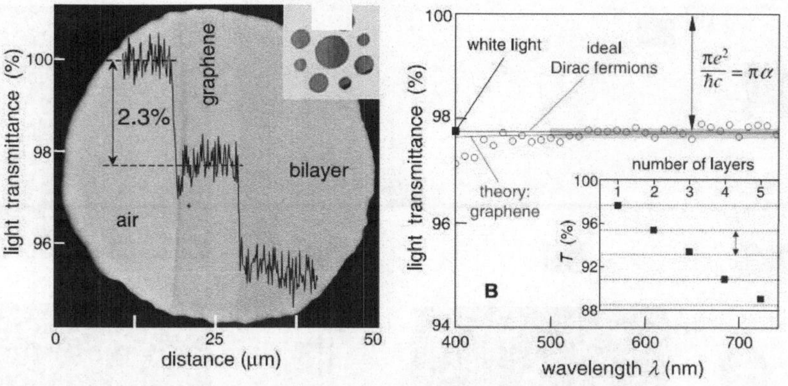

Figure 6.13

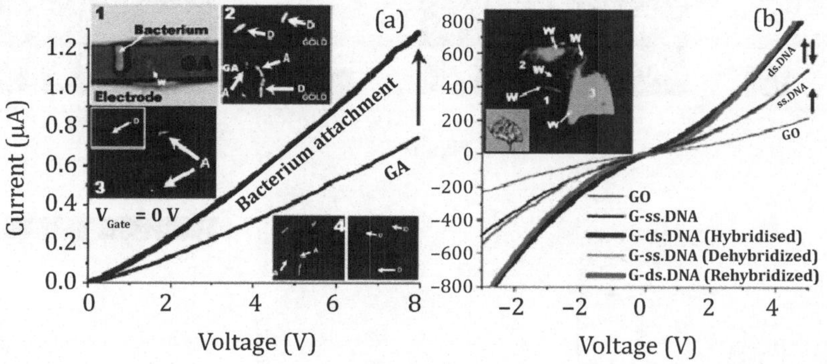

Figure 6.24

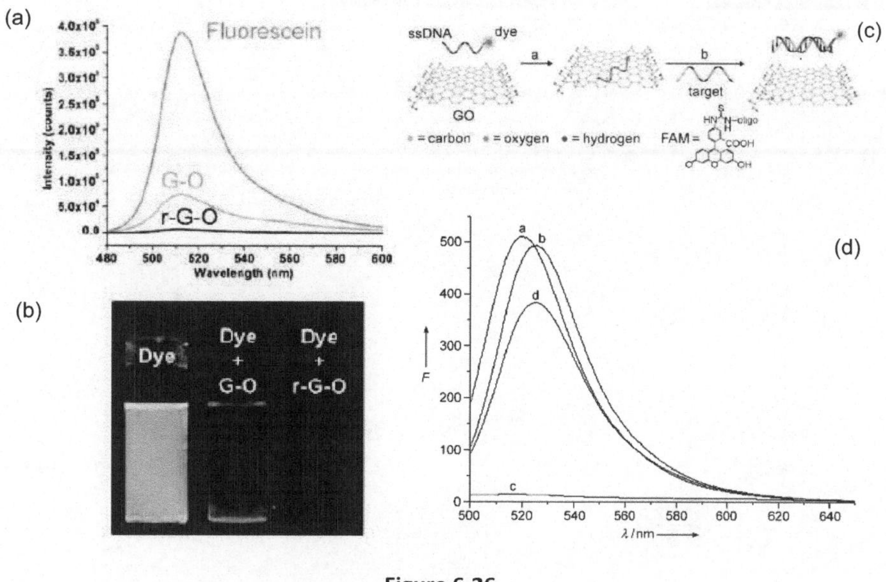

Figure 6.26

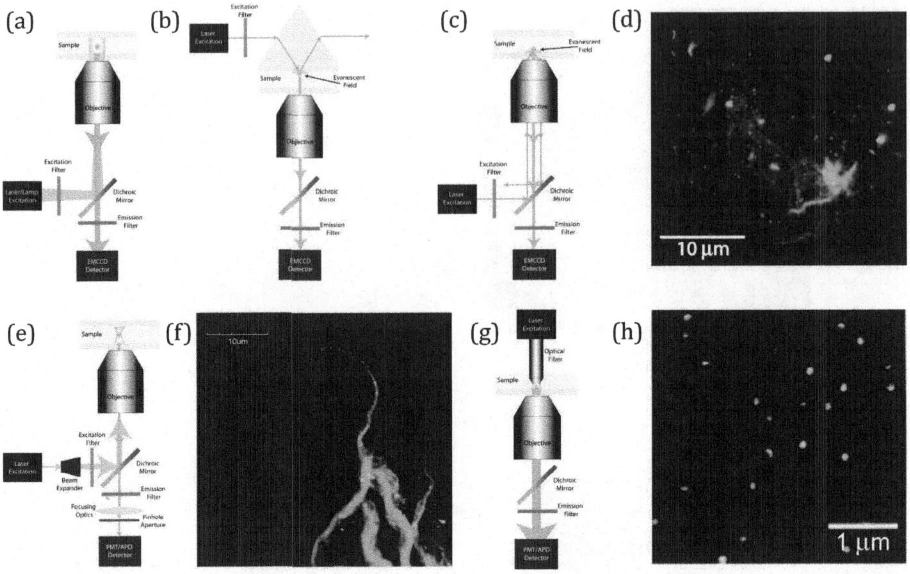

Figure 7.4

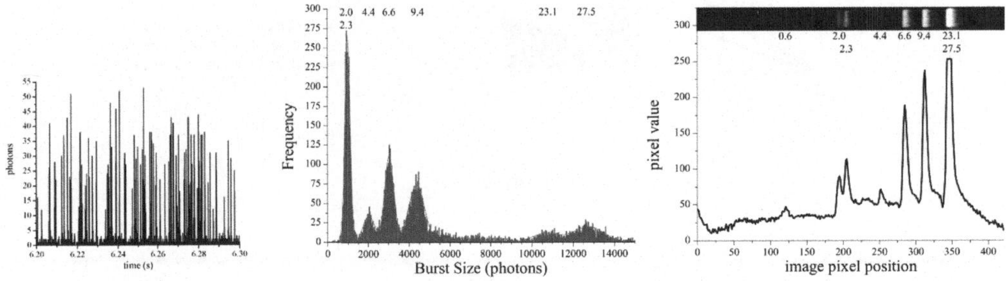

Figure 7.5

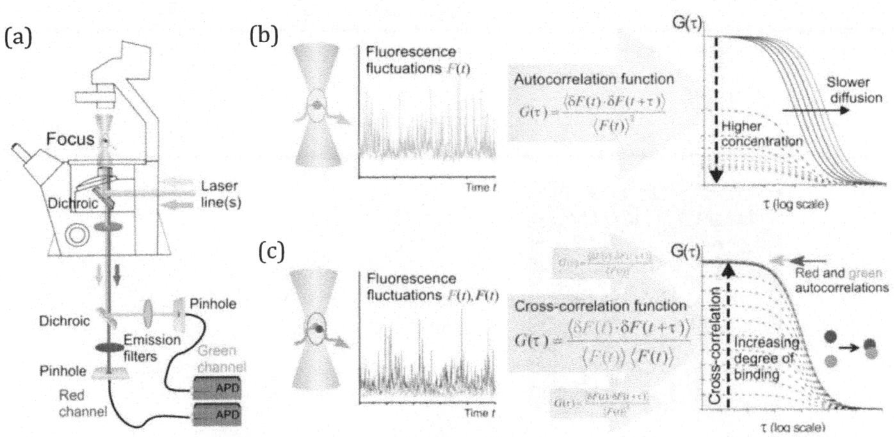

Figure 7.6

Figure 9.7

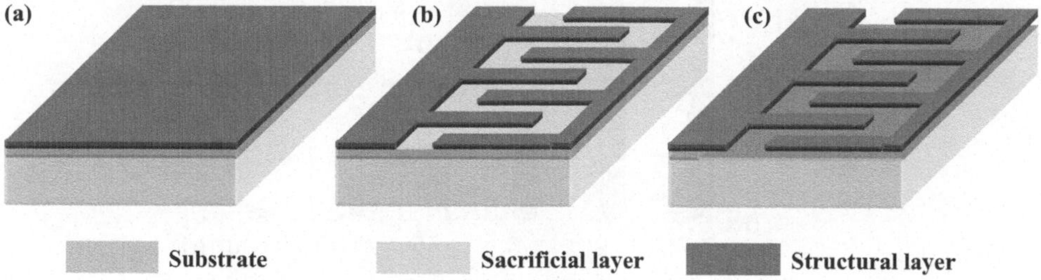

Figure 10.1

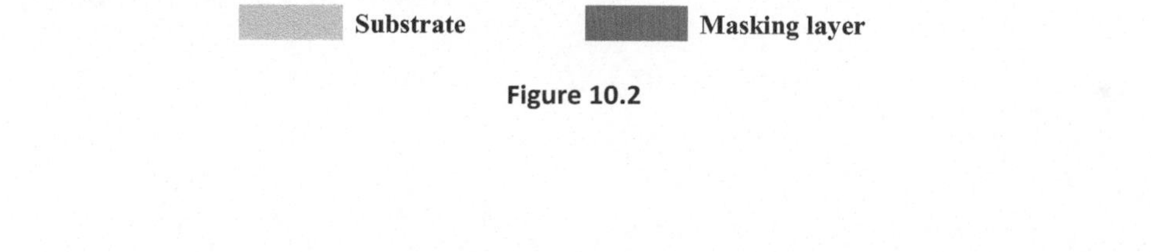

Figure 10.2

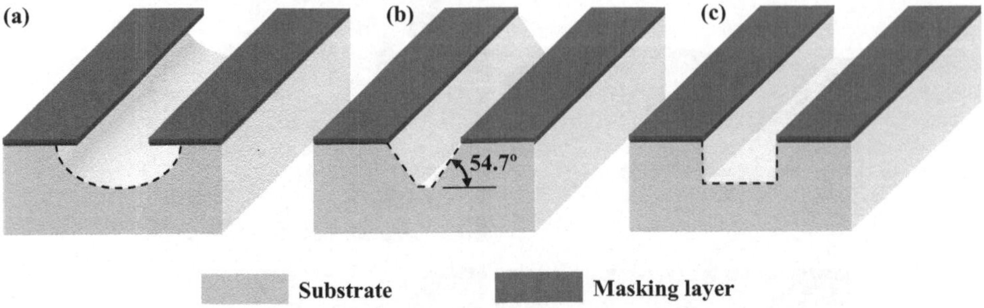

Figure 10.3

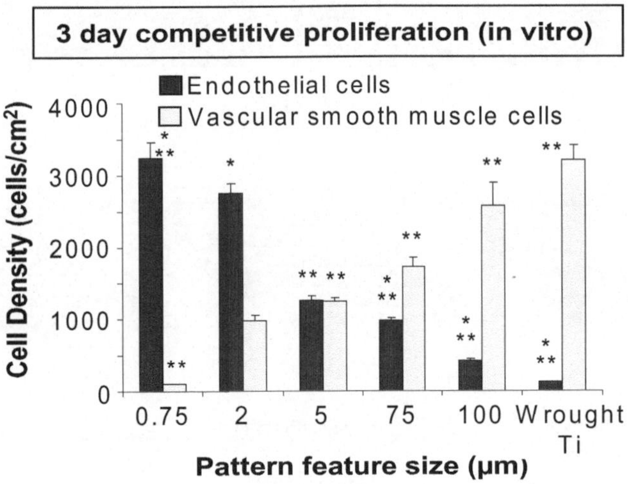

Figure 10.12